Mit Werkzeugen Mathematik und Stochastik lernen – Using Tools for Learning Mathematics and Statistics

Thomas Wassong · Daniel Frischemeier ·
Pascal R. Fischer · Reinhard Hochmuth ·
Peter Bender
Herausgeber

Mit Werkzeugen Mathematik und Stochastik lernen – Using Tools for Learning Mathematics and Statistics

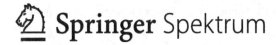

Springer Spektrum

Herausgeber

Thomas Wassong
Universität Paderborn, Deutschland
wassong@math.upb.de

Prof. Dr. Reinhard Hochmuth
Leuphana Universität Lüneburg, Deutschland
reinhard.hochmuth@leuphana.de

Daniel Frischemeier
Universität Paderborn, Deutschland
dafr@math.upb.de

Prof. Dr. Peter Bender
Universität Paderborn, Deutschland
bender@math.upb.de

Dr. Pascal R. Fischer
Universität Kassel, Deutschland
fischer@uni-kassel.de

ISBN 978-3-658-03103-9
DOI 10.1007/978-3-658-03104-6

ISBN 978-3-658-03104-6 (eBook)

Primary 97-02
Secondary 97U70, 97K70, 97U50, 97K60, 97U30, 97C70, 97D70, 97D20, 97D40, 97D50, 97D60, 97K80, 97M10

Die Deutsche Nationalbibliothek verzeichnet diese Publikation in der Deutschen Nationalbibliografie; detaillierte bibliografische Daten sind im Internet über http://dnb.d-nb.de abrufbar.

Springer Spektrum
© Springer Fachmedien Wiesbaden 2014

Planung und Lektorat: Ulrike Schmickler-Hirzebruch | Barbara Gerlach

Gedruckt auf säurefreiem und chlorfrei gebleichtem Papier.

Springer Spektrum ist eine Marke von Springer DE. Springer DE ist Teil der Fachverlagsgruppe Springer Science+Business Media
www.springer-spektrum.de

Vorwort

„Der beste Weg, die Zukunft vorauszusagen, ist, sie zu gestalten." Dieser Satz, der Willy Brandt zugeschrieben wird, ist mir durch den Kopf gegangen, als ich von meinem Kollegen und Freund Rolf Biehler das erste Mal das Wort „pro-aktiv" hörte. In vielfältigen Zusammenhängen habe ich in den zurückliegenden sieben Jahren eine Ahnung davon bekommen, was Rolf Biehler damit meinen könnte: einmal erkannte Probleme sach- und zielorientiert anzugehen, mit langem Atem und, da dies die Lösung eines Problems nicht selten verlangt, nach Möglichkeit in Kooperation. Die Beiträge im vorliegenden Band belegen, dass Rolf Biehler damit bemerkenswert erfolgreich war: In der Regel bereits zu einem frühen Zeitpunkt, dann, wenn sich neue Perspektiven und Möglichkeiten oder eben auch Problemlagen erst anzudeuten begannen, wurde (und wird) Rolf Biehler aktiv und gestaltete (und gestaltet) Entwicklungen entscheidend mit.

Dies trifft seit den 80er Jahren insbesondere auf Fragen zur Verwendung von Computerwerkzeugen beim Lernen der Stochastik in der Schule, der universitären Ausbildung und insbesondere auch in der Lehrerbildung zu. Dabei ist der Fokus in erster Linie nicht auf ein reines Anwenden von Werkzeugen gerichtet. Im Kern geht es in aller Regel um die klassische mathematikdidaktische Frage nach dem Lernen grundlegender mathematischer Konzepte und Ideen. Wie können die neuen Möglichkeiten genutzt werden, um das Verständnis von Lernenden im Hinblick darauf zu fördern und Lernbemühungen zu unterstützen? Wie kann man erreichen, dass sich die Lernenden Mathematik so aneignen, dass sie darüber im Sinne ihrer Lebensinteressen, und das umschließt nicht nur materielle, verfügen können? Dabei geht es (auch) um konkrete Fragen der Visualisierung komplexer mathematischer Konzepte sowie insbesondere des Experimentierens mit stochastischen Methoden und des Explorierens von Daten. So hat Rolf Biehler in diesem Bereich wesentlich dazu beigetragen, didaktische Potenziale neuer technischer Möglichkeiten zu heben, diese auszuloten und für das Handeln Lehrender zugänglich zu machen. Makar und Confrey weisen in ihrem Beitrag darauf hin, dass es dafür nicht ausreichte, neue Aufgaben zu erfinden, sondern dass dies auch eine neue Art und Weise statistischen Denkens und eines Denkens über Statistik erforderte. Mit anderen Worten: In der Analyse und Bewältigung der „concept-tool gaps" geht es nicht nur um die Bewertung und Ausgestaltung der „tools", sondern eben auch um die „concepts".

Ohne dieses wichtige Thema aus dem Auge zu verlieren, nahm Rolf Biehler in den letzten zehn Jahren Projekte, die im Bereich des Übergangs Schule-Hochschule angesiedelt sind, stärker in den Fokus. Bezogen auf diesen Übergang haben sich in dieser Zeit nicht nur in Deutschland eine Reihe von Randbedingungen stark verändert: In den Schulen geht es nun um andere mathematische Inhalte und Kompetenzen. Ein größerer Anteil eines Jahrgangs kommt an die Universitäten und möchte studieren. Viele Studierende treten ohne Abitur in die akademische Welt ein. Rolf Biehler hat früh erkannt, dass die Hochschulen hier aktiv werden müssen und dass Erstsemester mit den veränderten Bedingungen nicht alleine gelassen werden dürfen. Dabei geht es ihm nicht um ein schlichtes „Anpassen an" oder „sich Fügen in" Veränderungen an Schulen oder Hochschulen im Kontext des Bologna-Prozesses, die eigentlich teilweise kritisiert oder zumindest diskutiert werden müssten, sondern in erster Linie um deren Gestaltung sozusagen im „Hier und Jetzt" im Interesse derjenigen, die mit Hoffnungen und Erwartungen an die Universitäten kommen und ein Recht darauf haben, in ihren Bemühungen so unterstützt zu werden, dass sie eine Chance haben, die Anforderungen zu erfüllen, sich zu entwickeln und professionelle Handlungsfähigkeit zu erlangen.

Auch diese Bemühungen erfolgten nicht im stillen akademischen „Kämmerchen": Materialien und Maßnahmen zur Unterstützung der Studierenden wurden in enger Kooperation mit zahlreichen Beteiligten entwickelt, und von Beginn an wurden nahezu alle mathematikhaltigen Studiengänge in den Blick genommen. Mit relativ geringen Mitteln wurde so ein Prozess gestartet, der bis zum aktuellen bundesweit wahrgenommenen Projekt VEMINT und letztlich dann auch zum Kompetenzzentrum Hochschuldidaktik Mathematik (khdm) geführt hat. Zweifellos war zu Beginn im Jahr 2003 nicht abzusehen, dass es 2013 eine vom khdm organisierte bundesweite Tagung mit nahezu 300 Teilnehmern/innen zur Übergangsproblematik Schule/Hochschule geben würde.

In den letzten zwei Jahren ist noch ein weiterer bedeutender und großer Schwerpunkt von bundesweiter Bedeutung mit dem Fokus Lehrerfortbildung hinzugekommen: Der Auf- und Ausbau des Deutschen Zentrums für Lehrerbildung Mathematik (DZLM). Auch hier stehen konkrete Problemlagen im Fokus wie etwa die Erfordernis der Qualifizierung der zahlreichen Lehrkräfte, die fachfremd Mathematik unterrichten oder die Qualifizierung von Mathematikmoderator/innen.

Die Breite der in dem vorliegenden Band versammelten Beiträge zeigt, dass Rolf Biehler sowohl inhaltlich als auch personell in der gesamten Mathematikdidaktik zu Hause ist, und dies national wie international. Neben dem bereits erwähnten Charakter des Visionären betonen einige Beiträge die große Ernsthaftigkeit der Bemühungen, die immer wieder dazu führt, dass Problemlagen zunächst einmal genauer beschrieben und analysiert werden, statt schnelle und dann häufig nur einem ersten kritischen Blick standhaltende scheinbar endgültige Antworten zu produzieren.

Überhaupt: Müsste man die Frage beantworten, welche kurze Formulierung Rolf Biehlers Bemühungen geeignet zusammenfassen würde, so könnte dies der Ausbau der „Didactics of Mathematics as a Scientific Discipline" sein. Aus den wenigen Bemerkungen ist sicher schon deutlich geworden, dass damit kein Rückzug in den sog. universitären Elfenbeinturm angesprochen ist, sondern gewissermaßen im Gegenteil die feste Überzeugung, dass sich „Wissenschaftlichkeit" und das Anliegen, das Lernen von Schülern/innen, Studierenden und Lehrkräften konkret zu unterstützen, nicht nur vertragen, sondern insbesondere in der heutigen Zeit wechselseitig voraussetzen und erst in einem Miteinander wirklich produktiv werden können. Es entspricht dem Naturell von Rolf Biehler, dabei den unvermeidlich auftretenden Unsicherheiten und Widersprüchen nicht aus dem Weg zu gehen, sondern in der Arbeit mit anderen fruchtbar werden zu lassen. Vielleicht ist dies „das" oder zumindest „ein" Geheimnis des großen Erfolges von Rolf Biehler.

Es war für uns eine große Freude zu erleben, wie viele nationale und internationale Kollegen/innen sich spontan bereit erklärt haben, einen wissenschaftlichen Beitrag zu diesem Band zu liefern. Dabei entstand folgende breite Palette von Themen:

    I. Didaktik der Mathematik – Didactics of Mathematics
    II. Modellieren mit Funktionen – Modeling with Functions
    III. Didaktik der Stochastik – Didactics of Statistics
    IV. Stochastik in der schulischen Ausbildung – Statistics in School
    V. Stochastik in der Lehrerbildung – Statistics in Teacher Education
    VI. Stochastik in der universitären Ausbildung – Statistics in Higher Education
    VII. Hochschuldidaktik der Mathematik – University mathematics education

Wir meinen, dass eine interessante und lesenswerte Mischung aus Beiträgen von international renommierten Experten/innen und im engeren Sinne „jungen" Schülern/innen von Rolf Biehler entstanden ist, so dass dieser Band über den konkreten Anlass hinaus einen guten Überblick zu ausgewählten aktuellen Entwicklungs- und Forschungsfragen der Mathematikdidaktik liefert und damit einen Gewinn für die mathematikdidaktische Community darstellt. Das wäre unseres Erachtens zumindest ganz im Sinne von Rolf Biehler.

Reinhard Hochmuth im Namen der Herausgeber

Inhaltsverzeichnis

Kapitel 1
Abstrakte Mathematik und Computer

Willi Dörfler

Universität Klagenfurt

Abstract Jede Philosophie der Mathematik muss sich mit dem Phänomen befassen, dass es sehr erfolgreich und effizient möglich ist am beziehungsweise mit dem Computer, ausgestattet mit geeigneter Software, mathematische Tätigkeiten durchzuführen. Diese Problematik betrifft vor allem den ontologischen und auch epistemologischen Status der mathematischen Objekte. Werden diese als abstrakte (Platonismus) oder mentale (Intuitionismus) Objekte interpretiert, so entsteht eine schwer auflösbare Spannung zur Materialisierung der Mathematik am Computer. Ein semiotischer Fokus auf die Zeichenebene und das regelgeleitete Operieren mit Zeichen und Symbolen wie bei Peirce und Wittgenstein scheint demgegenüber eine rationale Erklärung einer Mathematik am Computer zu ermöglichen. Die Auflösung des Spannungsverhältnisses zwischen Abstraktheit und Computer erfolgt damit dadurch, dass die artifizielle Trennung zwischen mathematischer Tätigkeit und ihren Gegenständen und deren fehlende Vermittlung in einem integrativen und holistischen Ansatz überwunden werden.

Zugang zum Abstrakten

Im allgemeinen Verständnis gilt Mathematik als die abstrakteste Wissenschaft. Wenn auch die Bedeutung des Begriffes „abstrakt" eher vage ist, so ist damit jedenfalls verbunden, dass Abstraktes den Sinnen nicht oder nicht direkt zugänglich ist. Vielleicht auch aus diesem Grund hat „abstrakt" jedenfalls für die Mathematik die Konnotation „schwierig". Die Mathematik handelt ja von vielen abstrakten Objekten wie da sind: Mengen, Zahlen, Funktionen, geometrische Figuren, diverse Räume, etc. Damit die Mathematik etwas über diese abstrakten, den Sinnen nicht zugänglichen Objekte aussagen kann, bedienen sich die Mathematiker sogenannter Darstellungen oder Repräsentationen der mathematischen Objekte. Diese spielen dann insbesondere in der Didaktik und Methodik des Mathematikunterrichts eine große Rolle: mit ihrer Hilfe sollen die Lernenden Wissen über die anders ja nicht zugänglichen mathematischen Gegenstände erwerben oder entwickeln. Diese Darstellungen sind klarerweise sinnlich wahrnehmbar und am Papier oder am Bildschirm als Graphen, Diagramme, Formeln oder Zeichnungen auch manipulierbar durch Rechnen, Konstruieren, Umformen u.a. Interessanterweise wird durch diese „Versinnlichung" oder Materialisierung die Sichtweise des abstrakten Charakters der Mathematik nicht beeinträchtigt, wahrscheinlich vor allem deswegen, weil die Darstellungen weiterhin bloß als Verweis, als eine Art von (relationalen) Bildern, auf die eigentlichen mathematischen Objekte und Gegenstände aufgefasst werden. Die Darstellungen sind in diesem Sinne nur ein Hilfsmittel, ein sinnlicher Weg ins (vorgegebene) Abstrakte. Bei dieser epistemologischen und didaktischen Sicht bleibt es meines Erachtens aber vollkommen unklar, wie die Darstellungen die ihnen zugeschriebene Rolle erfüllen können, also wie sie den kategorialen Unterschied zum Abstrakten überbrücken können. Insbesondere stellt sich die Frage, woher die Mathematiker oder Didaktiker wissen, dass sie die „richtige" Darstellung für ein mathematisches, also abstraktes Objekt gewählt haben. Andererseits ist es ein eher mystischer Prozess, in dem Lernende abstrakte Objekte erfassen, begreifen und verstehen sollen, besonders wenn man noch meint, dass diesen eine mentale oder sogenannte interne Repräsentation entsprechen soll. In der kognitivistisch orientierten Didaktik ist aber genau dies eine grundlegende These oder Annahme.

Computer als mathematisches Werkzeug

Ist wie skizziert die notorische Abstraktheit der Mathematik sowohl ein epistemologisches wie auch ein didaktisches Problem, so ergeben sich aus dieser Interpretation des Charakters der Mathematik auch Fragen und Spannungen im Verhältnis der Mathematik zum Computer. Es muss hier nicht weiter erläutert werden, dass mit den heute verfügbaren Programmen große Teile der Mathematik am Computer ausgeführt werden können. Das gilt für numerische und algebraische Rechnungen und viele andere Kalküle ebenso wie für die Geometrie. Computerprogramme werden eingesetzt in komplexen Beweisen („Vier Farben Satz" der Graphentheorie als prominentes Beispiel), zum Finden von Beweisen in axiomatisch-deduktiven Systemen und selbst zum Entdecken von neuen Sätzen oder Formeln (etwa in der Kombinatorik). Der Computereinsatz in der Katastrophentheorie und in der Theorie der Fraktale ist ebenso gut bekannt. In der Didaktik der Mathematik finden der Computer bzw. geeignete Programme vielfältige Verwendung oder es wird eine solche jedenfalls vorgeschlagen und auch konzipiert. Dabei ist eine Grundidee, dass die oben schon erwähnten Darstellungen am Computer implementiert werden und dadurch die Lernenden eine Lernumgebung angeboten bekommen, in der sie durch Exploration der jeweiligen Darstellungen den erwünschten Zugang zu den abstrakten Objekten und Konzepten der Mathematik erhalten. Am Grundkonzept des Erwerbs abstrakter Begriffe oder Objekte durch die Lernenden ändert sich also beim Computereinsatz eigentlich nichts. Der Computer und seine Software werden als viel leistungsfähigere Mittel zur Ermöglichung dieses Lernprozesses verstanden als dies die klassischen Mittel von Papier und Bleistift sein können. Ob diese Versprechungen auch eingelöst werden, soll trotz berechtigter Skepsis - auch auf Basis der bisherigen praktischen Erfahrungen - hier nicht weiter diskutiert werden. Die von Rolf Biehler für die Statistik entworfenen

Konzepte zur Nutzung des Computers im Lernprozess und auch im Prozess des mathematischen Tuns und Problemlösens (auch in den Anwendungen) stellen dagegen aus meiner Sicht sehr positive Beispiele dar. Beispielhaft seien erwähnt die Arbeiten Biehler (1991; 1993; 1997). Und sie sind dies aus meiner Sicht vor allem deswegen, weil sie nicht oder nicht vordringlich den Darstellungscharakter im Hinblick auf Abstraktes im Auge haben. Etwas oberflächlich ausgedrückt: Biehler verfolgt nicht ein vielleicht gar nicht erreichbares Ziel, sondern nutzt den diagrammatischen (im Sinne von Peirce) Charakter der verwendeten „Darstellungen", die erst in ihrer Verwendung in einer entsprechenden Praxis allgemeine und „abstrakte" Beziehungen und Relationen für die Lernenden quasi erzeugen (dies im Sinne von Krämers „symbolischer Konstitution" mathematischer Objekte, Krämer 1991 und auch 1988). Die epistemologische, kognitive und didaktische Richtung des Entwicklungsprozesses wird gleichsam umgedreht: von den Diagrammen (im Sinne von Peirce) oder traditionell gesprochen von den Darstellungen zu den Abstrakta (erstere als Vorlage und Modell, letztere als Ergebnis). Und zwar werden die Abstrakta als Verwendungsformen der Darstellungen angesehen, sie liegen damit nicht mehr außerhalb oder vor der mathematischen Tätigkeit und Praxis, sondern werden in ihr als deren allgemeine Form konstituiert. Abstrakta sind dann nicht mehr die nicht greifbare und nicht verstehbare Vorlage, das nichtsinnliche Vorbild für die Darstellungen, sondern eine Weise des Sprechens über letztere und ihre Verwendungen in der Mathematik.

Philosophische Fragen

Trotz aller großartigen Erfolge – oder vielleicht gerade wegen dieser - beim Einsatz von Computern und von Software im Bereich der Mathematik bleibt es eine philosophische Frage (oder auch: man kann begründet eine solche aufwerfen), wie es denn möglich ist, mit einem letztlich technisch-physikalischen Gerät abstrakte Objekte zu bearbeiten oder zu untersuchen. Natürlich stellt sich diese Frage auch beim Rechnen und Beweisen am Papier, aber Leistungsfähigkeit und Materialität des Computers verschärfen aus meiner Sicht die Brisanz dieser Frage. Im Computer jedenfalls kann es keine abstrakten Objekte geben, dort gibt es je nach Betrachtungsebene Bits oder Bytes oder Magnetisierungszustände oder Pixel, etc. Als Analogie kann man sagen, so wie man durch Anatomie keinen menschlichen Geist finden kann, findet man im Computer auch keine abstrakten Gegenstände, jedenfalls nicht durch Analyse der Abläufe dort. In einem gewissen Sinn bleibt es ein Mirakel, wie wir mit dem Computer Mathematisches entdecken, erfinden oder konstruieren können, jedenfalls wenn man dieses als genuin abstrakt auffasst. Denn mit dem Computer verlassen wir grundsätzlich nie die Ebene der sogenannten Darstellungen, der Computer kann nur diese manipulieren und untersuchen und uns dann das Ergebnis zeigen. Wir beobachten also den Computer beim Arbeiten mit Darstellungen (wovon allerdings?) oder besser mit Diagrammen im Sinne von Peirce. In traditioneller Manier könnte man hier nun einfach sagen, dass die Mathematiker dann aus diesen Beobachtungen ihre Schlüsse auf die abstrakten Objekte ziehen. Sie verstehen eben die Computerergebnisse und Produkte geeignet als Beschreibungen von abstrakten Objekten und deren Eigenschaften (vergleiche dazu etwa Brown 1998). Eine solche Sicht möchte ich aber im Folgenden mit zwei alternativen Sichtweisen auf die Mathematik kontrastieren, die eine weniger metaphysische Erläuterung der hier aufgezeigten philosophischen Fragen nahelegen. Vorwegnehmend sei gesagt, dass in beiden (hier nur grob skizzierbaren) Positionen, also bei Peirce und bei Wittgenstein, mathematische Tätigkeiten als Zeichentätigkeit beschrieben werden, wobei der Bezug, die Referenz dieser Zeichen auf von ihnen bezeichnete Objekte weitgehend als irrelevant angesehen wird. Dies gilt sowohl für die Referenz auf materielle Gegenstände und Situationen (wie etwa in den Anwendungen), wie auch für die Funktion der Zeichen als Darstellungen abstrakter Objekte. Priorität in jeder Hinsicht haben für beide Autoren die Zeichen selbst und die auf sie bezogene und durch sie erst ermöglichte Tätigkeit. Klarerweise gibt es markante Unterschiede, aber hier ist nicht der Platz für eine vergleichende Untersuchung der Philosophien der Mathematik bei Peirce und Wittgenstein. Trotz der Fokussierung auf die Zeichen, Symbole, Diagramme und der mit ihnen ausgeführten und ausführbaren mathematischen Handlungen ist für beide Philosophen andererseits deren Verwendung als Modell (Anwendungen) jedenfalls eine wichtige Legitimierung für Mathematik als menschliche Praxis überhaupt.

Peirce und Diagramme

Charles Sanders Peirce (1839-1914) war ein amerikanischer Mathematiker, Logiker und Philosoph, bekannt als einer der Begründer und Vertreter des Pragmatismus und der modernen Semiotik, etwa durch Begriffe wie Ikon, Index, Symbol. Als diesbezügliche Literatur sei besonders verwiesen auf Hoffmann (2005), für die Konzepte Diagramm und diagrammatisches Denken auch auf die Arbeiten von Dörfler (2006b; 2008; 2010). Wir konzentrieren uns hier auf diese beiden Konzepte von Peirce, weil sie besonders relevant erscheinen im Hinblick auf eine rationale Aufklärung des (scheinbar) antagonistischen Verhältnisses von Abstraktheit der Mathematik und von deren so effizienter Durchführung am und mit dem Computer. Als mögliches Ergebnis sei vorweggenommen: durch die weitgehende Identifizierung von mathematischer und diagrammatischer Tätigkeit wird es sehr plausibel, dass Mathematik verstanden als Zeichentätigkeit auf einer zeichenverarbeitenden Maschine ausgeführt werden kann. Am Beginn soll ein Originaltext von Peirce stehen (1976, S. 47f):

> By diagrammatic reasoning, I mean reasoning which constructs a diagram according to a precept expressed in general terms, performs experiments upon this diagram, notes their results, assures itself that similar experiments performed upon any diagram constructed according to the same precept would have the same results, and expresses this in general

terms. This was a discovery of no little importance, showing, as it does, that all knowledge without exception comes from observation.

Oder etwas anders formuliert: „In der Logik sind Diagramme seit Aristoteles Zeiten ständig verwendet worden; und keine schwierige Schlussfolgerung kann ohne sie gezogen werden. Die Algebra hat ihre Formeln, die eine Art von Diagrammen sind. Und wozu dienen diese Diagramme? Um Experimente mit ihnen anzustellen. Die Ergebnisse dieser Experimente sind oft ganz überraschend. Wer hätte vorher vermutet, dass das Quadrat über der Hypotenuse eines rechtwinkeligen Dreiecks gleich der Summe der Quadrate über den Schenkeln ist?" (zitiert nach Hoffmann 2000, S. 40).

Für das Folgende sollen Inskriptionen das sein, was Mathematiker aufschreiben, oder was gedruckt erscheint, oder am Bildschirm ersichtlich ist. Ein Diagramm im Sinne von Peirce kann dann beschrieben werden als eine Inskription mit einer wohl definierten Struktur, meistens festgelegt durch Beziehungen zwischen Teilen oder Elementen, zusammen mit Regeln für Umformungen, Transformationen, Kompositionen, Kombinationen, Zerlegungen, etc. Diagramme sind dabei stets in Systemen organisiert und ein Diagrammsystem ergibt eine Art von Kalkül (vergleiche dazu unten Wittgenstein), zu dem eine Praxis des Operierens mit den Diagrammen gehört. Ein Blick in die Mathematik liefert viele Beispiele: Arithmetik der Dezimalzahlen, Bruchrechnung, euklidische Geometrie, elementare Algebra, Algebra der Polynome, formale Logiken, etc. Der Peircesche Gesichtspunkt ist dabei der, dass die mathematische Tätigkeit sich auf die Diagramme und damit auch auf die Inskriptionen konzentriert und im Sinne der obigen Zitate diese Diagramme experimentell „erforscht". Der Untersuchungsgegenstand der Mathematik sind also nicht abstrakte Objekte, sondern (die allerdings in einem anderen Sinne abstrakten Formen der) Diagramme und die sich aus den Operationsregeln ergebenden Konsequenzen. Die einfachsten Fälle sind etwa Rechnungen mit Dezimalzahlen, mit algebraischen Formeln, mit Polynomen, aber auch Beweise gehören dazu, mit dem Prototyp der euklidischen Beweise. Weitere instruktive Beispiele wären die komplexen Zahlen als $a+bi$ oder als Matrizen und analog die Quaternionen $a+bi+cj+dk$. Sieht man sie als Diagramme, als ein Diagrammsystem, so sind die Inskriptionen nicht mehr „Darstellungen" oder Zeichen abstrakter Zahlen, sondern sie selbst sind im diagrammatischen Sinne (das heißt bedingt und bestimmt durch die Operationen mit ihnen) die mathematischen Objekte. Ähnliches gilt für die Matrizenrechnung, siehe dazu Dörfler (2007). Im Konzept des Diagramms ist somit nicht die Referenz der Zeichen bedeutungsstiftend, die Diagramme haben Bedeutung durch ihre Struktur und die Regeln für ihre Transformationen. Auch Peirce merkt an, dass das Objekt eines Diagramms (in seiner Zeichen-Triade) fiktiv sein kann bzw. seine „Form" ist, die wiederum wechselseitig durch die Operationsregeln bestimmt wird. In anderen Worten: Diagramme „erzeugen" ihre Objekte (Referenz) selbst; oder besser: die Praxis des diagrammatischen Operierens in einem Diagrammsystem konstituiert die mathematischen Objekte. Etwa: im System der komplexen „Zahlen" wird i durch das

Rechnen mit dem Symbol zur Zahl. Man vergleiche dazu wieder Krämer (1988; 1991).

Zur weiteren Erläuterung einige Anmerkungen zum diagrammatischen Denken oder Schließen im Anschluss an Peirce. Diagrammatisches Denken ist regelbasiert innerhalb eines Diagrammsystems und des damit verbundenen Kalküls, der aber im Allgemeinen in jedem Schritt viele alternative Wege offenlässt (Unterschied zum Algorithmus). Damit ist diagrammatisches Denken oft konstruktiv und kreativ, es ist ein einfallsreiches Manipulieren von Diagrammen zur Aufdeckung noch verborgener Beziehungen und Eigenschaften (der Diagramme!). In der Geometrie müssen so Hilfslinien gezogen werden, in der Algebra Formeln kreativ umgeformt werden, in vielen Beweisen der Analysis muss geschickt mit Ungleichungen operiert werden; oft muss für einen Beweis einfach ein neues Diagramm erfunden werden. In der zitierten Literatur finden sich dafür viele Beispiele, für die Analysis siehe etwa Dörfler (2010). Wichtig erscheint dabei, dass sich in dieser Sicht Mathematik auf dem Papier ereignet bzw. ausgeführt wird. Der Gegenstand und das Mittel dieses Denkens und Schließens sind die Diagramme selbst und ihre Eigenschaften; potentielle Referenten spielen dabei keine Rolle, außer vielleicht die der Motivation für dieses Tun. Erfolgreiches diagrammatisches Denken erfordert dementsprechend intensive und extensive Erfahrung und Praxis im Umgang mit den jeweiligen Diagrammen, deren Transformationen und Kombinationen. Ein reichhaltiger und verfügbarer „Vorrat" an Diagrammen, ihren Eigenschaften und Beziehungen unterstützt und ermöglicht die kreative Manipulation, Erfindung und Beobachtung der Diagramme. Insbesondere muss Lernen von Mathematik umfangreiches und effektives „diagrammatisches Wissen" (altmodisch: Formelwissen) und diagrammatische Erfahrung beinhalten (durch Rechnen, Umformen, Ableiten, Beweisen, aber auch durch Beobachten, also Lesen von diagrammatischen Prozessen bei anderen Mathematik Betreibenden). Mathematische Tätigkeit als diagrammatisches Denken ist damit eine beobachtbare, nachvollziehbare und nachahmbare Tätigkeit: Mathematik verlagert sich vom Kopf auf das Papier und wird dadurch auch von einer individuellen zu einer sozial teilbaren und vermittelbaren Tätigkeit. Die Lehrbarkeit der Mathematik ist bei einem Verständnis als Wissenschaft von abstrakten Objekten (vermittelt nur durch ihre Darstellungen) ein sehr mystischer Vorgang. In diagrammatischer Sicht kann dieses Lernen noch immer schwierig und komplex sein, jedoch sind die mathematischen Objekte als Diagramme nun möglicher Inhalt und Gegenstand von Kommunikation (etwa durch Vorlesungen als „Vorschreibungen"). Man kann auch so sagen: man nehme den Darstellungen ihren Darstellungscharakter und betrachte sie als Diagramme. Und ich meine, das ist de facto auch das, was die Mathematiker in ihrer professionellen Arbeit tun, auch wenn dort dann wieder von den abstrakten Objekten die Rede ist.

Welche Schlussfolgerungen ermöglicht dies nun im Hinblick auf den Computer? Die Diagramme der Mathematik sind nach Regeln strukturierte Inskriptionen, die wiederum nach Regeln manipuliert werden können, einerseits um in einem allgemeinen Sinn ein Ergebnis auszurechnen oder um andererseits noch unbekannte Zusammenhänge aufzudecken und Eigenschaften feststellen zu können.

Nicht alle diese Diagramme werden unmittelbar und ohne zusätzliche Annahmen und Auflagen am Computer geeignet darstellbar sein, aber durch geschickte Schematisierung und Standardisierung sollte dies in den meisten Fällen machbar sein. In den oben genannten Beispielen jedenfalls ist dies bereits heute gegeben und auch die Transformationsregeln für die Diagramme sind ohne prinzipielle Schwierigkeit implementierbar, so dass damit dann das ganze jeweilige Diagrammsystem am Computer basierend auf der geeigneten Software verfügbar und handhabbar ist. Überzeugende Beispiele dafür sind meines Erachtens die heute verfügbaren Computer Algebra Systeme (CAS). Die von Rolf Biehler vorgeschlagenen Software-Systeme zur Statistik sind hier ebenfalls zu erwähnen, weil sie in der Interpretation dieses Artikels dem Nutzer Diagramme und ihre Manipulationen zur Verfügung stellen (Biehler 2007; Biehler et al. 2011). Ein schon klassisches Beispiel sind die Systeme zur Dynamischen Geometrie (DGS) wie Cabri oder Geogebra. Dort sind Grundregeln der euklidischen Geometrie implementiert, um damit euklidische Diagramme explorativ zu manipulieren. Alle diese Beispiele machen deutlich, dass das Arbeiten mit Diagrammen jeder Art vorwiegend nicht algorithmisch ist, sondern durch die Regeln eines Kalküls gesteuert und ermöglicht wird. Am Computer laufen dabei natürlich im Hintergrund vielfältige Algorithmen, aber auf der Benutzeroberfläche entsteht ein Kalkül mit im Prinzip uneingeschränkten Operationsmöglichkeiten innerhalb des Regelsystems. Andere Beispiele sind denkbar oder naheliegend: Diagramme zu nichteuklidischen Geometrien, zu endlichen Geometrien oder zu diversen algebraischen Systemen. Schon die Implementierung am Computer als bloßes Gedankenexperiment lenkt die Aufmerksamkeit auf die jeweiligen Diagramme und macht so deutlich, dass diese jedenfalls der materialisierte Gegenstand der Mathematik sind, für dessen Bearbeitung und Untersuchung die Referenz auf abstrakte Objekte nicht notwendig und auch nicht hilfreich ist. Dieser narrative Überbau über die Kalküle der Diagrammsysteme hinaus, also das explizite Sprechen von den abstrakten mathematischen Objekten (mit den Diagrammen dann als deren Darstellungen), gehört allerdings zur Praxis der Mathematik. Eine Analyse mittels der Begrifflichkeit von Peirce zeigt aber die getrennten Ebenen, was durch die Umsetzung am Computer noch drastisch unterstrichen wird: am oder im Computer oder in der Software kann es keine abstrakten Objekte geben und dennoch funktioniert dort die Mathematik problemlos. So könte der Computer auf diesem Wege eine epistemologische und ontologische Positionsänderung hinsichtlich Mathematik bewirken.

Die Identifizierung des Diagrammatischen mit dem Mathematischen, siehe auch Stjernfelt (2000), kann aber nicht ohne Schwierigkeiten durchgängig aufrecht gehalten werden. So gibt es ja weite Teile der Mathematik einschließlich komplexer Beweise, die jedenfalls nicht direkt oder explizit Diagramme (auch im weiten Sinne von Peirce) verwenden, wie dort, wo verbal und begrifflich argumentiert wird (auch schon bei Euklid). Eine bei Peirce angedeutete Lösung bestünde darin, auch sprachliche Ausdrücke als Diagramme aufzufassen, die wiederum nach den vereinbarten Regeln manipuliert werden. Aber diese Sicht führt dann schon sehr nahe an Positionen von Ludwig Wittgenstein heran, die aus meiner Sicht eine plausiblere Beschreibung mathematischer Tätigkeiten anbieten.

Wittgenstein, Spiele und Regeln

Obwohl sich Ludwig Wittgenstein (1889-1951) vielleicht intensiver und extensiver als die meisten anderen Philosophen mit der Mathematik auseinander gesetzt hat, werden seine Sichtweisen und Positionen wahrscheinlich wegen ihrer Radikalität in der Philosophie der Mathematik meist ebenso radikal abgelehnt und kritisiert oder gleich gar nicht zur Kenntnis genommen. Das liegt teilweise auch an Wittgensteins Stil, denn es gibt keinen zusammenhängenden Text, der seine Sicht auf die Mathematik klar und übersichtlich darstellt. Diese hat sich auch von Anfängen im Tractatus logico-philosophicus hin zur Spätphase seines Denkens gravierend verändert und liegt in Form postum herausgegebener Schiften mit teilweise fragwürdigen Auslassungen vor (Bemerkungen zu den Grundlagen der Mathematik, Philosophische Untersuchungen, Philosophische Grammatik, Philosophische Bemerkungen u.a. in der Werkausgabe bei Suhrkamp, Bd. 1-8). Für eine gründlichere Auseinandersetzung mit Wittgensteins Philosophie (der Mathematik) verweise ich beispielsweise auf Mühlhölzer (2010) oder Kienzler (1997). Hier kann es nur darum gehen, die für unser Thema relevanten Konzepte bei Wittgenstein in aller Knappheit – und damit notwendig verkürzt – vorzustellen.

Nicht nur für die Mathematik sondern für seine gesamte Sprachphilosophie grundlegend ist die Idee des Sprachspieles, in der wiederum zum Ausdruck kommt, dass sprachliche Bedeutung von Wörtern und Zeichen ganz allgemein in deren Gebrauch innerhalb der Praxis dieser Sprachspiele besteht. Dies steht im schroffen Gegensatz zur sonst üblichen Sicht, dass Zeichen allgemein, und somit auch die Zeichen und Symbole der Mathematik, ihre Bedeutung durch den Verweis auf das von ihnen Bezeichnete (etwa materielle oder abstrakte Gegenstände) erhalten. Ein extremer Vertreter dieser Sichtweise war Gottlob Frege, vergleiche dazu etwa seine Grundlagen der Arithmetik (Frege 1987), wo die (natürlichen) Zahlen als (nichtsinnliche) Gegenstände angesehen werden, für die die Zahlzeichen stehen. Das Bezeichnete (bei Frege eben die Zahlen) reguliert in dieser Sicht von Bedeutung als Referenz auch die Verwendung der Zeichen und Wörter. Ein Sprachspiel oder für meine Zwecke hier noch allgemeiner ein Zeichenspiel (wie etwa in der Mathematik) verwendet dagegen ein System aus Zeichen nach (oft impliziten und konventionellen) Regeln, die konstitutiv für die nun interne, primär und wesentlich operative Bedeutung der Zeichen sind. Diese entsteht also im Handeln mit den Zeichen, in der Zeichentätigkeit (also etwa in der Praxis des Sprechens oder Rechnens) und ist nicht bloß ein statischer, kontemplativer Bezug auf irgendwelche Gegenstände außerhalb von und unabhängig vom Sprach-/Zeichenspiel. Zur Verdeutlichung verwendet Wittgenstein mehrfach die Analogie zum Schachspiel, wie dies schon beispielsweise die Vertreter der formalen Arithmetik, E. Heine und J. Thomae, gemacht haben (vgl. Epple 1994). Dazu Zitate von Wittgenstein, wobei „Kalkül" im Sinne von Sprachspiel oder Zeichenspiel zu lesen ist:

> Für Frege stand die Alternative so: Entweder wir haben es mit Tintenstrichen auf dem Papier zu tun (etwa mit Zahlzeichen, W.D.), oder diese Tintenstriche sind Zeichen von

etwas, und das, was sie vertreten, ist ihre Bedeutung. Dass diese Alternative nicht richtig ist, zeigt gerade das Schachspiel: Hier haben wir es nicht mit den Holzfiguren zu tun, und dennoch vertreten die Figuren nichts, sie haben in Freges Sinn keine Bedeutung. Es gibt eben noch etwas Drittes, die Zeichen können verwendet werden wie im Spiel. (zitiert in Kienzler 1997, S.201)

Eine Gleichung ist eine syntaktische Regel. […] [Dies] macht auch die Versuche der Formalisten begreiflich, die in der Mathematik ein Spiel mit Zeichen sehen. (Philosophische Bemerkungen, S. 143, Werkausgabe Bd.2; zitiert nach Mühlhölzer 2010)

Die Mathematik ist immer eine Maschine, ein Kalkül. Der Kalkül beschreibt nichts. […] Der Kalkül ist ein Abakus, ein Rechenbrett, eine Rechenmaschine; das arbeitet mit Strichen, Ziffern etc. (Wittgenstein und der Wiener Kreis, S.106, Werkausgabe Bd. 3; zitiert nach Mühlhölzer 2010)

In der Mathematik ist alles Algorithmus, nichts Bedeutung; auch dort, wo es so scheint, weil wir mit Worten über die mathematischen Dinge zu sprechen scheinen. Vielmehr bilden wir dann eben mit diesen Worten einen Algorithmus. (Philosophische Grammatik, S.468, Werkausgabe Bd. 4; zitiert nach Mühlhölzer 2010)

Man hat mich in Cambridge gefragt, ob ich denn glaube, dass es die Mathematik mit den Tintenstrichen auf dem Papier zu tun habe. Darauf antworte ich: In genau demselben Sinn, wie es das Schachspiel mit den Holzfiguren zu tun hat. Das Schachspiel besteht nämlich nicht darin, dass ich Holzfiguren auf dem Brett herumschiebe. … Es ist egal, wie ein Bauer aussieht. Es ist vielmehr so, dass die Gesamtheit der Spielregeln den logischen Ort des Bauern ergibt. (Zitiert nach Epple 1994)

Wenn in diesen Zitaten auch zum Ausdruck kommt, dass die konkrete Gestalt von Zeichen für das Zeichenspiel sekundär ist, bleibt festzuhalten, dass es dennoch solche Zeichen geben muss und zwar als materielle Gegenstände, quasi zum Angreifen und Bewegen, die wir dann nach den entsprechenden Regeln verwenden und mit ihnen operieren. Allerdings sei angemerkt, dass die Beliebigkeit der Zeichen (das heißt hier der Figuren) beim Schachspiel zwar gegeben sein mag und auch auf die Wörter der natürlichen Sprache zutrifft, dass aber in der Mathematik die Form einer Notation weitreichenden Einfluss auf die Möglichkeiten des jeweiligen Zeichenspiels hat, das sich gleichsam mit der Notation verändert. Dies wird etwa durch die algebraische Notation überzeugend belegt. Aber dies ändert nichts an der Grundposition, dass die Bedeutungen der Zeichen in deren operativem Gebrauch im Zeichenspiel liegen. Das Handeln mit den Zeichen (in der Mathematik also das Rechnen, das Konstruieren, das Umformen, das Beweisen, etc.) steht dabei autonom für sich selbst und es ist nicht sinnvoll, eine Bedeutung der Zeichen als irgendwie von den Operationen mit den Zeichen „erzeugt" anzusehen, die dann außerhalb der Zeichentätigkeit liegt.

Im Grunde ist die mathematische Tätigkeit das Operieren in einem Zeichenkalkül, jedenfalls kann Wittgenstein in diesem Sinne interpretiert werden. Das heißt jetzt wiederum nicht, dass ein solches Zeichenspiel (wie das der Arithmetik, der Bruchrechnung, der komplexen Zahlen, der Matrizenrechnung, des Infinitesimalkalküls, etc.) nicht „angewendet" werden kann, das heißt zur Beschreibung von Situationen und Abläufen außerhalb des jeweiligen Spiels und innerhalb und außerhalb der Mathematik verwendet werden kann. Wittgenstein sieht in dieser Möglichkeit gerade die Rechtfertigung (relevance) der Mathematik, aber eben nicht ihre Bedeutung (meaning). Wittgenstein sagt sinngemäß: In der Mathematik ist nichts Bedeutung (Referenz bzw. Beschreibung), und alles Kalkül. Dabei spielt „Kalkül" eine ähnliche Rolle wie „Sprachspiel" und diese Äußerung bezieht sich natürlich nur auf die „reine" Mathematik. Auch ist diese Wittgensteinsche Position vom Formalismus etwa im Sinne von Hilbert klar zu unterscheiden. In den formalen Systemen des Formalismus wird nämlich nicht konkret und aktuell operiert, also eigentlich keine Mathematik betrieben, denn sie dienen ganz anderen Zwecken (Beweistheorie, Metamathematik), siehe auch Mühlhölzer (2012).

Diese doch recht radikale Sicht Wittgensteins sei an Beispielen kurz illustriert. Der „logische Ort" der Null als Zahl in der Arithmetik ist durch die auf sie bezogenen Rechenregeln festgelegt ($0+a=a$ oder $0*a=0$), was besonders deutlich wird beim Blick etwa auf die Algebra der Ringe. Die „Bedeutung" der Null ist dabei nicht die Referenz auf ein metaphysisches „Nichts". Auch „sehr große" natürliche Zahlen entbehren zwangsweise eines Referenten, etwa als Anzahlen. Analoges gilt für die negativen ganzen Zahlen, insbesondere -1, und ganz besonders deutlich für das notorische i der komplexen Zahlen. Dieses ist ja durch die Ersetzungsregel $i*i=-1$ bestimmt, und für den ganzen Kalkül genügt dies auch. Noch deutlicher wird diese exklusiv operative Festlegung der Rolle von Zeichen in einem Zeichenspiel oder Kalkül bei den imaginären Einheiten der Quaternionen (siehe http://de.wikipedia.org/wiki/Quaternion). Genau genommen ist dieses Charakteristikum auch schon in den „Elementen" des Euklid gegeben, in denen ja Konstruktionsregeln als Postulate (Axiome) an den Anfang gestellt werden. Dabei ist ferner wichtig, dass es sich bei diesen und vielen anderen möglichen Beispielen nicht um künstliche Formalisierungen inhaltlich gegebener mathematischer Bereiche handelt, sondern um Standardkalküle bzw. übliche mathematische Systeme. So kann auch der Leibnizsche Kalkül mit Infinitesimalien durch Regeln für diese, nun aufgefasst als Symbole ohne Referenten, festgelegt werden, worauf schon Leibniz selbst angesichts der unlösbaren Probleme mit geeigneten Referenten für die dx und dy etc. hingewiesen hat, siehe Krämer (1991).

Diese Festlegung der Rolle von Zeichen im jeweiligen Zeichenspiel oder Kalkül nach Wittgenstein durch Operationsregeln ist es dann auch, die eine Durchführung und Implementierung am Computer durch geeignete Software ermöglicht bzw. plausibel macht. Bräuchte man notwendigerweise für (inner-)mathematische Tätigkeiten eine referentielle Bedeutung der Zeichen, so erschiene eine Mathematik am Computer undenkbar, denn am Computer gibt es tatsächlich keine (referentielle) Bedeutung und dort ist wirklich alles Kalkül. Ein anderer Aspekt der operativen Festlegung des „logischen Ortes" der Zeichen ist, dass dadurch eine weitgehende Ersetzbarkeit, Austauschbarkeit der als Zeichen dienenden Inskriptionen ermöglicht wird, worauf auch schon Wittgenstein in seiner Schachspiel-Metapher aufmerksam gemacht hat. Dies ermöglicht dann die Codierung der mathematischen Zeichen in Computer-interne Zeichen, die nach entsprechend übertragenen Regeln manipuliert werden können.

Wie im obigen schon angeklungen ist, spielt der Begriff der Regel und des Regelfolgens bei Wittgenstein in seiner Auseinandersetzung mit der Mathematik eine große Rolle und dies hat auch für unser Thema einige Relevanz. So sagt er etwa sinngemäß, dass Mathematik keine Beschreibungen von Sachverhalten liefert (im Gegensatz zu Naturwissenschaften oder Sozialwissenschaften), sondern Regeln für Beschreibungen aufstellt und ableitet. Dies wird auch gegen den prima facie Eindruck gesagt, dass mathematische Aussagen meist als Aussagen über mathematische Objekte formuliert werden, also in durchaus physikalistischer Manier. Dies kann jedoch als der Effekt einer fast notwendigen Pragmatik gesehen werden: die Regeln für Beschreibungen werden als Beschreibungen formuliert. In modernen Axiomensystemen wird dieser Regelcharakter sehr deutlich, wo (Grund-) Objekte ja nur mehr als Variable, als Zeichen für undefinierte Grundelemente auftreten, deren Rolle im jeweiligen Zeichenspiel (=mathematische Theorie) durch Operationsregeln festgelegt wird. Genau genommen sind diese Regeln vorwiegend oder ausschließlich Regeln für den Gebrauch der Zeichen in dem jeweiligen Kalkül (Rechenregeln in der Arithmetik und Algebra aber auch in der Analysis, Konstruktionsregeln in der Geometrie) und Wittgenstein spricht von grammatischen Regeln zur Abgrenzung von jeder Sicht, in der Mathematik Beschreibungen von Sachverhalten liefert. Jeder neue Satz, jede neue Formel ist für ihn dann eine neue Regel, die die Grammatik der verwendeten Zeichen und Begriffe betrifft und nicht einen speziellen Inhalt beschreibt: so wollen wir die Zeichen in Zukunft verwenden. Das ergibt nun eine gute Möglichkeit, die notorische Wahrheit, Sicherheit und Zweifellosigkeit sowie die Unzeitlichkeit mathematischer Aussagen (vgl. zu diesem Thema Heintz 2000) nüchtern zu klären: Regeln können im Unterschied zu Beschreibungen weder wahr noch falsch sein, sie sind bestenfalls mehr oder weniger praktikabel oder fruchtbar. Damit im Einklang steht auch, dass im historischen Ablauf mathematische Aussagen nie falsifiziert wurden, wie dies für Aussagen aus der Physik eher die Regel ist. Das heißt natürlich nicht, dass es in Beweisen keine „Fehler" gibt, aber das ist keine Falsifizierung, die ja durch einen vom Satz zu beschreibenden Sachverhalt geleistet werden muss (was es nach Wittgenstein in der Mathematik eben nicht gibt). Wittgenstein wendet seinen Regelbegriff auch auf Sprachspiele an: auch Wörter werden vorwiegend nach Regeln verwendet. Damit kann die Mathematik auch dort, wo sie auf den ersten Blick nicht diagrammatisch im Sinne von Peirce vorgeht, also wo sie ihre Aussagen und Beweise sprachlich mit Wörtern und nicht mit Formeln und Figuren formuliert, als ein „System von Normen und Regeln" beschrieben werden, vergleiche dazu Ramharter und Weiberg (2006) sowie Kroß (2008). In diesem Sinne sind dann etwa (verbale) Definitionen Regeln zum Gebrauch gewisser Wörter und nicht Beschreibungen oder Konstruktionen von Objekten: Es wird die „Grammatik" der Wörter festgelegt und nicht irgendeine referentielle Bedeutung. In diesem Sinne spricht beispielsweise die Mathematik auch nicht über Unendliches, sondern regelt den Gebrauch von „unendlich" und untersucht die Konsequenzen der vereinbarten „Sprechregeln" (für eine drastische platonistische Gegenposition vergleiche Deiser 2010). Damit könnte aber auch der Computer über Unendliches „sprechen", wir müssen nur die Regeln dafür entsprechend implementieren. Damit sei jedoch wiederum nicht abgestritten, dass für die Mathematikerin in ihrer Arbeit mit und innerhalb der Regeln eine Fülle von Vorstellungen, Intuitionen, Bildern und dergleichen motivierende und auch kreative Funktionen haben kann. Aber letztendlich muss sie sich regelkonform verhalten, was keineswegs im Gegensatz zu Kreativität und Einfallsreichtum steht. In der Formulierung von Regeln gibt es in der Mathematik eine große Vielfalt (Formeln aller Art, Gleichungen, Figuren und Diagramme, sprachliche Sätze, etc.) und es wäre eine lohnende Aufgabe, den Regelcharakter verschiedenster mathematischer Zeichensysteme zu analysieren. Den spezifischen Zusammenhang zwischen der Regelhaftigkeit in der Mathematik und dem allgemeinen Konzept eines Computers untersucht Heintz (1993).

Zusammenfassung

Dieser Beitrag geht aus von einer zumindest konzeptuellen Spannung zwischen der vorherrschenden Interpretation der Mathematik als einer Wissenschaft von abstrakten Objekten und den vielfältigen erfolgreichen Verwendungsformen des Computers bzw. geeigneter Software bei der Durchführung von relevanten mathematischen Tätigkeiten wie Rechnen, Beweisen, Darstellen, Simulieren, etc. Diese letzteren bestehen am Computer in der mechanischen Manipulation physikalisch materialisierter Symbole, wogegen abstrakte Objekte im diesbezüglichen philosophischen Diskurs außerhalb von Raum und Zeit und jenseits jeder sinnlichen Erfahrung angesiedelt werden. Die damit zumindest philosophisch gegebene Problematik eines Spannungsverhältnisses entsteht somit wesentlich dadurch, dass mathematische Tätigkeit einerseits und ihre Gegenstände andererseits artifiziell getrennt und sozusagen in verschiedenen Sphären angesiedelt werden. Es liegt mit der üblichen Sichtweise der abstrakten Objekte eine zum Kartesischen Dualismus sehr ähnliche Situation vor, in der sich ebenfalls die klassischen Fragen der Vermittlung zwischen den Sphären stellen. Mit dieser Spannung kann nun unterschiedlich verfahren werden. Eine Möglichkeit wäre, auf das Konzept der abstrakten Objekte ganz einfach zu verzichten (vergleiche dazu Dörfler 2006a), bzw. diese nur als eine Sprechweise, eine facon de parler, anzusehen, wie dies im Nominalismus und im Fiktionalismus getan wird. Ein anderer Ausweg ist das Konzept der Darstellungen: Wir verwenden mathematische Zeichen und Notationen quasi als strukturelle Bilder der abstrakten Objekte und gewinnen über letztere Erkenntnisse durch Untersuchung der ersteren (so denkt Brown in Brown 1998). Das Problem dabei ist, dass wir uns ja nie sicher sein können, ob wir die richtigen Darstellungen gewählt haben, also ob sie auf die abstrakten Objekte „passen", also diese getreu wiedergeben und beschreiben. Dennoch scheint diese Lösung implizit auch im verbreiteten Verhältnis zum Computer-Einsatz vorzuliegen: die Mathematiker interpretieren die wahrnehmbaren Computer-Resultate als Einsichten in eine mathematische Welt. Durch die Konzentration auf die Tätigkeiten statt auf die Objekte können die philosophischen Ansätze bei Peirce und Wittgenstein mit

Vorteil zur Lösung der genannten Spannung eingesetzt werden: am Computer werden nicht mathematische Objekte direkt simuliert oder dargestellt, sondern es werden mathematische Tätigkeiten (diagrammatisches Denken, Manipulation von Zeichen nach Regeln) nachgebildet. Die Regelhaftigkeit dieser Tätigkeiten macht ihre Simulation am Computer einerseits möglich und erklärt andererseits diese Möglichkeit. Auch hier kann abschließend die Schachspielmetapher vielleicht helfen: ein Schachprogramm simuliert nicht die Schachfiguren sondern die Spielregeln. Käme es beim Schach auf die Figuren selbst an, so könnte ein Computer nicht Schach spielen. Mathematik als formale Zeichentätigkeit kann ebenso am Computer ausgeführt oder simuliert werden, und in dieser Sicht verschwinden die genannten metaphysischen Probleme.

Literaturverzeichnis

Biehler, R. (1991). Computers in probability education. In R. Kapadia, & M. Borovcnik (Eds.), *Chance encounters: probability in education* (S. 169-211). Dordrecht: Kluwer Academic Publishers.

Biehler, R. (1993). Software tools and mathematics education: The case of statistics. In C. Keitel, & K. Ruthven (Eds.), *Learning from computers: Mathematics education and technology* (S. 68–100). Berlin, Germany: Springer

Biehler, R. (1997). Software for learning and for doing statistics. *International Statistical Review*, 65(2), 167-189.

Biehler, R. (2007). TINKERPLOTS: Eine Software zur Förderung der Datenkompetenz in Primar- und früher Sekundarstufe. *Stochastik in der Schule*, *27(3)*, 34-42.

Biehler, R., Hofmann, T., Maxara, C., & Prömmel, A. (2011). *Daten und Zufall mit Fathom - Unterrichtsideen für die SI und SII mit Software-Einführung*. Braunschweig: Schroedel.

Brown, J. R. (1998). *Philosophy of Mathematics: Introduction to a World of Proofs and Pictures*. London: Routledge.

Deiser, O. (2010). *Einführung in die Mengenlehre*. Berlin: Springer.

Dörfler, W. (2006a). Keine Angst - Mathematik ist nicht (nur) abstrakt. In *Beiträge zum Mathematikunterricht 2006* (S. 71-74). Hildesheim: Franzbecker.

Dörfler, W. (2006b). Diagramme und Mathematikunterricht. *Journal für Mathematikdidaktik 27*, 200-219.

Dörfler, W. (2007). Matrizenrechnung: Denken als symbolisches Handwerk. In B. Barzel, T. Berlin, D. Bertalan, & A. Fischer (Hrsg.), *Algebraisches Denken. Festschrift für Lisa Hefendehl-Hebeker* (S. 53-60). Hildesheim: Franzbecker.

Dörfler, W. (2008). Mathematical Reasoning: Mental Activity or Practice with Diagrams. In M. Niss (Hrsg.), *ICME 10 Proceedings, Regular Lectures* [CD-Rom]. Roskilde: IMFUFA, Roskilde University.

Dörfler, W. (2010). Mathematische Objekte als Indizes in Diagrammen. Funktionen in der Analysis. In G. Kadunz (Hrsg.), *Sprache und Zeichen. Zur Verwendung von Linguistik und Semiotik in der Mathematikdidaktik* (S. 25-48). Hildesheim: Franzbecker.

Epple, M. (1994). Das bunte Geflecht der mathematischen Spiele. *Mathematische Semesterberichte 41*, 113-133.

Frege, G. (1987/1884*). Die Grundlagen der Arithmetik*. Stuttgart: Reclam.

Heintz, B. (1993). *Die Herrschaft der Regel. Zur Grundlagengeschichte des Computers*. Frankfurt: Campus Verlag.

14   Willi Dörfler

Heintz, B. (2000). *Die Innenwelt der Mathematik. Zur Kultur und Praxis einer beweisenden Disziplin.* New York: Springer.

Hoffmann, M. (2000). Die Paradoxie des Lernens und ein semiotischer Ansatz zu ihrer Auflösung. *Zeitschrift für Semiotik 22*(1), 31-50.

Hoffmann, M. (2005). *Erkenntnisentwicklung. Ein semiotisch-pragmatischer Ansatz.* Philosophische Abhandlungen, Bd. 90. Frankfurt am Main: Vittorio Klostermann.

Kienzler, W. (1997). *Wittgensteins Wende zu seiner Spätphilosophie 1930-1932.* Frankfurt: Suhrkamp.

Krämer, S. (1988). *Symbolische Maschinen. Die Idee der Formalisierung im geschichtlichen Abriss.* Darmstadt: Wissenschaftliche Buchgesellschaft.

Krämer, S. (1991). *Berechenbare Vernunft.* Berlin: de Gruyter.

Kroß, M. (Hrsg.). (2008). *Ein Netz von Normen. Wittgenstein und die Mathematik.* Berlin: Parerga.

Mühlhölzer, F. (2010). *Braucht die Mathematik eine Grundlegung? Ein Kommentar des Teils III von Wittgensteins „Bemerkungen über die Grundlagen der Mathematik".* Frankfurt: Vittorio Klostermann.

Mühlhölzer, F. (2012). On live and dead signs in mathematics. In M. Detlefsen & G. Link, (Hrsg.), *Formalism and Beyond. On the Nature of Mathematical Discourses.* Frankfurt: Ontos Publisher.

Peirce, Ch. S. (1976). *The New Elements of Mathematics I-IV* (C.Eisele, Hrsg.). Den Haag: de Gruyter.

Ramharter, E., & Weiberg, A. (Hrsg.). (2006). *Die Härte des logischen Muss. Wittgensteins Bemerkungen über die Grundlagen der Mathematik.* Berlin: Parerga.

Stjernfelt, F. (2000). Diagrams as Centerpiece of a Peircean Epistemology. *Transactions of the C.S. Peirce Society 36*, 357-392.

Wittgenstein, L. (1984). *Werkausgabe in 8 Bänden* (Bde. 1-8). Frankfurt: Suhrkamp.

Kapitel 2
Der Body-Mass-Index – von Quetelet zu Haldane

Hans Niels Jahnke

Universität Duisburg-Essen

Abstract Die Idee dieser Arbeit entstammt einer Vorlesung für Studierende des Lehramtes für die Sekundarstufe I, in der gezeigt werden sollte, wie man schon mit elementarer Mathematik grundlegende Phänomene unserer Umwelt modellieren und daraus überraschende und einschneidende Erkenntnisse gewinnen kann. Phänomene in Natur und Technik legen es nahe, den Quotienten aus Volumen und Oberfläche zu bilden. K. Menninger und daran anknüpfend H. Winter bezeichnen diesen Quotienten suggestiv als *Massigkeit*. Man sieht nun leicht, dass unter den üblichen beim Body-Mass-Index (BMI) gemachten Annahmen Massigkeit und BMI proportional zueinander sind. Der Aufsatz zeigt, dass sich aus dieser Proportionalität eine Reihe von inhaltlichen Deutungen für den BMI ergeben. Insbesondere wird geometrisch plausibel, warum in der Definition des BMI das Körpergewicht durch das Quadrat der Körperlänge geteilt wird – eine Definition, die üblicherweise aus statistischen Daten abgeleitet wird. Der Aufsatz geht insbesondere der Frage nach, ob derartige Deutungen auch schon für Adolphe Quetelet (1796 – 1874), dem man die Erfindung des BMI zuschreibt, eine Rolle gespielt haben.

1 Massigkeit

Die folgenden Überlegungen sind aus einer Vorlesung über Inhalte des Mathematikunterrichts der Sekundarstufe I hervorgegangen. In einem einleitenden Kapitel sollte exemplarisch gezeigt werden, wie man schon mit elementarer Mathematik grundlegende Phänomene unserer Umwelt modellieren und daraus überraschende und einschneidende Erkenntnisse gewinnen kann. Konkret ging es um die Beziehung zwischen Umfang und Inhalt von ebenen Figuren bzw. zwischen Oberfläche und Volumen von Körpern. Martin Wagenschein hat dafür die glückliche Bezeichnung „Kern und Schale" gefunden, die Winter übernimmt.

Aus Winters reichem Material (Winter 1997, S. 56-66) hat der Autor für die Vorlesung zwei Themenbereiche ausgewählt. Zunächst geht es um die „relative Unabhängigkeit von Kern- und Schalenmaß". Während das naive Denken annehmen mag, dass eine Schale umso größer ist, je größer der Kern ist, den sie um-

schließt, zeigt der analysierende Blick des Mathematikers, dass ein Kern von vorgegebenem Inhalt (aber natürlich veränderlicher Gestalt) von Schalen unterschiedlicher Größe umgeben sein kann. Für die idealen Objekte der Mathematik gilt sogar, dass zu vorgegebenem Flächeninhalt Figuren mit beliebig großem Umfang
gefunden werden können. Das dazu duale isoperimetrische Problem führt bekannt-
lich auf die Aussage, dass bei vorgegebenem Umfang der Kreis die Figur mit dem
größten Flächeninhalt ist. Im Dreidimensionalen gelten analoge Aussagen. Natur
und Technik machen von dieser „relativen Unabhängigkeit von Kern- und Schalenmaß" reichlichen Gebrauch. Der Polarfuchs etwa hat eine gedrungene, möglichst kugelnahe Gestalt, um die Abgabe von Körperwärme an seine kalte Umgebung zu minimieren. Dagegen hat sein Artgenosse aus tropischen Regionen das
umgekehrte Problem zu lösen und verfügt über einen fein gegliederten Körper mit
wesentlich größerer Oberfläche. Ebenso konstruiert der Ingenieur einen Wärme-
tauscher (Heizung, Kältemaschine) zu vorgegebenem Volumen so, dass seine
Oberfläche möglichst groß wird: Natur und Ingenieur nutzen also die Tatsache,
dass die Kapazität zum Wärmeaustausch proportional zur Oberfläche ist.

Ein zweiter Themenschwerpunkt erwächst aus der Frage, wie Kern und Schale
sich bei maßstäblicher Größenänderung verhalten. Wenn bei zwei ähnlichen Figu-

ren F_1 und F_2 zwei sich entsprechende Seiten l_1 und l_2 im Verhältnis $\dfrac{l_1}{l_2} = k$ ste-

hen, dann verhalten sich ihre Flächeninhalte wie k^2. Entsprechend gilt für Körper,
dass sich die Oberflächen wie k^2 und die Volumina wie k^3 verhalten, wenn die Li-
neardimensionen im Verhältnis k stehen. Man illustriert dies z.B. durch die Feststellung, dass ein bis auf halbe Höhe gefülltes (kegelförmiges) Sektglas nur zu

$$\left(\frac{1}{2}\right)^3 = \frac{1}{8}$$ mit Sekt gefüllt ist.

Man kann diese Aussagen allgemein beweisen, wozu im Dreidimensionalen
streng genommen schon die Integralrechnung erforderlich ist. Man sollte sie aber
auch an konkreten Formeln für Flächen- und Rauminhalte demonstrieren, um den
Studierenden zu zeigen, dass diese Aussagen sich auch im Bau dieser Formeln
ausdrücken. Formeln, die Flächeninhalte zu berechnen gestatten, sind Funktionen
2. Grades, Formeln für Volumina solche 3. Grades.

Es kann kein Zweifel bestehen, dass der Begriff der Ähnlichkeit wegen seiner
fundamentalen Bedeutung für die Anwendung der Mathematik in Technik und
Naturwissenschaften eine zentrale Rolle im allgemeinbildenden Mathematikunterricht spielen sollte. Dasselbe gilt für die unterschiedlichen Größenverhältnisse von
Längen, Flächen und Volumina bei ähnlichen Figuren und Körpern.

Wir stellen uns nun zwei Personen vor, die gleiches Gewicht haben, uns in ihrer Konstitution aber ganz ungleich vorkommen. Die eine Person mag klein und
rundlich sein, die andere groß und schlank, die eine ein „Fässchen", die andere eine „Bohnenstange". Solche Gegensätze findet man überall: der massige Elefant
und die sperrige Giraffe, die wuchtige romanische Basilika und der filigrane gotische Dom, die massive Steinbrücke und die luftige Hängebrücke (Menninger
1958, S. 181ff).

Kann man diesen qualitativen Unterschied mathematisch quantifizieren, so dass man den binären Gegensatz „dick vs. dünn" in graduelle Unterschiede „dicker" bzw. „dünner" auflösen kann? Offenbar kann die Angabe des Volumens (oder der Masse) allein dieses Problem nicht lösen. Man müsste das Volumen irgendwie zur „Ausgedehntheit" ins Verhältnis setzen. Für die Ausgedehntheit reicht aber nicht allein die Länge (Größe). Die romanische Basilika und der gotische Dom etwa mögen gleiche Längenausdehnung haben, sie unterscheiden sich aber in der „Gegliedertheit" ihrer Oberfläche.

Die letztere Beobachtung führt darauf, einen neuen Begriff einzuführen und als „*Massigkeit*" eines Körpers das Verhältnis seines Volumens zu seiner Oberfläche zu definieren (Menninger 1958, S. 182). Die Massigkeit M_K eines Körpers K ist also definiert als

$$M_K := \frac{V_K}{O_K},$$

wobei V_K das Volumen und O_K die Oberfläche von K bedeuten.

Offenbar trifft diese Definition in allen angeführten Beispielen das Gewünschte. Der Elefant wird massiger sein als die Giraffe, der kugelige Polarfuchs massiger als der fein gegliederte, schlanke Fuchs aus den tropischen Regionen.

Ein Würfel mit der Kantenlänge 10 cm hat ein Volumen von 1000 cm³. Seine Oberfläche ist $6 \cdot 100 \text{ cm}^2 = 600 \text{ cm}^2$, seine Massigkeit also

$$\frac{1000 \text{ cm}^3}{600 \text{ cm}^2} = \frac{5 \text{ cm}^3}{3 \text{ cm}^2} \approx 1,67 \frac{\text{cm}^3}{\text{cm}^2}.$$

Eine Holzleiste von 10 m Länge und einem Querschnitt von 1 cm mal 1 cm hat hingegen wieder ein Volumen von 1000 cm³, aber eine Massigkeit von

$$\frac{1000}{4 \cdot 1000 + 2 \cdot 1^2} \frac{\text{cm}^3}{\text{cm}^2} = \frac{1000}{4002} \frac{\text{cm}^3}{\text{cm}^2} \approx 0,25 \frac{\text{cm}^3}{\text{cm}^2}.$$

Die Massigkeit des Würfels beträgt also etwa das 6,67 - fache der Massigkeit der gleich voluminösen und gleich schweren Holzleiste.

Allgemein ist die Massigkeit eines Würfels W mit der Kantenlänge a

$$M_W = \frac{a^3}{6a^2} = \frac{a}{6}.$$

Wenn man sich die Oberfläche eines Körpers in der Ebene ausgebreitet denkt (manchmal geht das nur angenähert) und das Volumen des Körpers gleichmäßig auf die Oberfläche verteilt, dann gibt die Massigkeit die Dicke dieser Volumenschicht an. Anders gesagt: *Die Massigkeit gibt an, wie viel Volumen eines Körpers auf eine Flächeneinheit seiner Oberfläche entfällt.* Man kann sie sich als die Höhe einer Säule über dieser Flächeneinheit vorstellen. Das drückt sich auch darin aus, dass die Massigkeit die Dimension einer Länge hat.

Geht ein Körper K' aus einem Körper K durch Streckung mit dem Faktor k hervor, dann ist die Massigkeit von K' das k-fache der Massigkeit von K, denn

$$M_{K'} = \frac{V_{K'}}{O_{K'}} = \frac{k^3 \cdot V_K}{k^2 \cdot O_K} = k \cdot M_K$$

Auf die Flächeneinheit der Oberfläche des k-mal so großen Körpers K' entfällt also im Vergleich zum ursprünglichen Körper K das k-fache Volumen; demnach ist die Säule, die die Massigkeit von K' darstellt k-mal so hoch wie die von K. Diese Eigenschaft der Massigkeit ist fundamental, wir wollen sie die Skalensensitivität der Massigkeit nennen.

Auf der Suche nach Übungsaufgaben, in denen ein mathematischer Begriff angewendet werden kann, verfallen Lehrende heutzutage gerne auf vermeintlich Populäres, und, wie allgemein bekannt, ist der Body-Mass-Index (BMI) in dieser Hinsicht ein heißer Kandidat. So kam dem Autor geradezu unvermeidlich der BMI in den Sinn, und, siehe da, der BMI erweist sich als eine direkte Anwendung des Begriffs der Massigkeit. Der BMI ist definiert als

$$\text{BMI} := \frac{G}{l^2},$$

wobei G das Körpergewicht in kg und l die Körpergröße in m bedeuten. Unter den üblichen, beim BMI gemachten Voraussetzungen nimmt man für den menschlichen Körper eine einheitliche Dichte an. Dann ist aber das Gewicht eines Körpers proportional zu seinem Volumen ($G \sim V$), und seine Oberfläche ist proportional zum Quadrat der Körpergröße ($l^2 \sim O$). *Folglich ist der BMI proportional zur Massigkeit* des jeweiligen Menschenkörpers,

$$\text{BMI} \sim \frac{V}{O} = M.$$

Der „Body-Mass-Index" (BMI) ist ein einfaches Verfahren aus der Anthropometrie zur vergleichenden Einschätzung des Ernährungszustandes des Menschen (Bohlen 2010, S. 1). Als solcher muss er in gewissen Grenzen allgemein anwendbar sein. Die WHO (World Health Organisation) gibt Vergleichswerte für erwachsene Menschen an. Deren Größe variiert etwa zwischen 1,60 m und 2,10 m. Die Definition

$$\text{BMI} := \frac{G}{l^2}$$

setzt also voraus, dass im Intervall [1,60m; 2,10m] für normalgewichtige Personen mit einem idealen BMI der Größe b die Proportionalität

$$G = b \cdot l^2$$

besteht. Um den BMI als angemessenen Maßstab benutzen zu können, müsste also gezeigt werden, dass eine solche Gesetzmäßigkeit den Zusammenhang von G und l^2 hinreichend gut beschreibt.

Durch die Proportionalität von BMI und Massigkeit werden inhaltliche Deutungen des BMI nahegelegt, auf die weiter unten eingegangen wird. Es ist zudem eine naheliegende Frage, ob solche Deutungen, also letztlich geometrische Überlegungen, in der Genese des BMI eine Rolle gespielt haben. Dem soll im Folgenden nachgegangen werden.

2 Leibniz' Tempel

Als kleines Präludium der folgenden Geschichte wurden die Studierenden in der oben erwähnten Vorlesung an eine Überlegung des Philosophen und Mathematikers G. W. Leibniz (1646 - 1716) erinnert. Sie hat nichts mit dem Body-Mass-Index zu tun, führt aber in die Problematik auf wunderbar anschauliche Weise ein.

Im Zusammenhang eines philosophisch-mathematischen Versuchs, den geometrischen Begriff der Ähnlichkeit zu definieren, stellte Leibniz das folgende Gedankenexperiment an:

> Denken wir uns, es seien zwei Tempel oder Gebäude in der Weise eingerichtet, dass sich in dem einen nichts finden lässt, was sich nicht auch in dem anderen vorfände: dass also das Material durchweg dasselbe, etwa weißer parischer Marmor ist, dass ferner die Wände, die Säulen und alles übrige beiderseits genau dieselben Verhältnisse zeigen, die Winkel in beiden gleich sind usw. Wird nun jemand mit verbundenen Augen nacheinander in diese beiden Tempel geführt, und wird ihm erst nach dem Eintritt die Binde abgenommen, so wird er, wenn er in ihnen umhergeht, an ihnen selbst kein Merkmal entdecken, an dem er sie unterscheiden könnte. Trotzdem aber können sie der Größe nach voneinander verschieden sein …(Leibniz o.J./1966, S. 72).

Leibniz schließt daraus, dass zwei Objekte ähnlich sind, wenn sie ununterscheidbar in allen Eigenschaften außer ihrer Größe sind. Denkt man den Beobachter auf ein punktförmiges Auge reduziert, erscheinen ihm die beiden Tempel als gleich. Wenn der Beobachter hingegen seinen Körper in die Betrachtung einbezieht, dann verfügt er über einen Maßstab, durch den er die beiden Tempel unterscheiden kann. Erst bei Einbeziehung eines Maßstabes ist dann ein 40 cm hoher Tempel von einem 40 m hohen Tempel verschieden.

Die Studierenden wurden auch an das Buch *Gullivers Reisen* (1726) von Jonathan Swift erinnert, in dem zu Zwecken politischer Satire mit einer ähnlichen Vision wie der von Leibniz' Tempeln gespielt wird. Bis auf die Größe sind das Land Liliput und seine Einwohner vom England des frühen 18. Jahrhunderts nicht unterscheidbar.

Leibniz' Gedankenexperiment und Swifts dichterische Vision sind Fiktionen. Aber worin besteht das Fiktionale? Handelt es sich um gedanklich mögliche Vorstellungen, die die Natur nur zufällig nicht realisiert hat, oder um Ideen, die fundamentalen Gesetzmäßigkeiten unserer Welt widersprechen? Es zeigt sich, dass Letzteres der Fall ist. Der Grund, warum zwei ähnliche Welten, die in ihren Größen wesentlich verschieden sind, nicht existieren können, ist so fundamental, dass man ihn als „quasi-mathematisch" bezeichnen könnte. Der Leibnizsche 40-cm-Tempel und sein 40-m-Gegenstück sind durch einen versteckten geometrischen Parameter charakterisiert, der es ausschließt, dass der eine Tempel möglich ist, wenn der andere existiert. Wie im Folgenden gezeigt wird, ist die Massigkeit dieser Parameter.

3 Quetelets „Soziale Physik"

Kehren wir zum Body-Mass-Index zurück. Als sein Erfinder gilt der belgische Mathematiker Adolphe Quetelet (1796 – 1874). Quetelet nahm schon in jungen Jahren eine zentrale Stellung im Wissenschaftsbetrieb des neu gegründeten Staates Belgien ein. So gründete er 1826 die Sternwarte in Brüssel und wurde 1834 ständiger Sekretär der Brüsseler Akademie der Wissenschaften. Er verfolgte das Ziel, die Sozialwissenschaften als eine der Physik vergleichbare strenge Wissenschaft zu entwickeln und zu begründen. Sein Hauptwerk (Quetelet 1835/1921) enthält im Titel programmatisch den Begriff „Soziale Physik". In Quetelets Soziologie ist der Begriff des „mittleren Menschen" („homme moyen") zentral. Quetelet nimmt an, dass Eigenschaften der Menschen wie Körpergröße, „Neigung zum Verbrechen" etc. im Allgemeinen normal verteilt sind. Dann kann man sich hypothetisch den Menschen vorstellen, der in Bezug auf diese Eigenschaften jeweils den Mittelwert repräsentiert und diesen den „mittleren Menschen" nennen. Der hypothetische mittlere Mensch ist das Produkt „wesentlicher Ursachen". Die Eigenschaften der realen Menschen oszillieren aufgrund von „zufälligen Ursachen" jeweils um die des mittleren Menschen. Umgekehrt kann man den mittleren Menschen dadurch konstruieren, dass man aus statistischen Daten die Verteilungen der jeweiligen Eigenschaften und ihre Mittelwerte bestimmt. Quetelet wurde so zum Begründer der modernen Sozialstatistik und Anthropometrie.

Im 2. Band der „Sozialen Physik" untersucht Quetelet zunächst die Verteilungen von physischen Eigenschaften des Menschen wie Größe, Gewicht, Kraft, Schnelligkeit, Beweglichkeit. Man vergleiche zu den folgenden Ausführungen auch die Arbeit (Mamerow 2012). Zu seiner Zeit lagen Quetelet nur wenige und unsystematische empirische Daten über Größen und Gewichte von Menschen vor. Er versuchte, diesen Mangel zu beheben, indem er Daten aus unterschiedlichen Bereichen zusammenführte. So verschaffte er sich Messwerte aus Brüsseler Kliniken, Schulen, Waisenhäusern und aus Unterlagen für Musterungen zum Militär. Das erlaubte es ihm, für die belgische Bevölkerung Tabellen aufzustellen, die Größe und Gewicht von Männern und Frauen in Abhängigkeit vom Alter darstellen. Aus diesen Daten versuchte er Gesetze über das Wachstum der Menschen abzuleiten.

Seine Überlegungen über den Zusammenhang von Gewicht und Größe beim Menschen leitet Quetelet im Kapitel „Verhältnis zwischen Gewicht und Größe" mit einer allgemeinen Formulierung seines Ergebnisses ein. Er sagt:

> Würde der Mensch gleichmäßig nach allen Dimensionen wachsen, so würden sich die Gewichte in den verschiedenen Lebensaltern wie die Kubikzahlen der Größe verhalten. In Wirklichkeit macht man aber eine andere Beobachtung. Die Zunahme des Gewichts geschieht weniger schnell, ausgenommen im ersten Jahr nach der Geburt, in welchem das

angegebene Verhältnis ziemlich regelmäßig beobachtet wird. Nach dieser Zeit aber und bis gegen das Pubertätsalter entspricht die Zunahme des Gewichts fast den Quadraten der Größe. Die Entwicklung des Gewichts wird noch sehr schnell zur Zeit der Pubertät und steht dann ungefähr mit 25 Jahren still. Im allgemeinen weicht man wenig von der Wahrheit ab, wenn man annimmt, *dass sich während dieser Entwicklung die Quadrate des Gewichts in den verschiedenen Lebensaltern wie die fünften Potenzen der Größe verhalten*; was, vorausgesetzt, daß das spezifische Gewicht dasselbe bleibt, von selbst zu der Folgerung führt, dass das Wachstum des Menschen in die Breite geringer ist als das in die Länge (Quetelet 1835/1921, S. 89).

Also: wenn der Mensch so wächst, dass der größere Mensch ähnlich zum kleineren ist (er wächst „gleichmäßig in allen Dimensionen"), dann verhalten sich die die Gewichte wie die Kubikzahlen der Größen. Die Gewichte sind proportional zu den Volumina, und letztere wachsen in der dritten Potenz. Das sei, so Quetelet, im ersten Jahr nach der Geburt noch richtig, danach beobachte man aber etwas anderes, dass nämlich die Zunahme der Gewichte den Quadraten der Größen entspreche. In den verschiedenen Phasen des Wachstums (vor, während und nach der Pubertät) sind die Wachstumsgeschwindigkeiten noch unterschiedlich, so dass man insgesamt mit der $2\frac{1}{2}$-ten Potenz wenig von der Wahrheit abweiche. Folglich wachse der Mensch nicht in allen Dimensionen gleich, und der größere Mensch sei nicht ähnlich zum kleineren. Stattdessen sei das Wachstum des Menschen in die Breite geringer als in die Länge. Gemessen am Maßstab der Ähnlichkeit seien größere Menschen also im Allgemeinen schlanker als kleinere und umgekehrt seien kleinere korpulenter als größere.

Unter Hinzunahme von Beobachtungen an gleich alten, aber unterschiedlich großen Menschen entscheidet sich Quetelet dafür, *dass insgesamt die Gewichte sich verhalten wie die Quadrate der Größen.*

Wie begründet Quetelet diese letztere Aussage? Der Leser, der nun umfangreiche statistische Rechnungen an Daten zu Größen und Gewichten erwartet, sieht sich gründlich getäuscht. Stattdessen setzt Quetelet größte und kleinste Werte ins Verhältnis. Dies sei am Beispiel der Daten zu Männern erläutert, für Frauen ergibt sich mit anderen Werten ungefähr dasselbe Ergebnis (Quetelet 1835/1921, S. 90). Quetelet wählt aus seinen Daten die Größen und Gewichte der zwölf größten und zwölf kleinsten Männer aus und bildet Mittelwerte. Das führt ihn auf das Ergebnis, wie es in Tabelle 3.1 zu sehen ist.

Tab. 3.1 Mittelwerte Größen und Gewichte von Männern

	Größe	Verhältnis des Gewichts zur Größe
Mittelwert kleinste Männer	t = 1,511 m	p/t = 36,7
Mittelwert größte Männer	T = 1,822 m	T/P = 41,4

In Tabelle 3.1 bezeichnen t und T die Mittelwerte für die Größe sowie p und P die Mittelwerte für das Gewicht der größten und kleinsten „Individuen".

Bildet man den Quotienten der beiden Mittelwerte für die Größe, erhält man

$$t : T = \frac{1,511 \text{ m}}{1,822 \text{ m}} \approx 0,829 \approx \frac{5}{6}.$$

Ebenso ergibt sich für die beiden Quotienten aus Gewicht und Größe

$$\frac{p}{t} : \frac{P}{T} = \frac{36,7}{41,4} \approx 0,886 \approx \frac{5}{6}.$$

Offensichtlich wird hier kräftig gerundet. Da der Quotient in beiden Spalten gleich 5/6 ist (ein Ergebnis der Daten), darf man die linken Seiten gleichsetzen:

$$\frac{p}{t} : \frac{P}{T} = t : T \Leftrightarrow \frac{p}{P} = \frac{t^2}{T^2}.$$

Das ist das gewünschte Ergebnis: die *Gewichte verhalten sich wie die Quadrate der Größen.*

Eine populäre Version desselben Argumentes findet sich drei Seiten später (Quetelet 1835/1921, S. 93). Dort erwähnt er einen schwedischen „Riesen", der bei den Leibgardisten Friedrichs II. diente und der 2,52 m groß gewesen sein soll, und einen englischen „Zwerg" von 0,44 m. Dann sagt er, dass diese zwei Individuen sozusagen die Grenzen unserer Gattung bilden. Wenn sie nun zueinander im geometrischen Sinne ähnlich wären, dann würden sich ihre Gewichte wie 188 zu 1 verhalten. „Ein derartiges Missverhältnis hat gewiss niemals zwischen zwei Menschen bestehen können." (Quetelet 1835/1921, S. 93) Dies ist einfach ein Appell an die Anschauung (oder den gesunden Menschenverstand) und kann sicher nicht als wissenschaftliches Argument gewertet werden.

Ausdrücklich wendet sich Quetelet gegen den französischen Naturforscher Buffon (1707 – 1788), von dem er ohne Quellenangabe behauptet, dieser habe die Gewichte als proportional zu den dritten Potenzen der Größen betrachtet. Er wirft Buffon vor, keine Daten studiert und sich damit begnügt zu haben,

> „ein Ergebnis der Theorie auszusprechen, indem er voraussetzte, dass die Menschen im geometrischen Sinne unter sich ähnlich seien." (Quetelet 1835/1921, S. 94)

Es ist schwer zu beurteilen, wie intensiv und mit welchen Methoden Quetelet sich tatsächlich mit seinen Daten über Größe und Gewicht auseinandergesetzt hat. Über die oben vorgeführte Rechnung hinaus gibt es bei ihm keinerlei datenbezogene Diskussion. Insbesondere versucht Quetelet nicht, rechnerisch zu zeigen, dass der von ihm postulierte Zusammenhang

$$G \sim l^2$$

zwischen Gewicht G und dem Quadrat der Größe l^2 besser mit den Daten übereinstimmt als Buffons Annahme

$$G \sim l^3.$$

Das wäre auch nicht leicht gewesen. Wenn man nämlich mit modernen Methoden Quetelets Daten analysiert, ist es schwer zu sehen, wie sich daraus die Proportionalität $G \sim l^2$ ablesen lassen soll. Auf S. 91 von (Quetelet 1835/1921) findet man eine Tabelle, in der zu den Größen zwischen 0,60 m und 1,90 m, jeweils in Schritten von 10 cm, die zugehörigen Mittelwerte der Gewichte aus Quetelets Erhebungen angegeben sind. Berechnet man die Korrelationen zwischen G und l, G und l^2

sowie G und l^3, dann erhält man in allen drei Fällen hohe Werte oberhalb von 0,9, eine Bevorzugung des Verhältnisses von G und l^2 lässt sich aber nicht ablesen. Der Leser sollte sich, um die Situation angemessen zu bewerten, klar machen, dass Quetelet die statistischen Werkzeuge, die wir heutzutage benutzen, nicht zur Verfügung standen. Regression und Korrelation wurden erst am Ende des 19. Jahrhunderts von Galton und Yule entwickelt.

So ist es kein Wunder, dass Quetelet sich letztlich sehr vorsichtig ausdrückt, wenn er resümierend feststellt:

> „Nach den zahlreichen Untersuchungen, die ich über die Wechselbeziehungen zwischen den verschiedenen Größen und Gewichten der erwachsenen Menschen angestellt habe, glaubte ich den Schluss ziehen zu können, dass die Gewichte einfach wie die Quadrate der Größen sich verhalten." (Quetelet 1835/1921, S. 94)

Wie gesagt: welcher Art diese zahlreichen Untersuchungen sind, sagt er nicht.

Insgesamt ergibt die Lektüre von Quetelets „Sozialer Physik" das folgende Bild. Quetelet betrachtet drei mögliche Proportionalitäten, nämlich

$$(1)\ G \sim l^2 ,\ (2)\ G \sim l^{2,5}\ \text{und}\ (3)\ G \sim l^3 .$$

Den Fall (2) zieht er in Erwägung, schließt ihn aber unter Hinweis auf die Gesamtheit seiner Daten aus. (3) ist die These von Buffon. Sie beruht auf der Annahme, dass große und kleine „mittlere Menschen" zueinander im geometrischen Sinne ähnlich sind. Dem hält Quetelet entgegen, dass Menschen in der Länge stärker wachsen als im Querschnitt und dass von geometrischer Ähnlichkeit daher keine Rede sein könne. Seine Begründung für diese Aussage ist aber mehr phänomenologischer Art. In seinen Daten findet diese Aussage keine zwingende Stütze. Da er keine weiteren Begründungen anführt, ist es schwer zu sagen, warum Quetelet die Proportionalität (1) $G \sim l^2$ bevorzugt. Man kann nicht ausschließen, dass es sich hierbei um eine Meinung handelt, die zu seiner Zeit auch andere Wissenschaftler vertreten haben. Aussagen dazu gibt es bei Quetelet allerdings nicht.

Der Leser sollte sich klarmachen, dass Quetelet den Quotienten $\frac{G}{l^2}$, also den eigentlichen Body-Mass-Index nicht gebildet hat, geschweige denn, dass er auf die Idee gekommen wäre, diesen Quotienten als Beurteilungsgröße für den Ernährungszustand von Menschen anzuwenden. Quetelet ist ausschließlich an der Formulierung eines Gesetzes interessiert, das die Körpergröße mit dem Gewicht des „mittleren Menschen" verknüpft.

Eine Geschichte des Body-Mass-Index ist noch nicht geschrieben. In der Anthropometrie am Ende des 19. und beginnenden 20. Jahrhunderts wird mit verschiedenen Indizes experimentiert. Auch der Quotient $\frac{G}{l^3}$, der auf Buffons Annahme der Proportionalität $G \sim l^3$ beruht, spielte als „Ponderalindex" weiter eine Rolle. Die Arbeit (Keys et al. 1972) scheint die entscheidende Wende zugunsten des Index $\frac{G}{l^2}$ eingeleitet zu haben. In dieser Arbeit wurden mehrere Studien mit großen

Anzahlen von Probanden ausgewertet, in denen für jede Versuchsperson die Größe, das Gewicht, die Dicke der subkutanen Fettschicht und die Körperdichte gemessen worden sind. Subkutane Fettschicht und Körperdichte erlauben eine direkte Beurteilung des Ernährungszustands, Gewicht und Körpergröße sind leicht
messbar. Die Autoren der Studie stellen nun die einleuchtenden Forderungen auf,
dass ein aus G und l gebildeter Index zur Beurteilung des Ernährungszustandes einerseits von der Länge unabhängig sein sollte, andererseits sollte er möglichst gut
mit den Parametern ‚subkutane Fettschicht' und ‚Körperdichte' korrelieren. Verglichen werden die drei Möglichkeiten

$$(1)\ \frac{G}{l}, \qquad (2)\ \frac{G}{l^2}, \qquad (3)\ \frac{G}{l^3}.$$

Aus den Daten ergibt sich eine deutliche Präferenz von (2) gegenüber (3) und eine
leichte Präferenz von (2) gegenüber (1), genauer:

> Judged by the criteria of correlation with height (lowest is best) and to measures of body
> fatness (highest is best), the ponderal index [das heißt $\frac{G}{l^3}$] is the poorest of the relative
> weight indices studied. The ratio of weight to height squared, here termed the body mass
> index, is slightly better in these respects than the simple ratio of weight to height. The
> body mass index seems preferable over other indices of relative weight on these grounds
> as well as on the simplicity of the calculation and, in contrast to percentage of average
> weight, the applicability to all populations at all times. (Keys et al. 1972, S. 341)

4 Haldanes Prinzip

Der Body-Mass-Index $\frac{G}{l^2}$ setzt sich aus Größen zusammen, die leicht messbar

sind. Da er mit anderen Größen, die eine direkte Beurteilung des Ernährungszustands erlauben, statistisch gut korreliert, ermöglicht er eine praktikable Gewin-
nung von Daten zur Beurteilung des Ernährungszustands. Wie angemessen und relevant dieser Indikator tatsächlich ist, kann in einem mathematischen Aufsatz
natürlich nicht diskutiert werden und ist unter Medizinern und Physiologen in der
Diskussion.

Die Eignung des BMI wird statistisch begründet. Wir wollen uns nun mit der
Frage einer möglichen inhaltlichen Deutung beschäftigen. Dazu knüpfen wir an
einen berühmten Aufsatz des britischen Biologen John Burdon Sanderson Haldane
(1892 – 1964) an, der in den 1920er Jahren geschrieben wurde und den Titel *On
Being the Right Size* trägt. Haldane war theoretischer Biologe und Genetiker und
einer der Begründer der Populationsgenetik. Vorweg sollte gesagt werden, dass
Haldane nicht an Indices für Fettleibigkeit interessiert war, sondern an der Formulierung einer grundlegenden biologischen Gesetzmäßigkeit.

Die Hauptaussage dieses Aufsatzes findet sich in den einleitenden Zeilen.
Haldane sagt, dass die offensichtlichsten Unterschiede in der Natur solche der

Größe seien und dass die Zoologen dieser Tatsache bisher nur wenig Beachtung geschenkt hätten. In Lehrbüchern der Zoologie werde nicht thematisiert, dass der Adler größer sei als der Sperling oder das Nilpferd größer als der Hase. Aber, sagt Haldane, es könne leicht gezeigt werden, dass ein Hase nicht so groß sein könne wie ein Nilpferd oder ein Wal so klein wie ein Hering. Dem liege das Prinzip zugrunde, *dass es für jedes Lebewesen eine angemessene Größe gebe und dass eine beträchtliche Änderung der Größe unvermeidlich eine Änderung der Form nach sich ziehe.* Diese Aussage wird in der Literatur als das *Prinzip von Haldane* bezeichnet.

Den Kern seines Arguments stellt Haldane in amüsanter Form am Beispiel von menschenähnlichen Riesen dar. Doch hören wir ihn selbst:

> Let us take the most obvious of possible cases, and consider a giant man sixty feet high - about the height of Giant Pope and Giant Pagan in the illustrated *Pilgrim's Progress* of my childhood. These monsters were not only ten times as high as Christian, but ten times as wide and ten times as thick, so that their total weight was a thousand times his, or about eighty to ninety tons. Unfortunately the cross-sections of their bones were only a hundred times those of Christian, so that every square inch of giant bone had to support ten times the weight borne by a square inch of human bone. As the human thigh-bone breaks under about ten times the human weight, Pope and Pagan would have broken their thighs every time they took a step. This was doubtless why they were sitting down in the picture I remember. But it lessens one's respect for Christian and Jack the Giant Killer. (Haldane 1927/1985, S. 1)

Pilgrim's Progress war ein weit verbreitetes Erbauungsbuch des britischen Baptistenpredigers und Schriftstellers John Bunyan, das 1678, also 40 Jahre vor Swifts *Gullivers Reisen*, erschienen ist. Auf seinem Weg zur Seligkeit trifft der Held Christian auf alle möglichen Feinde des rechten Glaubens, die er besiegen muss und unter denen –in einem anglikanischen Buch!- der Papst (‚giant pope') nicht fehlen darf. Unterstellt man, dass Riesen im geometrischen Sinne zu Menschen ähnlich sind, dann hat der zehnmal so große Riese ein tausendmal so großes Volumen und eine tausendmal so große Masse, die von seinen Knochen zu tragen ist, deren Querschnitt nur hundertmal so groß ist. Also hat die Flächeneinheit des Oberschenkels von Giant Pope das zehnfache Gewicht zu tragen, das ein menschlicher Oberschenkel tragen muss. Wenn der Knochen des Riesen in seiner Konsistenz sich vom menschlichen Knochen nicht unterscheidet, muss das zum Bruch führen.

Generell gilt, dass die Bruchfestigkeit eines Balkens proportional zu seinem Querschnitt ist. Wenn man sich zu einem beliebigen Lebewesen eine zehnmal so große Kopie vorstellt, dann hat diese ein tausendmal so großes Gewicht, aber nur eine hundertmal so große Tragfähigkeit. Das Skelett wird unter dieser Last brechen. Nun ist das Gewicht proportional zum Volumen V, also proportional zur dritten Potenz der Größe l, d. h. $G \sim V \sim l^3$, der Querschnitt Q ist proportional zum Quadrat der Größe und damit auch zur Oberfläche O, also $Q \sim l^2 \sim O$. Die physikalischen Eigenschaften der Knochen des betrachteten Lebewesens erfordern, dass Gewicht und Querschnitt der Knochen in einem angemessenen Verhältnis stehen. Dieses Verhältnis ist proportional zu $\dfrac{l^3}{l^2}$ bzw. zu $\dfrac{V}{O}$. Letzterer Quotient war als Massigkeit definiert worden. Die oben erwähnte Skalensensitivität der Massigkeit hat also zur Folge, dass das Skelett eines mit dem Faktor k vergrößerten Lebewesens (k hinreichend groß) keine hineichende Tragfähigkeit mehr hat. Genau das wird in Haldanes Prinzip ausgesagt. Ein Lebewesen mit der organisch-physiologischen Ausstattung des Menschen muss grosso modo auch die Größe von Menschen haben. Stellt man sich einen größeren Menschen vor, hat das notwendig eine Veränderung der Form bzw. seiner organisch-physiologischen Ausstattung zur Folge. Das ist es, was möglicherweise auch Quetelet schon wusste oder ahnte, als er so selbstgewiss gegen Buffon sagte, dass der Mensch, wenn er in der Länge wächst, nicht im selben Maße auch in die Breite wachse.

Mit dieser Überlegung ist auch gezeigt, dass die Leibnizsche Vision der unterschiedlich großen, aber geometrisch ähnlichen Tempel unmöglich ist, wenn das Material jeweils dasselbe ist. Der 40 cm hohe, aus einem gewissen Material hergestellte Tempel wird zusammenbrechen, wenn man aus demselben Material einen 40 m hohen Tempel bauen wollte. Umgekehrt gilt: die Statik eines Gebäudes kann nicht an einem verkleinerten Modell getestet werden, wenn man für das Modell dasselbe Material nimmt wie für das zu bauende Objekt.

Haldane dekliniert sein Prinzip an einer Fülle aufschlussreicher zoologischer Beispiele durch. Man stelle sich etwa vor, dass die graziöse Gazelle, die über lange Beine verfügt, beträchtlich größer wird. Dann würden ihre Knochen brechen, wenn sie nicht eines von zwei Dingen tut. Zum einen könnte sie ihre Beine kurz und dick machen wie die eines Rhinozerus, so dass jedes Kilo Gewicht immer noch von derselben Querschnittsfläche Knochen getragen würde, oder sie könnte ihren Körper komprimieren und ihre Beine schräg stellen wie eine Giraffe. Rhinozerus und Giraffe sind von derselben Größenordnung und mechanisch sehr erfolgreich, weil bemerkenswert schnelle Läufer. (Haldane 1927/1985, S. 1/2) Entsprechend Haldanes Prinzip hat also die Änderung der Größe notwendig eine Änderung der Form zur Folge.

Stürzt ein Mensch in einen tausend Meter tiefen Schacht, würde er sterben, ein Pferd würde zersplittern, aber eine Maus würde dies überleben. Denn der Luftwiderstand ist proportional zur Oberfläche. Wenn man ein Lebewesen in allen Dimensionen auf ein Zehntel seiner ursprünglichen Größe verkleinert denkt, dann reduziert sich sein Gewicht auf ein Tausendstel, seine Oberfläche aber nur auf ein Hundertstel. Folglich verzehnfacht sich sein Luftwiderstand.

Ein letztes Beispiel: Säugetiere sind darauf angewiesen, ihre Körpertemperatur auf einer konstanten Höhe zu halten. Der Körper produziert Wärme und gibt sie über die Haut an die kühlere Umgebung ab. Wärmeproduktion und –abgabe müssen im Gleichgewicht sein. Die Wärmeproduktion ist proportional zum Volumen des Körpers, die Wärmeabgabe proportional zur Oberfläche, durch die sie erfolgt. Man stelle sich wieder eine Maus vor, bei der beides im Gleichgewicht ist. Verzehnfacht man in Gedanken die Größe der Maus, dann vertausendfacht man ihr Volumen und die Menge der produzierten Wärme. Die Oberfläche und damit die

Kapazität zur Wärmeabgabe würden aber nur verhundertfacht. Die Körpertemperatur würde entsprechend steigen, die zehnmal vergrößerte Maus wäre nicht lebensfähig.

Insgesamt ergibt sich aus diesen und vielen weiteren Beispielen, die man bei Haldane und in der daran anschließenden Literatur findet, folgendes Bild:

1. Der Quotient $\frac{G}{l^2} \sim M = \frac{V}{O} \sim \frac{l^3}{l^2}$ ist charakteristisch für einen physiologischen bzw. statischen *Gleichgewichtszustand* bei Lebewesen und materiellen Objekten. Bei Lebewesen geht es um den Wärmehaushalt, die zu leistende Arbeit bei der Fortbewegung und den Einfluss der Schwerkraft auf die Lebensbedingungen. Bei unbelebten Objekten steht die Bruchfestigkeit im Mittelpunkt.

2. Wegen der Skalensensitivität der Massigkeit können dieses Gleichgewichte bei Größenänderung nur bestehen bleiben, wenn sich die Form ändert, wenn also die Größenänderung nicht in allen Dimensionen in gleichem Maße stattfindet, oder, mathematisch ausgedrückt, wenn sie keine Ähnlichkeitstransformation ist.

3. Aus derselben Problematik heraus sucht man heute bei der Modellierung und Simulation technischer Objekte und Prozesse nach sogenannten *dimensionslosen Kennzahlen*. Nur solche Kenngrößen können vom „kleinen" Modell auf „große" Objekte übertragen werden. Eine solche dimensionslose Kennzahl heißt dann auch *Ähnlichkeitszahl*.

4. In gewissen Grenzen sind die betrachteten belebten und unbelebten Objekte gegen Schwankungen des Gleichgewichts tolerant. Die Statik ist so ausgelegt, dass eine maßvoll höhere Belastung kein Problem dargestellt, Störungen im Wärmehaushalt können durch Körperreaktionen ausgeglichen werden, etc.

5. Es hängt also von weiteren spezifischen Eigenschaften des betrachteten Objektes ab, welche Größenänderung zu einer tatsächlichen Beeinträchtigung bzw. Gefährdung seiner Existenz führt.

6. Im Hinblick auf den BMI ist daher die statistische Analyse von Daten zum Ernährungszustand und zur Physiologie des Menschen unabdingbar, um seine Aussagekraft abschätzen zu können. Es könnte ja durchaus sein, dass die Gleichgewichtstörungen, die durch Haldanes Prinzip prognostiziert werden, erst bei Größenänderungen relevant werden, die außerhalb der Bandbreite der bei Menschen üblichen Größendifferenzen liegen.

5 Verständnisprobleme beim proportionalen Denken

Es gibt eine Reihe von Publikationen, in denen Haldanes Prinzip für den (Mathematik- und Naturwissenschaftsunterricht aufbereitet worden ist. Wir verweisen exemplarisch auf den Abschnitt „Wie im Kleinen, so nicht im Großen" in (Glaeser 2008), die Einleitung zu (Sexl et al. 1990) sowie auf (Schlichting und Rodewald 1986) und (Apolin 1996).

Auf Details unserer Vorlesung zu diesem Thema soll nicht eingegangen werden. Die erheblichen Verständnisprobleme vieler Studierender sind allerdings lehrreich. Der sachliche Kern des ganzen Problems steckt in der folgenden Aufgabe:

a) Ein Mann mit 1,60 m Körpergröße habe eine Masse von 50 kg. Welche Masse hat ein anderer Mann mit *ähnlicher* Statur, der 2 m groß ist?
b) Ein Mann mit 1,60 m Körpergröße habe eine Masse von 50 kg. Welche Masse hat ein anderer Mann mit demselben Body-Mass-Index, der 2 m groß ist?
c) Erklären Sie den Unterschied der Ergebnisse in a) und b) mit Hilfe des Begriffs der Massigkeit.

Der Teil a) der Aufgabe stammt aus (Glaeser 2008, S. 53), die Teile b) und c) wurden vom Autor hinzugefügt. Zur Beantwortung wird etwa folgende Lösung erwartet:

a) Der 2 m-Mann mit ähnlicher Statur geht aus dem 1,60 m-Mann durch eine Streckung

mit dem Faktor (5/4) hervor. Sein Volumen ist daher das $\left(\dfrac{5}{4}\right)^3 = \dfrac{125}{64} \approx 2$ -fache des

1,60 m-Mannes. Gleiche Gewebedichte vorausgesetzt, verdoppelt sich folglich auch das Gewicht auf ca. 100 kg.

b) Der BMI des 1,60 m - Mannes beträgt

$$BMI_{1,60\,\mathrm{m}} = \frac{50\ \mathrm{kg}}{1,60^2\ \mathrm{m}^2} \approx 19,5\ \frac{\mathrm{kg}}{\mathrm{m}^2}\ .$$

Daraus folgt für das Gewicht $M_{2\mathrm{m}}$ des 2 m-Mannes

$$M_{2\mathrm{m}} = 19,5\ \frac{\mathrm{kg}}{\mathrm{m}^2} * 4\ \mathrm{m}^2 = 78\ \mathrm{kg}\ .$$

c) Der BMI ist proportional zur Massigkeit. Folglich muss er nach dem Ergebnis in a) bei dem 2 m-Mann „mit ähnlicher Statur" um den Faktor 5/4 wachsen. Umgekehrt beträgt das Gewicht des 2 m-Mannes, der denselben BMI hat, nur 4/5 des Gewichts des „ähnlichen" 2 m-Mannes. Es ist

$$100*(4/5) = 80 \approx 78.$$

(Hätten wir genau gerechnet, wäre es auch genau herausgekommen.)
Folglich: wenn ein größerer Mensch denselben BMI wie ein kleinerer haben soll, muss er bei gleicher Gewebedichte deutlich schlanker sein. Dies ist eine Folge der Skalensensitivität der Massigkeit und des BMI.

Bei Glaeser steht die Aufgabe a) im Zusammenhang mehrerer Übungen, in denen die Einsicht, dass sich Volumina mit k^3 ändern, in verschiedenen Kontexten angewandt werden soll. Eine Aufgabe fragt z.B. nach der Volumenänderung einer Pyramide bei maßstäblicher Vergrößerung. Die Mehrzahl unserer Studierenden hatte große Hemmungen, den allgemeinen Sachverhalt tatsächlich anzuwenden und das ursprüngliche Volumen einfach mit k^3 zu multiplizieren. Stattdessen zogen sie es vor, mit der „sicheren" Formel für das Volumen einer Pyramide zu rechnen.

Im Falle des maßstäblich auf 2 m vergrößerten Mannes gibt es allerdings keine Formel. Man ist also gleichsam „gezwungen", den allgemeinen Sachverhalt zu nutzen. Das haben viele Studierende aber nicht getan, sondern „proportional" gerechnet. Sie multiplizierten das Gewicht des 1,60 m – Mannes nicht mit $\left(\frac{5}{4}\right)^3$,

sondern mit $\left(\frac{5}{4}\right)$, und waren sich ziemlich sicher, dass das auch zu einem richtigen Ergebnis führt.

Der Kontrast schließlich zwischen den Gewichten, die sich aus den verschiedenen Ansätzen in a) und b) ergeben, war für viele Studierende eine große Überraschung. Der Unterschied von etwa 20 kg ist in der Tat beträchtlich. Nur wenige Studierende konnten diesen Sachverhalt durch Hinweis auf die Skalensensitivität der Massigkeit, aus der sich auch die Skalensensitivität des zu ihr proportionalen BMI ergibt, angemessen erklären. Für die meisten Studierenden ist der BMI intuitiv ein Parameter, der die Güte einer Gestalt misst. Er soll ja Fettleibigkeit von Schlankheit unterscheiden. Folglich erfasst er (analog zum „Goldenen Schnitt") so etwas wie die „richtigen Proportionen" einer Gestalt. Daher sollte ein gut proportionierter „großer" Mensch denselben BMI haben wie ein ähnlich gut proportionierter „kleiner" Mensch.

Den Studierenden wurde in diesem Zusammenhang das folgende Gedankenexperiment vorgeschlagen. Man stelle sich vor, dass der in Florenz stehende David des Michelangelo in Fleisch und Blut von seinem Podest herabsteigt. Der David ist 5,17 m groß, er sei ähnlich zu einem lebenden Mann von 1,80, der über einen „idealen" BMI von 23 verfügt. Was kann man über den BMI des David sagen? Auch hier fiel es schwer, die einfache Rechnung anzustellen, dass der lebende David etwa 2,8 mal so groß wäre wie die Vergleichsperson und demgemäß einen krankhaften BMI von etwa 64 hätte. Dieses Gedankenexperiment ist selbstverständlich nur eine weitere Variation von Haldanes Überlegungen zu Giant Pope und Christian.

Insgesamt zeigt sich, dass Proportionalität und Ähnlichkeit für Studierende der Anfangssemester ein durchaus schwieriges Begriffsfeld darstellen. Um sich in diesem Begriffsfeld zu bewegen, ist ein erhebliches Abstraktionsvermögen erforderlich. So ist es schon nicht selbstverständlich zu verstehen, dass man bereits über Information verfügt, wenn man weiß, dass zwei Größen proportional sind. Was bedeutet es, dass eine Größe proportional zum Quadrat einer anderen Größe ist? Letzteres ist fundamental für ein Verständnis des BMI. Gleichzeit besteht eine Tendenz zur proportionalen Übergeneralisierung. Zur Vielschichtigkeit des Proportionalitätsbegriffs vergleiche man den Überblick in (Jahnke und Seeger 1986). Dieses Begriffsfeld bedarf dringend einer Klärung spätestens beim Studienbeginn, denn seine Beherrschung stellt eine Schlüsselqualifikation für ein erfolgreiches Arbeiten im Bereich der MINT - Fächer dar.

Literaturverzeichnis

Apolin, M. (1996). Von Gullivers Reisen bis Jurassic Park. Größenordnungen in der Natur. *Plus Lucis*, 2, 29-32.

Bohlen, A. (2010). *Der „Body Mass Index" – Eine bibliometrische Analyse* (Dissertation). Medizinische Fakultät Charité, Berlin.

Galilei, G. (1973). Unterredungen und mathematische Demonstrationen über zwei neue Wissenszweige, die Mechanik und die Fallgesetze betreffend. (A. von Oettingen, Hrsg. & Trans.). Darmstadt: Wissenschaftliche Buchgesellschaft. (Originalwerk veröffentlicht 1638).

Glaeser, G. (2008). *Der mathematische Werkzeugkasten. Anwendungen in Natur und Technik* (3. Aufl.). Heidelberg: Spektrum Akademischer Verlag.

Haldane, J. B. S. (1985). On Being the Right Size. In J. B. S. Haldane, *On Being the Right Size and other essays* (S. 1-8). Oxford & New York: Oxford University Press. (Originalwerk veröffentlicht 1927).

Hankins, F. H. (1908). *Adolphe Quetelet as statistician*. New York: AMS Press.

Jahnke, H. N., & Seeger, F. (1986). Proportion. In G. v. Harten, H. N. Jahnke, T. Mormann, M. Otte, F. Seeger, H. Steinbring, & H. Stellmacher (Hrsg.), *Funktionsbegriff und funktionales Denken* (S. 35-83). Köln: Aulis.

Keys, A., Fidanza, F., Karvonen, M.J. Kimura, N., & Taylor, H. L. (1972). Indices of relative weight and obesity, *Journal of chronical diseases, 25*, 329-343.

Leibniz, G. W. (1966). De analysi situs. In G. W. Leibniz, *Hauptschriften zur Grundlegung der Philosophie* (Bd.1, übersetzt von A. Buchenau, durchgesehen und mit Einleitungen und Erläuterungen herausgegeben von E. Cassirer, 3. Auflage, S. 69-76). Hamburg: Felix Meiner. (Originalwerk ohne Jahresangabe).

Mamerow, Th. (2012). *Der BMI bei Quetelet.* (Schriftliche Hausarbeit im Rahmen der Ersten Staatsprüfung für das Lehramt GHR mit Schwerpunkt HRGe) Universität Duisburg-Essen.

Menninger, K. (1958). *Mathematik in deiner Welt. Von ihrem Geist und ihrer Art zu denken.* Göttingen: Vandenhoeck&Ruprecht.

Quetelet, A. (1921). *Soziale Physik oder Abhandlung über die Entwicklung der Fähigkeiten des Menschen* (Band 2, Übersetzung von 1869). Jena: Gustav Fischer. (Originalwerk von 1835).

Schlichting, H. J., & Rodewald, B. (1986). Energiehaushalt und Körperbau. Unterrichtsmodell für die Sekundarstufe 1. (9./10. Schülerjahrgang). *Unterricht Biologie, 120*, 20-24.

Sexl, R., Raab, I., & Streeruwitz, E. (1990). *Das mechanische Universum. Eine Einführung in die Physik* (1. Band). Frankfurt: Diesterweg & Aarau: Sauerländer.

Stigler, S. M. (1986). *The History of Statistics. The Measurement of Unvertainty before 1900.* Cambridge/Mass & London: Harvard University Press.

Winter, H. (1997). Mathematik als Schule der Anschauung, oder: Allgemeinbildung im Mathematikunterricht des Gymnasiums. In R. Biehler, & H. N. Jahnke (Hrsg.), *Mathematische Allgemeinbildung in der Kontroverse. Materialien eines Symposiums an der Universität Bielefeld.* Occasional Paper des IDM der Universität Bielefeld, Bd. 163. IDM Bielefeld.

Kapitel 3
Unterrichtsgestaltungen zur Kompetenzförderung: zwischen Instruktion, Konstruktion und Metakognition

Stanislaw Schukajlow

Westfälische Wilhelms-Universität Münster

Werner Blum

Universität Kassel

Abstract   Im Beitrag werden unter Rückgriff auf verschiedene Lehr-Lern-Konzeptionen mögliche Unterrichtsgestaltungen zur Förderung des Kompetenzerwerbs diskutiert, konkret zum Erwerb von mathematischer Modellierungskompetenz. Zentral erscheint für einen fachlichen ertragreichen Unterricht, auf die Ausgewogenheit der drei folgenden Leitprinzipien zu achten: (1) fachkundige Instruktion durch die Lehrperson, (2) eigenständige Konstruktion von Lösungselementen durch die Lernenden und (3) metakognitive Reflexionen beim Lernprozess. Eine geeignete Orchestrierung dieser drei Faktoren im Mathematikunterricht der Jahrgangsstufe 9 wird mit Bezug auf das DISUM-Projekt (Didaktische Interventionsformen für einen selbständigen aufgabengesteuerten Unterricht am Beispiel Mathematik) am Beispiel der so genannten „methoden-integrativen" Lernumgebung demonstriert.

1 Einleitung

Entwicklung und Evaluation von Instruktionsmodellen und speziell von Unterrichtsgestaltungen sind zentrale Aufgaben fachdidaktischer und psychologisch-pädagogischer Forschungen. Drei wichtige Fragen in diesem Umfeld können wie folgt formuliert werden:

- Welche Ziele sollten in einem guten Unterricht verfolgt werden?
- Welche Elemente bilden den Kern eines guten Unterrichts und sollten im alltäglichen Lehrerhandeln beachtet werden, um die Unterrichtsziele zu erreichen?

- Wie können diese Elemente in die Unterrichtsplanung implementiert und dann auch in die Praxis umgesetzt werden?

In den Phasen der Planung und Durchführung von Unterricht sowie bei der Beurteilung der Effektivität von Lernumgebungen sind die Ziele des Unterrichts von entscheidender Bedeutung. Die möglichen Unterrichtsziele weisen eine große Breite und Heterogenität auf und sind schwer alle zugleich zu erreichen. Emotionales Wohlbefinden und unterrichtsbezogene Motivation gehen nicht immer mit kognitiven Fortschritten einher, so dass auch die zentralen Eigenschaften eines guten Unterrichts durchaus differieren können. Es wird zum Beispiel vermutet, dass ein kognitiv anspruchsvoller, aufgabenbasierter Unterricht längerfristig sogar zu motivational-emotionalen Einbußen bei Lernenden führen kann (Helmke und Weinert 1997). Das Verfolgen eines bestimmten Ziels befördert somit nicht unbedingt die anderen Intentionen beim Lernprozess. Im Hinblick auf die intendierten Unterrichtsziele wird in der Forschung versucht, wichtige Leitprinzipien qualitätvoller Unterrichtsgestaltungen zu identifizieren. Bei der Bestimmung solcher handlungsleitenden Elemente greift man auf die Lehr-Lerntheorien zurück, die einen allgemeinen Rahmen für die Instruktionsmodelle bilden. Der Lernbegriff umfasst dabei Änderungen in kognitiven, sozialen und emotional-affektiven Bereichen und damit zusammenhängende Verhaltensmodifikationen. Die Elemente oder Leitprinzipien der Unterrichtsgestaltungen müssen bei der Unterrichtsplanung konkretisiert und in die Praxis umgesetzt werden. Diese äußerst anspruchsvolle Aufgabe bildet den schwierigsten, zugleich aber auch den interessantesten Teil der Lehrtätigkeit. Vielfältige Einflussfaktoren seitens der Lernenden, der Lehrperson sowie der äußeren Rahmenbedingungen erschweren einen programmatischen Umgang mit der Entwicklung von Unterrichtsgestaltungen und erfordern situative Anpassungen bei der Unterrichtsplanung und -durchführung. In diesem Beitrag stehen primär kognitive, auf den fachlichen Lernzuwachs gerichtete Unterrichtsziele im Fokus. Wir konzentrieren uns hierbei auf drei wichtige Prinzipien bei der Gestaltung von Lernumgebungen: Instruktion, Konstruktion und Metakognition. Diese betrachten wir zuerst im Kontext von Lehr-Lerntheorien (Kap. 2). In einer für das Primat des jeweiligen Unterrichtsprinzips prototypischen Lehr-Lernform werden diese Prinzipien analysiert (Kap. 3) und ihre ausgewogene Umsetzung wird am Beispiel einer operativ-strategischen Lernumgebung aus DISUM vorgestellt (Kap. 4).

2 Instruktion, Konstruktion und Metakognition in Lehr-Lerntheorien und -konzeptionen

2.1 Instruktion

Lerntheoretisch wird oft zwischen instruktions- und konstruktionsorientierten Theorien unterschieden (Shuell 1996). Diese Einteilung hängt mit der historischen Entwicklung der Lerntheorien vom Behaviorismus zum Kognitivismus und Konstruktivismus zusammen. Klassisches und später auch operantes Konditionieren (Skinner 1971), die den Kern des Behaviorismus bilden, haben sich mit der Erforschung von Instruktionsmaßnahmen und ihrem Einfluss auf Verhaltensänderungen beschäftigt. Bekannt sind im Bereich schulischen Lernens die so genannten Verstärkungspläne, die eine gewünschte Verhaltensänderung hervorrufen sollen (vgl. Steiner 2001). Gemäß der operanten Lerntheorie kann das gewünschte Verhaltensmuster durch äußere Anreize verstärkt werden. Ein schrittweises Abstellen der Belohnung führt zur Stabilisierung der intendierten Verhaltensweisen und zum dauerhaften Lernen des neuen Verhaltensmusters. Die aktive Rolle des Lernenden im Lernprozess wurde dabei kaum beachtet. Nachfolgende kognitionsorientierte Theorien haben zwar den Lernenden in den Blick genommen, ihre Hauptaufmerksamkeit galt jedoch weiterhin der Instruktion. Die Aufgabe des Lehrers wurde darin gesehen, eine geeignete Darbietung für den Unterrichtsstoff zu finden, der gelernt werden soll. So beschreibt Gagné (1962, S. 88) drei grundlegende Schritte bei der Entwicklung einer Instruktion im Rahmen seiner Lerntheorie wie folgt:

1. identifying the component task of final performance,
2. insuring that each of these component tasks is fully achieved,
3. arranging the total learning situation in a sequence which will insure optimal mediational effects from one component to another.

Die nähere Betrachtung der genannten Aktivitäten macht deutlich, dass eine Analyse des Lerngegenstandes, seine Sequenzierung und Vermittlung und somit die Tätigkeit des Lehrenden im Mittelpunkt der genannten Theorie steht. Der Lernende kommt hier in einer passiven Rolle als Rezipient des zu vermittelnden Wissens vor.

Eine andere instruktionsorientierte Lerntheorie ist das Lernen am Modell von Bandura (1976). Wesentliche Lernelemente dieser Theorie sind Beobachtung eines Verhaltens und seine anschließende Nachahmung. Diese Theorie setzte sich von behavioristischen Lerntheorien dadurch ab, dass die Belohnung bzw. Bestrafung nicht unmittelbar von den lernenden Personen erlebt werden muss. Untersuchungen von Banduras Arbeitsgruppe zeigten, dass das *Beobachten* von Verhalten und dessen Folgen alleine schon Lernprozesse stimulieren und zu Verhaltensänderungen führen kann. Die Entscheidungen des Lehrenden bei der Entwicklung einer Instruktion gemäß dem Lernen am Modell umfassen vor allem Fragen, welche Elemente der Experte den Lernenden demonstrieren und wie dies geschehen soll.

Die beschriebenen Beispiele der instruktionsorientierten Lerntheorien könnten den falschen Eindruck erwecken, dass diese Theorien veraltet seien oder gar ausgedient hätten. Vielmehr kann man jedoch von einer Erweiterung der Sichtweise auf das Lernen sprechen. Die instruktionsorientierten Theorien stützen sich auf die Annahme, dass Lernprozesse bei allen Individuen einer Lerngruppe ähnlich ablaufen. Diese Auffassung des Lernens ermöglicht es, Lern- und Instruktionsmodelle aufzustellen, die für alle Lernende gültig sein sollen und diese dann empirisch zu prüfen. Solche Modelle müssen nicht wie bei radikalen Behavioristen die Existenz von kognitiven Prozessen verneinen. Gerade in der Kognitionspsychologie sind Gedächtnismodelle verbreitet, die den Prozess der Informations-Aufnahme, -Verarbeitung und -Speicherung beschreiben. Die Forschungen zur Informationsverarbeitung werden in neuer Zeit mit den Erkenntnissen und Forschungsmethoden aus der Neurowissenschaft verknüpft, um weitere Indikatoren für Lernprozesse zu finden.

2.2 Konstruktion

In konstruktivistisch orientierten Ansätzen wird die individuelle Komponente des Wissenserwerbs betont. Shuell (1996, S. 744) schreibt in diesem Zusammenhang: „The constructive nature of meaningful learning implies that no two students have exactly the same perception of the instructional situation or end up with exactly the same understanding of the material being learned." Die Art der Auseinandersetzung mit dem Lernmaterial und die ablaufenden kognitiven Prozesse beeinflussen die Lernergebnisse und nicht die Instruktion an sich. Man kann somit annehmen, dass der Einfluss der Instruktion auf die Lernergebnisse über die individuellen kognitiven Prozesse vermittelt wird.

Die Geschichte der konstruktivistischen Lerntheorien geht auf die Gestaltpsychologie zurück und wurde ausführlich von Aebli (1980) nachgezeichnet. Nach der Auffassung von Aebli sollen alle Instruktionsmaßnahmen im Hinblick auf den Lernenden konzipiert werden. Der Prozess des Wissensaufbaus erfolgt schrittweise von der Handlung über eine Operation zum Begriff und besteht im Verknüpfen, Verdichten und Strukturieren von Wissen (siehe Anwendungsbeispiele bei Steiner 2001). Die genannten Prozesse des Wissensaufbaus laufen individuell ab. Die Aeblische Lerntheorie fokussiert auf das Verstehen und nicht etwa auf Verhaltensänderungen oder auf Abspeichern und Abrufen von Informationen (Hasselhorn und Gold 2006). Zentral ist für Aebli die Tätigkeit des Problemlösens, die als Vehikel der Denkprozesse verstanden wird (Schukajlow 2011).

Die konstruktivistischen Lerntheorien wurden unter Akzentuierung der Selbstregulation und Metakognition weiterentwickelt. Die Modelle des selbstregulierten Lernens können zu einer eigenen Gruppe von Lerntheorien zusammengefasst werden.

2.3 Metakognition

Theorien des selbstregulierten Lernens suchen eine Antwort auf die Frage, wie Expertise in eigenen Lernprozessen erworben werden kann. Sie stützen sich auf die Annahme, dass Lernende ihren Wissenserwerb aktiv steuern. In der Tradition der konstruktivistisch orientierten Lehr-Lerntheorien stehen Lernende in Theorien der Selbstregulation im Mittelpunkt des Lernprozesses. „This approach views learning as an activity that students do for themselves in a proactive way, rather than as a covert event that happens to them reactively as a result of teaching experience" (Zimmerman 2001, S. 1). Die Orientierung an der Selbstregulation im Lernprozess hat Folgen für die Konzeption des Lehrens. Instruktionsmaßnahmen werden als Lernumgebungen angesehen, die individuelle Strategien von Lernenden fördern und fordern sollen. Der Grad der Selbstregulation eines Lernenden ist variabel und hängt vom Individuum sowie von der Instruktion ab. Lernende, die selbstreguliert lernen, fokussieren ihre Gedanken, Gefühle und Handlungen auf die Ziele im Lernprozess. Speziell für den Wissenserwerb im kognitiven Bereich ist Metakognition von entscheidender Bedeutung. Die metakognitiven Lernprozesse umfassen Planung, Kontrolle und Regulation eigenen Lernens. Der Lernprozess beginnt idealtypisch mit der Setzung der Ziele und der Planung der Arbeitsschritte. Die Arbeitsschritte werden laufend kontrolliert. Die Ergebnisse der Kontrolle können Regulationsstrategien aktivieren, welche ggf. einen Wechsel in kognitiven Strategien anleiten (Schukajlow und Leiss 2011). Einschränkend muss man betonen, dass die drei Gruppen von Lehr-Lerntheorien zwar verschiedene Schwerpunkte setzen, jedoch keinesfalls disjunkt sind. Lehren, Lernen und Selbstregulation finden ihren Platz in jeder dieser Lehr-Lern-Theorien, wenn auch mit je unterschiedlicher Gewichtung (siehe exemplarisch zur Selbstregulation Zimmerman 2001). Nach der Einordnung der Lehr-Lerntheorien in diese Gruppen folgt nun ein kurzer Überblick über Unterrichtsdesigns, die unter Rückgriff auf diese Theorien entwickelt wurden.

3 Empirische Ergebnisse zur Wirksamkeit von Unterrichtsdesigns mit den Schwerpunkten auf Instruktion, Konstruktion und Metakognition

Die Einschätzung von Lerneffekten kann mit Hilfe von verschiedenen Forschungsparadigmen wie Persönlichkeits-, Prozess-Produkt- oder Experten-Paradigma vorgenommen werden. Bisher wurden in fast allen Metaanalysen, die sich mit der zusammenfassenden Untersuchung der Wirksamkeit verschiedener Faktoren befassen, die Effekte der Instruktion im Rahmen des Prozess-Produkt-Paradigmas evaluiert (Seidel und Shavelson 2007). Der Unterrichtsprozess beeinflusst gemäß diesem Modell das Produkt des Lernens. Am Produkt des Lernens – also der emotionalen, motivationalen oder kognitiven Entwicklung von Lernenden

– wird der Erfolg von Interventionsmaßnahmen gemessen. Kontextmerkmale wie Alter oder Geschlecht, aber auch die allgemeinen Rahmenbedingungen des Lernens, z.B. sozioökonomischer Status der Familie, werden in die Auswertungen mit einbezogen. In neueren Metaanalysen werden Bestrebungen deutlich, die Perspektive der Lernenden noch stärker zu berücksichtigen und metakognitive, selbstregulative Elemente in den Vordergrund zu stellen. Folgende Aspekte des Lehrens werden dabei herausgestellt: Konstruktivismus, Domänenspezifität, Soziales Lernen, Zielorientierung, Evaluation und Regulation (Seidel und Shavelson 2007). In Unterrichtsdesigns wurden in der letzten Dekade selten nur ein spezielles Merkmal oder eine Lerntheorie implementiert. Viel häufiger wurden mehrere einzelne Aspekte zusammengefügt und komplexe Unterrichtsmuster evaluiert (Borko 2004). Aufgrund der integrativen Zusammensetzung solcher Unterrichtsmuster ist es kaum möglich, eine in der Praxis erprobte Lernumgebung einer einzigen Lehr-Lerntheorie eindeutig zuzuordnen. Nimmt man eine solche Anordnung vor, kann direkte Instruktion eher als eine instruktionsorientierte Lehr-Lernform bezeichnet werden. Konstruktivistisch orientierte Lehr-Lernformen sind vor allem entdeckendes (genauer: entdeckenlassendes) und problembasiertes Lernen. Konstruktivistisch orientierte Lernumgebungen mit metakognitiven, selbstregulativen Elementen sind Unterrichtsdesigns, in denen die Vermittlung von Strategien im Vordergrund steht.

3.1 Direkte Instruktion

Unter direkter Instruktion wird eine Unterrichtsmethode verstanden, bei der das Unterrichtsgeschehen vergleichsweise stark von außen her – durch Lehrer oder auch Computer – gesteuert wird. Weinert (1996) benennt folgende Elemente der direkten Instruktion:

- Die Lehrperson legt angemessene Lernziele fest und zerlegt den Lernstoff in kleine, sinnvolle Lerneinheiten,
- Diese Lerneinheiten werden im fragend-entwickelnden Unterrichtsgespräch vermittelt,
- Es werden Aufgaben verschiedener Schwierigkeit gestellt, die Lernende bearbeiten können,
- Die Lehrperson sorgt für ausreichende Übungen, kontrolliert individuelle Lernfortschritte der Schüler und hilft bei der Überwindung von Schwierigkeiten im Lernprozess.

Wenn man verschiedene Beschreibungen der direkten Instruktion miteinander vergleicht, unterscheiden sich diese in der Zusammensetzung und in der Auswahl ihrer einzelnen Elemente. Gelegentlich wird z.B. auch Lernen am Modell mit Vor- und Nachmachen als ein Bestandteil der direkten Instruktion aufgeführt (Hattie 2009, S. 265). Die Beschreibungen der direkten Instruktion zeigen zugleich eine

Gemeinsamkeit: Handlungen werden von Lernenden in dieser Lehr-Lernform stets durch die Lehrkraft gesteuert.

Die Lehr-Lernform der direkten Instruktion wurde vielfach erfolgreich evaluiert. In Mathematik wurden bei der direkten Instruktion mittlere Effektgrößen – Cohens-d von 0,50 bis 0,65 – beobachtet (Baker et al. 2002; Haas 2005). Die Anwendung der direkten Instruktion führt in diesen Studien also zu einem Leistungsunterschied von mehr als der Hälfte einer Standardabweichung im Vergleich zu Kontrollgruppen. Diese Effekte sind etwas größer bei Schülern mit Lernschwierigkeiten, bleiben aber auch für andere Leistungsgruppen bedeutsam.

3.2 Entdeckenlassendes und problembasiertes Lernen

Eine Lehr-Lernform, die sich auf den ersten Blick konsequent an den Prinzipien des Konstruktivismus orientiert und wo auf die Steuerung des individuellen Lernprozesses von außen verzichtet wird, ist das endeckende Lernen. Betrachtet man aber diese Lernumgebung in ihrer Reinform genauer, erweist sich die Konstruktion von Wissen für Lernende in dieser Lernumgebung oft als nicht möglich und die Prämisse „fehlende Steuerung von außen führt zur Maximierung der Eigenkonstruktion" als falsch. Die Reduktion der Steuerung auf ein Minimum überfordert Lernende im Allgemeinen, behindert ihre Eigenkonstruktion und entspricht somit nicht dem Leitprinzip des Konstruktivismus. Aus den genannten Gründen gelten „geleitetes endeckendes Lernen" oder problembasiertes Lernen als prototypische Formen von konstruktivistischen Lernumgebungen. Die Vorzüge dieser Lehr-Lernformen werden immer wieder hervorgehoben und kontrovers diskutiert (Hmelo-Silver et al. 2007; Mayer 2004). Wir stellen hier problembasiertes Lernen als ein Beispiel einer konstruktivistisch-orientierten Lernumgebung vor. In dieser Lehr-Lernform werden Eigenkonstruktionsprinzipien in die Praxis umsetzt, und gerade in Mathematik wurde diese Form öfters evaluiert. In Anlehnung an Barrows (1996) charakterisieren Gijbels et al. (2005, S. 6) problembasiertes Lernen wie folgt:

* learning needs to be student-centered
* learning has to occur in small student groups
* the presence of a tutor as a facilitator or guide
* authentic problems are primarily encountered in the learning sequence, before any preparation or study has occurred
* the problems encountered are used as a tool to achieve the required knowledge and the problem-solving skills necessary to eventually solve the problem
* new information needs to be acquired through self-directed learning.

Diese Charakterisierung von Aktivitäten zeigt, dass die konstruktive Eigentätigkeit der Lernenden im Mittelpunkt des problembasierten Lernens steht. Die bisherigen Untersuchungen konstatieren eine breite Streuung der Effektivität dieser Lehrumgebung. Im Mittel wurden in diesen Studien Effekte des problembasierten Lernens von lediglich 0,15 nachgewiesen (Hattie 2009). Insbesondere die Vermittlung von Wissen gelingt im problembasierten Lernen nur schwer. Die Wirkungen sind jedoch bedeutsam (bis zu 0,75), wenn allgemeine Prinzipien des Problemlösens oder die Anwendung und Konsolidierung vorhandenen Wissens als Lernziele gesetzt und überprüft werden.

3.3 Training von Selbstregulation und Strategien

In Lehr-Lernformen mit dem Fokus auf dem Training von Selbstregulation und Strategien wird angenommen, dass Lernende ihren Lernprozess mit Hilfe von Strategien selber steuern. Die Reflexion über eigene Lernprozesse auf der Metaebene ist ein wichtiges Element solcher Lernumgebungen, welche die Vermittlung, Auswahl und anschließende produktive Nutzung der neuen Strategien ermöglichen sollen. In Mathematik orientiert sich die Vermittlung von Strategien oft an dem Vierphasenmodell von Polya (1948). Kramarski und Mevarech (2003, S. 283) sehen die eigenständige Formulierung und Beantwortung von Fragen zu drei folgenden Aspekten als gemeinsames Element solcher Lernumgebungen:

* the nature of the problem or task
* construction of relationships between previous and new knowledge
* the use of strategies appropriate for solving the problem or task.

Da es ein breites Spektrum von Strategien gibt, die in verschiedenen Studien vermittelt wurden, variiert die Effektivität des Trainings von Strategien und Selbstregulation von 0,22 bis 0,85 und erreicht für Lernumgebungen mit dem Schwerpunkt auf Metakognition im Mittel immerhin den Wert 0,69 (Hattie 2009). Eine Regulation eigener Aktivitäten kann bereits in der Grundschule durch geeignete Interventionsmaßnahmen angeleitet und signifikant verbessert werden (Kramarski et al. 2010).

Der Vergleich der empirischen Wirksamkeit der drei beschriebenen Unterrichtsdesigns zeigt, dass ihre Implementation zu unterschiedlich starken Leistungssteigerungen bei Lernenden führt. Am effektivsten von den drei betrachteten Lernumgebungen erscheinen die direkte Instruktion und das Training von Selbstregulation und Lernstrategien. Das entdeckende und das problembasierte Lernen sind weniger effektiv und zeigen ihre Wirkungen in der Regel, wenn zudem die Unterstützung von Lernenden durch Instruktionsmaßnahmen in das Unterrichtsskript aufgenommen wird. Diese Befunde bestätigen zunächst die lerntheoretische Erkenntnis, dass Konstruktion ohne Instruktion nur in seltenen Fällen möglich ist. Die Integration von metakognitiven Elementen in die Lernumgebungen ist ein weiterer wichtiger Baustein, der die Wirksamkeit von Lernprozessen steigern kann. Im problembasierten Lernen und auch bei der Vermittlung von Lernstrategien spielt die metakognitive Komponente eine unentbehrliche Rolle. Wie Instruktion, Konstruktion und Metakognition in einer Lernumgebung zusammengeführt werden können, soll ein Beispiel aus dem Forschungsprojekt DISUM illustrieren.

4 Methoden-integratives Design: Aufbau, Prinzipien und empirische Erkenntnisse

Im Rahmen des Forschungsprojekts DISUM wurde untersucht, wie Schüler und Lehrer mit Modellierungsaufgaben umgehen (Leiss 2007; Schukajlow 2011) und welche Lehr-Lernformen die Entwicklung der Modellierungskompetenz begünstigen. In der ersten Phase des Projekts zeigten sich stärkere förderliche Wirkungen eines selbständigkeitsorientierten, so genannten „operativ-strategischen" Unterrichts, auf Schülerleistungen und motivational-affektive Merkmale von Lernenden im Vergleich zum „direktiven", stärker instruktionsorientierten Unterricht (siehe den Überblick in Schukajlow et al. 2012). Den methodischen Kern des „operativ-strategischen" Unterrichts bilden erstens die ko-konstruktive Gruppenarbeit und zweitens Reflexionsphasen im Plenum. Die ko-konstruktive Gruppenarbeit besteht aus drei Schritten: (1) Einzelarbeit, in der erste Lösungsansätze entwickelt werden, (2) Austausch zu Schwierigkeiten und Lösungswegen in der Gruppe, unterstützt durch möglichst adaptive Lehrerinterventionen (Leiss 2010), und (3) individueller Aufschrieb der Lösung (siehe die Arbeitskarte in Schukajlow et al. 2011). Inhaltlich geht es in DISUM um die Vermittlung von Modellierungskompetenz zu den Inhaltsbereichen „Satz des Pythagoras" (SdP) und „Lineare Funktionen" (LF). Lehr-lerntheoretische Überlegungen und eine Analyse der Realisierung der beiden genannten Lernumgebungen im Feld lassen vermuten, dass eine Ergänzung des operativ-strategischen Unterrichts mit direktiven Instruktionselementen und mit einem metakognitiven Instrument, dem Lösungsplan für Schüler, die beobachteten positiven Effekte verstärken kann (siehe eine Version des Lösungsplans in Schukajlow et al. 2011). Der Lösungsplan (siehe Blum 2010) ist eine verkürzte Version des bekannten Modellierungskreislaufs und stellt eine Art Fahrplan für die Bearbeitung von Modellierungsaufgaben dar, der prozedurales Metawissen über das Modellieren systematisiert und vermittelt. Zugleich enthält der Lösungsplan strategische Hilfen für Lernende (z.B. „Stell dir die Situation konkret vor!"), wie das in selbständigkeitsorientierten Konzeptionen angelegt ist. Nach einer positiven Evaluation des Lösungsplans in einer Laborstudie (Schukajlow et al. 2010) wurde der methoden-integrative Unterricht entsprechend den drei Leitprinzipien Instruktion, Konstruktion und Metakognition für zehn Unterrichtsstunden (fünf Doppelstunden) wie in Tab. 4.1 konzipiert.

In jeder Doppelstunde dieser methoden-integrativen Unterrichtseinheit sind Instruktion, Konstruktion und Metakognition als Leitprinzipien vertreten. Die Instruktion bereitet die Eigenkonstruktion des Wissens vor und sichert die Unterstützung bei Schwierigkeiten im Konstruktionsprozess. Auch die metakognitive Reflexion zu Strategien und ihrer Anwendung wird durch die Instruktion der Lehrperson in Gruppen- und Plenumsphasen initiiert und gesteuert. Wichtig erscheint in dieser komplexen Interaktion von Lehr-Lernprinzipien, die Orientierung jedes einzelnen Schüler auf seine eigenen Lernprozesse zu stärken, statt nur auf die Produktion von Ergebnissen in Form richtiger Lösungen zu achten (Schukajlow 2011). Erste äußerst ermutigende Hinweise zu positiven Wirkungen des methoden-integrativen Unterrichts im Vergleich zum operativ-strategischen Unterricht liegen bereits vor (Krämer et al. 2011). Die Lernfortschritte der beiden Pilotierungsklassen lagen im Mittel bei über einer Standardabweichung. Videoanalysen des Unterrichts und Detailvergleiche von Schülerbearbeitungen lassen darauf schließen, dass insbesondere die metakognitiven Elemente wie erhofft für einen besseren Transfer gesorgt haben und deshalb die Lösungsquoten im Nachtest deutlich höher lagen als in den vorangehenden Studien.

Tab. 3.1 Ablauf des methoden-integrativen Unterrichts

Unterrichts-stunde	Ablauf
1./2.	Sicherung des Vorwissens zu den beiden involvierten Stoffgebieten mit dem Arbeitsblatt „Kannst du das lösen?", Instruktion zur Einführung der operativ-strategischen Gruppenarbeit und ihre Einübung an einer einfachen Aufgabe zum Satz des Pythagoras.
3./4.	Bearbeiten der mittelschweren Modellierungsaufgabe „Kletterwald" (SdP) in Gruppenarbeit; Thematisieren von Schwierigkeiten und Strategien von Lernenden in Gruppenarbeits- und Reflexionsphasen als Vorbereitung zur Einführung des Lösungsplans. Lehrerinstruktion zur Verwendung des Lösungsplans durch das Vormachen der Bearbeitung der mittelschweren Modellierungsaufgabe „Der Berg" (SdP) an der Tafel entsprechend dem „Lernen am Modell".
5./6.	Fragend-entwickelnder Unterricht zur Aufgabe „Reiterhof" (LF), mit der Fokussierung auf die Erkenntnis, dass der Lösungsplan auch beim Modellieren zum Inhaltsbereich Lineare Funktionen tragfähig ist. „Operativ-strategische" Bearbeitung der Aufgaben „Sportstudio" (LF), „Gemeindefest" (SdP) und „Abkürzung" (SdP) unter Verwendung des Lösungsplans und Einbezug von Checklisten zur Verwendung einzelner Stationen des Lösungsplans durch die Schüler. Reflexion auf Metaebene zu Aktivitäten beim Modellieren mit dem Lösungsplan als Folie für die Beschreibung der Bearbeitungsprozesse.
7./8.	Bearbeitung der Aufgaben „Wäscheleine" (SdP), „Fahrschule" (LF) und „Zuckerhut" (SdP) gemäß der operativ-strategischen Gruppenarbeit. Lehrerinterventionen in den Gruppenphasen nach Möglichkeit unter expliziter Verwendung des Lösungsplans. Reflexion auf Metaebene zur Verwendung des Lösungsplans und zu Schwierigkeiten und Strategien von Lernenden.
9./10.	Übung in Einzelarbeit anhand der Aufgaben „Zirkel" (SdP), „Drucker" (LF) und „Feuerwehr" (SdP) mit Reflexionsphasen im Plenum.

Die Effektivität der Integration von Instruktion, Konstruktion und Metakognition bestätigen auch andere Studien, die alle drei Leitprinzipien miteinander kombinieren (Swanson und Hoskyn 1998). Die Realisierung dieser Prinzipien in den Unterrichtsdesigns sowie ihre optimale Gewichtung und Orchestrierung im Hinblick auf verschiedene Unterrichtsziele bleiben weiterhin offene Forschungsfragen, deren Bearbeitung viele neue Erkenntnisse verspricht.

Literaturverzeichnis

Aebli, H. (1980). *Denken: das Ordnen des Tuns*. Stuttgart: Klett-Cotta.

Baker, S., Gersten, R., & Lee, D. S. (2002). A synthesis of empirical research on teaching mathematics to low-achieving students. *Elementary School Journal 103*(1), 51-73.

Bandura, A. (1976). *Lernen am Modell. Ansätze zu einer sozial-kognitiven Lerntheorie*. Stuttgart: Klett.

Barrows, H. S. (1996). Problem-based learning in medicine and beyond. In L. Wilkerson, & W. H. Gijselaers (Eds.), *Bringing problem-based learning to higher education: Theory and practice. New directions for teaching and learning* (S. 3-13). San Francisco: Jossey-Bass Inc. Publishers.

Blum, W. (2010). Modellierungsaufgaben im Mathematikunterricht – Herausforderung für Schüler und Lehrer. *PM - Praxis der Mathematik in der Schule, 52*(4), 42-48.

Borko, H. (2004). Professional development and teacher learning: Mapping the terrain. *Educational Researcher, 33*(8), 3-15.

Gagné, R. M. (1962). Military training and principles of learning. *American Psychologist, 17*(2), 83-91.

Gijbels, D., Dochy, F., Van den Bossche, P., & Segers, M. (2005). Effects of Problem-Based Learning: A Meta-Analysis from the Angle of Assessment. *Review of Educational Research, 75*(1), 27-61.

Haas, M. (2005). Teaching Methods for Secondary Algebra: A Meta-Analysis of Findings. *NASSP Bulletin, 89*(1), 24-46.

Hasselhorn, M., & Gold, A. (2006). *Pädagogische Psychologie. Erfolgreiches Lehren und Lernen*. Stuttgart: Kohlhammer.

Hattie, J. (2009). *Visible learning : a synthesis of meta-analyses relating to achievement*. London [u.a.]: Routledge.

Helmke, A., & Weinert, F. E. (1997). Bedingungsfaktoren schulischer Leistungen. In F. E. Weinert (Hrsg.), *Enzyklopädie der Psychologie. Themenbereich D. Psychologie des Unterrichts und der Schule* (S. 71-176). Göttingen: Hogrefe.

Hmelo-Silver, C. E., Duncan, R. G., & Chinn, C. A. (2007). Scaffolding and Achievement in Problem-Based and Inquiry Learning: A Response to Kirschner, Sweller, and Clark (2006). *Educational Psychologist, 42*(2), 99-107.

Kramarski, B., & Mevarech, Z. R. (2003). Enhancing mathematical reasoning in the classroom: Effects of cooperative learning and metacognitive training. *American Educational Research Journal, 40*(1), 281-310.

Kramarski, B., Weisse, I., & Kololshi-Minsker, I. (2010). How can self-regulated learning support the problem solving of third-grade students with mathematics anxiety? *ZDM, 42*(2), 179-193.

Krämer, J., Schukajlow, S., Blum, W., Messner, R., & Pekrun, R. (2011). Mit Vielseitigkeit zum Erfolg? Strategische Unterstützung von Lernenden in einem "methoden-integrativen" Unterricht mit Modellierungsaufgaben. In R. Haug, & L. Holzäpfel (Hrsg.), *Beiträge zum Mathematikunterricht 2011* (S. 479-482). Münster: WTM Verlag.

Leiss, D. (2007). *„Hilf mir es selbst zu tun". Lehrerinterventionen beim mathematischen Modellieren*. Hildesheim: Franzbecker.

Leiss, D. (2010). Adaptive Lehrerinterventionen beim mathematischen Modellieren - empirische Befunde einer vergleichenden Labor- und Unterrichtsstudie. *Journal für Mathematik-Didaktik, 31*(2), 197-226.

Mayer, R. E. (2004). Should there be a three-strikes rule against pure discovery learning? The case for guided methods of instruction. *American Psychologist, 59*(1), 14-19.

Pólya, G. (1948). *How to solve it a new aspect of mathematical method*. Princeton, N.J.: Princeton University Press.

Schukajlow, S. (2011). *Mathematisches Modellieren. Schwierigkeiten und Strategien von Lernenden als Bausteine einer lernprozessorientierten Didaktik der neuen Aufgabenkultur.* Münster u.a.: Waxmann.

Schukajlow, S., Blum, W., & Krämer, J. (2011). Förderung der Modellierungskompetenz durch selbständiges Arbeiten im Unterricht mit und ohne Lösungsplan. *Praxis der Mathematik in der Schule, 38*, 40-45.

Schukajlow, S., Krämer, J., Blum, W., Besser, M., Brode, R., & Leiss, D. (2010). Lösungsplan in Schülerhand: zusätzliche Hürde oder Schlüssel zum Erfolg? In A. Lindmeier & S. Ufer (Hrsg.), *Beiträge zum Mathematikunterricht 2010* (S. 771-774). Münster: WTM Verlag.

Schukajlow, S., & Leiss, D. (2011). Selbstberichtete Strategienutzung und mathematische Modellierungskompetenz. *Journal für Mathematikdidaktik, 32*(1), 53-77.

Schukajlow, S., Leiss, D., Pekrun, R., Blum, W., Müller, M., & Messner, R. (2012). Teaching methods for modelling problems and students' task-specific enjoyment, value, interest and self-efficacy expectations. *Educational Studies in Mathematics, 79*(2), 215-237.

Seidel, T., & Shavelson, R. J. (2007). Teaching Effectiveness Research in the Past Decade: The Role of Theory and Research Design in Disentangling Meta-Analysis Results. *Review of Educational Research, 77*(4), 454–499.

Shuell, T. J. (1996). Teaching and learning in the classroom context. In D. C. Berliner, & R. C. Calfee (Hrsg.), *Handbook of educational psychology* (S. 726-764). NY: Simon & Schuster Macmillan.

Skinner, B. F. (1971). *Analyse des Verhaltens.* München: Urban und Schwarzenberg.

Steiner, G. (2001). Lernen und Wissenserwerb. In D. H. Rost (Hrsg.), *Handwörterbuch Pädagogische Psychologie* (S. 139-204). Weinheim: Beltz.

Swanson, H. L., & Hoskyn, M. (1998). Experimental Intervention Research on Students with Learning Disabilities: A Meta-Analysis of Treatment Outcomes. *Review of Educational Research, 68*(3), 277-321.

Weinert, F. E. (1996). Lerntheorien und Instruktionsmodelle. In F. E. Weinert (Hrsg.), *Psychologie des Lernens und der Instruktion* (S. 1-48). Göttingen: Hogrefe.

Zimmerman, B. J. (2001). Self-regulated learning and academic achievement: Theoretical perspectives. In B. J. Zimmerman, & D. H. Schunk (Hrsg.), *Theories of self-regulated learning and academic achievement: An overview and analysis* (S. 1-37). Mahwah, NJ: Erlbaum.

Chapter 4
Low Achievers' Understanding of Place Value – Materials, Representations and Consequences for Instruction

Petra Scherer

University of Duisburg-Essen

Abstract A sound understanding of place value is a crucial competence for successful mathematics learning and the flexible handling of bigger numbers. Firstly, the contribution presents in detail the specific meaning of place value as a fundamental idea as well as results of empirical studies with respect to low achievers. For diagnosing the competencies of low achievers dealing with numbers up to thousand, specific test items were designed that cover different levels of representation and that can be used in interviews or as a paper and pencil test. The contribution presents results of a small case study with fifth- and sixth-graders with special needs using the developed test instrument followed by consequences and indications for classrooms activities. These indications touch aspects like diagnosis with respect to basic competencies, the selection of adequate problems or linking different levels of representation. For low achieving students the topic of place value and its didactical implementation should take a more prominent role for didactical proposals and classroom practice.

1 Introduction: understanding place value – a fundamental idea for learning mathematics

Place value represents one of the fundamental ideas of mathematics (Winter 2001), in particular ideas of arithmetic (Wittmann 1994; see also Hart 2009). A firm understanding of the place value concept is necessary for understanding our decimal number system in general, for coping with bigger numbers and number sense (e.g. estimating). In detail, the importance becomes clear for developing effective calculation strategies (e.g. to replace one-by-one finger counting), for understanding the written algorithms or for extending integers to fractions. Moreover, place value relates to measurement with decimal and non-decimal structures. So, the relevance for different fields of school mathematics can be stated (see also van de Walle 1994, p. 154).

Research shows that especially low achievers, even in higher grades, have great difficulties with this mathematical topic. In this contribution different examples with respect to the type of problem and the level of representation will be analysed and discussed. For this purpose a small case study will be presented that looked for both a better understanding of the difficulties of low achievers and consequences for teaching and learning processes. The concrete problems and activities described can be used on the one hand for a diagnostic instrument and on the other hand for activities in classroom.

2 Exemplary research results with respect to low achievers

Several studies show that place value belongs to one of the difficult topics in mathematics:

Moser Opitz (2007) found in her study with 5th- and 8th-graders who showed learning disabilities in mathematics, that understanding of the decimal system belongs to one of the most difficult basic topics. The study included among other activities counting in steps bigger than 1, addition and subtraction, doubling and halving, multiplication & division, word problems, decimal system, estimating, informal & written calculation.

Kamii (1986) asked primary students (1st to 5th grade) to determine the number of given counters by estimating and counting, the latter with a self-chosen strategy as well as by counting in tens. She found that the understanding of place value was not completely developed even in higher grades.

Hart (2009) investigated with African students the competencies concerning the written algorithms: She found an increasing error-rate in written additions and subtractions with numbers up to 1,000 when carry-overs were included while the number size remained nearly unchanged. Handling carry-overs requires a firm understanding of place value. Moreover, the general test scores of 3rd- and 4th-graders were lower than in 1st and 2nd grade which she ascribes to the place value concept required for solving the test items.

MacDonald (2008) found students' misconceptions when extending whole numbers to decimals. The students assumed a symmetry with respect to the decimal point (see example in Fig. 4.1). »Interestingly, when applied correctly, the mirror metaphor actually explains the structure of decimal place value, however, the important distinction to be made is that the symmetry is based around the ones column and not the decimal point« (MacDonald 2008, p. 13).

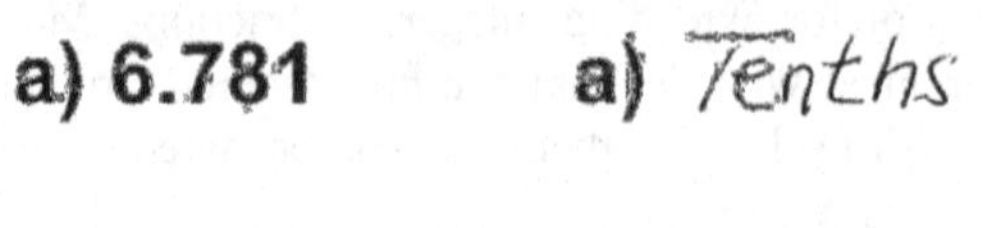

Find the place value of 8

a) 6.781   a) *Tenths*

Fig. 4.1 Misconceptions within decimal numbers (see MacDonald 2008, p. 12)

Fuson (1990) stressed the irregular structure of number names in different languages, especially compared to many Asian languages. This would complicate understanding of place value. Moreover, linguistic similarities and confusions with numbers like ›fourteen‹ and ›forty‹ or in combination with decimals ›hundreds‹ and ›hundredths‹ my cause further difficulties (see Resnick et al. 1989). Comparable linguistic similarities also exist in German language.

Fuson et al. (2000) found differences in different instructional approaches. Comparing traditional concepts to the »EM-concept« (Everyday Mathematics) brought up advantages for the experimental groups especially when solving problems that focus on concepts (like place value) in comparison to calculations competences.

3 Results from a small case study

In the following section, selected items of an instrument for diagnosing competencies concerning place value will be presented and discussed (see Scherer 2009). The students' documents are taken from a small case study with students with special education needs. The study was carried out with 12 low achieving pupils who attend a special school for learning disabled (5th and 6th grade, see Table 4.1). Although in Germany the methods for diagnosing special needs are not standardized, difficulties in mathematics are one of the main reasons for attending special schools (see Langfeldt 1998) and this was also true for the participating students. For some of the students German language was not their first language.

Table 4.1 Overview of the participating students

-	girls	boys	average age (years; months)
5th grade	1	3	10;9
6th grade	3	5	12;3

The ideas and concepts of place value usually are dealt with in classroom when introducing numbers up to 100 (grade 2 in primary school and usually in grade 3 in special education). This topic then develops over several grades until the extension to decimal numbers (see also van de Walle 1994, p. 154). The special education students taking part in the study should have worked on place value with two- and three-digit numbers for at least one to two school years. The test items used ask for basic competencies that form the foundation for further mathematical topics.

The examples discussed in the following will illustrate the problems of those students as well as give some indications for classroom activities that will be described in section 4.

For this instrument, specific tasks with numbers up to 1,000 have been designed. The items could be used for paper-and-pencil tests as well as for interviews to get information for both, oral and written competencies of the students. For the case study presented here a combination was used: During a one-to-one interview situation the students at first should write down their spontaneous solution or give it verbally. Afterwards, the interviewer had the opportunity to question and discuss this solution to get further information on the students' underlying ideas.

On the one hand, the test-items cover different levels of representations and touch the basis for calculation strategies. On the other hand, unknown formats or challenging problems (e.g. referring to the specific role of zero) which cannot be solved in a mechanistic way have been chosen rather than only standard items. Usually, these types of problems do not appear in textbooks and therefore may not been addressed in the classroom yet.

The whole test covers the following topics:

- counting forward/backwards (different starting points to pass the tens or different hundreds)
- counting in steps (forward/backward in steps of 2, 10, 100)
- splitting up numbers into place values (different number values, partly with vacant tens or units)
- identifying numbers in the place value table; operative variations in the place value table
- identifying place values of digits in given numbers (symbolic level)
- composing numbers on iconic (hundred-squares, tens and units) and on symbolic level (different number values, partly with vacant tens or units; order of place values varying)
- solving simple addition and subtraction problems (type of problems HTU, HT, HU as addends, without passing a place value)

A first analysis of the spontaneous solutions showed a certain understanding of place value for all pupils but also revealed a variety of difficulties, especially with the non-standard items. In this contribution the focus will be on two item-groups, namely *splitting up numbers in place values* and *composing numbers*, the latter items were offered on the *iconic level* as well as on the *symbolic level*.

For *splitting up numbers into place values*, the (intended) standard solution (e.g. to notate $378 = 300 + 70 + 8$) appeared very rarely. In contrast, some students drew hundred-squares, bars of 10 und single dots to represent the decomposition (Fig. 4.2). Others marked the letters H (the German word »Hunderter« for hundreds), Z (the German word »Zehner« for tens) and E (the German word »Einer« for units) above the corresponding place values (Fig. 4.3). When asked about the meaning, this student could explain that the number contains »*Two hundreds, zero tens and nine units*«. With other students it remained as an open question whether a real understanding of the vacant place value exists or if the place values were just notated in a rather mechanistic way according to the well-known order. »It is quite easy to attach words […] without realizing what the materials and symbols are supposed to represent« (van de Walle 1994, p. 155).

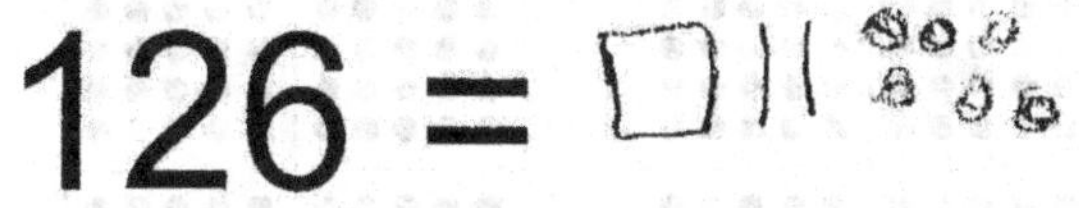

$$126 = \square \,||\, \substack{\text{\tiny ⬦⬦⬦}\\\text{\tiny ⬦⬦⬦}}$$

Fig. 4.2 Student 7 (6th grade) using an iconic representation

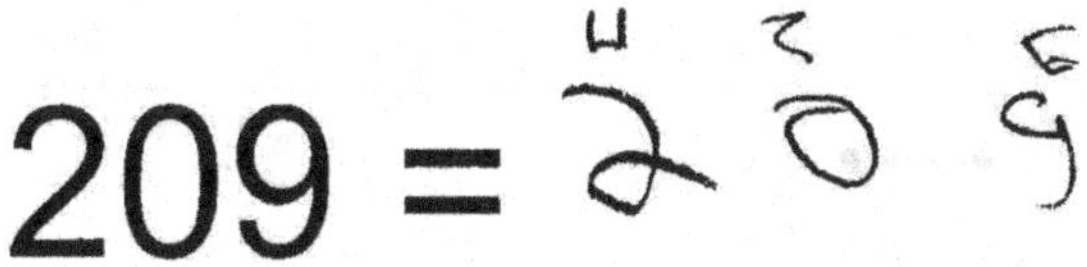

$$209 = \overset{\text{H}}{2}\ \overset{\text{Z}}{0}\ \overset{\text{E}}{9}$$

Fig. 4.3 Student 12 (6th grade) marking the place values

Interesting notations occurred for the test items with vacant place values: notation as addition with 0 (Fig. 4.4a), or in a listed form with 00 according to the position of the tens (Fig. 4.4b). These specific solutions and notations give a first indication for instruction: students should work on such non-routine problems and reflect on the different notations. One could assume that the students experienced examples with vacant place values on the enactive and iconic level but were unsure how to handle these vacant positions on the symbolic level. Here, the importance of activities of an intermodal transfer (Bruner 1977), the flexible transfer between enactive, iconic and symbolic level becomes obvious (see also section 4).

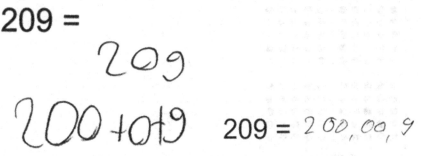

Fig. 4.4 Left: 4.4a, Right: 4.4b. Solutions of Student 1 (5th grade) and Student 10 (6th grade) for splitting up the number

Solving the items for *composing numbers on the iconic level* for some of the students was an easy task, whereas others used rather wasteful strategies: One student noted down the single hundreds, the tens and ones entirely and worked out the complete result by using the written algorithm (Fig. 4.5). He did so not only with a rather huge number for the different place values (e.g. five tens or six hundreds) but also for reasonable numbers like 120 as shown in Fig. 4.6.

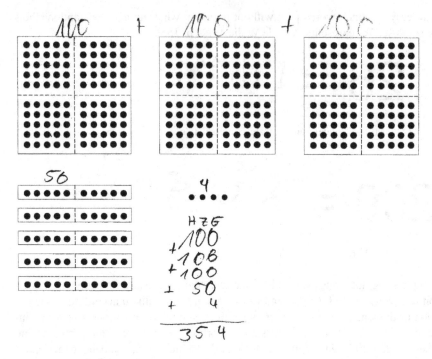

Fig. 4.5 Student 8 (6th grade) using the written algorithm

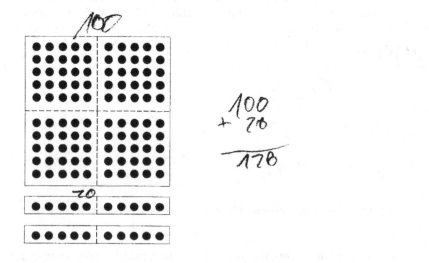

Fig. 4.6 Student 8 (6th grade) using the written algorithm even for ›easy‹ tasks

For another student the task appeared more demanding as she did not identify the hundred on first sight but firstly used the sub-structure of 25 (Fig. 4.7).

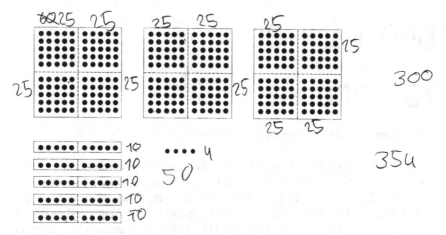

Fig. 4.7 Student 4 (5th grade) using a sub-structure for the field of hundred

The students participating in this study had different school biographies and as a consequence one cannot definitely say if array structures, like the field of dots, had been experienced in former grades (field of 20 dots or field of 100 dots) so that they might be expected to be familiar with this type of representation. On the other hand, it would be desirable that such a material with a common structure (see also base ten material; base ten blocks or Dienes blocks, arrays; van de Walle 1994) could be used effectively. This also belongs to basic competencies and could appear as test items in comparative studies.

For *composing a number on the symbolic level* many students spontaneously used the written algorithm, whereas some others tried to figure it out by mental calculation. With this method, sometimes a structural aid was used (see Fig. 4.8).

Fig. 4.8 Structural aid used by Student 11 (6th grade)

For these items the vacant place values as well as the changing and unfamiliar order of place values played an important role: For example, composing a number out of the term 70+200+3 led to the number 723 whereas a more or less standard item like 300+50+4 yielded a correct solution. Also *Student 6* (6th grade) had no problems to solve the routine problem 300+50+5. Moreover, a vacant place value and the role of zero did not cause a problem and yielded a correct solution for 100+20 and 400 + 8. In contrast to this, the varying order lead to bigger problems (see Fig. 4.9)

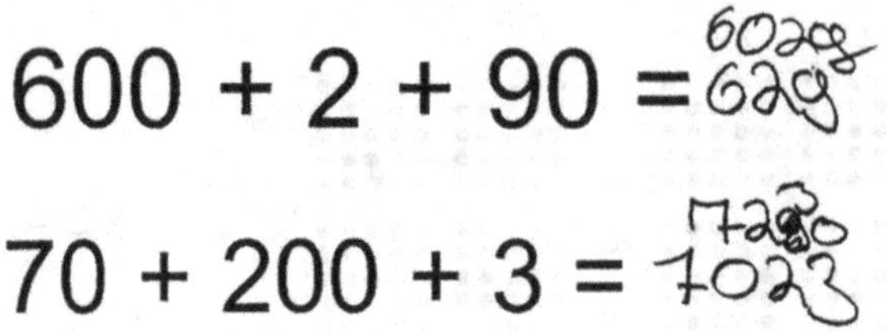

$$600 + 2 + 90 = 6\cancel{29}\,{}^{6029}$$

$$70 + 200 + 3 = \cancel{7023}\,{}^{7230}$$

Fig. 4.9 Student 6 (6th grade) solving the non-standard-items

The written algorithm did not automatically guarantee a correct solution. *Student 1* (5th grade) was successful also with the non-standard-item as he ordered the place values by height (given item: 600+2+90, see Fig. 4.10), whereas *Student 8* was not successful in every case: The written algorithm only functioned well with the hundreds given in first position (Fig. 4.11) but did not when the given problem started with the tens (Fig. 4.12). In his first attempts he started with 70 and did not succeed in writing the next addends below in the correct position. The algorithm as such was worked out correctly. Only when the interviewer suggested to start with the biggest number the student solved the item correctly (third solution in Fig. 4.12).

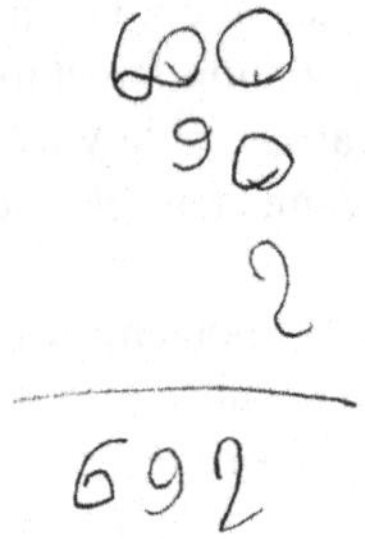

$$
\begin{array}{r}
600 \\
90 \\
2 \\
\hline
692
\end{array}
$$

Fig. 4.10 Written algorithm used by Student 1 (5th grade) with the correct order of the place values

$$600 + 2 + 90 = 692$$

$$
\begin{array}{r}
600 \\
2 \\
+\ 90 \\
\hline
692
\end{array}
$$

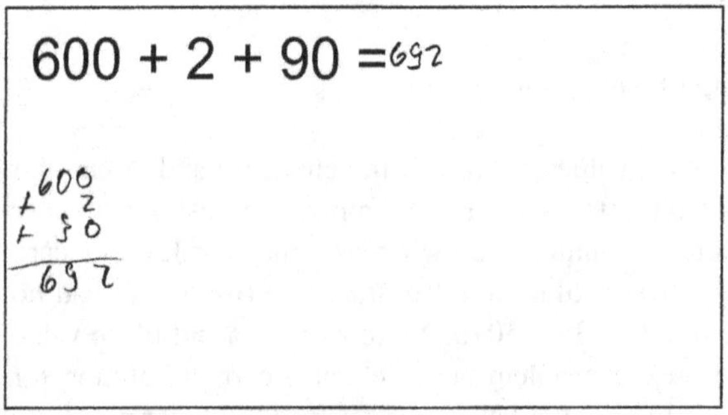

Fig. 4.11 Written algorithm used by Student 8 (6th grade) without changing the order of the place values

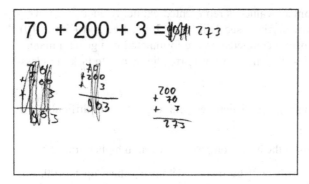

Fig. 4.12 Written algorithm used by Student 8 (6th grade) leading to an incorrect solution in his first attempts

Summarizing the results of this specific type of problem it is important to look more precisely at the solution process: When students used spontaneously the written algorithm the interviewer encouraged them to work out the next item mentally. But most of the students refused and used the algorithm again. If pupils are not able to compose three place values automatically but need the written algorithm (Fig. 4.10, 4.11 and 4.12) or a structural aid (Fig. 4.8) this might influence the development and their choice of calculation strategies.

4 Indications for classroom activities

One of the important results of the presented study was that many students found correct results, including for calculations, even though basic competencies were missing. Especially the non-standard or challenging items helped to identify the deficits and missing basics of some students. One could see that solving routine items may give an impression of understanding.

Consequences for learning and instruction on the one hand touch future topics in higher grades (e. g. numbers up to 10 000 or up to one million). On the other hand, the results show evidence that topics dealt with earlier should be reflected critically in a preventive way (see also van de Walle 1994, pp. 154). E.g., when dealing with numbers up to 100 the instruction should stress the understanding of place value and the students' learning processes and difficulties should be diagnosed carefully.

The topic ›place value‹ appears in the German Standards for primary level under two central ideas (›Leitidee‹): For the central idea ›number and operations‹ the concrete competence »understanding composition of the decimal place values system« (Kultusministerkonferenz [KMK] 2005, p. 11; translation PS) is formulated. For the central idea ›patterns and structure‹ the competence »understanding and using structural representations of numbers (e.g. 100 square)« (KMK 2005, p. 12; translation PS) can be found. It remains questionable whether the central meaning of the mathematical topic ›place value‹ is represented explicitly enough. In every case it assumes a competent teacher (see Hill et al. 2005; Lee and Kim 2010). Moreover, the required students' competences are formulated in a general manner and not in concrete activities. So, they do not mirror the concrete tasks in classroom practice.

Important aspects for assessing difficulties and coping with difficulties in mathematics education are

- the diagnosis and support of the basic competences even in higher grades

Although place value usually will be treated within numbers up to 100, this topic should be a continuing theme especially with respect to decimal numbers. Hart emphasizes that teachers have to differentiate between the »fundamental« and the »trivial« (Hart 2009, p. 27) and also Barker reports on the positive effect of receiving extended instructional time on the composition and decomposition of ten and its applications which lead to a higher efficiency at computation and a greater number sense (Barker 2009, p. 338; see also van de Walle 1994). In the curriculum, the introduction of the ›hundreds‹ and ›thousands‹ (within numbers with three and more digits) might be assumed to follow after having addressed the ›tens‹ (within numbers up to 100). In every case, specific difficulties or misconceptions with smaller numbers have to be analysed carefully and considered for further activities (see Scherer 1999; Scherer 2003).

- a suitable selection of problems for diagnostic procedures as well as for instruction

The non-routine items could detect many existing difficulties and those items should be an integral part of everyday lessons (as they are also integrated in comparative studies like TIMSS; see Walther et al. 2008). Moreover, activities should be integrated that require mental operations and thinking abilities (cf. operating with hidden place values, Scherer 2003).

- linking the different levels of representation and not only focusing on calculations and working on the symbolic level (e.g. Scherer 1995; Scherer and Häsel 1995)

For introducing place value concepts with numbers up to 100 usually different models are used. You find base ten material (cubes, bars and squares), field of dots or the place value table, the latter with numbers or sometimes with money (for a critical analysis concerning money as a model for place value see Scherer and Moser Opitz 2010, pp. 148; Steinbring 1997). With different models and materials the students then work for a certain time on the enactive or iconic level and are expected to connect the models with the symbolic expressions. When extending numbers up to 1,000 those models are used again but normally just for a short time of introduction. The main focus is the work on the symbolic level.

Adequate activities, even in the higher grades with numbers up to 1,000 or beyond that, should cover both, the intermodal transfer as well as intramodal transfer

(Bruner 1977) to ensure that students really understand the position system and especially the role of zero (see also Scherer 2011). In Fig. 4.13a and 4.13b the number given in the place value table has to be transferred on the iconic level (intermodal) as well as on the symbolic level outside the place value table (intramodal). With the example in Fig. 4.13b the specific role of zero in the table at first position could be made a subject of discussion. With the self-chosen number in Fig. 4.14 the extension to higher place values becomes obvious and could be discussed.

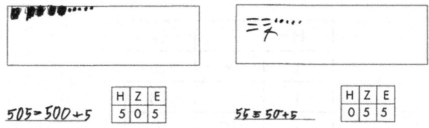

Fig. 4.13 Left 4.13a, Right 4.13b. Numbers represented on different levels

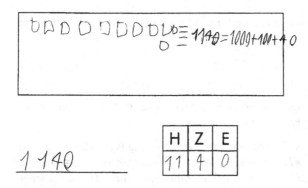

Fig. 4.14 Self-chosen number of a student represented on different levels

Moreover, equivalent representations should take a prominent role and those activities should be treated on the different representation levels (see Scherer and Moser Opitz 2010, pp. 145; Scherer and Steinbring 2004, p. 166; van de Walle 1994, p. 163): The number 123 can be shown with base ten material in different ways as well as in the place value table (Fig. 4.15).

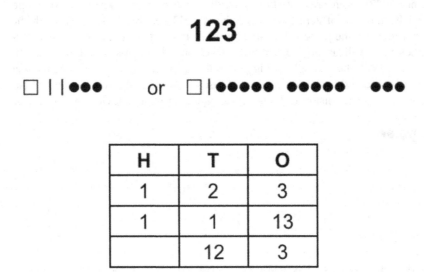

Fig. 4.15 Equivalent representations of one number with base ten material or in the place value table

Last but not least,

- the importance of language and linguistic specifics (e. g. Scherer and Steinbring 2003; Thompson 1997)

should be considered. As there exist similarities in language when speaking about whole numbers (e. g. »*Two hundred*« for 200) in contrast to speaking about the place value in the place value table (»*Two hundreds*« for 2 H), the teacher should be aware of and require a precise verbalization.

5 Final remarks

The benefit of the assessments described here and corresponding classroom practice is to meet the individual needs of the students, to get advice for organising differentiation in classroom, to identify specific problems that could be made a subject of discussion for the whole group and to lead to a deeper understanding of mathematical topics. In general, classroom practice should require more than getting the correct result or being able to perform the algorithm but explaining and reasoning about solution strategies and these solutions strategies and reasoning should be considered carefully. Teachers »need to know how to use pictures or diagrams to represent mathematics concepts and procedures to students, provide students with explanations for common rules and mathematical procedures, and analyze students' solutions and explanations« (Hill et al. 2005, p. 372).

For low achieving students the topic of place value and its didactical implementation should take a more prominent role for didactical proposals and classroom practice.

References

Barker, L. (2009). Ten Is the Magic Number! *Teaching Children Mathematics*, *15*(6), 336-345.

Bruner, J. (1977). *The process of education*. Cambridge, MA: Harvard University Press.

Fuson, K. C. (1990). Issues in place value and multidigit addition and subtraction learning and teaching. *Journal for Research in Mathematics Education*, *21*(4), 273-280.

Fuson, K. C., Carroll, W. M., & Drueck, J. V. (2000). Achievement Results for Second and Third Graders. Using the Standards-Based-Curriculum Everyday Mathematics. *Journal for Research in Mathematics Education*, *31*(3), 277-295.

Hart, K. (2009). Why do we expect so much? In J. Novotná, & H. Moraova (Eds.), *SEMT 2009. International Symposium Elementary Maths Teaching. August 23 – 28, 2009. Proceedings: The Development of Mathematical Understanding* (pp. 24-31). Prague: Charles University.

Hill, H. C., Rowan, B., & Loewenberg Ball, D. (2005). Effects of teachers' mathematical knowledge for teaching on student achievement. *American Educational Research Journal*, *42*(2), 371-406.

Kamii, C. (1986). Place Value: an explanation of its difficulty and educational implications for the primary grades. *Journal for Research in Childhood Education*, *1*(2), 75-86.

Kultusministerkonferenz [KMK] (2005). *Bildungsstandards im Fach Mathematik für den Primarbereich Beschluss vom 15.10.2004*. München: Wolters Kluwer.

Langfeldt, H.-P. (1998). *Behinderte Kinder im Urteil ihrer Lehrkräfte. Eine Analyse der Begutachtungspraxis im Sonderschul-Aufnahme-Verfahren*. Heidelberg: Edition Schindele.

Lee, J.-E., & Kim, K.-T. (2010). Positional System: Pre-service Teachers' Understanding and Representations. *The Mathematics Educator*, *12*(2), 81-102.

MacDonald, A. (2008). »But what about the oneth?« A Year 7 student's misconception about decimal place value. *Australian Mathematics Teacher*, *64*(4), 12-15.

Moser Opitz, E. (2007). *Rechenschwäche/Dyskalkulie. Theoretische Klärungen und empirische Studien an betroffenen Schülerinnen und Schülern*. Bern: Haupt.

Resnick, L. B., Nesher, P., Leonard, W., Magone, M., Omanson, S., & Peled, I. (1989). Conceptual bases of arithmetic errors: The case of decimal fractions. *Journal for Research in Mathematics Education*, *20*(1), 8-27.

Scherer, P. (1995). Ganzheitlicher Einstieg in neue Zahlenräume – auch für lernschwache Schüler?! In G. N. Müller, & E. C. Wittmann (Eds.), *Mit Kindern rechnen* (pp. 151-164). Frankfurt/M.: Arbeitskreis Grundschule.

Scherer, P. (1999). *Entdeckendes Lernen im Mathematikunterricht der Schule für Lernbehinderte – Theoretische Grundlegung und evaluierte unterrichtspraktische Erprobung (2nd edition)*. Heidelberg: Edition Schindele.

Scherer, P. (2003). *Produktives Lernen für Kinder mit Lernschwächen: Fördern durch Fordern. Band 2: Hunderterraum/Addition & Subtraktion*. Horneburg: Persen.

Scherer, P. (2009). Diagnose ausgewählter Aspekte des Dezimalsystems bei lernschwachen Schülerinnen und Schülern. In M. Neubrand (Ed.), *Beiträge zum Mathematikunterricht 2009* (pp. 835-838). Münster: WTM.

Scherer, P. (2011). Übungsformen zur Förderung des Stellenwertverständnisses – Sicherung von Basiskompetenzen. *Grundschulmagazin*, *4*, 29-34.

56   Petra Scherer

Scherer, P., & Häsel, U. (1995). Ganzheitlicher Einstieg in den Tausenderraum – Zum Einsatz des Tausenderbuches im 4. Schuljahr einer Schule für Lernbehinderte. *Die Sonderschule*, *40*(6), 455-462.

Scherer, P., & Moser Opitz, E. (2010). *Fördern im Mathematikunterricht der Primarstufe*. Heidelberg: Spektrum.

Scherer, P., & Steinbring, H. (2003). The professionalisation of mathematics teachers' knowledge – teachers commonly reflect feedbacks to their own instruction activity. In M. A. Mariotti (Ed.), *Proceedings of CERME 3*. Pisa: Universita di Pisa.

Scherer, P., & Steinbring, H. (2004). Übergang von halbschriftlichen Rechenstrategien zu schriftlichen Algorithmen – Addition im Tausenderraum. In P. Scherer, & D. Bönig (Eds.), *Mathematik für Kinder – Mathematik von Kindern* (pp. 163-173). Frankfurt/M.: AK Grundschule.

Steinbring, H. (1997). »… zwei Fünfer sind ja Zehner…« – Kinder interpretieren Dezimalzahlen mit Hilfe von Rechengeld. In E. Glumpler, & S. Luchtenberg (Eds.), *Jahrbuch Grundschulforschung*. (Vol. 1, pp. 286-296). Weinheim: Deutscher Studienverlag.

Thompson, I. (1997). Mental and written algorithms: can the gap be bridged? In I. Thompson (Ed.), *Teaching and learning early number* (pp. 97-109). Buckingham: Open University Press.

Van de Walle, J. A. (1994). *Elementary and Middle School Mathematics: Teaching Developmentally (2 ed.)*. New York: Addison Wesley Longman.

Walther, G., Selter, C., Bonsen, M., & Bos, W. (2008). Mathematische Kompetenz im internationalen Vergleich: Testkonzeption und Ergebnisse. In W. Bos, M. Bonsen, J. Baumert, M. Prenzel, C. Selter, & G. Walther (Eds.), *TIMSS 2007. Mathematische und naturwissenschaftliche Kompetenzen von Grundschulkindern in Deutschland im internationalen Vergleich* (pp. 49-85). Münster: Waxmann.

Winter, H. (2001): *Inhalte mathematischen Lernens*. Retrieved September, 8th 2012 from http://grundschule.bildung-rp.de/lernbereiche/mathematik/wissenschaftliche-artikel/inhalte-mathematischen-lernens.html.

Wittmann, E. C. (1994). Teaching aids in primary mathematics: Less is more. In L. Bazzini, & H.-G. Steiner (Eds.), *Proceedings of the Second Italian-German Bilateral Symposium on the Didactics of Mathematics* (Vol. 39, pp. 101-111). Bielefeld: IDM.

Chapter 5
Visual integration with stock-flow models: How far can intuition carry us?

Peter Sedlmeier, Friederike Brockhaus, Marcus Schwarz

Chemnitz University of Technology

Abstract Doing integral calculus is not easy for most students, and the way it is commonly taught in schools has attracted considerable criticism. In this chapter we argue that stock–flow models have the potential to improve this state of affairs. We summarize and interpret previous research and the results of some of our own studies to explore how an intuitive understanding (and teaching) of integral calculus might be possible, based on such stock–flow models: They might be used for doing "visual integration" without calculations. Unfortunately, stock-flow tasks themselves seem to be quite difficult to solve for many people, and most attempts to make them more intuitive and easily solvable have not met with much success. There might, however, be some potential in using animated representations. In any case, a good starting point for students to eventually be able to perform visual integration in an intuitive way and to arrive at a deeper understanding of integral calculus seems to be to present flows as a succession of changes in stocks.

Introduction

Doing integral calculus is not easy for most students; and the way it is commonly taught in schools has attracted considerable criticism. From a rather general perspective, it has been lamented that teaching often focuses on the calculation itself while neglecting whether students really comprehend what they do (e.g. Borneleit et al. 2001; Hußmann and Prediger 2010). As a consequence, students might not develop adequate mental models and process tasks solely automatically without a deeper understanding (Danckwerts and Vogel 1992; Davis and Vinner 1986; Bürger and Malle 2000; Blum 2000; Hahn and Prediger 2004; Tall 1992; Tall and Vinner 1981). Based on these considerations there have been some suggestions to improve the traditional math education by emphasizing the importance of basic mental representations (Biehler 1985; Davis and Vinner 1986; Hußmann and Prediger 2010; Hahn and Prediger 2008). It might, for instance, be much easier and more intuitive to do integral calculus *not* with numbers and formulas but rather by imagining the changes over time for the quantities in question. And this *visual in-*

tegration might work better the more similar the representations used to convey an integration task are to processes and objects students are already familiar with.

In this chapter, we deal with a specific approach to visual integration, using so-called *stock-flow models* (Sterman 2000). Stock-flow models have been examined in cognitive science and decision research; in our view, they also have the potential to improving the teaching of integral calculus. The main aim in this chapter is to summarize and interpret previous research and our own studies in regard to how an intuitive understanding (and teaching) of integral calculus might be possible, based on such stock-flow models. We begin by discussing the potential uses of intuition in math education. Then we briefly explain what stock-flow models are and proceed to summarize previous attempts to make such stock-flow models intuitively understandable to students. After that we report results from our own studies and finish with some potential implications of this research for teaching integral calculus.

Math education and intuition

The well-known *heuristics and biases* research program initiated by Tversky and Kahneman (1974) claimed that the intuitions we have about solving mathematical tasks are not always helpful and demonstrated numerous "fallacies" based on wrong intuitions. Meanwhile, however, a consensus seems to have been reached that there are both valid and invalid intuitions and that an important precondition for the development of valid intuitions is the opportunity to learn the regularities that exist in the environment (Kahneman and Klein 2009). In recent years, the idea that intuitions can be a powerful basis for good judgments and decisions has been widely publicized by several best-selling books (e.g., Gigerenzer 2008; Gladwell 2007). This idea, that valid intuitions can be very helpful in solving problems has also repeatedly been propagated in the context of math education (e.g., Brown and Campione 1994; Senge et al. 2000). For instance, Ebersbach and Wilkening (2007) found that young children already possess intuitive knowledge about the characteristics of nonlinear growth, long before these functions are taught in school. As another example, Fischbein (1994) demonstrated that multiplication can be performed in an intuitive and in a non-intuitive way. He constructed text problems in which the solution was to multiply two numbers, 15 and 0.75. If the formulation of the problem asked for multiplying 0.75 by 15, about 75% of his fifth-, seventh- and ninth grade students solved it correctly, whereas the solution rate for multiplying 15 by 0.75 was only about 25%. Fischbeins's explanation for the discrepancy was that, in the first case (in contrast to the second), students could use an intuition about multiplication: *If a number is multiplied it becomes larger*. Fischbein (1995) argued that there exist intuitions that are especially useful in learning about probabilities. And indeed, there is ample evidence that such valid statistical intuitions do exist (Sedlmeier 1999; 2007). Fischbein (1975) also claimed that schooling, with its overwhelming emphasis on deterministic explanations about the world might considerably weaken childrens' valid intuitions about probabilities. Engel and Sedlmeier (2005) took this claim to the test and found evidence consistent with it: Irrespective of type of school (Hauptschule, Realschule, and Gymnasium), solution rates decreased with amount of schooling (see Green 1983; 1991 for similar results). However, there is also good news: valid statistical intuitions can be re-activated even in adults by using suitable external representations (e.g., Sedlmeier 2000; Sedlmeier and Gigerenzer 2001; Sedlmeier and Hilton 2012).

What kinds of external representations are these? Norman (1993) argues that such representations must follow the *naturalness principle*: „Experiential cognition is aided when the properties of the representation match the properties of the thing being represented" (p. 72). In many cases, the essential manipulation to evoke valid intuitions is to use a suitable visual or graphical format that represents the task in hand well. But changing the order in which numbers are presented (such as in Fischbein's "multiplication study" mentioned above) may already have strong effects.

There are basically two explanations for why representation influences cognition. The first is an evolutionary account: over millennia, our perception has become sensitive to the structure of the environment and therefore, external representations that represent well the part of the environment dealt with in a given task automatically evoke intuitive ways to deal with the respective problems (e.g., Cosmides 1989; Cosmides and Tooby 1996). The second explanation for the facilitating effects of external representations rests on learning processes (Frensch and Rünger 2003). Such an explanation has, for instance, been presented in the form of a computational model (a neural network) for intuitions about probabilities (Sedlmeier 1999; 2002; 2007). Assuming that intuitions arise at least partly as a result of learning processes might be regarded as the more convincing explanation because this assumption allows us to explain intuitions as a result of the experiences one has within one's life-time (Sedlmeier 2005). For instance, Bayes' formula, presented in its usual format, is totally non-intuitive for most students but is intuitive for most statisticians because they were exposed to it and thought about it in varied ways and circumstances for many years: expertise can make originally non-intuitive representations intuitive (Hogarth 2001). Such an explanation is also consistent with "street-math" results that show that South American street vendors and carpenters can perform relatively complex calculations with extensive exposure to suitable external representations but with little or no formal schooling (Nunes et al. 1993).

Sometimes, like in the case of valid statistical intuitions, this expertise is acquired in our daily routines, by monitoring the relative frequencies with which things and events occur (Sedlmeier 2007). This might also hold for processes that can be represented by stock-flow models.

Stock-flow models

In everyday life, we are often faced with judgments about changing quantities where the underlying processes can be represented by simple *stock-flow models*. Let us illustrate this kind of model with some examples: An obvious one is a bank account. Here, the *stock* consists of a certain sum of money, which over time can be increased by some *inflow* (e.g., salary, interest rates) and decreased by some *outflow* (e.g., expenses, bank fees). Another example for a stock is the concentration of carbon dioxide in the atmosphere, which is responsible for global warming. In this case, the inflow consists of naturally occurring emissions of carbon dioxide and of man-made emissions (e.g. from burning oil, gas or coal), and the outflow consists of the absorption of carbon dioxide by biomass (e.g. plants and trees), and the oceans. But we don't have to move outside ourselves to encounter stock-flow systems: Our weight can be considered a stock that is increased by the intake of calories (inflow) and decreased by the energy consumption of our body (outflow).

To make judgments about the patterns in the quantities over time, we need to integrate the quantity at the beginning of the time interval in question and quantities that are added and subtracted to the initial one over a given time span. Stock-flow tasks usually differ somewhat from the way integral-calculus tasks are presented in school: textbooks usually only present the netflow, that is, the difference between in- and outflow, and only ask for the stock at the end of the process (e.g., the bank account, the carbon dioxide concentration, or the body weight *after* some period of time). However, distinguishing between the two kinds of flow and examining intermediary states might be quite informative for students, too, because of the ubiquity and everyday-relevance of stock-flow tasks. Because we are confronted with such tasks so frequently, one might expect that we have developed some ability that lets us make these judgments at least partly in an intuitive way. If there are such valid intuitions, they might prove to be very helpful in teaching integral calculus in schools.

Intuition and stock-flow models: previous results

If students have valid intuitions about change and accumulation, then presenting integral-calculus tasks in a format that corresponds to these intuitions should facilitate their understanding. One might argue that stock-flow tasks as used in the literature do exactly this: they "translate" abstract integration tasks into a format that should be more easily understood by problem solvers. Before we briefly discuss the results obtained in previous research, let us look at a typical stock-flow task, a variant of the *bathtub-task* and its usual representation (e.g., Sterman and Booth Sweeney 2005).

The task is usually introduced by presenting some form of bathtub such as in Figure 5.1. The Figure depicts the stock, that is, the amount of water in the tub and it shows the two potential flows: the inflow indicated by the tap at the upper left and the outflow indicated by the drain at the lower right of Figure 5.1.

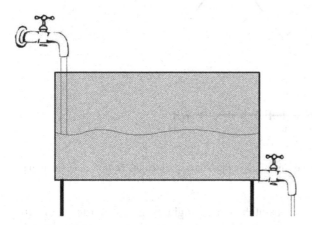

Fig. 5.1 Representation of the general setup of a typical stock-flow task, the bathtub task

After (or with) this figure, usually a *flow diagram* is presented, such as the one in Figure 5.2. The present flow diagram indicates a constant outflow of 50 liters per minute but a variable inflow of 75 liters per minute for the first four minutes that diminishes to 25 liters per minute for the next four minutes and repeats this pattern once.

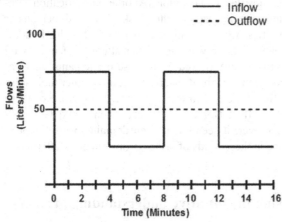

Fig. 5.2 Flow diagram for the bathtub task

From that information, participants are to draw the alteration of the quantity of water in the tub over time, starting, say, from 100 liters. The correct solution is shown in Figure 5.3.

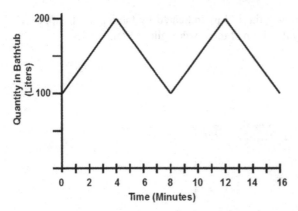

Fig. 5.3 Correct solution for the version of the bathtub task specified by the flow diagram in Fig. 5.2

Everybody is familiar with a bathtub, so one might expect the existence of valid intuitions for solving tasks of this kind. However, initial results were quite discouraging: the usual solution rates for tasks of this kind, even for mathematically trained participants were often markedly below 50% correct answers (e.g., Booth Sweeney and Sterman 2007; Sterman and Booth Sweeney 2005). Following these disappointing results, several studies tried modifications of the task representations. One modification was to offer other, possibly more easily understandable representations such as bar graphs or tables depicting the pattern of both in- and outflows instead of the commonly used line graph. Another modification concerned the task contents: for instance, instead of water, tasks dealt with persons or cars, that is, with countable entities (which might be easier to deal with). In addition to the representational changes, there were also other attempts at improving solution rates: monetary incentives were offered to increase participants' motivation, and in some studies, participants received feedback about whether their solutions were correct and were offered the possibility to change them. None of these manipulations had a noticeable effect (for overview see Cronin and Gonzalez 2007; Cronin et al. 2009). Moreover, it seems that even domain-specific experience does not make a difference in these kinds of tasks (Brunstein et al. 2010).

Our attempts at increasing the intuitive understanding of stock-flow tasks

The rather surprising outcomes of studies that attempted to improve the understanding of stock-flow tasks lead us to try further modifications concerning the representation of such tasks that might make them more intuitive. We report the results from three selected kinds of studies that examined (i) the use of animated representations, (ii) possible inconsistencies between the parts that are commonly used to present stock-flow tasks, and (iii) a way of reconciling potential inconsistences in the usual task representation.

Does animation help?

Stock-flow tasks usually involve some kind of change or movement over time. However, the usual way to represent such tasks (e.g., Figures 5.1 to 5.3) only involves static graphical representations. Previous attempts to optimize the representation of stock-flow tasks (e.g. bar graphs, tables etc.) might have failed because these task representations might not have triggered adequate mental representations in participants. In three studies (Schwarz and Sedlmeier 2013) we showed participants animated simulations of stock-flow systems before we had them work on commonly used stock flow tasks. The studies were conducted with different samples, partly different stock-flow manipulations, and with a variety of tasks. The results, however, were quite comparable: Figure 5.4 shows that, over all, seeing an animated ("Simulation") instead of a static ("Control") stock-flow task did not consistently increase solution rates in subsequently presented stock-flow tasks.

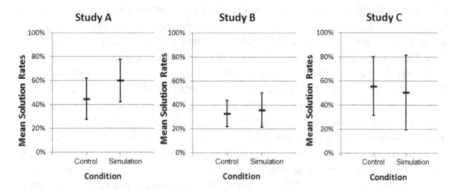

Fig. 5.4 Mean solution rates of animated simulations and a control condition. Error bars represent 95% confidence intervals.

These results should, however, not be taken to be the last word on the potential facilitating effect of animated representations. After all, there might still be possibilities to improve the animations. Moreover, there seem to be some gender differences concerning the impact of animated representations (more beneficial for males); and substituting animated for static representations also increases the amount of information to be processed by participants. So the non-impact of animated representations on solution rates shown in Figure 5.4 might at least partly be due to gender effects and information overload. We are currently exploring both potential explanations.

Might less actually be more?

In most previous studies, flow diagrams, showing both in- and outflow, such as the one in Figure 5.2 have been used. Participants had to infer changes in stock over time, using the information shown in the flow diagram, and to draw these changes into a (then empty) solution diagram such as the one shown in Figure 5.3. At a first glance, the *frames* for Figures 5.2 and 5.3 look quite alike and participants might implicitly conclude that these two diagrams are to contain similar kinds of information. Of course, this is not the case: the flow diagram contains rates (e.g. liters per minute) whereas the stock diagram contains quantities (e.g., liters). Previous results indeed indicate that quite a number of participants might not make a distinction between the two different kinds of information and seem to use an invalid intuition by drawing changes in stock in analogy to changes in flows – a result often termed "correlation heuristic" (e.g., Cronin et al. 2009). The simplest way to prevent participants' use of such a correlation heuristic would be to just omit the flow diagram and instead describe the flows only verbally. However, omitting the flow diagram also reduces the amount of information available to participants and violates Mayer's (2001) well-known *multi-media principle*, according to which people learn better from a combination of words and pictures than from words alone. Therefore, without the diagram, solution rates should decrease if participants use the information contained therein properly. In contrast, if participants tend to use the correlation heuristic, solution rates should increase if the flow diagram is omitted. Figure 5.5 shows the results from a study (Brockhaus and Sedlmeier 2013), in which one group obtained only a verbal description of the flows ("No graph", $n = 32$) and the other obtained the usual flow diagram ("Graph", $n = 32$). In this study, seven criteria were used to judge the degree to which the solution was correct (solutions were, for instance, judged as partly correct if participants recognized the points at which the stock decreased or increased but drew wrong quantities). Omitting the flow diagram ("No graph") increased the solution rates, indicating that less information (no flow diagram) seems to have had beneficial effects in that this manipulation prohibited the use of the correlation heuristic. In the condition without a flow diagram, 45% of participants solved the task completely correctly as compared to 28% in the (commonly used) condition with a flow diagram.

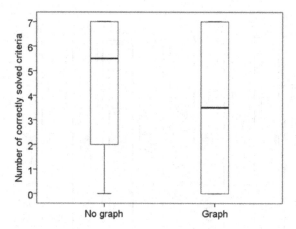

Fig. 5.5 Boxplots that compare the performances of the groups with ("Graph") and without ("No graph") a flow diagram

This result indicates that the presentation of both rates and quantities in a similar graphical way might be a serious impediment to understanding stock-flow tasks intuitively. But even if the flow diagram is omitted, the solution rates (45% completely correct solutions) do not indicate an intuitive task understanding.

Can flow be understood as a succession of stocks?

If participants have difficulties understanding the difference between rates and quantities, as indicated in the previous paragraph, one might try to find some kind of connection or transition between the two. One might, for instance, conceive of flow as a succession of stocks. This idea is illustrated in Figure 5.6. Here, the "flow," both in and out, is successively accumulated over some (discrete) time steps (one minute each); and the result of this stepwise accumulation is shown as a *change* in the stock (here contained in a quite narrow "tub"). For instance, after the 1st minute, the stock has increased by 9 units (upper row of stocks – "Water flowing into the tub"), and at the same time it also has decreased by 6 units (lower row of stocks – "Water flowing out of the tub") yielding an accumulated net-flow of 3 units. In the second time step, in- and outflow remain constant, so after 2 minutes, the net-flow has accumulated to 6 units, etc. If participants can imagine (and understand) these changes well, they should be able to arrive at the correct stock after the 12 minutes used in this task.

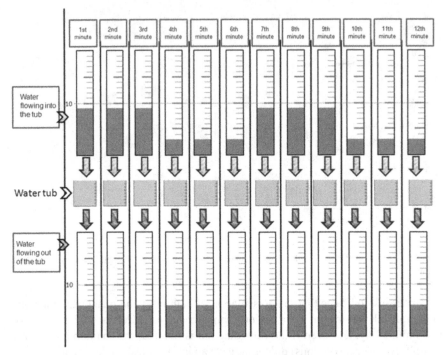

Fig. 5.6 The modified flow diagram (as a succession of changes in stocks). The amount of the in-flow and outflow after every minute is shown in the tub itself.

Participants' understanding of the task was assessed using again a succession of stocks that was depicted in the same format (Figure 5.7). Here, an initial amount of water in the tub had to be modified according to the in- and outflows depicted in Figure 5.6.

The modified version of the bath-tub task (both diagrams presented as a succession of stocks) was compared to the usual baseline condition (see Figure 5.2) in a further study (Brockhaus and Sedlmeier 2013). In this study, a scoring system was used that yielded 2 points for a completely correct solution. As the boxplots in Figure 5.8 show, the median number of points in the modified condition is 2 (indicating a completely correct solution) as compared to a median number of 0 points in the baseline condition. In the modified condition, 83.3% of participants (25 out of 30) solved the tasks completely correctly as compared to only 30% (9 out of 30) in the baseline condition.

Presenting the flow as a succession of stocks was clearly helpful and seems to have prevented the use of an invalid intuition, that is, to use the pattern in the flow diagram as the basis for one's response in the solution diagram (the correlation heuristic). The present manipulation seems to have enabled participants to determine the net-flow step by step and to update the stock accordingly by imagining the respective changes and then adding up the changes in the stock. In other words: participants performed visual integration successfully.

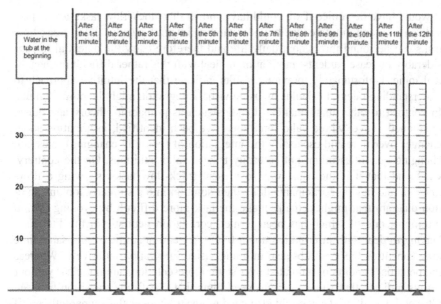

Fig. 5.7 Modified solution diagram to draw the changes in stock over time as a succession of stocks

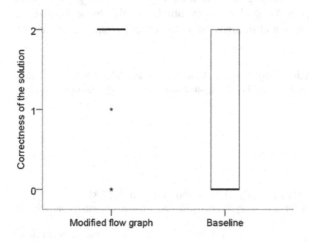

Fig. 5.8 Boxplots that compare the performance in the condition with the modified representation ("Modified flow graph") and the classical flow diagram with a line graph ("Baseline"). In the condition with the modified flow graph, the solutions of 2 and 3 participants were scored as having 1 and 0 points, respectively (depicted as extreme outliers in the left boxplot).

How Far Can Intuition Carry Us?

It would have been nice if we could have shown that stock-flow tasks *per se* provide an easy and intuitive way to understanding integral calculus. Such stock-flow tasks can be found throughout daily life and therefore carry the potential to considerably increase students' motivation to deal with this rather difficult mathematical topic. Unfortunately, previous results indicate that the intuitive approach to understanding integral calculus by using visual integration might be rather limited. In contrast to other mathematical domains such as probability theory and others mentioned above, where valid intuitions have been identified, information about changes over time and especially the integration of (possibly changing) rates into changing quantities might not so easily trigger valid intuitions. On the contrary, we found that it is the usual way to present stock-flow tasks, showing a flow-diagram and having participants draw their solution into a similarly looking solution diagram that might actually make the task more difficult by evoking *invalid* intuitions, one of which has been termed correlation heuristic in the literature. Having to deal with information about both rates and quantities at the same time may constitute the main obstacle in doing visual integration intuitively. Whereas the potential benefits of animated over static representations are not clear yet, our results indicate that students might need hints that clarify the transition between rates and quantities. They might also need more structure as they apparently profit from a step-by-step representation of the results of in- and outflows if one of these flows changes over time. Presenting flow as a succession of changes in stocks might be a good starting point for students to eventually be able to perform visual integration in an intuitive way and to arrive at a deeper understanding of integral calculus.

Acknowledgments We thank Juli Eberth, Tim Erickson, Rob Gould, Juliane Kämpfe, Thomas Schäfer, and Isabell Winkler for their very helpful comments on a previous version of this chapter.

References

Biehler, R., (1985). Graphische Darstellungen. *Mathematica Didactica*, *8*, 57-81.
Blum, W. (2000). Perspektiven für den Analysisunterricht. *Der Mathematikunterricht, 46*(4-5), 5-17.
Booth Sweeney, L. B., & Sterman, J. D. (2007). Thinking about systems: student and teacher conceptions of natural and social systems. *System Dynamics Review*, *23*(2-3), 285-312.
Borneleit, P., Danckwerts, R., Henn, H.-W., & Weigand, H.-G. (2001). Expertise zum Mathematikunterricht in der gymnasialen Oberstufe. *Journal für Mathematik-Didaktik*, *22*(1), 73-90.
Brockhaus, F., & Sedlmeier, P. (2013). *Is intuition the key to understanding stock and flow systems? The influence of modifying the systems' representation format on the stock flow performance* (Unpublished manuscript). University of Technology, Chemnitz.

Brown, A., & Campione, J. (1994). Guided discovery in a community of learners. In K. McGilley (Ed.), *Classroom lessons: Integrating cognitive theory and classroom practice* (pp. 229-279). Cambridge; MA: MIT Press.

Brunstein, A., Gonzalez, C., & Kanter, S. (2010). Effects of domain experience in the stock–flow failure, *System Dynamics Review, 26*(4), 347–354

Bürger, H., & Malle, G. (2000). Funktionsuntersuchungen mit Differentialrechnung. *Mathematik lehren, 103*, 56-59.

Cosmides, L. (1989). The logic of social exchange: Has natural selection shaped how humans reason? Studies with the Wason selection task. *Cognition, 31*(3), 187–276.

Cosmides, L., & Tooby, J. (1996). Are humans good intuitive statisticians after all? Rethinking some conclusions from the literature on judgment under uncertainty. *Cognition, 58*(1), 1–73.

Cronin, M., & Gonzalez, C. (2007). Understanding the building blocks of system dynamics. *System Dynamics Review, 23*(1), 1–17.

Cronin, M., Gonzalez, C., & Sterman, J. D. (2009). Why don't well-educated adults understand accumulation? A challenge to researchers, educators and citizens. *Organizational Behavior and Human Decision Processes, 108*(1), 116–130.

Danckwerts, R., & Vogel, D. (1992). Quo vadis, Analysisunterricht? *Der mathematisch-naturwissenschaftliche Unterricht, 45*, 370-374.

Davis, R. B., & Vinner, S. (1986). The Notion of Limit: Some Seemingly Unavoidable Misconception Stages. *Journal of Mathematical Behavior, 5*, 281-303.

Engel, J., & Sedlmeier, P. (2005). On middle-school students' comprehension of randomness and chance variability in data. *ZDM, 37*(3),168-179.

Ebersbach, M., & Wilkening, F. (2007) Children's intuitive mathematics: The development of knowledge about nonlinear growth. *Child Development, 78*(1), 296-308.

Farrel, S., & Lewandowsky, S. (2010). Computational models as aids to better reasoning in psychology. *Psychological Science, 19*(5), 329-335.

Fischbein, E. (1994). The interaction between the formal, the algorithmic, and the intuitive components in a mathematical activity. In R. Biehler, R. W. Scholz, R. Sträßer, & B. Winkelmann (Eds.), *Didactics of mathematics as a scientific discipline* (pp. 231-245). Dordrecht: Kluwer.

Fischbein, E. (1975*). The intuitive sources of probabilistic thinking in children.* Dordrecht: D. Reidel.

Frensch, P. A., & Rünger, D. (2003). Implicit learning. *Current Directions in Psychological Science, 12*(1), 13-18.

Gigerenzer, G. (2008). *Gut Feelings: The Intelligence of the Unconscious.* London: Penguin Books.

Gladwell, M. (2007). *The power of thinking without thinking.* New York: Back Bay Books.

Green, D. R. (1983). A survey of probability concepts in 3,000 students aged 11–16 years. In D. R. Grey, P. Holmes, V. Barnett, & G. M. Constable (Eds.), *Proceedings of the First International Conference on Teaching Statistics* (pp. 766-783). Sheffield, UK: University of Sheffield: Teaching Statistics Trust, University of Sheffield.

Green, D. R. (1991). *A longitudinal study of pupil's probability concepts.* Loughborough, UK: Loughborough University.

Hahn, S., & Prediger, S. (2004): Vorstellungsorientierte Kurvendiskussion. Ein Plädoyer für das Qualitative. In A. Heinze, & S. Kuntze (Eds), *Beiträge zum Mathematikunterricht 2004* (pp. 217-220). Hildesheim: Franzbecker.

Hahn, S., & Prediger, S. (2008). Bestand und Änderung – Ein Beitrag zur didaktischen Rekonstruktion der Analysis. *Journal für Mathematikdidaktik, 29*(3/4), 163-198.

Hogarth, R. (2001). *Educating intuition.* Chicago: University of Chicago Press.

Hußmann, S., & Prediger, S. (2010). Vorstellungsorientierte Analysis – auch in Klassenarbeiten und zentralen Prüfungen. *Praxis der Mathematik in der Schule, 31*, 35-38.

Kahneman, D., & Klein, G. (2009). Conditions for intuitive expertise: A failure to disagree. *American Psychologist, 64*(6), 515–526.

Mayer, R. E. (2001). *Multimedia learning.* New York: Cambridge University Press.

Norman, D. A. (1993). *Things that make us smart.* Cambridge, MA: Perseus Books.

Nunes, R., Schliemann, A. D., & Carraher, D. W. (1993). *Street mathematics and school mathematics.* New York: Cambridge University Press.

Schwarz, M., & Sedlmeier, P. (2013). *It is not that easy: animated representation formats and instructions in stock-flow tasks* (Unpublished manuscript). University of Technology, Chemnitz.

Sedlmeier, P. (1999). *Improving statistical reasoning: Theoretical models and practical implications.* Mahwah: Lawrence Erlbaum Associates.

Sedlmeier, P. (2000). How to improve statistical thinking: Choose the task representation wisely and learn by doing. *Instructional Science, 28*(3), 227-262.

Sedlmeier, P. (2002). Associative learning and frequency judgments: The PASS model. In P. Sedlmeier, & T. Betsch (Eds.), *Etc. Frequency processing and cognition* (pp. 137-152). Oxford: Oxford University Press.

Sedlmeier, P. (2005). From associations to intuitive judgment and decision making: Implicitly learning from experience. In T. Betsch, & S. Haberstroh (Eds), *Experience based decision making* (pp. 83-99). Mahwah: Erlbaum.

Sedlmeier, P. (2007). Statistical reasoning: valid intuitions put to use. In M. Lovett, & P. Shah (Eds.), *Thinking with data* (pp. 389-419). New York: Lawrence Erlbaum Associates.

Sedlmeier, P., & Gigerenzer, G. (2001). Teaching Bayesian reasoning in less than two hours. *Journal of Experimental Psychology: General, 130*(3), 380–400.

Sedlmeier, P., & Hilton, D. (2012). Improving judgment and decision making through communication and representation. In M. K. Dhami, A. Schlottmann, & M. Waldmann (Eds.), *Judgment and decision making as a skill: Learning, development and evolution* (pp. 229-258). Cambridge: Cambridge University Press.

Senge, P., Cambron-McCabe, N., Lucas, T., Smith, B., Dutton, J., & Kleiner, A. (2000). *Schools that learn: A fifth discipline fieldbook for educators, parents, and everyone who cares about education.* New York: Doubleday.

Sterman, J. (2000). *Business dynamics: Systems thinking and modeling for a complex world.* New York: McGraw Hill.

Sterman, J. D., & Booth Sweeney, L. (2005). Managing complex dynamic systems: Challenge and opportunity for naturalistic decision-making theory. In H. Montgomery, R. Lipshitz, & B. Brehmer (Eds.), *How professionals make decisions.* (pp. 57-90). Mahwah: Erlbaum.

Tall, D. (1992, August). *Students' Difficulties in Calculus.* Paper presented at the ICME-7, Québec, Canada.

Tall, D., & Vinner, S. (1981). Concept Image and Concept Definition in Mathematics with Particular Reference to Limits and Continuity. *Educational Studies in Mathematics, 12*(7), 151-169.

Tversky, A., & Kahneman, D. (1974). Judgment under uncertainty: Heuristics and biases. *Science, 185*, 1124–1131.

Chapter 6
Games, Data, and Habits of Mind

William Finzer

KCP Technologies, Emeryville, CA, USA

Abstract This paper describes three simple, data-producing computer games and the means by which students are expected to analyze the data to produce a model with which they can improve their game-playing strategy. The first game is an in-out machine with linear models. In the second game students must perceive an underlying conditional probability model. And in the third game students invent ways of scoring. Lessons learned from classroom field tests of activities based on these games are described. The activities belong to the family of model eliciting activities in that they are open ended, allow for multiple problem-solving methods, and result in a model whose utility can be evaluated objectively. Habits of mind for working with data are discussed in the context of each game: Graph data and look for patterns; turn patterns into models; use models to help solve problems; automate time-consuming processes; be critical of models; and use formulas with weights to develop ratings.

1 Introduction

When you play a computer game, you generate data consisting of your moves, other players' moves, the computer's moves, and events that occur during the game. In most such games the data evaporate when the game ends, and you are left with a summary of what happened, often simply a single number, a score. In the Data Games Project[1] we are exploring the consequences of making the game data available to players during and after game play. We are doing so in the context of ordinary U.S. middle and secondary mathematics classrooms to explore the question: To what extent can students build mathematical models with which they can improve their game-playing strategies?

This paper describes some of the games and activities we have developed, the data analysis environment in which the games are embedded, and some of the ob-

[1] This material is based upon work supported by the National Science Foundation under: KCP Technologies Award ID: 0918735 and UMass Amherst Award ID: 0918653. Any opinions, findings, and conclusions or recommendations expressed in this material are those of the author and do not necessarily reflect the views of the National Science Foundation.

servations we have gathered from field-testing. The reader is invited to try out the games at http://www.ccssgames.com.

2 Cart Weight—A Simple In-Out Machine Game

A cart appears, bearing some bricks. You enter a guess for its weight. The cart gets weighed and you get some points if your guess is close to the actual, as shown in Figure 6.1. An exactly correct guess is worth 100 points. Five carts constitute a game. A high enough score allows you to go on to the next level. There are six levels, increasing in difficulty. Notice that what makes this a game rather than an exercise: You get a score based on how well you did.

Cart Weight is simple. Once you figure out how much each brick weighs and how much the cart weighs, you have a model with which you can correctly predict the weight of a cart. (But higher levels introduce complexities—variation in brick weight and two sizes of bricks, for example.)

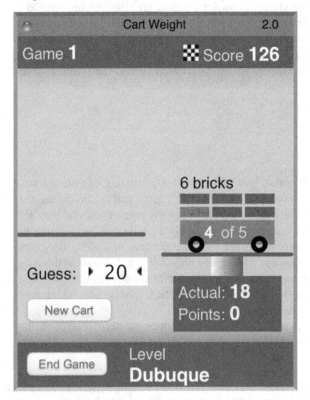

Fig. 6.1 The game of Cart Weight with 4 out of 5 carts completed

The data generated in a game of Cart Weight includes the number of bricks and the actual weight. For all but the simplest level you would find it important to keep track of these numbers. But, in fact, the game does not exist solely on its own but is embedded in a data analysis environment with tools for working with the data. Figure 6.2 shows Cart Weight, a table listing data from the four moves, and a scatterplot whose points, by appearing to lie on a line, suggest the linear relationship between weight and bricks. In fact, at this first level of the game, the relationship is strictly proportional, with the cart itself having no weight.

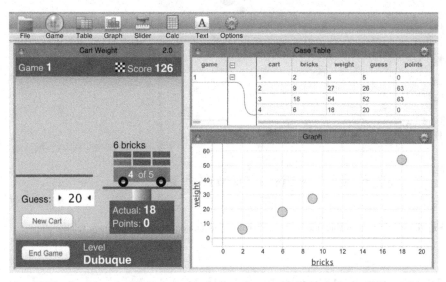

Fig. 6.2 The game is embedded in a data analysis environment with tools for exploring and modeling.

Data Games runs as a web application entirely within a browser. It is implemented using only HTML5 and JavaScript, and it requires no installation. By signing up for a free account, users can save and restore Data Games documents just as they would word processing documents. The data analysis capabilities are limited, but quite adequate to support the level of modeling required by the games themselves.

Fathom Dynamic Data® Software (Finzer 2006) is the inspiration for the data analysis tools, so, as in *Fathom*, data is dynamically linked between representations, objects such as axes, sliders, and movable lines are dynamically manipulable, and when data points are dragged all derivative visualizations and quantities are dynamically updated. Figure 6.3 shows a movable line that has been dragged to an approximate fit for Cart Weight data. Notice the highlighted point is not exactly on the line—the relationship is noisy!

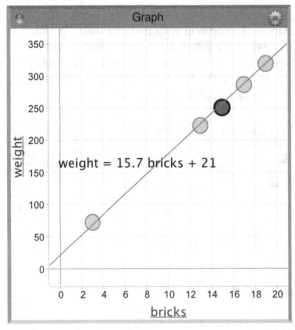

Fig. 6.3 Placing a movable line in the scatterplot yields an equation for the linear relationship

3 Classroom and Curriculum Contexts

Before moving to a discussion of what we observed in field test classrooms, we provide a bit of helpful context. The games and associated activities were designed to be usable within a single classroom period of an hour or less. For each activity we provide two student videos 1–2 minutes long that explain the game and show how to use the most relevant features of the environment; e.g. the movable line in the scatterplot. There is also a 5-minute video introducing teachers to the activity, the mathematics students will encounter, and tips for conducting the activity. An editable student worksheet gives teachers an easy default activity structure and a record of student work.

Each game and activity is designed to fit into an area of the mathematics curriculum as it is taught in the U.S. Cart Weight, for example, fits nicely into the study of linear relationships. A given teacher may choose to use a Data Games activity as an introduction to new concepts, as an experience to deepen understanding, or as a capstone to a unit in which students put newly mastered mathematics to use.

4 Cart Weight in the Classroom

Cart Weight was developed late in the project as the result of discussions with field test teachers. They wished for a game with simpler mathematics than any we had developed up to that time, one that requires no equation solving. Once students have the equation of a movable line, they can accurately predict the weight of a cart from the number of bricks. Somewhat to our surprise, this did not make the game itself easy.

To explain the difficulties students have playing Cart Weight, we need to describe the first three (of six) levels. In the first level the relationship is *weight* = 3 * *bricks*, with no change from game to game. Because of the small integer coefficient, students typically determine this rule quickly without need of table or graph. In the second level, *weight* = 4 * *bricks* + 8, again with no change from game to game. Many students will use the table for this level and will get high scores after a couple of trials during which they accumulate enough data to "eyeball" the equation. Though few students have difficulty with this level, those that do can get a solution from their neighbors because the relationship is the same for everyone.

The trouble (or the opportunity for mathematical learning) occurs in the third level, in which the relationship, again with small integer coefficients, *changes* with each game. At this level, students must use the first two carts in a game to determine the linear relationship in order to score well enough with the remaining three carts to be allowed to continue to the next level. In 4 classroom field tests at grade levels ranging from 8 to 12, less than 50% in each class were able to do so within a 50-minute class period. And yet *all* students in all these classes had studied linear relationships, some within recent days or weeks and others in previous years.

Our attempts to explain these results center on students attitudes toward data and graphs. While students are willing and anxious to look at individual rows of data in a *table*, they are much less interested in observing and making use of spatial patterns of data as they appear in a *graph*. We frequently saw students who had not made a graph in spite of explicit instruction to do so, and, just as frequently we saw students who had made a graph (in which data points lay along a line) but were studying numerical values in the table. Many students appeared puzzled that we wanted them to make use of the graph.

Conjecture: Students' encounters with graphs center on the task of making a graph, or of reading values from graphs. They seldom have an encounter in which graphs are used as tools for helping to solve a problem. Graphs are tasks, not tools. In a Data Games activity there is no task associated with the graph—it is, after all, already constructed. Consequently most students ignore the graph even though using it as a tool is the most effective way to get a high score for a third-level game. This conjecture points to an instructional remedy, a question to pose to students: "How can you use this graph to improve your score?"

5 Habits of Mind

More important than any particular set of algebraic skills that might be improved by playing Cart Weight is that students develop ways of thinking about how to work with data. These are what Cuoco et al. (1996) call *habits of mind*, "a repertoire of general heuristics and approaches that can be applied in many different situations." An activity based on the game of Cart Weight offers an opportunity for students to develop, and put into practice, habits of mind that pertain to building data models. Students generate data in real time, the tools for data exploration and analysis are immediately accessible, and the problem—how to increase one's score enough to advance to the next level—is clearly defined.

It is a habit of mind, when confronted with data, to seek to visualize that data in a graph. Once having obtained a graph, it is a habit of mind to study it to find patterns. If a pattern is observed, it is a habit of mind to attempt to describe it as a mathematical relationship. Finally, armed with a relationship, it is a habit of mind to make use of it to help understand the situation and solve problems associated with that situation. For Cart Weight the data represent the turns in the game; the graph is a scatterplot of weight versus number of bricks; the pattern is that the points in the scatterplot lie on a straight line; the relationship or model is a linear equation; and this equation is a tool for accurately predicting the weight of the cart from the number of bricks.

At the risk of over-simplification, this description of habits of mind associated can be turned into a concise list of injunctions as follows:

- Graph data and look for patterns.
- Turn patterns into models.
- Use models to help solve problems

6 Markov—An Encounter with Conditional Probability

Your objective in the game of Markov is to save Madeline the dog from the evil doctor Markov by beating him at Rock-Paper-Scissors. You and Markov each choose rock (R), paper (P), or scissors (S). Rock beats scissors; scissors beats paper; and paper beats rock. In Markov's cruel version of the game, you lose if you both make the same choice. When you lose, Madeline the dog goes down, and when you win, she goes up. Get her all the way up and she goes home with you. Otherwise Markov keeps her in his lab. Figure 6.4 shows the basic setup.

As with Cart Weight, you won't get far without studying a graphical representation of the data. The particular graphical representation students work with is the dot chart shown in Figure 6.5. It's a bit like a 2-way table, and it takes a while to master reading it. This graph and other cues strongly lead the player to attempt to predict Markov's next move from his previous two. For example, in Figure 6.5 the player should conclude that if Markov's previous two moves were SS, his next move will almost certainly be R, so the smart move for the player is P.

Fig. 6.4 You won't succeed at helping Madeline the dog to the exit without careful analysis of Doctor Markov's moves in the game of Rock-Paper-Scissors.

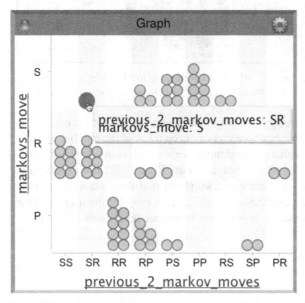

Fig. 6.5 Each data point in the dot chart represents Markov's previous two moves (on the x-axis) and the move he made next (on the y-axis). It's not totally random; e.g. SS is always followed by R.

Once a player can properly interpret the dot chart, it becomes a fairly mechanical operation to save Madeleine. The fun of playing would be over were it not for the ability of the player set an automated strategy, the interface for which is shown in Figure 6.6. The strategy shown there suits the data from Figure 6.5.

Notice that three of the panels do not yet have a response set, which makes sense because there is still too little data for those situations. Notice also that 'weights' have been distributed so that the panels with the most certain and frequent outcomes get the most weight. The weight associated with a panel determines how far Madeleine will move up or down when that panel's situation applies. By heavily weighting a panel that has a certain outcome and occurs frequently, a strategy can be made to win games in a small number of moves. Using an automated strategy, a game that takes many minutes to play manually can be finished in just a few seconds.

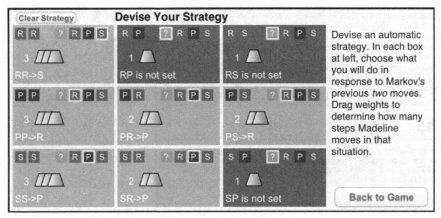

Fig. 6.6 In each box, the strategy specifies what to do in response to Markov's last two moves. The number of 'weights' in a box determine how far Madeline will move, either up or down.

Once you can specify a strategy and play games very quickly, it becomes possible to compare strategies. For example you can keep track of the number of turns it takes to rescue Madeline with a given strategy. How much better does a strategy that uses weights do compared to one that does not? Figure 6.7 shows the results of an experiment to answer just that question. (Note that the Data Games environment treats the game data hierarchically so that it is easy for the user to make graphs of either turn level cases as in Figure 6.5 or game level cases as in Figure 6.7.)

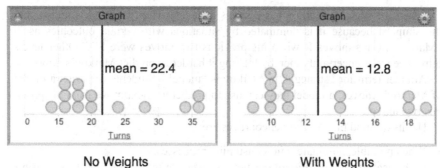

No Weights   With Weights

Fig. 6.7 Each dot plot shows a distribution of number of turns it took to win a game of Markov. The strategies for the games in the left-hand plot did not distribute the weights whereas those in the right-hand panel used the weights shown in Figure 6.6. The games played with weights take many fewer turns.

7 Markov in the Classroom

As you would expect from the preceding discussion of Cart Weight, students resisted trying to figure out how to read and use the dot chart. We endeavored to improve their chances by renaming attributes, providing explanatory tooltips, and adding some explanations to student worksheets. But we also felt that struggles with the representation were beneficial and would lead to improved understanding of conditional probability and two-way tables in future encounters. Nearly all students in field test classrooms were successful in using the dot chart to save Madeline from Markov.

Markov makes a brief sound for each move, and, because of this, the shift from manual play to autoplay is a dramatic event in a computer lab. The first time a rapid-fire game is heard, heads turn, and students want to know how to do it. The change from figuring out individual moves to setting an entire strategy involves a profound cognitive shift. The strategy embodied in the panel shown in Figure 6.6 becomes an object of study in itself. Students have to explain what they see happening during autoplay in terms of their strategy; e.g. that occasionally Markov wins very quickly because a low probability event triggers a rapid descent, or that a strategy takes a very long time to win because weights haven't been effectively used. Students often became competitive with their strategies, tuning them to produce wins with the least number of moves and arguing over whether it should be the average number of moves or the least number of moves a strategy produces that determine how good a strategy is.

In field test classrooms where there was sufficient time for students to experiment with weights in the strategy panel, we heard astute discussions about the need to balance certainty of outcome with frequency of occurrence when considering which moves to weight heavily and which to weight lightly.

Like Cart Weight, Markov has a series of levels. The first level is undeniably the simplest because it is dominated by situations with certain outcomes as in "Markov always moves R when his previous two moves were SS." Other levels introduce more uncertainty thereby making it harder to predict Markov's moves.

Another term for "strategy" in Markov is "model." Students construct a model of Markov's moves, a model they can use to predict his next move from his previous two moves.

Habits of mind that Markov encourages are:

- When possible, automate time-consuming processes.
- Be critical of models. Figure out how to measure their accuracy. Devise ways to improve them.

8 Lunar Lander—Defining a Measure

By the criteria put forward earlier, Lunar Lander should not be considered a game because no score is assigned. A player attempts to land by firing thrusters to increase or decrease rate of descent, but no indication is given as to whether the landing was 'successful' or not. Students are asked to make as good a landing as they can, which leads to the question, "How do I know how good the landing is?" As shown in Figure 6.8, two students can play at the same time, which still begs the question of which one produced the better landing.

Fig. 6.8 In Lunar Lander a player controls the thrust on a rocket headed for the moon's surface.

Figure 6.9 shows the default graph, a plot of altitude versus time, for two landings. By studying these and other graphs, students generally arrive at the conclusion that low impact, short time, and more fuel are all good. But they discover this is still a qualitative description and leaves open the question of how to compare one landing that has a shorter time than another but a higher impact. The important concept for students to come to grips with is that numeric measures have advantages over qualitative descriptions in comparison situations.

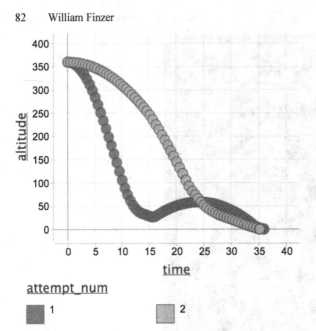

attempt_num

☐ 1   ☐ 2

Fig. 6.9 Which of these two traces shows a better landing?

Attribute Name:	score
Formula:	fuel_remaining−impact−total_time

Cancel   Apply

Fig. 6.10 This score formula at least has the right dependencies on fuel, impact, and time, but it probably counts fuel more heavily than it should.

Students are asked to write a formula that computes the score for a landing, a simple example of which is shown in Figure 6.10. With a formula for score in place, the numeric values appear in the table of game values as shown in Figure 6.11.

attempt_num	total_time	impact	fuel_remaining	score
1	35.94	3.8	11.54	-28.2
2	35.19	4.57	63.96	24.2

Fig. 6.11 The scores are computed for the two landings. Should we allow negative scores?

The important work of Lunar Lander, once there are several candidate scoring formulas, is to critically evaluate them with actual landings and to choose a formula that everyone can agree to use to determine a winner at the game.

9 Lunar Lander in the Classroom

In one field test class we found that the idea of writing a formula to express something, especially a formula in which the variable names are whole words instead of single letters, to present a considerable hurdle. Watching this class made us as observers realize the complexity involved in asking students to shift between landing the rocket, making sense of the altitude versus time graph, designing a formula, and comparing how well different formulas capture the idea of a good landing. For this class, Lunar Lander did not fit well within a single class period.

In other classes students were able to spend significant time trying various formulas for the score and deciding which ones worked better than others. Students were able in these classes to deal with the vagueness of the challenge and to appreciate that the decision about which formula to use is driven by qualitative concerns. The activity introduces students to a way of thinking they may not often encounter in mathematics classes, but one that has broad application in statistics and science. Here it is expressed as a habit of mind:

- To evaluate how good something is, consider designing a formula that combines several attributes with appropriate weights.

10 Discussion

Computer games, whether designed for classroom use or not, are often cited as vehicles for learning. James Gee, in fact, has described how learning to play many of the complex computer games involves much of the same skill set as learning to think scientifically (Gee 2003). These include observing, interpreting, refining techniques, and making and testing conjectures.

Data games differ from most education games in that the concepts to be learned are not embedded in the context of the games, but in act of engaging in data modeling. The three games discussed here illustrate ways that data analysis can be combined with learning mathematics. The games are sources of data. By embedding them in a data analysis environment, there is a tight loop: generate data, make conjectures, construct a quantitative model, evaluate the model back in the game to generate more data.

As Erickson (2012) has discussed, the design constraints on the games are rather strict. Among other things, the data produced by a game has to be easily useful in improving the player's score, and, conversely, the game must not be easily susceptible to winning by trial and error practice.

Data Games activities belong loosely to the genre of "model eliciting activities" described by Lesh et al. (2008) and Doerr and English (2003). The problem of getting a high score on a data game is fairly open-ended, and no sequence of steps, or previously worked example, is provided. There is typically no "best" strategy, and, when students' strategies are compared, there is often no way to say that one is better than another.

Though the activities fit within a single class period, some teachers have extended activities to last two or three periods. The levels in the games give students plenty of room for further exploration while allowing entry for everyone.

11 Next Steps

The games and activities were released for free use in Fall, 2012. We expect to learn from this release how well they can be adapted by teachers who have not been part of the field test. The Data Games environment and its application programming interface (API) for creating components that can live within it are now being used and extended as a platform for the InquirySpace project at the Concord Consortium (see http://concord.org/projects/inquiryspace).

Meanwhile, ongoing collaborative research and development under the direction of Cliff Konold has resulted in additional games and activities currently being tested in classrooms. These new materials are scheduled for release as part of the Data Games web site by Fall, 2013.

Acknowledgments   The Data Games project is a team effort. The results described in this paper would not have been possible without the excellent games designed and prototyped by Tim Erickson. Thanks to Cliff Konold who heads our collaborative research project at the University of Massachusetts, Amherst; to Kirk Swenson for his tireless work on the data analysis environment, to Vishakha Parvate, Jaimie Stevenson, and Rick Gaston for organizing and conducting the piloting and field testing, and to the thirteen middle and high school teachers who tested the activities and gave us useful feedback.

References

Cuoco, A., Goldenberg, E. P., & Mark, J. (1996). Habits of mind: An organizing principle for mathematics curricula. *Journal of mathematical behavior*, *15*(4), 375-402.

Doerr, H. M., & English, L. D. (2003). A modeling perspective on students. *Journal for research in mathematics education*, *34*(2), 110-36.

Erickson, T. E. (2012, July). Designing games for understanding in a data analysis environment. Paper presented at IASE 2012 Roundtable Conference. Technology in Statistics Education: Virtualities and Realities, Cebu, Philippines.
http://icots.net/roundtable/docs/Tuesday/IASE2012_Erickson.pdf

Finzer, W. (2006). *Fathom dynamic data software* [Computer Software]. Emeryville, CA: Key Curriculum Press.

Gee, J. P. (2003). *What video games have to teach us about learning and literacy*. New York: Palgrave Macmillan.

Lesh, R., Middleton, J. A., Caylor, E., & Gupta, S. (2008). A science need: Designing tasks to engage students in modeling complex data. *Educational Studies in Mathematics*, *68*(2), 113-130.

Chapter 7
Caging the Capybara: Understanding Functions through Modeling

Tim Erickson

Epistemological Engineering, Oakland, California, USA

Abstract How do advances in technology expand and improve the ways we can teach about mathematical functions? In addition to indispensable tools for dealing with stochastics, technology gives us modeling tools: tools that enhance data analysis by letting students graph functions with their data, create new computed variables, and control model functions dynamically by varying the functions' parameters. In this paper, we will use an extended example—an optimization task we will call the "Capybara Problem"—to show how we can use these tools to address common difficulties students have with functions. This paper describes seven different approaches to this problem, beginning at the concrete end of the spectrum—using physical materials to represent the problem and its constraints—and then gradually introducing abstraction in the form of variables and functions. Technology supports students throughout this process, helping them understand the nature of variables, and helping them learn to construct symbolic functions and to meaning in their forms and parameters.

Introduction

Functions are among the most important concepts in mathematics—and not just for mathematicians. As lay people, we deal with functions to make observations and decisions large and small, personal and public. We often look at functions of *time*: my rent is going up. Unemployment is going down. Sometimes, we're looking at how one quantity varies with another: The faster I drive, the worse my gas mileage. The more coffee I drink in the evening, the harder time I have getting to sleep. In any case, functions are relevant because they're about *relationships*.

Traditionally, learning about functions has been an abstract and formal endeavor, rooted in algebra. Readers of this volume (especially older readers) learned about elementary functions—linear, quadratic, polynomial, power-law, exponential, logarithmic, trigonometric—from an algebraic perspective. We learned how to solve for quantities in symbolic expressions, and came to understand how combinations of symbols encoded features of functions—for example, how a positive leading coefficient in a quadratic means that the curve opens "up," or how a number added to a sine function raises the wave. Graphs helped us understand these principles, but we did not use graphs to solve problems: they were too hard to draw. We saw some data, although not very much of it. More often, we struggled with abstract symbolic representations, and learned their properties.

We can improve this. It is now easy to make graphs, and easy to find and incorporate data into problems and investigations. We should use these new tools to make learning about functions more accessible, concrete, and effective.

Developers at the postsecondary level (e.g., Engel 2010) have addressed this problem under the aegis of applied mathematics. At the school level, it is currently popular in the US to call this a *modeling* approach. Fortunately, there has recently been a greater call for more modeling in mathematics education (e.g., in the "Common Core" State standards, NGA 2010). Because of improvements in technology, it is now practical to do genuine modeling in the secondary classroom. (See also Erickson 2005 for examples.)

Let us take as given that we believe that being rooted in data is a good idea, and that graphs—created through technology—can give you more insight than a typical algebraic expression (Kaput 1989).

We will spend most of this paper dissecting an example in detail. During this journey, we will see how two ingredients—dynamic graphs and data—combine to help students make sense of functions. The graphs give us insight into the functions; the data gives us realism and rich, interesting contexts. We will also see how data, the epitome of concreteness, helps us on the road to abstraction.

The Capybara Problem

Consider this typical introductory calculus problem:

> The Queen wants you to use a total of 100 meters of fence to build a Circular pen for her pet Capybara and a Square pen for her pet Sloth. Because she prizes her pets, she wants the pet pens paved in platinum. Because she is a prudent queen, she wants you to *minimize* the total area.
> What are the dimensions of the Queen's two pet pens?

In a problem like this one, students often have trouble setting up the function to be minimized, in particular:

- Choosing a suitable *single* independent variable.
- Confusing the side length of the square with its perimeter.
- Maintaining the 100-meter constraint.
- Finding the area of the circle given its circumference.

There are some chain-rule challenges in taking the derivative, but if you can't make the original equation, no amount of differentiating will get you the right answer. In the traditional calculus class, moreover, the instructor is focused on the calculus, not on how to build the right function.

Seven approaches to the capybara problem

To see how modeling and data can help with this, let us approach this problem from different points along a "continuum of abstraction," gradually leaving behind physical materials and the construction of specific circles and squares as we gradually introduce variables, functions, and generalizations. (We can also view it as a developmental sequence.) These different approaches will help students understand different parts of the formula-creation process.

1. Each pair of students gets 100 centimeters of string. They cut the string in an arbitrary place, form one piece into a circle and the other into a square, measure the dimensions of the figures, and calculate the areas. They glue or tape the shapes to pieces of paper. The class makes a display of the shapes and their areas, organizes them—perhaps by the sizes of the squares—and draws a conclusion about the approximate dimensions of the minimum-area enclosures.

   This approach is the most concrete; even elementary students can use it. It is modeling even though it does not use functions. The experienced teacher can highlight the way the problem constrains the total amount of fence: To make a small total area, why can't you just make a little circle and a little square? Because you have to use all the string.

2. Same as above, but we plot the data on a graph. To do this, we have to decide on an independent variable—a number that tells us which of the versions of fence is which. We probably choose the side length of the square to "name" a configuration. (The dependent variable is easier; here we would use the sum of the areas.) We can estimate the dimensions and the minimum area from a sketch of a curve through the points—an informal function, if you will.

3. We enter the data into dynamic data software, plot the points, guess that they fit a parabola, and enter a quadratic in vertex form, adjusting its parameters to fit the data. This approach introduces symbolic mathematics, and needs technology in order to be practical in the classroom.

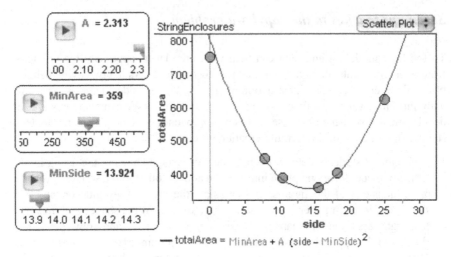

Fig. 7.1 Fitting a quadratic function informally using Approach (3), using Fathom

The Figure 7.1 shows what this looks like using Fathom (Finzer 2007); we've plotted total area (`totalArea`) against side. The sliders at left are the parameters that control the curve. These vertex-form parameters are the side of the "best" square (`MinSide`), the minimum total area (`MinArea`), and (A), the quadratic's leading coefficient. The function's formula here is

$$\texttt{totalArea = MinArea + A (side - MinSide)}^2.$$

You could do (1), (2), and (3) in rapid succession, to help students see connections among the string diagrams, the measurements, the graph of data points, and the graph of the function on the computer.

4. We use diagrams: instead of making the shapes with string, we draw them on paper. An individual or a small group can more easily draw several different sets of enclosures. This requires more calculation. For example, if we draw the square first, we have to calculate how big the circle must be before we draw it.
   This is a big step towards abstraction. The string doesn't help us keep track of the fence-length constraint, so we ourselves need to ensure that the sum of the perimeters is constant.
5. We make a diagram again, but use a variable for the length of a side and calculate expressions (instead of numbers) for all the other quantities. So we create no specific cases, we no longer measure, we no longer plot data at all. Instead, we use our expressions for the areas of the figures and plot their sum as a function of the side length—and read the minimum off the graph.

Specifically, if the square's side length is x, its perimeter will be $4x$. The circle's circumference will have to be $(100 - 4x)$, and from that we can find the radius. We use those to find the total-area function $A(x)$:

$$A(x) = x^2 + \pi r^2 = x^2 + \pi \left(\frac{100 - 4x}{2\pi} \right)^2$$

This problem setup is the same as for calculus, but to find the minimum, we use graphing software and approximation. Thus this approach includes all of the abstraction that many students find so difficult.

6. We use algebraic techniques (including completing the square) to convert this expression to vertex form, from which we read the exact solution. Using this approach, we might not even plot the function.
7. We use calculus, and avoid some messy algebra in (6). Interestingly, completing the square in (6)—which students traditionally learn before calculus—is *much* harder than taking a derivative, setting it to zero, and solving.

How data, modeling, and technology help

You can see that these different approaches gradually introduce more abstraction. We conjecture that they will "scaffold" students as they work to write that $A(x)$ function (above). Let us look in detail at the roles that data and dynamic graphs play in that process.

From data to table to scatter plot

First let us look at what data the students need to record. In the first three approaches (1–3), it is useful if they write the (measured) side of the square and the diameter of the circle; possibly the radius of the circle; the two (calculated) areas[1]; and their sum. Although the way students record data spontaneously may be informal and incomplete, this is a good chance to help them decide to organize the data into a table, with columns like a spreadsheet; this is how it will go into the computer, after all.

---

[1] One could have students make the figures on grid paper and count squares to determine the area, but we'll skip that here for simplicity.

StringEnclosures

	side	diameter	S_area	C_area	totalArea
=			$side^2$	$\left(\dfrac{diameter}{2}\right)^2 \pi$	S_area + C_area
1	15.5	12.4	240.25	120.763	361.013
2	10.4	19.0	108.16	283.529	391.689
3	18.3	9.4	334.89	69.3978	404.288

Fig. 7.2 In this illustration, the students measured the side of the square and the diameter, but computed S_area (the square area), C_area (circle area), and totalArea. They did not need columns for perimeter and circumference; that will be essential in approach (4).

In approach (3), students enter this data into a table on the computer. This brings up a question: of these columns, which ones did you *measure* and which did you *compute*? And if you computed them, how did you do it? Students tend to calculate area, for example, using a calculator, and then type the result into the table cell. But instead, we can encourage them to have the computer do the calculations for the columns they compute.

For example, they probably divide the (measured) diameter by 2 to get the radius; they square that and multiply by π to get the area of the circle (see Figure 7.2). Asking them to write these as formulas so that the computer does the calculation accomplishes two things: it alerts students to the fact that they actually know how to perform *parts* of the calculations, and it separates what will become a very complicated formula into manageable chunks.

Adding the dynamic (and empirical) function to the graph

After students graph the data in approach (3), we introduce a dynamic function: we put a function on the graph and adjust it to fit using slider-parameters. At this level, students need to know how to write a formula for a parabola in vertex form. They need to parameterize it appropriately as well, so they can enter it into the software.

The vertex form is perfectly suited to dynamic graphing software: students see how changes in parameter values change the function. When they drag the sliders, they get a visceral feel for how the transformations work. When a residual plot is present as well, students see the parameters' effects there, for example, how changing the principal coefficient *straightens* the residuals (See Erickson 2008 for details).

Yet the symbolic form of this function will *not* look like the one that arises when we do the calculus problem—it is an *empirical* function, and therein is one of the delicate aspects of this approach. We will soon see how students come to discover the more "theoretical" form.

Having the computer calculate as much as possible

In (4), when students begin with a diagram instead of string, students have to use even more symbolic tools: they have to deal with the 100-meter constraint, and figure out the diameter of the circle given its circumference.

It helps students to have a detailed, many-columned table like the one they used above. But they now increase the number of columns, and write formulas for more of them. If they start with the side of the square, for example, they can write a formula for its perimeter. And if they know the perimeter, they can calculate the circle's circumference. They may not feel comfortable writing formulas for *all* of the derivable quantities, but that is our ultimate goal. If they start by choosing the length of the side of the square (or any particular quantity except total area), they can calculate *every* other quantity in the table—and make the graph.

Enclosures

	side	perim	circ	diameter	S_area	C_area	totalArea
=		$4side$	$100-perim$	$\dfrac{circ}{\pi}$	$side^2$	$\left(\dfrac{diameter}{2}\right)^2\pi$	$S\_area + C\_area$
1	10	40	60	19.0986	100	286.479	386.479
2	20	80	20	6.3662	400	31.831	431.831
3	15	60	40	12.7324	225	127.324	352.324

Fig. 7.3 The student is using Approach (4) in this table, computing everything based only on the side of the square. In contrast to the table in Figure 7.2, nothing is actually measured.

In making these formulas, students become aware of the repeated calculations they have been making; they see how to express those calculations symbolically in formulas for columns; and most important, they see that it is worth the effort to make the formulas. When they realize that they don't need to actually measure anything in the diagram to determine the two areas, they are ready to move on.

Doing without the data points

Approach (5) is the modeling payoff. Instead of an empirical function that fits the data, we create the function from the geometry. The function is quadratic because it arises from area calculations, not simply because the data *look* quadratic. How does technology help students make that leap in understanding?

One way is this: in (4), once every column is computed from the side of the square, you can enter any side length you like, and the software will compute the sum of the areas. No measurement necessary. The string was a simulation of a fence; now we're using calculation to simulate the string. It is all numbers now. So we can blanket the domain with values, and plot the resulting data points, none of which depend on sliders. The points lie on a parabola, of course—the small points in Figure 7.4. If we could plot them all, we would have the function itself: the infinite set of points that describe the relationship.

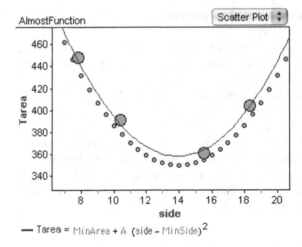

$$\text{— Tarea} = \text{MinArea} + \text{A} \ (\text{side} - \text{MinSide})^2$$

Fig. 7.4 In this graph, the small points are values computed from the side of the square based on a total length of 100; the function is the student's best fit to the data (the large points).

How do we turn an infinite set of points into a single curve, a single formula? We combine the formulas we have already made for each individual column. This is the spot where students need to know how to substitute—arguably a critical algebraic skill. If it is hard for them, they can do it step by step, eliminating one column at a time, and verifying that the simulation—for that is what this has become—gives the same results.

Now students can simply plot the function and find the coordinates of its minimum. They can even compare it to the vertex form; they can do it graphically, or, if they have the skills, they can put both formulas in the same algebraic form and see how well they match.

Reflecting on the process

This modeling approach addresses problems students have with creating the formula for total area. It also gives us good approximate answers to the original question—what are the dimensions of the pens?

We help students use abstraction by starting, sensibly enough, with something concrete: the string. Then we gradually introduce representations and abstractions as they become useful. This problem and these approaches are certainly not the entirety of learning about functions or about modeling in secondary mathematics, but the basic ideas should apply to other problems:

- A concrete simulation requires the least abstraction and often makes it unnecessary to model difficult-to-represent relationships.
- There are many approaches between that concrete model and the purely abstract, traditional mathematical function.

- An empirical graph—one made by measuring specific examples rather than by analyzing the generalized situation—often helps students understand underlying functions, and can lead to an approximate solution.
- An approximate solution may be good enough.
- Writing easy formulas for "bite-sized" calculations can lead to a more comprehensive solution.
- Dynamic data analysis software helps students organize their data, visualize it with graphs, create functions, and make those "bite-sized" calculations.

Finally, we should not think of using data, or graphs, or even string, as less sophisticated or desirable than using calculus. Consider one of the most confusing things about the problem—that there is in fact a *minimum* area. Most area problems using a fixed amount of fence *maximize* area. How can there be a minimum? A successful calculus student might say, "because the coefficient of the first term is positive." But a student who used the string might say, "because the shape you can make with the whole fence is much bigger than two shapes, each made with half the fence." One can make a case that the "string" answer shows more insight.

Additional Notes and Observations

Confusing the Data with the Model.

It is important that students (and teachers) be clear why the data appear as points and the model as a curve. The model is an ideal, a fantasy that we are proposing for consideration. It exists for any possible value. In contrast, data is reality, and exists only at specific values.

We can use data as a check on models (just like in science) rather than simply assuming that a formula must be better than a measurement. When we measure string figures, we will get points that do not lie on the curve. This discrepancy helps students connect math to reality. There are any number of reasons the curve might not go through the point. It is fair to blame the data if there is a measurement mistake (in our case, the string may have been a bit long). But the model may have missed because we modeled the ideal situation, leaving other aspects of reality *unmodeled*: our circles were not perfectly round, nor our squares square.

Habits of Mind.

We want students to develop good mathematical practices; modeling activities like this one offer opportunities. Here are two:

When we ask students to make separate columns for each quantity, and write simple formulas, we're doing two things: we're helping them encode their knowledge in chunks they understand, and learn to combine them; and we're also helping them learn to *identify and name variables*. By having them use the names of intermediate variables in formulas—instead of plugging a numerical value in, using a calculator—we're helping them see the advantage of "keeping a calculation in letters as long as possible."

Another is to *check limiting cases*. Even when students are making shapes with string, a teacher can ask, "what if we put all of the fence into the square? What would the areas be then?" Students can see that the maximum square side is 25 meters, and the sum-of-areas is 625. This point is an anchor for our curve, one we can be sure of theoretically. We can do the same on the other side, with the circle—which also reminds students how to find area if they know circumference.

Looking for New Questions.
Rich approaches like these open up new questions, often suitable for more experienced students that finish early. Here is one that we can investigate if we have a table with many columns: suppose we plot the side of the square against the diameter of the circle (see Figure 7.5, left). Why is it a straight line? Why is the slope -1.27?

If that is too easy, plot the area of the circle against the area of the square (Figure 7.5, right). What function models *that*?

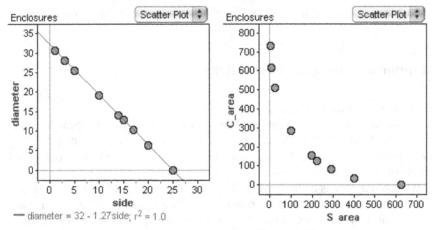

Fig. 7.5 Two additional relationships students can investigate

Power and insight from graphs and data—especially as things get realistic.
An abstract, symbolic solution often gives an exact answer when a model or simulation will give us only an approximation. But sometimes an approximation is what we really need.

A task like the Capybara Problem, after all, is unrealistic and idealized. It is not that a Queen would not pave her pens in platinum; but a real optimization problem would not be so clean. We would have to consider where the pens had to be placed, the slope of the ground, the cost of posts, size and cost of gates, plumbing to bring water to the area, and so forth. An idealized model, like the proverbial spherical cow, is *useful* because it captures a mathematical essence. But we can't depend on it for the details. In our case, when we look at the graph, we can see that the curve, the parabola, is flat on the bottom, and that the total area varies no more than 10% when the side of the square is between 10 and 18 meters.

A graph tells you this vital piece of information immediately. The exact solution ($side = 100/(4 + \pi)$) does not.

In a modeling curriculum, we will still see these familiar and pristine problems, but we should see more and more realistic problems as well. We will see problems with more data and fewer clean and artificial constraints. We will see different kinds of functions (Thompson 1994), categorical data instead of just numerical, and data with inherent variability. Answers will become ranges instead of single numbers. We will need to cope with uncertainty in our conclusions. In short, we will need to understand stochastics and statistics.

In those problems, we will need clean, pure functions to serve as models: idealized relationships that approximate reality in some essential way. We will express them symbolically and explore their properties. But we will also have to be aware of their limitations—they are only models, after all—and use graphs, data, and all the tools of modeling to find real, practical answers and to make informed decisions.

Acknowledgments The author is grateful to thoughtful reviewers whose comments helped improve this paper; and to the editors for the hard work of organizing this "herd of cats."

Of course, all of us are also indebted to Rolf Biehler, for ongoing inspiration through his work, and for providing the occasion for collaborating on this volume.

References

Engel, J. (2010). *Anwendungsorientierte Mathematik: Von Daten zur Funktion*. Heidelberg: Springer.

Erickson, T. (2008). A Pretty Good Fit. *The Mathematics Teacher 102*(4), 256–262.

Erickson, T. (2005). Stealing from Physics: Modeling with Mathematical Functions in Data-Rich Contexts. *Teaching Mathematics and its Applications, 25*(1), 23–32.

Finzer, W. (2007). *Fathom Dynamic Data™ Software* (Version 2.1). Emeryville, CA: Key Curriculum Press. Retrieved from http://www.keypress.com/fathom.

Kaput, J. J. (1989). Linking representations in the symbol systems of algebra. In S. Wagner, & C. Kieran (Eds.), *Research issues in the learning and teaching of algebra* (Research agenda for mathematics education, Vol. 4, pp. 167–194). Reston, VA: NCTM.

National Governors Association (NGA) Center for Best Practices, Council of Chief State School Officers (2010). *Common Core State Standards for Mathematics*. Washington D.C: National Governors Association Center for Best Practices, Council of Chief State School Officers.

Thompson, P. W. (1994). Students, functions, and the undergraduate curriculum. In E. Dubinsky, A. H. Schoenfeld, & J. Kaput (Eds.), *Research in Collegiate: Mathematics Education I* (Issues in Mathematics Education, Vol. 4, pp. 21–44). Providence, RI: American Mathematical Society.

Kapitel 8
Visualisieren – Explorieren – Strukturieren: Multimediale Unterstützung beim Modellieren von Daten durch Funktionen

Markus Vogel

Pädagogische Hochschule Heidelberg

Abstract Phänomene aus der erlebten natürlichen, technischen und sozialen Umwelt lassen sich über Daten abbilden. Kern der Datenanalyse ist, im Rauschen der Daten Muster ausfindig zu machen, sie mit mathematischen Mitteln zu modellieren und aus diesen Modellen je nach Situation und Anforderung deskriptive, prognostische oder verallgemeinernde Aussagen abzuleiten. In der unterrichtlichen Umsetzung werden solche Modellierungsaktivitäten durch den Einsatz von Multimedia sehr gut unterstützt: dynamische Visualisierungsmöglichkeiten, die rechnerische Bewältigung auch größerer Datenmengen und computergestützte Simulationsmöglichkeiten eröffnen Möglichkeiten, welche die Leitidee Daten und Zufall für die Schülerinnen und Schüler „augenscheinlicher" und „greifbarer" werden lassen. Dieser Mehrwert ergibt sich jedoch nicht per se. Wenn nicht Einsichten der Forschung zum Lehren und Lernen mit Multimedia berücksichtigt werden, können im ungünstigsten Fall die Schülerinnen und Schüler durch die Notwendigkeit, computergestützte multiple Repräsentationen verarbeiten zu müssen, auch überfordert werden. Bei der unterrichtlichen Implementation ist daher vor einer isoliert stoffinhaltlichen Betrachtung und vor einem sorglosen Multimediaeinsatz zu warnen. Beim computergestützten Modellieren von Daten durch Funktionen sind die Möglichkeiten der technischen Unterstützung zu analysieren und didaktisch-methodisch zu reflektieren.

Modellieren von Daten durch Funktionen

Einführung und curriculare Legitimation

Mit den Veröffentlichungen von TIMSS und PISA hat die mathematikdidaktische Entwicklung in Deutschland dahingehend eine Neuakzentuierung erfahren, dass der mathematikdidaktische Aspekt der Anwendungsorientierung stärker in den Vordergrund getreten ist. Hierzu hat das paradigmatische Grundkonzept mathematical literacy der PISA-Untersuchungen (z.B. Deutsches PISA-Konsortium 2001), welches sich explizit auf den Mathematikdidaktiker Hans Freudenthal beruft, sicherlich nicht unwesentlich beigetragen. Im Mittelpunkt dieser didaktischen Sichtweise steht die phänomenologische Verankerung mathematischen Lernens. Einerseits sollen die Schülerinnen und Schüler so zur handlungsorientierten Umwelterschließung durch Mathematik befähigt werden, andererseits sollen auf diese Weise belastbare mentale Modelle für mathematische Begriffe ausgebildet werden (vgl. Freudenthal 1983). In dieser Hinsicht ist das Modellieren von Daten durch Funktionen ein geeignetes Thema für den Mathematikunterricht: Mit Daten lassen sich Phänomene der natürlichen, technischen und sozialen Umwelt vermitteln. Bei der Modellierung geht es nicht nur allein um die Erschließung der Datenstruktur und der dahinter stehenden phänomenologischen Gesetzmäßigkeit, sondern auch um die Handhabung der verwendeten Funktionen als mathematische Werkzeuge. Durch den kontext- und zweckgebundenen Gebrauch können sich diese in ihrer mathematischen Begrifflichkeit weiter erschließen. Entsprechend können beim datenbasierten Modellieren wesentliche Inhalte der verschiedenen Leitideen *Daten und Zufall, Messen* und *Funktionaler Zusammenhang* in den KMK-Standards (2003 und 2012) "zusammengedacht und auf der Basis von prozessbezogenen Kompetenzen wie z.B. „mathematisch Modellieren" oder „mathematische Darstellungen verwenden" vernetzt werden.

Grundgleichung der Datenmodellierung

Betrachtet man reale Daten als kontextualisierte Zahlen eines Phänomens, besteht ein wesentlicher Teil des Datenverstehens darin, eine den Daten zugrunde liegende Struktur (Trend) aufzudecken und in einem mathematischen Modell zu beschreiben. Statistische Daten hängen in ihrem Entstehungsprozess stets vom Zufall (genauer gesagt: erklärtermaßen als vom Zufall verursacht) ab. Daraus ergibt sich die Variabilität der Daten, die den Kern stochastischen Denkens darstellt. In der Einsicht, dass die Welt nicht streng deterministisch ist, sondern aus einem Mix an Variation und Struktur besteht, gleicht die Datenanalyse einer Suche nach dem Muster in der Variation. Eine funktionale Beschreibung, mit der die Struktur abgebildet werden soll, „fällt dabei nicht vom Himmel", sondern erwächst aus dem Bemühen, sich im Rauschen der Daten zurechtzufinden. Gleichzeitig ergibt sich aus diesem Suchvorgang, dass geeignete, weil handhabbare Funktionen die Daten nicht exakt abbilden, sondern lediglich einen Trend wiedergeben. Ob bei der Mathematisierung Techniken der Interpolation oder parametrische Standardmodelle zum Einsatz kommen, ob die Daten geglättet oder mittels Differenzen- bzw. Differenzialgleichungen lokale Änderungsraten analysiert werden, um zur Funktionsbeschreibung zu gelangen – dies ist eine Frage, deren Beantwortung sich an der jeweiligen Problemstellung und den spezifischen Lösungserfordernissen einerseits sowie dem mathematischen Vorwissen und verfügbaren mathematischen Modellierungstechniken der modellierenden Person orientiert. Unabhängig davon wird

aber in jedem Fall akzeptiert, dass geeignete Funktionen die Daten nicht exakt abbilden, sondern lediglich einen Trend wiedergeben, Abweichungen werden bewusst in Kauf genommen.

Aus dem, was Borovcnik (2005) Strukturgleichung nennt, kann man für das
Mathematisieren von funktionalen Abhängigkeiten aus Daten die Grundgleichung
der Datenmodellierung ableiten (vgl. z.B. Vogel 2006; Eichler und Vogel 2009):

Daten = Funktion + Residuen.

Diese einfache Grundgleichung beschreibt den Zusammenhang zwischen Daten, Funktion und Residuen grundlegend. Am einfachen Beispiel eines Datensatzes, welcher den linearen Zusammenhang (Gesetz von Gay-Lussac) zwischen
Temperatur und Druck eines konstanten Volumens Luft in einem luftdichten Me-
tallbehälter abbildet, lässt sich dieser Zusammenhang mit multimedialer Unterstützung für Schülerinnen und Schüler visuell zugangsnäher darstellen (Abb. 8.1).

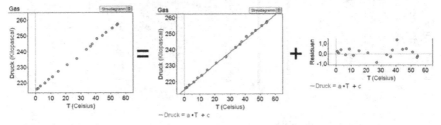

Abb. 8.1 Modellierung von Daten über eine deterministische und eine stochastische Komponente

Diese Grundgleichung der Datenmodellierung ist ein pragmatisches Konstrukt,
um mit der omnipräsenten Variabilität von Daten fertig zu werden: Der Teil der
Datenvariabilität, der erklärt werden kann, wird mit einer deterministischen Komponente, der Funktion, zum Ausdruck gebracht. Das, was übrig bleibt, der uner-
klärte Anteil an Variabilität, der sich in den Residuen zeigt, wird als nicht deterministisch erklärbar betrachtet und als zufallsbedingt modelliert. Damit kommt
eine stochastische Komponente in die Modellierung der Daten. Mathematisch lässt
sich dies im bivariaten Fall so formalisieren: Bei einer Datenmenge, welche den
Zusammenhang zwischen zwei Variablen x und y abbildet, handelt es sich um eine Menge von n Datenpunkten (x_1, y_1), (x_2, y_2), ..., (x_n, y_n). Ziel der Modellierung
ist es, eine Funktion f zu finden, welche erlaubt eine Menge von n Punkten $(x_1,
f(x_1))$, $(x_2, f(x_2))$, ..., $(x_n, f(x_n))$ zu ermitteln, die für die Originaldaten als Modell
stehen können. Ein Residuum r_i als Differenz zwischen Modell- und Originalwert
definiert ergibt sich dann formal zu $r_i = y_i - f(x_i)$. Um der geforderten Zufälligkeit
der Residuen modellierungstechnisch gerecht zu werden, können sie als Realisierungen von stochastisch unabhängigen und identisch verteilten Zufallsvariablen ε_i
betrachtet werden, für deren Erwartungswert $E(\varepsilon_i) = 0$ und deren Varianz $Var(\varepsilon_i)$
$= \sigma^2(x_i) > 0$ gilt. So werden die Responzwerte in einen deterministischen Teil
$f(x_i)$ und einen stochastischen Term r_i zerlegt. Der deterministische Wert gibt den
erwarteten Responzwert an der Stelle x an, der stochastische Term beschreibt den

trendfreien Einfluss aller weiterer Störgrößen einschließlich zufälliger Einflüsse und modelliert diejenige Variabilität in den Daten, die durch das deterministische Funktionenmodell f(x) nicht erfasst wird. Hinsichtlich der eingangs formulierten phänomenologischen Anknüpfung lässt sich die Grundgleichung modellierungstechnisch so in Beziehung setzen: Das interessierende Phänomen bildet sich in den Daten als Realmodell ab, die identifizierte Gesetzmäßigkeit wird in einem funktionalen Zusammenhang beschrieben und das, was durch das Funktionenmodell nicht erfasst wurde bzw. von diesem abweicht wird in den Residuen beschrieben. Der Modellbildungsprozess lässt sich wie in Abbildung 8.2 veranschaulichen.

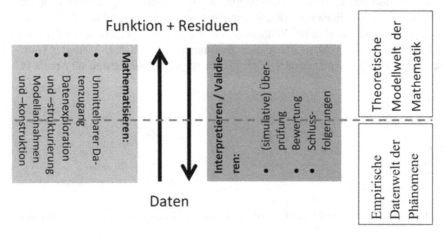

Abb. 8.2 Teilprozesse des Modellierens von Daten

Für den Mathematisierungsvorgang hat dies zur Konsequenz, dass es nicht genügt, nur darauf zu achten, dass die funktionale Anpassung möglichst gut den Datentrend wiedergibt. Die Residuen sollten nicht als „modellierungstechnischer Abfall" gering geschätzt werden (vgl. Vogel 2008a; Vogel und Eichler 2010): Aufgrund der additiven Struktur des funktionalen-stochastischen Modells sind sie ein Maß für die funktionale Anpassung. Daraus ergeben sich Bedingungen, die für Residuen gelten sollen: Sie sollen möglichst klein und zufällig im Sinne von trendfrei sein sowie sich in ihren Abweichungen nach oben und unten ausgleichen (vgl. Biehler und Schweynoch 1999). Das Überprüfen des mathematischen Modells besteht darin, den funktionalen Zusammenhang (z.B. durch die Deutung der Funktionsparameter) in der realen Modellebene zu reflektieren und sich mögliche Ursachen für das Rauschen in den Daten vor Augen zu führen.

Aspekte der multimedialen Unterstützung

Beispiele multimediagestützten Modellierens mit elementaren Funktionen

Im Schulunterricht begegnen den Schülerinnen und Schülern Fragen nach dem funktionalen Zusammenhang zweier Merkmale bereits im naturwissenschaftlichen Unterricht: Naturwissenschaftliche Phänomene, wie z.B. Gesetzmäßigkeiten beim radioaktiven Zerfall oder beim freien Fall lassen sich über Daten repräsentieren. Kern der Analyse des radioaktiven Zerfalls ist im Rauschen der Daten die exponentielle Abnahme zu spezifizieren, funktional zu beschreiben und dadurch wichtige Informationen über die Halbwertszeit zu erhalten, welches jedes radioaktive Nuklid eindeutig kennzeichnet. Die Auswahl einer Exponentialfunktion lässt sich durch theoretische Überlegungen begründen: Für die momentane zeitliche Änderung der radioaktiven Kerne gilt (N(t) ist die Anzahl der nach Ablauf der Zeit t noch nicht zerfallenen Kerne, λ ist die sog. Zerfallskonstante):

$$\frac{dN(t)}{dt} = -\lambda \cdot N(t).$$

Umsortieren nach Variablen und beidseitige Integration ergeben (N_0 ist die Anzahl der anfangs vorhandenen Kerne):

$$\left[\ln N(t)\right]_{N_0}^{N(t)} = -\lambda \cdot \left[t\right]_0^t.$$

Woraus sich nach weiteren Schritten schließlich das bekannte Zerfallsgesetz ergibt:

$$N(t) = N_0 \cdot e^{-\lambda \cdot t}.$$

Während für die Verwendung einer Exponentialfunktion noch theoretische Gründe sprechen, ist die Spezifizierung der Funktionsparameter ohne multimediale Unterstützung unterrichtspraktisch betrachtet nicht realisierbar. Bei der Modellvalidierung, zu der sowohl die Parameter- also auch die Residuenanalyse zu zählen ist, ist in diesem Beispiel noch zu vermerken, dass für die zunehmende „Verschlechterung" der Modell-Exponentialfunktion entscheidend ist, dass in den Residuen die so genannte „Nullrate" eingeht. Die Nullrate beschreibt die Aktivität der natürlichen Umgebungsstrahlung. Die Streuung der Nullrate wird als konstant großes und zufälliges Rauschen angenommen. In diesem Rauschen geht der schwächer werdende Trend der Aktivität des radioaktiven Nuklids zunehmend unter. So entsteht ein Funktionenmodell, hier beispielhaft skizziert für den radioaktiven Zerfall von metastabilem Barium Ba-137m (vgl. Abb. 8.3).

Beim Modellieren von Wahrscheinlichkeitsstrukturen und beim Arbeiten mit Wahrscheinlichkeitsverteilungen wird mathematisch betrachtet ebenfalls mit Funktionen gearbeitet, welche verschiedenen möglichen Realisierungen von Zufallsvariablen Wahrscheinlichkeiten zuordnen. Dies lässt sich am Schokolinsen-

Beispiel (z.B. Vogel und Eichler 2011) verdeutlichen: Wenn auf der Basis einer vorausgehenden Auszählung einer großen Anzahl von Schokolinsen-Tüten (im Beispiel n=100) festgehalten werden kann, dass in einer Tüte durchschnittlich 3 Linsen pro Farbe und insgesamt durchschnittlich 18 Linsen pro Tüte sind, und wenn dann noch die Modellannahmen der stochastischen Unabhängigkeit für den Befüllvorgang als plausibel angenommen werden können, dann sind alle Annahmen vorhanden, die die Modellierung der Wahrscheinlichkeit für das Auftreten der Anzahl von x Linsen einer bestimmten Farbe (z.B. r für rot) durch die Funktion einer Binomialverteilung erlauben:

$$B_{18,\frac{1}{6}}(r) = \binom{18}{r} \cdot \left(\frac{1}{6}\right)^r \cdot \left(1-\frac{1}{6}\right)^{18-r} \text{ mit } 0 \le r \le 18 \text{ und r ganzzahlig.}$$

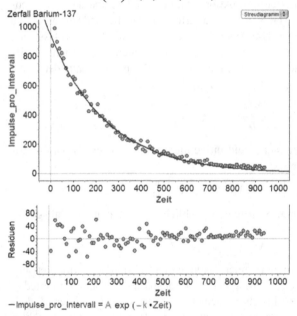

Abb. 8.3 Exponentielle Abnahme

Auch hier lassen sich mit multimedialer Unterstützung Residuenbetrachtungen anstellen: Die realen Daten werden hinsichtlich des Auftretens der interessierenden Frage in einem Diagramm mit relativen Häufigkeiten eingetragen. Als Differenz zu den Modellwerten der Binomialverteilungsfunktion sollten die Residuen den o.g. Kriterien hier ebenfalls genügen (vgl. Abb. 8.4 links). Eine weitere Möglichkeit, die sich hier im Unterricht anbietet, ist über Simulationen „virtuelle" Schokolinsen-Tüten herzustellen: Mit den o.g. Modellannahmen gleicht der Herstellungsprozess einer Tüte einem 18maligem Würfeln mit einem Laplace-Würfel. Mit Computerunterstützung lassen sich problemlos sehr viele solcher virtuellen Schokolinsen-Tüten herstellen. Werden diese ausgezählt und die jeweiligen relativen Häufigkeiten mit den entsprechenden Wahrscheinlichkeiten der Binomialverteilungsfunktion verglichen, dann erlauben computergestützte Simulationen, dass die Variabilität der Residuen und damit die Passung des Binomialverteilungsmodells weitergehend erforscht wird (vgl. Abb. 8.4 rechts). So lassen sich z.B. aus der gezielten Wahl von 100 Tüten und 10000 Tüten bei Betrachtung der Achsenskalierung gewissermaßen „mathematisch-phänomenologisch" Hinweise darauf hin ableiten, dass bei einem 100-fach größeren Stichprobenumfang die Residuen eine in der Größenordnung 10-fach kleinere Streuung aufweisen. Das gezielte Variieren von entsprechenden Stichprobengrößen kann dann im Unterricht eine Verallgemeinerung an weiteren Beispielen vorbereiten, die auf das 1/Wurzel-n-Gesetz führen.

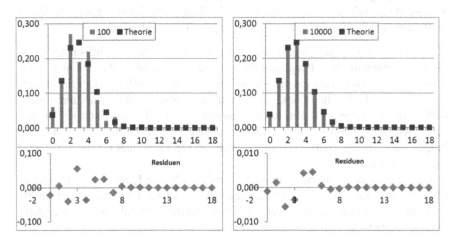

Abb. 8.4 Residuenbetrachtungen bei einer Wahrscheinlichkeitsanalyse

Mit Computerunterstützung lohnt sich in jedem Fall die Betrachtung von Residuen. Nicht schon zuletzt auch deshalb, weil auch über den Unterricht hinausreichende Konzepte der Statistik, wie z.B. die Testgröße eines Chi-Quadrat-Anpassungstests letztlich auf der Betrachtung von beobachteten und theoretisch erwarteten Werten beruht. Residuen enthalten wertvolle Informationen zu dem (mathematischen) Phänomen, welches den Daten zugrunde liegt, und können im Modellierungsprozess wichtige Hinweise für die sukzessive Datenanpassung geben. Sie stehen für das Spannungsverhältnis zwischen Daten und dem, was an Trend in den Daten zu entdecken ist. So kennzeichnen die Residuen die Anpassung der Daten durch Funktionen als das, was sie ist: eine Modellierung, die auf vereinfachenden Annahmen und Entscheidungen beruht, um im Überangebot der Dateninformation das wesentlich Erscheinende besser zu erkennen.

Multimediale Unterstützung von Arbeitsprozessen des Visualisierens, Explorierens und Strukturierens

Eine grundsätzliche forschende Vorgehensweise bei der Datenmodellierung erfordert Werkzeuge, die erlauben (vgl. Vogel 2006)

- bei der Datenarbeit einfach und leicht auf verschiedene Darstellungen zuzugreifen (Aspekt der Visualisierung),
- um Muster in der Zusammenschau dieser Darstellungen ausfindig machen zu können (Aspekt der Exploration) und
- diese strukturell zu erfassen und zu beschreiben (dabei können geeignete Visualisierungen wiederum hilfreich sein) (Aspekt der Strukturierung).

Computerunterstützung leistet hier sehr gute Dienste, da über einfachste Handhabung Streudiagramme, Funktionsgraphen und Residuenplots dargestellt werden können. Ein Simulationsmodell, wie es im Schokolinsen-Beispiel exemplarisch vorgestellt wurde, ist ohne Computerunterstützung praktisch überhaupt nicht realisierbar. Bei der Arbeit mit dem Rechner ist die intuitive Bedienbarkeit von besonderer Bedeutung: Werden keine nennenswerten kognitiven Ressourcen für die Programmbedienung benötigt, verbleibt mehr an Denkkapazität für die inhaltliche Arbeit (vgl. Vogel 2006).

Einzelne Aspekte, die zum Teil bereits in der Software *Fathom* (vgl. Biehler et al. 2011) implementiert sind und diese zu einem wichtigen Instrument der multimediagestützten Datenanalyse werden lassen, werden nachfolgend in einer Aufzählung erörtert, die sich noch ergänzen lässt:

- Wie schon aus den vorausgehenden Beispielen deutlich wurde, sind die wohl wichtigsten Werkzeuge der multimediagestützten Datenmodellierung das Streudiagramm und der Residuenplot. Das Streudiagramm spiegelt mit Blick auf den Datentrend eher eine „globale" Sicht auf die Modellierung wieder. Das Residuendiagramm, welches die Abweichungen zwischen Modellwerten und Datenwerten abbildet, ist demgegenüber eher durch eine „lokale" Sichtweise geprägt: es geht nicht primär um das „Ganze", sondern um den „Unterschied". In didaktischer Hinsicht lässt sich das Residuendiagramm mit der Metapher einer „Modellierungs-Lupe" erklären: Hier wird das vergrößert anvisiert, was bei der Modellierung übrig geblieben ist (vgl. Abb. 8.5). Dabei lassen sich zuweilen modellierungsdidaktisch wertvolle Beobachtungen machen (vgl. Vogel 2008a; Vogel und Eichler 2010). Durch die dynamische Verknüpfung von Streudiagramm und Residuenplot wird die rechnergestützte Datenanpassung dadurch erleichtert, dass bei der Spezifizierung der Parameter der gewählten Funktion sofort die Güte der Anpassung im Residuenplot sichtbar wird. Seine Wertschätzung für Residuen und den Residuenplot formuliert Tim Erickson (2005, S. 85) folgendermaßen: „How close is close enough? A real function never goes through all the points. It only comes close. There is no firm rule, but we will learn about a tool here - the residual plot - that may be the most important piece of data analysis machinery since the slide rule."

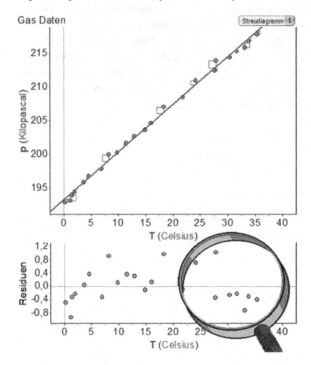

Abb. 8.5 Residuenplot als „Modellierungs-Lupe"

In Abbildung 8.5 sind Streudiagramm und Residuenplot anhand des o.g. Gesetzes von Gay-Lussac dargestellt. Ein weiteres multimediatechnisches Element, welches Fathom im Zusammenhang mit der Abweichung zwischen Funktionswerten und Daten darzustellen erlaubt, sind die Abweichungsquadrate: In didaktischer Hinsicht erschließt sich das didaktische Potential dieser Visualisierung in seinem propädeutischen Sinn, wenn die Schülerinnen und Schüler auf regressionsanalytische Probleme hingeführt werden sollen.

- Ein wichtiges Element hinsichtlich unmittelbarer Zugriffsmöglichkeit (im Sinn der Mensch-Maschine-Interaktion) und intuitiver Zugänglichkeit ist der Greif–und Zugmodus, mit dem beispielsweise Punkte bewegt, Schieberegler erstellt und bedient (vgl. auch vorgenanntes Zitat von Tim Erickson) oder Achsen umskaliert werden können. Ist die Programmbedienung über den Greif- und Zugmodus realisiert, so lassen sich z.B. Daten greifen und in einem Diagramm so ablegen, dass sie sofort visualisiert werden (vgl. Abb. 8.6, hier ist der erste Schritt der Erstellung des Streudiagramms zum o.g. Gas-Temperatur-Datensatz abgebildet). Mit dem Steuerungssymbol der Greifhand wird eine Operation der physischen Erfahrungswelt als Metapher für eine mentale Operation in einem nicht-haptischen Sachkontext verwendet. Dies kann ein Erklärungsmuster für den intuitiven Zugang zur Repräsentation darstellen. Durch das Schließen der Hand (repräsentiert im Drücken und Gedrückt-Halten der Maustaste) wird ein im wörtlichen Sinn verstandener Datenzugriff dargestellt, dem in der physischen Welt kein Äquivalent entspricht. Entsprechendes gilt für das Loslassen im Öffnen der Hand (Lösen des Maustastendrucks). Eine vergleichsweise aufwändige Diagrammerstellung, bei der mitunter aufgrund der langen Dauer das inhaltliche Ziel aus den Augen verloren wird, entfällt dadurch. Dies ist insbesondere hinsichtlich der vergleichsweise knappen Zeitressourcen in einer Unterrichtsstunde von unmittelbarer schulpraktischer Bedeutung.

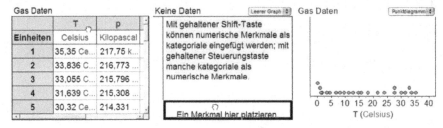

Abb. 8.6 Daten greifen und ablegen

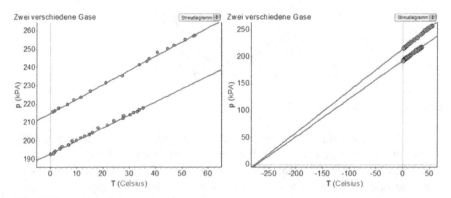

Abb. 8.7 Daten-Zooming

- Beim „Daten-Zooming" kann ein Datensatz in verschieden großen Bildausschnitten betrachtet werden (vgl. Abb. 8.7). Das ist bei der explorativen Datenanalyse dann von Belang, wenn Informationen der Funktionsanpassung erst ablesbar werden, wenn sie aus einer größeren Distanz betrachtet werden. Dies ist beispielsweise bei Fragen der Gültigkeit von Modellgrenzen von Belang: Wie weit kann die modellierende Funktion hinsichtlich ihrer Eigenschaften im Verhältnis zum Datenkontext sinnvoll interpretiert werden? In gleicher Weise wie das Aus-Zoomen, kann das Ein-Zoomen weitere Informationen bringen: Wie sehen Modellabweichungen im Detail aus, weisen die Daten in Teilbereichen bei fokussierter Betrachtung noch unerkannte Muster auf? Im Zusammenhang mit dem Gesetz von Gay-Lussac sind in Abbildung 8.7 zwei verschiedene Gase abgebildet. Die ausgezoomte Betrachtung der beiden linearen Temperatur-

Druck-Modelle führen darauf, dass die Schülerinnen und Schüler hier das Phänomen des tiefstmöglichen Temperatur, des sog. Temperaturnullpunkts, der bei ca. -273,15°C liegt (verallgemeinernd: gemeinsamer Abszissen-Schnittpunkt aller Funktionsgeraden zu verschiedenen Gasen) entdecken können.

- Wenn die Achsen eines Streudiagramms umskaliert werden können, lässt sich die Datenwolke aus unterschiedlich großer Distanz betrachten (s.o.). In Erweiterung dazu liegt ein Perspektivenwechsel auf die Daten dann vor, wenn die Datenpunkte logarithmiert betrachtet werden können (vgl. Abb. 8.8). Dies ist insbesondere für Modellierungstechniken interessant, bei denen eine nichtlineare Datenstruktur durch eine angemessene Transformation linearisiert werden soll, um auf die transformierten Daten Techniken der Geradenanpassung anzuwenden und schließlich das erhaltene Resultat wieder zurück zu transformieren (vgl. Vogel 2006; Engel 2009; Engel und Vogel 2012). Mit der Möglichkeit der logarithmierten Datenbetrachtung können die Erfolgsaussichten für das rechnerische Verfahren im Modellierungsprozess vorab abschätzbar werden. In Abbildung 8.8 ist ein Streudiagramm zum Zusammenhang von Körpergewicht und Hirnmasse von 62 verschiedenen Säugetieren als Beispiel dargestellt (vgl. Vogel 2006; Engel und Vogel 2012): Während bei sehr kleinen Tieren (z. B. Spitzmäusen) das Gehirn 10 % des Körpervolumens ausmacht (Datenpunkt mit dem kleinsten Körpergewicht in Abb. 8.8), stellt es bei sehr großen Säugetieren wie den Elefanten zwischen 0,1 % und 0,2% dar (afrikanischer und asiatischer Elefant als die zwei Datenpunkte mit dem größten Körpergewicht in Abb. 8.8). Die Werte für das Körpergewicht erstrecken sich über einen Bereich von sechs Zehnerpotenzen. Um Daten mit einem solch großen Wertebereich graphisch in einem Koordinatensystem besser überblicken zu können, bietet sich die logarithmierte Betrachtungsweise an (hier zur Basis 10), um die vielen kleineren Tiere im Streudiagramm besser sehen zu können und um den Zusammenhang (sogenannter negativ-allometrischer Wachstumszusammenhang, vgl. Roth 2000) besser studieren zu können.

Der Computer ermöglicht explorative Vorgehensweisen bei der Datenanalyse, die den Mathematikunterricht in seiner paradigmatischen Ausrichtung grundsätzlich bereichern kann. Es geht weniger um das Anwenden fertiger statistischer Verfahren auf Daten und die Überprüfung, was bei solchen standardisierten Verfahren aus den Daten herauszuholen ist. Die explorative Datenanalyse beginnt möglichst vorurteilsfrei bei den Daten (Tukey 1977; Biehler 1995) und versucht, die Daten in unterschiedlichen Darstellungen, Abständen und Perspektiven zu betrachten, um tatsächlich vorhandene Strukturen im Nebel der Datenwolke aufzuspüren, also nicht durch fertige Verfahren Strukturen von außen hineinzutragen. Gefundene Strukturen können hinsichtlich ihrer prognostischen Aussagekraft über Simulationen erforscht werden, wahrscheinlichkeitstheoretische Überlegungen lassen sich so empirisch unterlegen. Die vorgenannten Elemente der multimedialen Unterstützung sind bei diesen Arbeitsschritten von wesentlicher Bedeutung, sie machen den Computer zu einem unentbehrlichen Werkzeug bei der unterrichtlichen Arbeit mit Daten und Zufall.

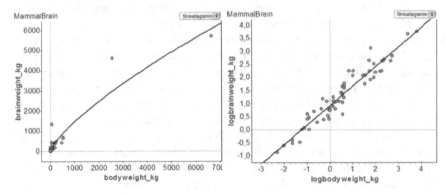

Abb. 8.8 Beispiel einer Datentransformation

Didaktisch-methodische Anmerkungen

Abschließend seien noch einige didaktische Punkte zu einer computergestützten unterrichtlichen Umsetzung der Leitidee Daten und Zufall angemerkt, welche bei der Datenanalyse beginnt:

- Im Sinne des Modellierungskreislaufs nach Blum et al. (2003) bietet sich für die Modellierung von Daten im Unterricht folgende Vorgehensweise an: Klärung des phänomenologischen Hintergrunds – Betrachtung der Daten in verschiedenen Repräsentationen, insbesondere im Streudiagramm – Überlegungen zur Funktionsanpassung der Daten (Auswahl eines Funktionstyps, Spezifizierung der Funktionsparameter) – Reflektieren der Datenmodellierung im Kontext (inhaltliche Deutung der Funktionsparameter, Ursachen für das Residuenverhalten, Modellgrenzen) und Beurteilung. Insbesondere die Modellgrenzen sind ins Bewusstsein der Schülerinnen und Schüler zu heben.
- Eine entscheidende Voraussetzung, um Daten mit elementaren Funktionstypen anpassen zu können, ist, dass diese schon bekannt sein müssen. Nur dann können sie als Modellmöglichkeit in Betracht gezogen werden. Mit der Verwendung dieser Funktionstypen ist das didaktische Ziel verbunden, dass sich ihr Verständnis durch die Arbeit im Datenkontext weiter vertieft.
- Den phänomenologischen Hintergrund miteinzubeziehen ist mitunter nicht nur Hilfe, sondern auch Notwendigkeit. So kann der Gefahr begegnet werden, dass die Datenanpassung nicht auf der grafisch-visuellen Oberfläche verbleibt (z.B. gedankenloses Hantieren mit Schiebereglern). Fragen nach der kontextuellen Bedeutung von verwendeten Funktionsparametern können dazu beitragen, dass der Datenkontext weiter erhellt und der verwendete Funktionstyp besser verstanden wird.

- Der Residuenplot kann dazu beitragen, dass die Schüler lernen, Daten als solche ernster zu nehmen: Streuungen gehören zu Daten und sind nicht als „Fehler" von vermeintlich „wahren realen Werten" zu betrachten. Der übliche Terminus „Fehlerrechnung" wirkt in diesem Zusammenhang didaktisch betrachtet etwas unglücklich.

- Im Sinne eines sukzessiven Aufbaus von Datenmodellierungs-Kompetenz wird zunächst sinnvollerweise mit Datensätzen gearbeitet werden, bei denen ein Trend noch gut erkennbar ist und das Rauschen weniger ins Gewicht fällt. So sollte es für die Schüler möglich sein, mit der parametrischen Modellierungstechnik sinnvoll zu arbeiten. Dabei geht es in erster Linie um das Nach-Entdecken bekannter Gesetzmäßigkeiten. Die parametrische Modellierung gerät dann an ihre Grenzen, wenn stark verrauschte Daten keinen bestimmten Trend erahnen lassen und der Datenkontext ebenfalls keine schlüssigen Argumente für die begründete Auswahl eines Funktionstyps liefert. Diese Erfahrung darf den Schülerinnen und Schülern nicht vorenthalten werden! Es werden dann neue Verfahren notwendig, wie z.B. das Anwenden von Glättungstechniken (z.B. Engel 2009), die zwar durchaus mit zum Teil elementaren mathematischen Mitteln umsetzbar sind, aber (bisher zumindest) im Unterricht der Sekundarstufen I und II kaum zum Einsatz kommen.

- Bei der Auswahl geeigneter Inhalte sind für die Lehrkraft zwei Fragen wesentlich: Welche Datensätze vermitteln Phänomene, die noch in der „Explorations-Reichweite" der Schülerinnen und Schüler liegen, und welche mathematischen Mittel stehen den Schülerinnen und Schülern als Modellierungswerkzeuge zur Verfügung? Im Bereich der Sekundarstufe bieten sich in diesem Zusammenhang insbesondere naturwissenschaftliche Themen wie Schallgeschwindigkeit (vgl. Vogel 2008b), freier Fall, Abkühlungsprozesse, aber auch der weltweite CO_2-Anstieg (vgl. Vogel 2008c) und viele andere mehr an. Möglichkeiten des fachintegrativen Zusammenarbeitens liegen hier nahe. Ungeachtet dessen können jedoch auch zunächst trivial erscheinende Dinge wie das Falten von Papier zu ansprechenden und gehaltvollen mathematischen Überlegungen führen (Biehler et al. 2007), bei denen die Modellierung von Daten und die Erarbeitung mathematisch-fundierter Aussagen deskriptiver wie prognostischer Art im Zentrum verschiedener mathematischer Aktivitäten steht.

Zusammenfassend ist festzuhalten: Sollen Schülerinnen und Schüler Daten wie in dem dargelegten grundsätzlich explorativen Zugang modellieren, dann sind die dazu konstitutiven Aspekte des Visualisierens, Explorierens und Strukturierens nur mit Computerunterstützung (auch in der unterrichtlichen Umsetzung) realisierbar. Der zentrale Punkt der multimedialen Unterstützung liegt in der Möglichkeit der multiplen Repräsentation von Streudiagramm und Residuenplot. Sind diese miteinander dynamisch verknüpft, werden Änderungen im Streudiagramm sofort im Residuenplot sichtbar. Dies ist für ein schülergerechtes exploratives Arbeiten mit den Daten von besonderer Bedeutung.

Literaturverzeichnis

Biehler, R. (1995). Explorative Datenanalyse als Impuls für fächerverbindende Datenanalyse in der Schule - Hintergründe und Besonderheiten ihrer Arbeitsweisen, Konzepte und Methoden. *Computer + Unterricht, 17*, 56-66.

Biehler, R., & Schweynoch, S. (1999). Trends und Abweichungen von Trends. *Mathematik lehren, 97*, 17-22.

Biehler, R., Hofmann, T., Maxara, C., & Prömmel, A. (2011). *Daten und Zufall mit Fathom - Unterrichtsideen für die SI und SII mit Software-Einführung*. Braunschweig: Schroedel.

Biehler, R., Prömmel, A., & Hofmann, T. (2007). Optimales Papierfalten – Ein Beispiel zum Thema „Funktionen und Daten". *Der Mathematikunterricht, 53*(3), 23-33.

Blum, W., Neubrand, M., Ehmke, T., Senkbeil, M., Jordan, A., Ulfig, F.. & Carstensen, C. H. (2003). Mathematische Kompetenz. In M. Prenzel, J. Baumert, W. Blum, R. Lehmann, D. Leutner, M. Neubrand, R. Pekrun, H.-G. Rolff, J. Rost, & U. Schiefele (Hrsg.), *PISA 2003. Der Bildungsstand der Jugendlichen in Deutschland – Ergebnisse des zweiten internationalen Vergleichs* (S. 47-92). Münster: Waxmann.

Borovcnik, M. (2005). Probabilistic and statistical thinking. In M. Bosch (Hrsg.), *Proceedings of the Fourth Congress of the European Society for Research in Mathematics Education* (S. 485-507). Fundemi IQS, Universitat Ramon Llull. http://www.ethikkommission-kaernten.at/lesenswertes/Upload/CERME_Borovcnik_Thinking.pdf. Abgerufen am 11.11.2012.

Deutsches PISA-Konsortium (Hrsg.) (2001). *Pisa 2000 - Basiskompetenzen von Schülerinnen und Schülern im internationalen Vergleich*. Opladen: Leske+Budrich.

Eicher, A., & Vogel, M. (2009). *Die Leitidee Daten und Zufall*. Wiesbaden: Vieweg+Teubner.

Engel, J. (2009). *Anwendungsorientierte Mathematik: Von Daten zur Funktion*. Berlin: Springer.

Engel, J., & Vogel, M. (2012). Vom Geradebiegen krummer Beziehungen: Zugänge zum Modellieren nichtlinearer Zusammenhänge. *PM - Praxis der Mathematik in der Schule, 44*, 29-34.

Erickson, T. (2005). *The model shop. Using data to learn about elementary functions*. Oakland: eeps media. (Third Field-test Draft).

Freudenthal, H. (1983). *Didactical phenomenology of mathematical structures*. Dordrecht: Reidel.

Kultusministerkonferenz (2003). *Beschlüsse der Kultusministerkonferenz. Bildungsstandards im Fach Mathematik für den Mittleren Schulabschluss [Beschluss vom 04.12.2003].* http://www.kmk.org/fileadmin/veroeffentlichungen_beschluesse/2003/2003_12_04-Bildungsstandards-Mittleren-SA.pdf. Abgerufen am 11.11.2012.

Kultusministerkonferenz (2012). *Bildungsstandards im Fach Mathematik für die Allgemeine Hochschulreife [Beschluss vom 04.12.2003].* http://www.kmk.org/fileadmin/veroeffentlichungen_beschluesse/2012/2012_10_18-Bildungsstandards-Mathe-Abi.pdf. Abgerufen am 11.11.2012.

Roth, G. (2000). *Das Gehirn und seine Wirklichkeit*. Frankfurt: Suhrkamp.

Tukey, J. W. (1977). *Exploratory data analysis*. Massachusetts: Addison-Wesley.

Vogel, M. (2006). *Mathematisieren funktionaler Zusammenhänge mit multimediabasierter Supplantation*. Hildesheim: Franzbecker.

Vogel, M. (2008a). "Reste verwerten" - Überlegungen zur didaktischen Wertschätzung von Residuen. In A. Eichler, & J. Meyer (Hrsg.), *Schulbuchkonzepte; Daten und Zufall als Leitidee für die Sekundarstufe I* (Anregungen zum Stochastikunterricht, Band 4, S. 159-168). Hildesheim: Franzbecker.

Vogel, M. (2008b). Wie schnell hört man eigentlich? Daten erheben, auswerten und interpretieren. *Mathematik lehren, 148*, 16-19.

Vogel, M. (2008c). Der atmosphärische CO2-Gehalt - Datenstrukturen mit Funktionen beschreiben. *Mathematik lehren, 148*, 50-55.

Vogel, M., & Eichler, A. (2010). Residuen helfen gut zu modellieren. *Stochastik in der Schule, 30*(2), 8-13.

Vogel, M., & Eichler, A. (2011). Das kann doch kein Zufall sein! Wahrscheinlichkeitsmuster in Daten finden (Basisartikel). *PM - Praxis der Mathematik in der Schule, 39,* 2-8.

Chapter 9
Change point detection tasks to explore students' informal inferential reasoning

Joachim Engel

University of Education, Ludwigsburg, Germany

Abstract Inferential statistics is the scientific method for evidence-based knowledge acquisition. However, the logic behind statistical inference is difficult for students to understand. Recent research has focused on learners' informal and intuitive ideas of inferential reasoning rather than on mastery of formal mathematical procedures. In this paper we consider a particular type of statistical decision problem. In change-point detection tasks, one must decide if a process is running smoothly or if it is out of control. It is argued that change-point detection tasks constitute a prime environment in which to investigate students' informal inferential reasoning. This claim is backed theoretically through reference to a framework by Zieffler et al. (2008) and demonstrated in an empirical study with second-year pre-service teachers.

1 Introduction

Most of our knowledge about the empirical world is based on careful generalizations from observations to a wider universe. Methods of statistical inference have a central role in evidence-based knowledge acquisition; they are used in all of the empirical sciences to ensure the legitimacy of new knowledge. The mathematical methods created for that purpose are based on advanced concepts of probability in combination with different epistemological positions grown in the history of scientific reasoning.

For several decades, as Zieffler, Garfield, and delMas (2008) explicate, psychologists and education researchers have studied and documented the difficulties people have making inferences about uncertain outcomes (see Kahneman et al. 1982). Researchers have pointed to various reasons for these difficulties including: the logic of statistical inference (e.g., Cohen 1994; Nickerson 2000; Thompson et al. 2007), students' intolerance for ambiguity (Carver 2006), and students' inability to recognize the underlying structure of a problem (e.g., Quilici and Mayer 2002). Other research has suggested that students' incomplete understanding of statistics such as distribution (e.g., Bakker and Gravemeijer 2004), variation (e.g., Cobb et al. 2003), sampling (e.g., Saldanha and Thompson 2002; 2006; Watson

2004), and sampling distributions (e.g., delMas et al. 1999; Lipson 2003) may also play a role in these difficulties.

Some recent studies (e.g., Haller and Krauss 2002) indicate that even empirical researchers do not always understand these methods or use them correctly. As far as teaching in school is concerned, the issue is not to teach complex formal methods of inferential statistics but to initiate a way of thinking that helps to make sound decisions in a world of uncertainty where progress in many areas is characterized by the analysis of empirical data. Today, introducing students at school to the basic concepts of inferential reasoning is regarded as a highly desirable goal of school education. Recent curriculum reforms worldwide, influenced by the Standards of the National Council of Teachers of Mathematics (NTCM), emphasize that "instructional programs from prekindergarten through grade 12 should enable all students to develop and evaluate inferences and predictions that are based on data" (NCTM 2000). More recently, the US Common Core State Standards for Mathematics (NGA 2010, p. 47) expect grade 7 students to "draw informal comparative inferences about two populations".

After summarizing some major aspects of statistical inference (Section 2) and referring to a framework for informal inferential reasoning (IIR) in Section 3, we argue that change point detection tasks constitute an ideal environment to explore students' IIR (Section 4). We then summarize results from a study based on videotaped interviews with pre-service teachers in which they reason with these types of problems (Section 5).

2 Statistical Inference

Drawing inferences from data is part of everyday life, and critically reviewing results of statistical inferences from research studies is an important goal for most students who enroll in an introductory statistics course (Zieffler et al. 2008). Statistical inference requires going beyond the data at hand, either by generalizing the observed results to a wider universe or by drawing a more profound conclusion about the relationship between the variables, e.g., determining whether a pattern in the data can be attributed to a real effect in a causal relationship.

From the viewpoint of logic, statistical inference attempts to generalize from a limited number of observations to more general statements about a whole. Because such a conclusion can never be done without errors, statistical inference is about how the uncertainty involved can be dealt with in a logically reasonable manner. Mathematics provides powerful tools to limit the negative consequences of this problem as much as possible and to quantify uncertainty. Besides the formal mathematics, different philosophical positions have developed their own logic of reasoning. Two predominant approaches are

- Classical Inference: Based on the classic or frequentist notion, probabilities represent objective properties of the attribute under investigation. The observations at hand are determined by the sampling distribution, whose parameters are fixed but unknown. Conclusions from the observation to the population can only be made as statements about the acceptability of certain parameter values in light of the observations made. Theoretical distributions are fixed objective but unknown quantities, which can be approximated by the parameter estimates.

- Bayesian inference: Referring to the subjectivist concept, probabilities are an expression of incomplete knowledge. They depend on the analyst. What we observe arises from the interplay of several random variables, to which we assign distributions according to our current knowledge. Since the distribution parameters are also assumed to be random variables, they may themselves also be assigned probability distributions. Starting from an a priori distribution and based on the observations at hand, probabilities are revised to an a posteriori distribution. Based on the obtained a posteriori distributions, probability statements about the parameters can be made.

The notion of chance variability is fundamental to statistical inference, taking into account that variation is everywhere and that conclusions are therefore uncertain. Because no inference can be done with certainty, inference requires an implicit or explicit probability model behind its reasoning. This is where methods of probability enter into statistics in order to keep the negative consequences of inductive reasoning under control.

3 Informal Inferential Reasoning

The recent literature (e.g., Pratt and Ainley 2008; Pfannkuch et al. 2012) discusses different concepts of an introduction to inductive reasoning, which precedes a formal treatment. They are based on the assumption that as early as in the middle grade classes (Biehler 2007; Watson 2008), basic concepts ("Grundvorstellungen") of statistical inference can be made accessible when students reflect on and evaluate arguments that are based on an analysis of data. The implied focus is on considerations of variability in data and an assessment of whether different characteristics are due to systematic effects or to random variability. Figure 9.1 (from Pfannkuch 2005) shows a diagram of the arrival times of a bus for the morning over the course of a month. Due to external factors (different traffic volumes, traffic positions, varying number of passengers, etc.) the bus is not at the bus stop at the exact same time every day. On day 11, the bus arrives substantially delayed. Is this an indication of an unusual event (road closure, accident, etc.) or can the actual delay still be explained by the daily random fluctuations? Informal inferential reasoning is neither about unambiguous yes / no conclusions nor about the exact calculation of error probabilities in a probabilistic model. Rather it is about building valid intuitions and heuristics that lead to judging the deviation just observed in the context of past deviations. The horizontal line in Figure 9.1 represents a threshold indicating that the viewer is no longer willing to attribute delays above that line to the inevitable daily fluctuations, but rather as an indication for an unusual disorder. This is a subjective choice, yet it is based on considerations of variability that tries to separate special causes (e.g., an accident, a closed road) from the daily fluctuations that are attributed to random influences.

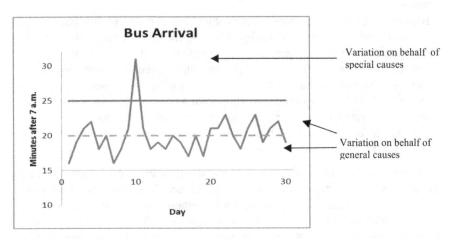

Fig. 9.1 Informal inference: Fluctuations of arrival times of a bus exceeding a certain threshold as indicator of an unusual disorder

Given the importance of understanding and reasoning about statistical inference, and the consistent difficulties students have with this type of reasoning, there have been attempts to expose students to situations that allow them to use informal methods of making statistical inferences (e.g., comparing two groups based on boxplots of sample data).

Research in this field begins to offer insight into the learners' inferential reasoning and how that thinking might be shaped more effectively by well-designed tasks. Rubin, Hammerman, and Konold (2006) define informal inferential reasoning as reasoning that involves the related ideas of properties of aggregates (e.g., signal and noise, and types of variability), sample size, and control for bias. Pfannkuch (2006) defines informal inferential reasoning as the ability to interconnect ideas of distribution, sampling, and center, within an empirical reasoning cycle (Wild and Pfannkuch 1999). Rossman (2008) has described informal inference as "going beyond the data at hand" and "seeking to eliminate or quantify chance as an explanation for the observed data" through a reasoned argument that employs no formal method, technique, or calculation.

Combining these perspectives, we follow Zieffler's et al. (2008) working definition of informal inferential reasoning as the way in which students use their informal statistical knowledge to make arguments to support inferences about unknown populations based on observed samples.

Furthermore, Zieffler et al. (2008) propose a framework for informal inferential reasoning with the following components:

1. Making judgments, claims, or predictions about populations based on samples, but not using formal statistical procedures and methods (e.g., p-value, t tests);
2. Drawing on, utilizing, and integrating prior knowledge (e.g., formal knowledge about foundational concepts, such as distribution or average; informal knowledge about inference such as recognition that a sample may be surprising given a particular claim; use of statistical language), to the extent that this knowledge is available; and
3. Articulating evidence-based arguments for judgments, claims, or predictions about populations based on samples.

Reading (2007) specifies these ideas by pointing out that informal inferential reasoning also includes ideas of choosing between competing models, expressing a degree of uncertainty in making inference, and making connections between the results and problem context.

4 Change point detection and informal inference

Change point detection problems can be characterized as follows: Suppose one accumulates independent observations from a certain process that is in a certain state. Observations vary around a certain mean. After a while some of the observations seem to be a bit unusual, too high or too low. Has something occurred that altered the state of the system (a "breakdown"), or are these observations within the range of the expected when acknowledging random variation? Should one declare that a change took place ("raise an alarm") as soon as possible? A false alarm costs resources, credibility etc., but NOT raising alarm when, for example, in fact a new health hazard occurs, may be even more harmful. When we want to detect the change quickly, any detection policy gives rise to the possibility of frequent false alarms when there is no real change. On the other hand, attempting to avoid false alarms too strenuously leads to long delays between the time of occurrence of a real change and its detection.

Practical examples of this problem arise in areas such as health, quality control and environmental monitoring. For instance, consider surveillance for congenital malformations in newborn infants. Under normal circumstances, the percentage of babies born with a certain type of malformation has a more or less known value p_0. Should something occur (such as an environmental change, the introduction of a new drug to the market, etc.) the percentage may increase (e.g., the thalidomide episode of the 1960s). One would want to raise an alarm as quickly as possible after a change takes place, while trying to control the risk of a false alarm. Generally, this type of problem arises whenever surveillance is being done.

In most practical situations, the question about a possible change point is a sequential decision problem: while the data become sequentially available, a decision has to make a trade-off between the risks of false alarm, misdetection and a detection delay. The input is usually given "online", i.e., in piece-by-piece or serial fashion without having the entire data available from the start as in contrast to an offline decision problem where all the data from the beginning to the end are available and the decision regarding a change point is done in retrospect. From a formal mathematical and decision theoretic perspective, extensive research has been done in this field during the last few decades to derive optimal decision rules (see e.g., Pollack 1985 or Tartakovsky and Moustakides 2010 and the literature therein).

We consider change point detection tasks as an ideal scenario to investigate learners' IIR. These types of problems have been used previously by Rubin, Hammerstein and Konold (2006). Engel, Sedlmeier and Woern (2008) investigated responses of 179 pre-service teachers in a context that required them to determine whether a particular change in a process occurred over time or not. One example of these tasks is presented in Figure 9.2, which is further discussed below where a summary and analysis of videotaped interviews with learners is presented.

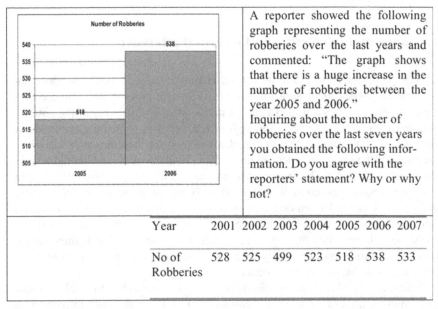

A reporter showed the following graph representing the number of robberies over the last years and commented: "The graph shows that there is a huge increase in the number of robberies between the year 2005 and 2006."
Inquiring about the number of robberies over the last seven years you obtained the following information. Do you agree with the reporters' statement? Why or why not?

Year	2001	2002	2003	2004	2005	2006	2007
No of Robberies	528	525	499	523	518	538	533

Fig. 9.2 Change point detection task "Robberies" to elicit students' informal inferential reasoning

At a first glance, students usually look at the graphical display and may be tempted to agree with the reporters' claim due to the fact that the bar's height more than doubled from 2005 to 2006. But the statistical issue here is an evaluation of the increase between 2005 and 2006 in light of the fluctuations of the number of robberies over a certain time period. Here – and this may be in contrast to formal procedures of classical hypothesis testing – context knowledge about what may cause or prevent robberies is also an important ingredient in the decision process. Eventually, a decision is asked for between the two models of a constant number of robberies over the years (e.g., the overall mean) and a jump that may have occurred around 2006, resulting in a piecewise constant function, see Figure 9.3.

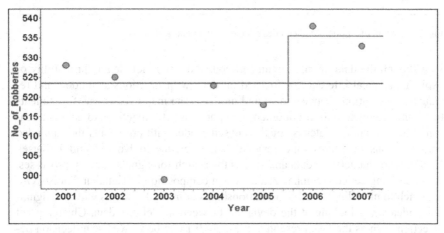

Fig. 9.3 Change point detection as a decision between two competing models: has there been a jump or is the increase in 2006 just due to random fluctuations?

This type of problem fits well into above framework for IIR because the task requires (1) making judgments and claims on the basis of the sample of observations over 7 years, (2) considering variation and the distribution of observed data (integrating prior knowledge including the recognition that a sample may be surprising) and (3) articulating evidence-based arguments for the claim. Further, the task requires a choice between competing models, connects the statistical aspect with the context and elicits multiple and partially contradictory arguments and also encourages expressing the degree of uncertainty in the decision process.

5 Students informal reasoning about change point detection problems

The task in Figure 9.2 is an example from a series of problems that were used to investigate students' progress in statistical reasoning after attending a class on data-based modeling of functional relationships. Participants were second-year pre-service students (preparing to be teachers for elementary and secondary schools) attending two different courses in applied mathematics. While one course had a more traditional syllabus (the control group), with no particular analysis of real data and only a moderate amount of explicit modeling, the other course (the treatment group) incorporated technology-supported modeling of functional relationships of real data as described in a recent textbook (Engel 2009) with a succession of topics as outlined in Figure 9.4.

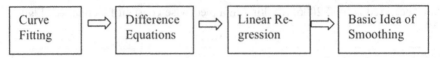

Fig. 9.4 Content of Applied Mathematics course for the treatment group

Although the data-oriented treatment course did not include any probability or explicit statistical inference, it focused on data, with functions as models and residuals as deviations between data and models. One major focus was for students to fit linear, polynomial, trigonometric, exponential etc. functions to data and reason about the appropriateness of the chosen model with arguments that included considerations of context and residuals. In reference to Konold and Pollatsek (2002), who characterize data analysis as the search for signals in noisy processes, the underlying concept emphasized data as a composition of signal and noise. Our approach to modeling functional relationships confronted students with the signal-noise idea when looking at the deviation between model and data. Change-point detection such as the task presented in Figure 9.1 can be viewed as a decision between two competing models for the time series of the number of robberies: a constant function, representing a fixed average number over the years, the actual number of robberies varying only randomly; OR a piecewise constant function with a systematic jump around the year 2006, see Figure 9.3.

The quantitative evaluation of the pre- and posttests with change point detection tasks has been reported in Engel et al. (2008). A sequence of items such as the one in Figure 9.2 and Figure 9.5 (below) was constructed that at a first sight seemed to have a jump or change point in the data. This impression was aggravated by starting the vertical scale high above the origin which made small changes look large. From a statistical point of view it is important to evaluate the sudden increase in the light of the variation of past data. In our examples, when taking into account the variation over the last several years, which was provided in tabular format, evidence for a change point became very weak. Students' responses were scored on a scale between 0 and 100 by two independent scorers according to a scheme that honored recognition of variation in the past data, enhanced by contextual consideration while attempts to search for specific reasons led to low scores. While in the pretest treatment and control group barely differed at all, the difference between the two groups in the posttest was highly significant. ($p=0.0031$) with an increase from a score of 28.9 to 43.8 or an effect size of 4.13 standard deviation for the treatment and from 29.5 to 29.8 or an effect size of 0.11 standard deviation for the control group. This effect is noteworthy, because – to restate again – the words probability or statistics were never mentioned throughout the whole class. Therefore, students had nothing but informal tools available to address the tasks. Engel et al. (2008) interpreted this result as sizable indicator that students are capable of transferring the signal-noise concept from the context of modeling functional relationships to broader statistical problems.

To obtain more in-depth information, seven video-taped interviews were recorded in the first week and after the last week of class on two problems similar to those in the written tests: "increase in robberies" (Figure 9.3) and "incidences of leukemia" (Figure 9.5).

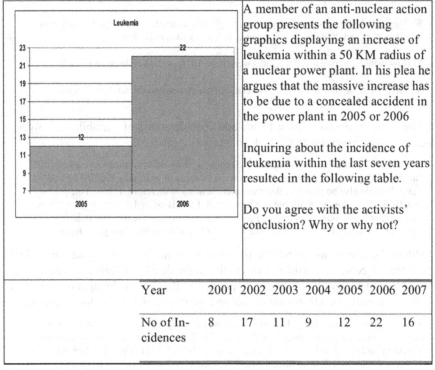

A member of an anti-nuclear action group presents the following graphics displaying an increase of leukemia within a 50 KM radius of a nuclear power plant. In his plea he argues that the massive increase has to be due to a concealed accident in the power plant in 2005 or 2006

Inquiring about the incidence of leukemia within the last seven years resulted in the following table.

Do you agree with the activists' conclusion? Why or why not?

Year	2001	2002	2003	2004	2005	2006	2007
No of Incidences	8	17	11	9	12	22	16

Fig. 9.5 Change point detection task "Leukemia" to elicit students' informal inferential reasoning.

All pre- and post-class interviews followed the same scheme: At the beginning of the interviews the graph with the jump between 2005 and 2006 was shown to the students together with the text explaining the context of the data. After confronting students with the tasks, the interviewer's job was to listen, to encourage the students to make judgments or predictions and to ask students to use and integrate prior knowledge to articulate evidence-based arguments. The interviewer also probed student responses and asked for explanations. The time series data with observations from the previous years and one following year in tabular format was not presented to the students from the onset, but a bit delayed in the course of the interview. This procedure was chosen to find out whether the students would ask for additional data by themselves or not.

In the pre-class interviews students tended to focus on the graph with the two robbery observations from 2005 and 2006 only, searching for arguments whether or not the jump from 518 to 538 was high enough to back the reporter's claim. They read the data in the graph adequately, and applied proportional reasoning when probed by the interviewer. However, they rarely asked by themselves for data preceding the year 2005, and tended to ignore the time series data even after the interviewer presented them to them, as illustrated in the following interview (S student, I interviewer):

> S.: The scale here is chosen inadequately, ranging from about 500 to 540. The difference between 2005 and 2006 isn't all that much as it looks like, only 20 more.

> I.: How would you rate it if the figures were from 1018 to 1038 or from 18 to 38?

> S.: From 18 to 38, that would be completely different, because the numbers have more than doubled then.

The same student was able to include considerations of variability 14 weeks later, when confronted with the leukemia task (Figure 9.5).

> S.: Okay, so the one who is giving this plea because, only refers to the two years of 2005 and 2006. May well be that in 2004, there were 18 cases of leukemia, then it'd be leveled again. It could also be possible, that there were fewer and therefore it rises. You only see that 10 children more were ill, but one cannot tell if this is normal or not, so to say. It may even be that the increase is not related to the nuclear plant, but the increase in leukemia is a pure coincidence. I'd like to know the number of incidences the three years before.

When the interviewer presented the table with the time series data the student discussed the context, considered variability in the data but finally resorted to a rather deterministic mode of thinking by pondering if previous fluctuations may also be due to an accident. He did not reflect on how uncertain his conclusion may be.

> S.: Ah, in 2002 the number of leukemia cases was also high, and one can ask whether there was something happening then. 2001 and 2004, the number was quite small and then it went up in 2006. Is this really due to an accident, hmm ? 8, 11, 9 cases of leukemia. The 17 (from 2002) doesn't fit in there at all. If the increase now (in 2006) is caused by the nuclear plant, then in 2002 here must have been an accident as well. …One should investigate, whether there was an accident in 2002. If this is the case, then I would include that there must have been an accident in 2006. If nothing has happened, then probably there was no accident later (in 2006).

Another student (A) pointed out in her pre-class interview that she wanted more factual information, reflecting her desire to search for more specific causes that may explain the increase between 2005 and 2006.

> A.: I wouldn't agree to the activists claim that given the graphical information one can conclude that there was an accident. For that conclusion, one would need much more data, e.g., on the level of radiation in the surrounding [area]. If that increased, yes, then I would agree that there was really an accident. More investigations in the plant are needed. Don't they have to record these things? Talk to people living in the neighborhood, find out if animals died. If there were an accident, then it also affected animals and plants.

When the interviewer presented her the data of the last 7 years, she started computing yearly changes.

> A.: Between 2001 and 2002 the number has risen by 9, now by 10. Aah, there was a similar case previously. The question is whether there has also been an accident at the plant. If I knew that there was also an accident in 2002, then I would believe that now there was also one. But otherwise, …

A third student (K) had a skeptical attitude towards the reporter in the robbery task. First he focused on the data display, but then turned towards considerations of variability in the time series.

K: Umm, the graphics were probably intentionally chosen to manipulate the reader to dramatize the increase, while in fact a rise from 518 to 538 in robberies over a year isn't really that much. I wouldn't agree with reporter's conclusion. It would be different if I had data over a period of 20 or 30 years, and it had always been to 518. Okay, then one could say this year, it's really been a dramatic increase.

The interviewer then presented data on the number of robberies over the past years. When the student immediately pointed to the yearly fluctuations, the interviewer asked for possible explanations.

K.: Maybe, people didn't report all of the robberies to the police in a certain year, ... or that in a certain year the potential robbers were jailed, the next year they were out again.

In the post-class interview (leukemia task) the same student asked right at the beginning for the time series.

K.: Graphics intends to demonstrate a sizeable increase in leukemia incidences. You have to look at the previous years to be able to make a judgment here. I need the trend over the last years. Some fluctuations are always possible.

6 Conclusions

To make the logic of inductive reasoning in empirical studies accessible to learners, recent research emphasizes the introduction of informal inferential reasoning before any mathematical procedures like formal hypothesis testing or confidence intervals. As in any other domain of mathematics learning, the acquisition of basic concepts has to precede formal treatments in order to lead to sustainable learning. When confronting students with tasks that require a judgment whether a given process continues to run smoothly or a breakpoint occurred, the framework of Zieffler et al. (2008) provided a very useful orientation to assess students' capacities for informal inferential reasoning. Furthermore, the interviews conducted with pre-service teachers confirmed earlier findings from a written assessment (Engel et al. 2008) that emphasizing data modeling in an applied mathematics course improves inferential reasoning even if probability and statistics are not included in the class' syllabus.

Acknowledgments I thank Tim Erickson, Gail Burrill and Jane Watson for helpful comments on a previous version of this chapter.

References

Bakker, A., Derry, J., & Konold, C. (2006). Technology to support diagrammatic reasoning about center and variation. In A. Rossman, & B. Chance (Eds.), *Working cooperatively in statistics education: Proceedings of the Seventh International Conference on Teaching Statistics, Salvador, Brazil* [CDROM]. Voorburg, The Netherlands: International Statistical Institute. Retrieved from http://www.stat.auckland.ac.nz/~iase/publications/17/2D4_BAKK.pdf.

Biehler, R. (2007). Denken in Verteilungen – Vergleichen von Verteilungen. [Considering distributions – comparing distributions] *Der Mathematikunterricht 53*(3), 3-11.

Engel, J. (2009). *Anwendungsorientierte Mathematik: Von Daten zur Funktion. Eine Einführung in die mathematische Modellbildung für Lehramtsstudierende* [Application oriented mathematics: from data to functions. An introduction into mathematical modeling]. Heidelberg: Springer.

Engel, J., Sedlmeier, P., & Woern, C. (2008). Modeling Scatterplot Data and the Signal-Noise Metaphor: towards Statistical Literacy for Pre-Service Teachers. In C. Batanero, G. Burrill, C. Reading, & A. Rossman (Eds.), *Joint ICMI/IASE Study: Teaching Statistics in School Mathematics. Challenges for Teaching and Teacher Education. Proceedings of the ICMI Study 18 and 2008 IASE Round Table Conference.* Monterrey, Mexico: International Commission on Mathematical Instruction and International Association for Statistical Education. Retrieved from http://iase-web.org/documents/papers/rt2008/T4P8_Engel.pdf.

Haller, H., & Krauss, S. (2002). Misinterpretations of significance: A problem students share with their teachers? *Methods of Psychological Research, 7*(1), 1–20.

Konold, C., & Pollatsek, A. (2002). Data analysis as the search for signals in noisy processes, *Journal for Research in Mathematics Education, 33*(4), 259-289.

NCTM – National Council of Teachers of Mathematics (2000). *Principles and Standards for school mathematics.* Reston, VA: National Council of Teachers of Mathematics.

NGA - National Governors Association Center for Best Practices, Council of Chief State School Officers (2010). *Common Core State Standards for Mathematics.* National Governors Association Center for Best Practices, Council of Chief State School Officers, Washington D.C. Retrieved from http://www.corestandards.org/.

Pfannkuch, M. (2005). Probability and statistical inference: How teachers can enable learners to make the connection? In G. Jones (Ed.), *Exploring probability in school: Challenges for teaching and learning* (pp. 267-294). New York: Kluwer/ Springer Academic Publisher.

Pfannkuch, M. (2006). Informal inferential reasoning. In A. Rossman, & B. Chance (Eds.), *Working cooperatively in statistics education: Proceedings of the Seventh International Conference on Teaching Statistics.* Salvador, Brazil. Voorburg, The Netherlands: International Statistical Institute. Retrieved from: http://www.stat.auckland.ac.nz/~iase/publications/17/6A2_PFAN.pdf.

Pfannkuch, M., Wild, C., & Parsonage, R. (2012). A conceptual pathway to confidence intervals. *Zentralblatt Didaktik der Mathematik, 44*(7), 899 – 911.

Pollak, M. (1985). Optimal Detection of a Change in Distribution. *Annals of Statistics, 13,* 206–227.

Pratt, D., & Ainley, J. (2008). Introducing the special issue on informal inferential reasoning. *Statistics Education Research Journal, 7*(2), 3-4. Retrieved from http://www.stat.auckland.ac.nz/~iase/serj/SERJ7(2)_Pratt_Ainley.pdf.

Reading, C. (2007). Cognitive development of reasoning about inference. Discussant reaction presented at the Fifth International Research Forum on Statistical Reasoning, Thinking and Literacy (SRTL-5), University of Warwick, UK.

Rossman, A. (2008). Reasoning about Informal Statistical Inference: One Statistician's View. *Statistics Education Research Journal, 7*(2), 5-19. Retrieved from http://www.stat.auckland.ac.nz/~iase/serj/SERJ7(2)_Rossman.pdf.

Rubin, A., Hammerman, J., & Konold, C. (2006). Exploring informal inference with interactive visualization software. In A. Rossman, & B. Chance (Eds.), *Working cooperatively in statistics education: Proceedings of the Seventh International Conference on Teaching Statistics, Salvador, Brazil* [CDROM]. Voorburg, The Netherlands: International Statistical Institute. Retrieved from http://www.stat.auckland.ac.nz/~iase/publications/17/2D3_RUBI.pdf.

Tartakovsky, A., & Moustakides, G. (2010). State-of-the-Art in Bayesian Changepoint Detection. *Sequential Analysis, 29,* 125–145.

Watson, J.M. (2008). Exploring beginning inference with novice grade 7 students. *Statistics Education Research Journal,* 7(2), 59-82. Retrieved from http://www.stat.auckland.ac.nz/~iase/serj/SERJ7(2)_Watson.pdf.

Wild, C., & Pfannkuch, M. (1999). Statistical thinking in empirical enquiry. *International Statistical Review, 3,* 223-266.

Zieffler, A., Garfield, J., & delMas, R. (2008). A framework to support research on informal inferential reasoning. *Statistics Education Research Journal,* 7(2), 40-58. Retrieved from http://www.stat.auckland.ac.nz/~iase/serj/SERJ7(2)_Zieffler.pdf.

Kapitel 10
Eine kleine Geschichte statistischer Instrumente: vom Bleistift über R zu `relax`

Hans Peter Wolf

Fakultät für Wirtschaftswissenschaften der Universität Bielefeld

Abstract Die Technologie hat in den letzten 50 Jahren viele Werkzeuge geschaffen, die stark das Lernen und Lehren statistischer Inhalte beeinflusst haben. Einige Meilensteine werden in diesem Kapitel aus Sicht des Autors dargestellt. Taschenrechner und FORTRAN erleichterten das Rechnen, *APL* half durch kompakte Notation und Interaktion, die Sprache S brachte mehr Verständlichkeit und viele statistische Rechen- und Graphik-Routinen. R als kostenlose S-Implementation garantiert die Verbreitung einer Vielzahl von R-Paketen. Zur Dokumentation von Datenanalysen, aber auch deren Reproduktion und Modifikation ist das Paket RELAX entstanden, welches weitere Werkzeuge zur Darstellung von R-Lösungen bietet und die statistische Ausbildung bereichern kann.

1 Technologie: Kosten und Wirkungen

Auch aus Gründen einer besseren Unterstützung von Lernprozessen denken zurzeit viele Hochschulen darüber nach, ihre IT-Infrastruktur zu modernisieren. Es wird nach integrierten Lösungen aus einem Guss gesucht und suggeriert, dass sich damit ein Paradies zum Lernen eröffnet. Diese Projekte werden in den nächsten Jahren an verschiedenen Universitäten siebenstellige Euro-Beträge verschlingen. Doch ist fraglich, ob sich solche Ausgaben für das Lernen und Lehren lohnen. Vor diesem Hintergrund wird nun ein Blick in die Vergangenheit geworfen. Dabei werden Werkzeuge für die universitäre Statistikausbildung und ihre Entwicklung in den letzten 50 Jahren aus der persönlichen Erfahrung mit wenigen Federstrichen skizziert.

Eine neue Mensa!
Betrachten wir ein kleines Beispiel. Im Rahmen der Renovierung der Universität Bielefeld wird eine neue Mensa gebaut, deren Ablauf genau überlegt sein will. Viele Dinge sind zu entscheiden, wie zum Beispiel die Organisationsform der Kassen. Hierbei ist eine interessante Frage, welche Auswirkungen Barzahlungen gegenüber Kartenzahlungen besitzen. Aus statistischer Sicht müssen Daten beschafft und ausgewertet werden. Flugs lassen sich Kassierdauern in Sekunden an vergleichbaren Kassen erheben.

- *Kartenzahlungen*: 13 6 8 3 14 1 9 10 10 13 8 10 22 7 19 5 3 5 11 17 10 17 7 12
- *Barzahlungen*: 13 25 11 20 16 24 19 17 28 25 19 64 30 36 22 12 19 13 31 26 24 22 28 14 17 31 15 22 26 32 9 25 25 22 28 29 38 30 13 14 15 14 16.

Was können wir von diesen Daten erfahren?

2 Eine kleine Zeitreise

Vor über 50 Jahren: Papier und Bleistift

Vor grob 50 Jahren hätten Studierende zusammenfassende Statistiken berechnet und mühevoll Histogramme erstellt. Das Vorgehen war Textbüchern zu entnehmen, Rechentricks wie:

$$d^2 = \overline{x^2} - \overline{x}^2$$

wurden bewiesen, und Arbeitstabellen wie in Abb. 10.1 mussten gefüllt werden.

```
------------------------------
        x_i      x_i^2
------------------------------
         13        169
          6         36
        ...        ...
------------------------------
sums    240       3034
------------------------------
```

Abb. 10.1 Typische Arbeitstabelle zur Berechnung von Statistiken

In dieser Zeit beherrschten Schemata und manuelle Rechenfertigkeiten das Geschehen. Als großes Manko wurde kostbare Zeit der Lernenden für zeitraubende Berechnungen statt für das Lösen von Problemen geopfert. Dafür *saßen* die Grundtechniken wie im Schlaf und liefern noch heute für die 24 Kartenzahler: $\bar{x} = 10$ und für die Barzahler: $\bar{x} = 22,77$.

70er Jahre: Taschen- und Großrechner

In den 70er Jahren kamen Taschenrechner auf, die sogar Funktionstasten zur Mittelwert- und Varianzberechnung besaßen. Natürlich wurden solche Hilfsmittel nicht für Klausuren zugelassen und noch heute müssen *programmierbare Taschenrechner* i.d.R. draußen bleiben. Andererseits bekamen Bielefelder Studierende schon in den 70er Jahren Zugang zum Großrechner TR 440 und konnten sich Berechnungen per FORTRAN und Lochkarten für umfangreichere Datensätze in Auftrag geben. Solche Programme besaßen, wie in Abb. 10.2 zu sehen, ein sehr sprödes Erscheinungsbild.

```
C       Berechnung Mittelwert
        DIMENSION X(100)
        DATA X/13.0,6.0,...
        N=24
        DO 10 I=1,N
        SUM=SUM+X[I]
 10     CONTINUE
        XMITTEL=SUM/N
 3333 FORMAT(F6.2)
        WRITE(6,3333) XMITTEL
C       ------------------ CCC
```

Abb. 10.2 FORTRAN-Code zur Mittelwertberechnung

Die technische Unterstützung zur Formelberechnung war folgerichtig und konnte Lernende, Lehrende und professionelle Statistiker von langweiligen Berechnungen befreien, und man gewann Zeit für anspruchsvollere Aufgaben. Gleichzeitig stellten sich zur Nutzung der Rechner-Ressourcen neue Herausforderungen ein. Es galt, die statistischen Verfahren zu programmieren. Das Schreiben langer FORTRAN-Programme erforderte eine algorithmische Art des Denkens, wobei die Maschinennähe der Sprache von den eigentlichen Problemen ablenkte und das Ringen mit dem Rechner viel Zeit kostete.

Mitte der 70er: EDA
Vor zirka 35 Jahren legte uns Tukey (1977) in seinem Buch *Exploratory data analysis* nahe, die Daten in den Mittelpunkt zu stellen: *Look at your data!* so lautete das Credo. Beim Einsatz rein automatischer Routinen können Ausreißer oder Datenfehler übersehen werden und führen dann ebenfalls automatisch zu falschen Interpretationen. Statt Hypothesentests ist oft eine Exploration der Beobachtungen für das Verständnis zweckmäßiger. So zeigen *stem and leaf displays* oft mehr als ein T-Test vermitteln kann. Solche semigraphische Darstellungen sind sehr einfach manuell zu erstellen. Für unsere Daten könnte ein Student aus den späten 70ern die Zusammenfassung in Abb. 10.3 notiert haben.

```
mit Karte              ohne Karte

 0  |  133              0  |  9
 0  |  55677889         1  |  12333444556677999
 1  |  000012334        2  |  02222445555668889
 1  |  779              3  |  0011268
 2  |  2                4  |
 2  |                   5  |
 3  |                   6  |  4
```

Abb. 10.3 Stem and leaf display – per Hand erstellt

Andererseits war es illusorisch, aus dem Stand ein FORTRAN-Programm zur Erstellung eines *stem and leaf display* zu schreiben. Waren Tukeys Vorschläge als Phobie gegen den wachsenden Einfluss von Rechnern zu werten? Sicher nicht, denn wenn wir Glättungsoperationen wie 3RSSH oder auch nur *letter value displays* betrachten, entdecken wir Konzepte der Informatik wie Iteration und Rekursion. Die Tiefen der ausgewählten Quantile (der *letter values*) zur Charakterisierung des Datensatzes berechnen sich z. B. nach der Formel:

$$depth(letter_i) = \frac{\lfloor depth(letter_{i-1})\rfloor + 1}{2}.$$

Insbesondere zur Überprüfung von Symmetrie-Fragen sind diese Displays gut geeignet. Oft reichen bereits fünf Werte bestehend aus Extrema, Hinges und Median für eine knappe, aber doch aussagekräftige Zusammenfassung eines Datensatzes aus. Wenn ein *stem and leaf display* vorliegt, können wir diese fünf charakteristischen Werte mit Hilfe der rekursiv berechneten Tiefen leicht bestimmen. Doch wie bekommen wir die Werte bei etwas größeren Datensätzen? Und wie können wir bei der Umsetzung der Aufforderung, näher hinzuschauen und explorativ vorzugehen, von den Fortschritten der Rechnerwelt profitieren?

Ende der 70er: APL für Vektoren, Matrizen und Interaktion
Interaktion lautet das Zauberwort: Wir benötigen eine Umgebung, in der wir flexibel mit Daten umgehen können! *APL* erfüllte diese Anforderung. Hatte ein Studierender Ende der 70er Jahre das Glück, Zugang zu einer *IBM 5110*[1] oder *IBM 5120*[2] zu bekommen, konnte er mit diesem Instrument rechnen, experimentieren und Daten aus verschiedensten Perspektiven betrachten. Diese Rechner, die im Nixdorf-Museum in Paderborn bewundert werden können, sind die unmittelbaren Vorgängermodelle des *IBM PC*, der die interne Bezeichnung *IBM 5150* besaß. Bemerkenswerterweise wurden diese Rechner entweder im Basic- oder im *APL*-Modus gestartet. *APL* steht für *A Programming Language* und ist eine Interpretersprache mit vielen seltsamen Zeichen, wie sie in Abb. 10.4 zu sehen sind.

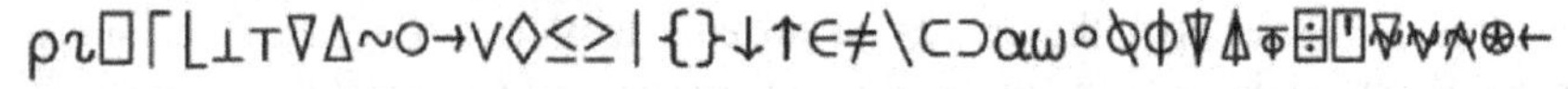

ρι□⌈⌊⊥⊤∇∆~○→∨◊≤≥ | {}↓↑∈≠\⊂⊃αω○⍉⌽∇⍋⍒⊟⍞⍢⍣⍟⊛←

Abb. 10.4 Einige Symbole aus dem APL-Zeichensatz

Viele der Zeichen sind Operatoren zum Sortieren, Runden oder für arithmetische Operationen, beispielsweise invertiert die Anweisung ⌹*MAT* die Matrix *MAT*. Wie bei allen Interpretern wird jede *APL*-Anweisung sofort ausgewertet und das Resultat ausgegeben. Die Sprache *APL* wurde von Iverson[3] zunächst als reine *Notationssprache* entwickelt, also um eleganter als mit FORTRAN mathematische

---

[1] Die IBM 5110 mit Zusatzgeräten war ab 1978 zum Preis von ca. 25.000,00 $ verfügbar, weitere Informationen sind über http://oldcomputers.net/ zu finden.

[2] Die *IBM 5120* gab es ab 1980 und kostete mit Diskettenlaufwerken knapp 10.000,00 $.

[3] Für weitere Infos siehe: http://en.wikipedia.org/wiki/Kenneth_E._Iverson.

Berechnungen ausdrücken zu können. Die Anweisungen in Abb. 10.5 berechnen beispielsweise zu dem Datensatz X die Werte für ein *five number display*.[4]

```
.5×+/X[⌊I],[1.5]X[⌈I←1ΦN,1,DH,DM,N+1-DH←.5×1+⌊DM←.5×1+N←ρX←X[⍋X]]
```

Abb. 10.5 Berechnung der five number displays der Kassierzeiten

Für unsere Kassierzeiten ergeben sich damit die Werte in Tab. 10.1.

Tab. 10.1 Five number displays der beiden Datensätze

data	minimum	lower hinge	median	upper hinge	maximum
Kartenzahler	1,0	6,5	10,0	13,0	22,0
Bargeldzahler	9,0	15,5	22,0	28,0	64,0

Jeder wird zustimmen, dass der Sport, Probleme in einer Zeile als sogenannte *one liner* zu lösen, eher zu unleserlichen Ergebnissen führt. Zu Recht gilt *APL* als *write only*-Sprache, und selbst *APL*-Experten gelang es oft nach kurzer Zeit nicht mehr, den eigenen Code zu verstehen. Es lässt sich festhalten, dass dieses Werkzeug für den Fachmann bestens für interaktive Datenanalysen[5] im Sinne von Tukey geeignet war. Kritisch bleibt festzuhalten, dass keine hochauflösenden Graphiken, sondern nur aus Textzeichen bestehende erstellt werden konnten.

80er Jahre: Graphik per PC für Datenanalysen
Im Buch von Tukey (1977) spielen Graphiken eine ganz zentrale Rolle. Verschiedene dieser Graphiken hat Biehler (1982) diskutiert. Explorativ analysieren bedeutet demnach: Daten sprechen lassen, Interaktion plus Graphik. Diese Kombination sollte angestrebt werden. Mit dem *IBM PC*[6] wurden Rechner zwar preiswerter, blieben aber für Studierende unerschwinglich. Doch für universitäre Einrichtungen bot die Kombination aus *PC* und neuen *APL*-Produkten mit hochauflösender Graphik eine interessante Alternative. Mit dieser Ausrüstung wurde es möglich, auch graphische Verfahren umzusetzen und Rechner gestützte Experimente zu entwerfen. In (Naeve et al. 1992) findet man hierzu Beispiele, die wesentlich in *APL*-Umgebungen entstanden sind. Solche Experimente konnten endlich mittels *fahrbarer PC*-Projektoren-Einheiten in den Hörsaal getragen werden – aus Dozentensicht ein Durchbruch. Was wollte man mehr? Leider blieben diese vielfältigen Möglichkeiten ausgewählten kleinen Gruppen vorbehalten.

Ende der 80er: The new S Language
Das *blaue* Buch von Becker, Chambers und Wilks (1988) mit dem Titel *The NEW S Language – a programming environment for data analysis and graphics* soll hier als nächster Meilenstein genannt werden: Becker, Chambers und Wilks beschreiben die dritte Version der Sprache S als eine Sprache von Statistikern (und

---

[4] Per Internetseite: http://tryapl.org/ lässt sich dieses überprüfen.

[5] *APL*-Funktionen zur Datenanalyse findet man in (Velleman und Hoaglin 1981).

[6] *IBM PC* war erhältlich ab September 1981 und kostete 1.500 bis 3.000 $.

nicht von Informatikern) für Statistiker. Als Zweck wird die Analyse von Daten hervorgehoben, für die graphische Verfahren inzwischen nicht mehr weg zu denken sind. Interaktion zeichnet die Umgebung aus sowie der Ansatz, seine Handlungen per Anweisungen zu beschreiben. Viele Konzepte der Sprache *APL* wie Interpreteransatz, Umgang mit Vektoren und Matrizen, funktionale Beschreibung von Abläufen wurden übernommen, zusätzlich bietet S Kontrollstrukturen, wie wir sie von der für didaktische Zwecke entwickelten Sprache PASCAL her kennen. Der große Vorrat an statistischen und graphischen Funktionen erlaubt vielfältige Datenanalysen. Der entscheidende Fortschritt besteht jedoch in der für Statistiker einleuchtenden Art, Operationsschritte zu beschreiben. So lässt sich schnell eine eigene Fünfzahlenzusammenfassung definieren und beispielsweise die mitgelieferte Medianfunktion überprüfen:

```
<* 1>=
 my.five.number.display <- function(x){
     x <- sort(x); n <- length(x)                  # sort data
     m.depth <- ( n + 1 )/2                         # find depth of
     h.depth <- (floor(m.depth) + 1)/2              # median/hinges
     idx <- c(1,h.depth,m.depth,(n+1-h.depth),n)    # find positions
     result <- (x[floor(idx)] + x[ceiling(idx)])/2  # find numbers
     return(result)
 }
```

Die S-Anweisungen bilden direkt die Schritte: Sortieren, Ermittlung der Tiefen und Extraktion der gesuchten Daten ab. Die Konstruktion von einfachen Graphiken erfordert ebenfalls wenig Sprachgewalt. Zur vergleichenden Darstellung der *five numbers* der Kassierdauern als *box and whisker plot* reicht nach einer Zusammenfassung der Daten als Liste ein einfacher Funktionsaufruf von boxplot() aus. Als Ergebnis erhalten wir die Boxplots von Abb. 10.6.

```
<* 1>=
 kassier.zeiten <- list("mit Karte"=z.karte, "per Bargeld"=z.bar)
 boxplot(kassier.zeiten, horizontal = TRUE)$stats
```

Der Weg von der Idee bis zur Umsetzung ist also gradlinig, und wir *sehen*, dass Kartenzahlungen deutlich zeitsparender als Barzahlungen sind.

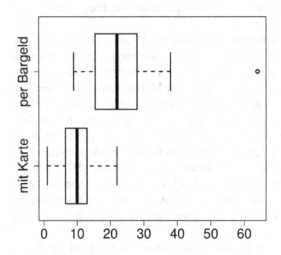

Abb. 10.6 Boxplots der Kassierdauern der Bar- und der Kartenzahler

1993: S-PLUS für breitere Nutzbarkeit

1990 gab es in Bielefeld unter Naeve erste Diplomanden, die eine S-Installation nutzen konnten. Doch war die Implementation der Sprache S mit einigen Schwierigkeiten verbunden. Dieses änderte sich, nachdem S unter dem Namen S-PLUS[7] als kommerzielles Produkt mit entsprechendem Support auf den Markt kam. Die erste WINDOWS-Version war 1993 erhältlich und mit universitären Campus-Lizenzen eröffnete sich ein Weg, S angemessen in der Lehre einzusetzen. Die Ausrichtungen des Werkzeugs auf die Bedürfnisse von Statistikern förderten den Einsatz, und es entstanden viele singuläre Lösungen. Hieraus ergaben sich neue Fragen: Wie lassen sich Datenanalysen und Verfahren verständlich dokumentieren und wie lassen sich Lösungen austauschen und weiter verwerten?

1995: Analyse-Reports zur Wiederbelebung

Donald E. Knuth ist einer der bedeutendsten Wissenschaftler der Informatik und vielen als Erfinder von TeX (Knuth 1986) bekannt. Zum Programmieren dieser vielschichtigen Software entwickelte er einen Programmierstil, der auf Problemzerlegung und unmittelbarer Dokumentation von Teillösungen basiert. Das Paradigma seines *literate programming* (Knuth 1992) lautet kurz: *Erkläre deinen Gedankengang zur Problemlösung, dann ergibt sich die Programmlösung automatisch.* Technisch wird dieser Ansatz durch WEB-Systeme unterstützt, die aus einem Quellfile sowohl das gewünschte Programm als auch eine wohl formatierte Dokumentation hervorzaubern. Mit den *APL*-Erfahrungen im Hinterkopf

---

[7] Anfangs wurde S-PLUS von StatSci vertrieben.

haben wir dankbar das neue Paradigma aufgegriffen, siehe bspw. (Naeve 1995), und setzen auch heute noch noweb[8] zur Dokumentation von Programmlösungen für unterschiedliche Sprachen ein.

Wie ein literates Programm besteht auch eine *explorative Datenanalyse* wesentlich aus einer Abfolge von Überlegungen und Umsetzungsschritten. Damit ist es zweckmäßig, diese Abfolgen ergänzt um Outputs zu Reports zusammenzufassen. Für den Prozess einer Datenanalyse wurde ein Mechanismus konstruiert, mit dem sich einzelne Anweisungen aktivieren und kommentieren sowie erschöpfende Report-Dokumente erstellen lassen. Die integrierten Befehlssequenzen motivieren dazu, sie erneut zu verwerten, sie also für weitere Zwecke wiederzubeleben. Mit der Möglichkeit, selektiv alte S-Anweisungen eines Reports zu aktivieren, war die Idee wiederbelebbarer Papiere bzw. der Reportschmiede (Wolf 1995) geboren. Mit dem Werkzeug konnte man ganze Analysen reproduzieren, aber auch alte nach neuen Ideen modifizieren oder aber weitere Analyseschritte zusammen mit Kommentaren ergänzen. Die neuen Reports ließen sich später ihrerseits wiederbeleben.

Das Konzept wiederbelebbarer Papiere verwendeten wir nicht nur für *literate Datenanalysen*, sondern auch für die Lehre. Bspw. ließen sich Skripte mit integrierten S-Anweisungen und Outputs spicken, so dass Dozenten in der Vorlesung und prinzipiell auch Studierende auf Basis der in den Reports abgelegten Anweisungen arbeiten und davon ausgehend Variationen probieren konnten (Wolf 1998). Damit war ein Instrument geschaffen, das Lehrende (in der Vorlesung) wie auch Lernende (beim Selbststudium) von lästigen und fehlerträchtigen Eintippprozessen befreite.

1999: offener S-Zugang durch R

Neben S-PLUS entstand Mitte der 90er die frei erhältliche Implementation R (CRAN 2012) der Sprache S. Auch wenn sich die internen Strukturen von S-PLUS und R unterscheiden, ließen sich sehr viele Anwendungen leicht portieren. Nach ersten Versuchen mit frühen R-Versionen (R-0.62*) wurde für die wiederbelebbaren Papiere eine R-Implementation geschaffen (Wolf 1999) und damit das Zugangsproblem prinzipiell gelöst.

Hingewiesen sei darauf, dass Leisch (2002) mit SWEAVE einen Automatismus vorgestellt hat, der S-Outputs in Text-Dokumente integrieren kann. Erforderlich ist als Input eine Quelldatei mit nach den Regeln des noweb-Systems integrierten S-Anweisungen. Aufgrund der gemeinsamen Syntax lassen sich wiederbelebbare Papiere als SWEAVE-Quelldokumente verwenden, und umgekehrt können SWEAVE-Dokumente von der Reportschmiede verarbeitet werden. SWEAVE gehört übrigens zur Grundausstattung von R und erfreut sich breiter Verwendung.

2000: per TCL/TK zum R-Paket relax

Im Juni 2000 erschien die R-Version 1.1.0, die als Besonderheit eine Schnittstelle zu TCL/TK besaß. TCL/TK ist ein Werkzeugkasten zum Bau von Oberflächen und

---

[8] Das WEB-System noweb wurde von Norman Ramsey (1994) vorgeschlagen.

ermöglichte es, für die wiederbelebbaren Papiere einen maßgeschneiderten Editor zu entwerfen. Zunächst wurde dieses Werkzeug nach dem Prinzip der Sparsamkeit nur mit den absolut notwendigen Funktionalitäten zur Erstellung von Reports und deren Reproduktion ausgestattet. Weitere Operationen wurden erst nach und nach ergänzt. Inzwischen hat sich die Reportschmiede mit `relax` der Version 1.3.8 (Wolf 2012b) zu einem mächtigen Werkzeug weiterentwickelt und ist als eines der über 4000 R-Pakete vom CRAN[9]-Server zu bekommen. Das aktuelle Aussehen des `relax`-Editors zeigt Abb. 10.7.

Abb. 10.7 Beispiel einer Ansicht des RELAX-Editors

Wir erkennen im oberen Bereich ein Arbeitsfeld und unten einen Ausgabebe-reich. Im Arbeitsfeld werden R-Anweisungen rot hervorgehoben, schwarze Schrift kennzeichnet textliche Abschnitte. Wie man sieht, können Ergebnis-Graphiken integriert werden. Wesentliche Management-Operationen werden über Knöpfe ausgelöst. Während der Arbeit mit `relax` schreibt der Bediener seine Gedanken und Aufträge an R ins Arbeitsfeld. Dann evaluiert er per Knopfdruck (`EvalRCode`) die Anweisungen und kann den Output anschließend ins Arbeitsfeld (`Insert`, `SavePlot`) übertragen.

Dank der Oberfläche kann der Anwender also seine Überlegungen und R-Anweisungen gemeinsam und direkt hantieren. Damit bekommt der Prozess der

---

[9] CRAN = *The Comprehensive R Archive Network*, (CRAN 2012)

Reporterstellung eine zentrale Rolle und aus dem *Software-Anwender* wird ein *literater* Datenanalytiker. Letztlich entstehen html- oder pdf-Dokumente wie das vorliegende, deren Quellfiles für weitere Bearbeitungen bereit stehen. Außerdem ist entsprechend der Grundidee der Reportschmiede die Reproduzierbarkeit gewährleistet.

Fazit und Verwendungen

Rechner sind inzwischen für die meisten Studierenden erschwinglich, mit R existiert ein für die Statistik ausgefeiltes und kostenloses Software-Produkt, dessen Verfügbarkeit über den CRAN-Server gesichert ist, und mit dem relax steht ein vielseitiges Instrument zur Unterstützung des Analyseprozesses zur Verfügung.

Was lässt sich nun alles mit relax anstellen? Für die Lehre haben wir Materialien mit kleinen Experimenten, Übungszettel mit R-Anweisungen und Ergebnissen bis hin zu Statistik-Klausuren erstellt. Weiter kann auf das Buch (Wolf et al. 2006) verwiesen werden, dessen integrierte R-Anweisungen rekonstruierbar sind. Naheliegend ist es, den Editor zur Erstellung von Folien oder Materialien mit R-Anweisungen zu nutzen, deren Beispiele mit höchster Sicherheit funktionieren. Am Rande sei bemerkt, dass das vorliegende Dokument ebenfalls in der relax-Umgebung geschrieben ist.

Im Rahmen solcher Anwendungen sind einige Nebenprodukte entstanden, wie die Funktion ConstructDemo(). Diese Funktion extrahiert aus einem Quellfile alle R-Abschnitte und erlaubt es dem Anwender, durch die *R-Chunks* zu *browsen*, ausgewählte Einheiten zu aktivieren, zu verändern usw. Mit dieser Technik hat ein Leser von (Wolf et al. 2006) leichten Zugriff auf alle R-Beispiele und kann so seiner Experimentierlust freien Lauf lassen. Darüber hinaus unterstützt die relax-Umgebung die Lösung und Dokumentation von Problemen, die kompliziertere Anweisungssequenzen erfordern. Dieses wird im folgenden Abschnitt demonstriert.

3 Eine kleine Simulation zu den Kassierzeiten

Dieser Abschnitt beschreibt eine im *literate programming style* geschriebene Simulation zur Frage, wie sich die Bezahlungsart auf die Verweildauern im Kassenbereich auswirkt. Dieses Beispiel enthält notwendigerweise vermehrt technische Elemente. Jedoch wird der Leser hoffentlich erkennen, wie Teillösungen erstellt werden und wie sich aus den Teilen die dokumentierte Gesamtlösung ergibt. Zur Abgrenzung ist dieses Beispiel in einer kursiven Schrift gesetzt, wobei R-Anweisungen in Schreibmaschinenschrift verfasst sind. Die Zeilen, die in spitzen Klammern eingefasst sind, adressieren Teillösungen, welche über die angehefteten Nummern verbunden sind bzw. gefunden werden können.

Simulation von Verweildauern

Besonders untersuchungswürdig sind Situationen an Kassen, in denen es zu Eng-
pässen kommt. Wir unterstellen für die Ankünfte der Kunden einen Poisson-
Prozess mit einer Rate, die 75% der Abfertigungsrate von den beobachteten Bar-
zahlern entspricht. Dann ist mit Warteschlangen zu rechnen, und wir können Verweildauern und Kassierzeiten graphisch gegenüberstellen. Zur Simulation entwerfen wir eine kleine Funktion, in der wir künstliche Ankunfts- und Bedienzeiten
generieren. Mit diesen Zeiten simulieren wir eine virtuelle Kasse. Die gemessenen
Kassierzeiten, die Kundenankunftsrate, die Anzahl der Kunden und der Start des
Zufallszahlengenerators sind über Argumente beeinflussbar.

```
<definiere simulate.cash 3>=
  cash.sim <- function(cash.times, rate.arrival, n.sim=20, zz=3){
    <ermittle Ankunfts- und Bedienzeiten 7>
    <setze Simulation um 8>
    return(result)
}
```

Nun lassen sich Simulationen per Funktionsaufruf starten. Für unsere Situation
lesen wir die Bedienzeiten ein, ermitteln hieraus eine Ankunftsrate, starten die Simulation und plotten die Ergebnisse mit Hilfe von Bagplots, vgl. Abb. 10.8.

```
<starte Simulation 4>=
  <lese Bedienzeiten z.karte und z.bar ein 5>
  rate <- .75*1/mean(z.bar)
  bar    <- cash.sim(cash.times=z.bar,    rate.arrival=rate, 1000)
  karte  <- cash.sim(cash.times=z.karte,  rate.arrival=rate, 1000)
  <plotte Ergebnisse als Bagplot 6>
```

Bagplots sind zweidimensionale Boxplots[10]. Im Zentrum ist der zweidimensionale Medien als Stern zu erkennen. Die innere Blase enthält 50% der Datenpunkte
und entspricht der Box eines Boxplots, die ganz außen liegenden Einzelpunkte
sind Ausreißer-Kandidaten. Der Unterschied zwischen den beiden Szenarien ist
gravierend, siehe Abb. 10.8. Barzahlungen kosten viel Zeit und damit Nerven der
Kunden. Kartenzahlungen führen zu deutlich kürzeren Warteschlangen bzw. War-
tezeiten, natürlich nur, wenn der Spareffekt später nicht durch verringerte Kas-
senanzahlen kompensiert wird.

---

[10] Zur Idee von Bagplots siehe Rousseeuw, Ruts undTukey (1999). Die Funktion bagplot()
stammt aus dem R-Paket aplpack (Wolf, 2012a), das übrigens auch eine Funktion zur Erstellung schöner *stem and leaf displays* enthält.

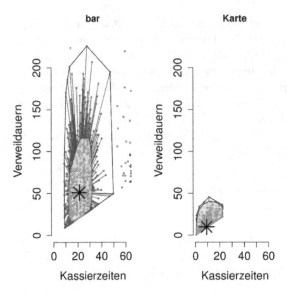

Abb. 10.8 Simulationsergebnisse als Bagplots dargestellt

Dateneingabe und Ploterstellung
Damit alles funktioniert, müssen wir noch einige Teillösungen definieren. So sind die beobachteten Bedienzeiten einzulesen.

```
<lese Bedienzeiten z.karte und z.bar ein 5>= is used in 4
 z.karte <- c(13,6,8,3,14,1,9,10,10,13,8,10,22,7,19,5,3,5,
              11,17,10,17,7,12)
 z.bar   <- c(13,25,11,20,16,24,19,17,28,25,19,64,30,36,22,
              12,19,13,31,26,24,22,28,14,17,31,15,22,26,32,
              9,25,25,22,28,29,38,30,13,14,15,14,16)
```

Die graphische Darstellung erfordert zwei Aufrufe der Bagplot-Funktion:

```
<plotte Ergebnisse als Bagplot 6>= is used in 4
 require(aplpack); par(mfrow=1:2)
 extr <- apply(rbind(bar,karte),2,range)
 extr <- extr[,c("t.service","t.response")]
 bagplot(bar[,"t.service"], bar[,"t.response"],
         xlab ="Kassierzeiten", ylab = "Verweildauern",
         xlim = extr[,1], ylim = extr[,2], main="bar" )
 bagplot(karte[,"t.service"], karte[,"t.response"],
         xlab = "Kassierzeiten", ylab = "Verweildauern",
         xlim = extr[,1], ylim = extr[,2], main="Karte" )
```

Ankunft und Bedienung
Nach Initialisierung des Zufallszahlengenerators ermitteln wir zufällig Quantile der Kassierzeitverteilung und addieren zu diesen kleine Störungen. Für die Ankunftszeiten verwenden wir aus einer Exponentialverteilung gezogene Zufallszahlen.

```
<ermittle Ankunfts- und Bedienzeiten 7>= is used in 3
 set.seed(zz)
 t.service <- jitter(quantile(cash.times,runif(n.sim),type=5),,,.5)
```

```
t.arrival <- cumsum(rexp(n.sim, rate.arrival))
```

Warteschlangensimulation

Die Verweilzeit eines Kunden ergibt sich aus der individuellen Bedienzeit sowie der Wartezeit. Zur Berechnung legen wir alle Ankunfts- sowie die Bedienende-Ereignisse in einer Ereignisliste (EL) ab. Aus diese Liste entnehmen wir in zeitlicher Reihenfolge Ereignisse und bearbeiten sie: Ein Ankunftsereignis erhöht die Anzahl der Elemente im System und induziert ein neues Bedienende-Ereignis, dessen Zeitpunkt erst festgelegt werden kann, wenn die entsprechende Person an der Kasse bedient wird. Ein Bedienende-Ereignis führt zur Entfernung einer Person aus dem System und die nächste Person kann an der Kasse bedient werden. Für die Auswertung wird der Exit-Zeitpunkt festgehalten.

```
<setze Simulation um 8>= is used in 3
 n.person <- 0; later <- Inf; arrival <- TRUE
 EL <- cbind(who=seq(t.arrival), Time=t.arrival, type=arrival)
 I.O.service <- cbind(t.arrival = t.arrival, t.exit = 0,
                      t.service = t.service)
 while( 0 < length(EL) ){
   # find next event ev -----------------------------------------
   idx  <- which.min(EL[,"Time"])   # find index of next event
   ev   <- EL[idx,]                  # extract event
   EL   <- EL[-idx,,drop=FALSE]      # remove event
   Time <- ev["Time"]; who <- ev["who"]; type <- ev["type"]
   if( type == arrival ){
     # manage ARRIVAL event ------------------------------------
     #  increment line counter, store new exit event
     exit.time <- if(0<n.person) later else Time + t.service[who]
     EL        <- rbind(EL ,c(who, exit.time, !arrival))
     n.person  <- n.person + 1
   } else {
     # manage EXIT event --------------------------------------
     #  decrement counter, set exit time in EL, store exit time
     n.person <- n.person - 1
     if(0 < n.person){
       idx <- which.min(EL[,"who"])
       EL[idx,"Time"] <- Time + t.service[EL[idx,"who"]]
     }
     I.O.service[who, "t.exit"] <- Time
   }
 }
 t.response <- I.O.service[,"t.exit"] - I.O.service[,"t.arrival"]
 result <- cbind( I.O.service, t.response)
```

Damit sind alle Teile der Simulation für R-Kenner ausreichend beschrieben. Andere Leser sollten zumindest Idee und Vorgehensweise verstehen können.

Reflexion

Die Graphik wird für einen Kassenverantwortlichen spannend sein. Dieser wird wahrscheinlich sofort vorschlagen, die Auswirkungen anderer Parametersetzungen zu untersuchen. Einfache Warteschlangensysteme lassen sich übrigens auch rein formal – also ohne Simulationen – analysieren. Jedoch hilft die Theorie nicht so richtig weiter, wenn das System komplizierter wird und ein Verantwortlicher die Auswirkungen *sehen* will.

Es entsteht der typische Wunsch zur Rekonstruktion und Modifikation, der sich mit dem Konzept der wiederbelebbaren Papiere und `relax` problemlos umsetzen lässt. Vielleicht möchte der Auftraggeber sogar das Simulationsexperiment selbst steuern können. Hierfür könnte eine kleine Oberfläche mit Schiebern zur Manipulation der Parameter hilfreich sein, wie sie mit der `slider()`-Funktion aus dem `relax`-Paket konstruierbar ist.

Mit neuen Möglichkeiten kommen also wieder neue Wünsche auf, die neue Herausforderungen für die Technologie darstellen und den Fortschritt im Gange halten. Ein Schritt in Richtung Interaktivität wird in folgendem Abschnitt beschrieben.

R-DemoBoxen: Simulation per Interaktion
In der Tat verwendet das Beispiel eher klassische Techniken. Auch ist es im Prinzip in anderen Welten, wie sie uns MAPLE, MATLAB oder MATHEMATICA bescheren, realisierbar. Deshalb soll zum Abschluss noch auf *R-Demo-Boxen* hingewiesen werden. Dieses zurzeit noch in der Entwicklung befindliche Werkzeug gestattet es, mit wenigen Handgriffen aus ausgewählten Anweisungen eines wiederbelebbaren Papiers ein Präsentationswerkzeug zu generieren. Dieses Werkzeug öffnet eine neue Oberfläche, in der sowohl formatierte Texte als auch R-Anweisungen angezeigt werden. Der Anwender kann die Anweisungen unverändert evaluieren lassen. Er kann aber auch Modifikationen der vorgegebenen Demos umsetzen. Zum Thema Simulation der Kassiersituation ist in Abb. 10.9 eine funktionstüchtige R-DemoBox zu sehen.

Abb. 10.9 Eine DemoBox zur Kassiersituation

Wir sehen im oberen Bereich eine Beschreibung des Gegenstandes und im unteren die zugehörigen Anweisungen. Per Klick auf *Eval* startet die Simulation und berücksichtigt vom Anwender vorgenommene Änderungen.

4 Alternative Sichten

Die in diesem Papier herausgestellten Meilensteine sind wesentlich auf das R-Paket `relax` ausgerichtet. Der Autor gibt zu, dass er dabei zum Teil einäugig die Vorteile hervorgehoben hat, die sich aus seiner Sicht ergeben. Natürlich gibt es viele andere Perspektiven, die aus anderen Erfahrungen erwachsen sind. Zum Beispiel ist es naheliegend, das Credo *Look at your data!* wörtlich zu nehmen und Zahlenrepräsentationen der Daten in den Vordergrund zu stellen.

In diese Richtung gehen Tabellenkalkulations-Lösungen. Sie zeigen dem Anwender permanent das Datenmaterial und erlauben ihm, Auswahlen zu treffen und Beziehungen von Variablen durch Klicks zu beschreiben. Die Definition von (neuen Verarbeitungs-) Prozessen steht nicht im Vordergrund, dagegen sind typische Operationen über Buttons oder Menüs anstoßbar. Daten werden weniger durch Namen repräsentiert und ein Anzeigen muss auch nicht per Funktionsaufruf eingeleitet werden. Weder Funktionsnamen noch Syntax-Fragen sind vom Anwender zu memorieren, er kann eben direkt mit den Daten hantieren. Zwar gibt es unter den vielen R-Paketen verschiedene, die Vorschläge in dieser Richtung anbieten, doch wird so schnell nicht zu schlagen sein, was uns Werkzeuge wie FATHOM schon seit einiger Zeit bieten (siehe Biehler et al. 2006).

Aus Sicht des Autors ist es wichtig, dass es verschiedene Ansätze gibt. Denn auch die Situationen zum Einsatz statistischer Verfahren sind höchst unterschiedlich. Für zeitlich sehr begrenzte Ausbildungseinheiten werden sich Produkte empfehlen, die sehr schnell bedienbar sind und die Grundoperationen für die Datenanalyse sehr offensichtlich anbieten. Ist es jedoch erforderlich, weniger standardmäßig explorative Datenanalyse zu betreiben oder Simulationsstudien durchzuführen, dürften mächtige Sprachen wie R und Konzepte, wie sie in diesem Papier hervorgehoben werden, die Nase vorn haben.

Literaturverzeichnis

Becker, R. A., Chambers, J. M., & Wilks, A. R. (1988). *The new S language – a programming environment for data analysis and graphics*. Pacific Grove, California: Wadsworth & Brooks.

Biehler, R. (1982). *Explorative Datenanalyse — Eine Untersuchung aus der Perspektive einer deskriptiv-empirischen Wissenschaftstheorie*. Bielefeld: IDM – Universität Bielefeld.

Biehler, R., Hofmann, T., Maxara, C., & Prömmel, A. (2006). *Fathom 2: Eine Einführung*. Berlin: Springer.

R Core Team. (2012). *R: A language and environment for statistical computing*. Wien: R Foundation for Statistical Computing. Abgerufen von http://www.R-project.org.

Ehrenberg, A.S.C. (1975). *Data reduction*. London: Wiley.

Knuth, D. E. (1986). *The TeX book*. Reading, Mass.: Addison-Wesley.

Knuth, D. E. (1992). *Literate programming*. Stanford, Calif.: CSLI.

Leisch, F. (2002). Sweave: Dynamic generation of statistical reports using literate data analysis. In W. Härdle, & B. Rönz (Hrsg.), *Compstat 2002 – Proceedings in computational statistics* (S. 575-580). Heidelberg: Physica Verlag.

Naeve, P. (1995). *Stochastik für Informatik*. München: Oldenbourg Verlag.

Naeve, P., Trenkler, D., & Wolf, H. P. (1992). Sehen und Glauben: über Experimente in der Statistikausbildung. In J. Andreß (Hrsg.), *Theorie, Daten, Methoden*. München: Oldenbourg Verlag.

Ramsey, N. (1994). Literate programming simplified. *IEEE Software, 11*(5), 97–105.

Rousseeuw, P. J., Ruts, I., & Tukey, J. W. (1999). The bagplot: a bivariate boxplot. *The American Statistician, 53*(4), 382–387.

Tukey, J. W. (1977). *Exploratory data analysis*. Reading, Mass.: Addison-Wesley.

Velleman, P. F., & Hoaglin, D. C. (1981). *Applications basics, and computing of exploratory data analysis*. Boston, Mass.: Duxbury Press.

Wolf, H. P. (1995*). Eine Reportschmiede für den Datenanalytiker* (Diskussionspapier No. 301). Bielefeld: Fakultät für Wirtschaftswissenschaften der Universität Bielefeld.

Wolf, H. P. (1998). *Ein wiederbelebbares Buch zur Statistik* (Diskussionspapier No. 407). Bielefeld: Fakultät für Wirtschaftswissenschaften der Universität Bielefeld.

Wolf, H. P. (1999). *RREVIVE – Funktionen zur Arbeit mit wiederbelebbaren Paieren unter R* (Diskussionspapier No. 415). Bielefeld: Fakultät für Wirtschaftswissenschaften der Universität Bielefeld.

Wolf, H. P. (2012a). *aplpack – another plotting package* (1.2.7) [Computer software]. Abgerufen von http://cran.r-project.org/web/packages/aplpack.

Wolf, H.P. (2012b). *relax – R Editor for Literate Analysis and lateX (1.3.12)* [Computer software]. Abgerufen von http://cran.r-project.org/web/packages/relax.

Wolf, H. P., Naeve, P., & Tiemann, V. (2006). *BWL-Crash-Kurs Statistik – aktiv mit R*. Konstanz: UVK Verlagsgesellschaft.

Chapter 11
Implications of technology on what students need to know about statistics[1]

Arthur Bakker

Utrecht University, Freudenthal Institute for Science and Mathematics Education

Abstract The availability of technology influences what people need to know about statistics, but do they need to know less, more or something different? As one piece in the jigsaw puzzle of this quest, this chapter focuses on the question of what student laboratory technicians in vocational education need to learn about statistics in the presence of technology. Through interviews with interns, intern supervisors and teachers, a questionnaire administered to interns, and workplace observations I have identified what statistical knowledge is taught and required. The knowledge required turned out to diverge across laboratories and to be highly influenced by the degree to which work is mediated by technology. For example, calibration and validation of measurement instruments is based on linear regression, but is often automated. Many computations are carried out on Excel sheets, but not all schools dedicate enough instruction time on spreadsheets. At least 30% of the interns (N=300) felt insufficiently prepared in terms of mathematics or statistics.

Introduction

Technology and statistics are two key themes in Rolf Biehler's work (Biehler 1997; Biehler et al. 2013). In this contribution I focus on the question: What is the impact of technology on what people need to know about statistics? More generally, the competences required in the workplace have changed over the past decades due to increasing use of information technology, computerisation (Felstead et al. 2007). Hoyles, Noss, Kent and Bakker (2010) investigated the impact of technology on what employees need to know about mathematics and statistics. They argued that there is increasing need for what they call *Techno-mathematical Litera-*

---

[1] This chapter is based on the following conference paper: Bakker, A., Wijers, M, & Akkerman, S. F. (2010). The influence of technology on what vocational students need to learn about statistics: The case of lab technicians. In C. Reading (Ed.), *Data and context in statistics education: Towards an evidence-based society. Proceedings of the Eighth International Conference on Teaching Statistics* [CD-ROM]. Voorburg: International Statistical Institute.

cies, mathematical knowledge mediated by technology and situated in specific contexts.

A first exploration of the question about this impact is puzzling. On the one hand, automation generally takes away computation from employees, which suggests that advanced calculations can be carried out by intermediate-level employees with the help of user-friendly software. In their case study of a pathology lab, Hoyles et al. (2002) indeed observed that "all the work of the laboratory is highly computerised and automated" (p. 62). One of the interviewees noted "that we do not do much maths in haematology, because the analysers [machines] do so much work for us" (p. 64). On the other hand, the advent of process improvement techniques, mostly statistically based, has meant that more and more employees, even at lower levels, are faced with handling data (Hoyles et al. 2007). This implies that many employees with vocational or little general education need some insight into data, variability, distribution, centre, spread, data trending etcetera (Hoyles et al. 2002). Many have to draw inferences from samples, for example about production processes (Bakker et al. 2008).

As a follow-up on this workplace research, I focused with other colleagues on an area of education that has been neglected in statistics education: vocational education. One interesting feature of this area is that it is here that school and work meet (Akkerman and Bakker 2012). The influence of workplace norms is big, and this also has consequences for both the technology used and what students need to know about statistics. In this chapter I focus on laboratory work, as it involves many calculations carried out by dedicated software. Common in industry, but also in health services and safety institutes, laboratories are statistically rich, hence interesting places to study the influence of technology on the statistical knowledge required by employees.

Vocational education is constantly confronted with the question of what future employees should learn. However, in their review of competence-based vocational education, Van den Berg and De Bruijn (2009) noted that most research in vocational education in the Netherlands, Germany, and Switzerland (countries with long vocational traditions) focuses on how-questions with regard to teaching, supervision, assessment and learning environments. Only a small minority of research studies address the question of *what* students actually need to learn. The importance of the what-question, however, is growing in a changing environment, as discussed above, hence the research question addressed in this chapter: *What do student lab technicians in vocational education need to learn about statistics given the computerisation trend?*

Background

About 40% of Dutch senior secondary students (16-17) attend general education or pre-university tracks; the remaining 60% enrol in senior secondary vocational education (MBO). From MBO level 4 (the highest level of four), students can move to higher professional education (HBO, bachelor level). In laboratory education, the percentage of level 4 students enrolling in is quite high: about 40% with an additional 30% hoping to gain a few HBO certificates.

MBO used to have attainment targets for each MBO occupation (e.g., hair dresser, baker, electrician, lab technician). For mathematics and statistics in many technical programmes this was a list of about fifty attainment targets (sinus, cosine, Pythagoras' theorem, etc.), but attainment targets that were judged less relevant for the occupation were ignored by mathematics teachers. Thus general subjects such as mathematics were generally considered separate from the occupation. With the more recent introduction in the Netherlands of competence-based vocational education, qualification files for 241 occupations were formulated in terms of what starting employees should be able to do. In the qualification file for lab technicians, references to the statistical knowledge required were scarce and broadly phrased (e.g., "basic knowledge of mathematics"; "care for quality"), with the effect that such knowledge was taught and assessed less often than about ten years ago. Teachers often had a hard time to convince their managers that students needed some disciplinary knowledge such as statistics in order to develop the competences in the qualification files (Bakker et al. 2011). In most cases, the number of teaching hours for such subjects decreased considerably.

Within this historical development the what-question has become urgent, even though it is sometimes considered unanswerable by established research methods (for a discussion of this issue see Van den Heuvel-Panhuizen 2005). I therefore address descriptive and evaluative questions that form the basis for answering my main normative question:

1. What do student lab technicians learn about statistics in various vocational laboratory schools?
2. How has laboratory work changed over the past decades according to experienced lab technicians and teachers?
3. What statistical knowledge do interns at MBO level need at work?
4. How well prepared do interns feel for the computational part of their work?

Methods

An answer to question 1 was sought by interviewing teachers at four different vocational laboratory schools (MBO, which is below Bachelor level) and studying their course materials. I also observed several lessons to get a sense of how course materials were used. For an answer to questions 2, 3 and 4, I conducted interviews with six supervisors of interns (students in work placement) in a variety of labs (in total 6:40 hrs), two managers (1:50 hrs), nine interns in the workplace (4:20 hrs), two interns who presented their work at school (2:16 hrs) and five teachers at four different schools (14:25 hrs). In addition, I undertook four workplace tours in different labs (2:10 hrs), spent a day of observation and interviewing in one lab, and collected several prototypical artefacts that represented how statistical knowledge was mediated (e.g., Standard Operating Procedures including calculations, graphs, data etc.). The interviews were transcribed verbatim and coded for statistical knowledge required during internships and general issues relevant to our questions (e.g., trends, influence of technology and transition from school to work).

Due to the large variability in findings from the interviews, I decided to administer a questionnaire to interns of ten laboratory schools to get a better image of how representative particular findings were. The questionnaire focused on the following questions:

1. Which calculations have you performed over the last weeks, for what purpose, and with which tools did you perform them?
2. What (other) mathematical and statistical knowledge did you need?
3. How well do you feel that school has prepared you in terms of mathematics and statistics?

I received 300 usable forms from nine schools. The national population of MBO lab students is 3,500 (16 schools); they spend about a third of their time as interns; hence I assume to have information from about 30% of the intern population of that time. When filling in the questionnaires, most students were about halfway through their internship. To get a better image of what statistical knowledge they needed at the end of their final nine months' internship in the fourth year, a teacher and the author also studied 31 final reports by interns on their research carried out in the workplace.

Results

1 What do students learn about statistics at laboratory school?

The topics addressed in the course materials are summarised in Table 11.1. There was considerable consistency across schools, partly because the same resource book was used (Raadschelders and den Rooyen 2005) in addition to the teachers' own materials. Several schools used jointly designed project materials that aim to develop high quality projects in which all competences required can be developed. Alongside these projects, general subject teachers address the general knowledge required in each project.

There were differences in when statistics was taught and for how long. In one school, statistics was taught during the one-day-at-school or block-release days during internships in the third and fourth year. In others, it was taught mainly in the first and second year. Topics such as the Dixon's test for outliers or the normal distribution were often introduced in one lesson. I observed one lesson in which students learned to make Shewhart control charts in Excel. One observation at another school was that significance tests were quickly introduced with their formulae and manually practised with three exercises. Given that these topics are not part of the general education curriculum and that vocational students typically find mathematics and statistics harder to learn than students from general education, I wonder not only how well these topics are understood, but also whether they learn the right things about these tests.

Table 11.1 Statistical topics addressed in the course materials

Descriptive statistics	Arithmetic mean, SD, outlier, range, coefficient of variation, variance
	Correlation, Linear regression for calibration
	Histogram, scatter plot
	Frequency distribution, normal distribution
Inferential statistics	t test, F test, Dixon's Q and Grubb's tests for outliers
	confidence intervals
Quality control	Youden plot: a graphical data analysis technique for carrying out an interlab comparison
	SPC (Shewhart or Levey-Jennings) and Westgard rules

2 How has laboratory work changed?

The changes over the past fifteen years differed per type of lab. The most "old-fashioned" lab I visited was a veterinary lab in which calculations were still mainly done manually or with calculators. As the supervisor said, "I'm computing all day long," but very few calculations went beyond means and standard deviations. As a contrasting case there was a quality control lab in which all procedures were standardized by means of SOPs (Standard Operating Procedures). The supervisor repeatedly said that the work of lab technicians at MBO level had become "like baking apple pie" and that this was common for regulated and accredited test labs. In line with Good Manufacturing Practices (GMP) all steps are prescribed and the acceptable range of possible outcomes is predetermined. All analyses are double-checked by employees with higher levels of education and there is hardly any room for judgement of results by MBO lab technicians. Their task has become to monitor computer outcomes and report anomalies to their supervisors. This seemed common for all laboratories that were highly regulated, either by GMP or health and safety rules (such as in hospitals). The drive to make the work error-free, one manager commented, has led to a situation where the younger generation often no longer knows what happens behind the screens. The paradoxical situation is that this does not lead to problems – those have been ruled out by the system – but I did hear concerns about this situation; many lab technicians found it important for interns to understand what they were doing and I have evidence from observations in one lab that blindly following procedures can lead to waste of materials and time.

The most apparent trend, the focus of this chapter, is that of computerisation. A typical consequence is that calibrating machines was in some cases currently done by means of pressing "calibrate" in a control panel with the result that MBO lab technicians did not encounter calibration graphs or the statistics behind them. In previous times, regression lines were made and correlation calculated.

3 Which statistical knowledge do interns need at work?

From the interviews and questionnaire I summarise the statistical topics encountered in laboratories. Because of the small number of labs visited, I only make a distinction between rare, common and pervasive use to give a sense of their frequency. 'BA' refers to examples carried out by a lab technician at BA level (higher professional education). A comparison of Tables 11.1 and 11.2 indicates a large amount of overlap and I indeed had the impression that teachers stayed in touch with workplaces – sometimes through their students – to see if they still taught the right topics. However, the question remains what these students need to understand about more complex statistics if it is carried out by a computer system. The fact that I saw t tests being used in a lab does not necessarily mean that an MBO lab technician should understand the formula or be able to perform one by hand, or even with a software package.

Table 11.2 Topics encountered in different laboratories. BA refers to being used by an employee at Bachelor level.

Descriptive statistics	Collecting and reporting data (pervasive)
	Arithmetic mean, standard deviation (pervasive, often automated)
	Relative standard deviation, coefficient of variance (common)
	Weighted mean when some measurements are more reliable than others and geometric mean in logarithmic scales (BA)
	Scatter plot, histogram, line chart (pervasive)
	Youden plot (rare)
	Correlation and linear regression for calibration, validation of instruments (pervasive, but often automated)
Inferential statistics	Interpreting confidence intervals (pervasive)
	Interpreting results of significant tests (common)
	Using tests for outliers: Grubbs, Dixon's Q (common)
	Carrying out with software: t test, F test, Fisher, analysis of variance, multivariate data analysis (BA)
Quality control	Statistical process control; Westgard rules for trends; difference between common-cause and special-cause variation (common)
	Design of experiment (BA – rare)

General	"Number sense": a feel for which numbers are correct, outliers (pervasive)
	Statistics in Excel (pervasive), mostly using standard worksheets
	Understanding to some extent options in the software (e.g., six types of standard deviation in Excel) and what software is doing (common)

I highlight a few general issues. First, I frequently heard about the need for "number sense", by which supervisors typically meant a feel for which numbers are correct and which are outliers. This is not just based on statistical insight (e.g., SD cannot be bigger than the range), but also on contextual knowledge (e.g., typical concentrations of particular substances).

Secondly, computations are not always taken away from employees. In some labs (according to the questionnaire results about 14%), interns reported that all calculations were automated in Excel sheets or dedicated computer software (such as LIMS: Laboratory Information Management System). In most other labs, a mix of computational tools (calculator, Excel, software) was reported to be used. The general impression from the interviews was that calculations had become easier over the years because of software and automated machines, but what lab technicians need to know has not become less, only different; for example fluency in Excel has become more important. Not surprisingly, a new book for the laboratory sector focuses on doing statistics with Excel (Klaessens 2009).

Thirdly, with computations outsourced to software, it becomes important to know something about the software and what it is doing. A simple example illustrates this. When using Excel to compute a SD, one is faced with a choice between several types: STDEV, STDEVA, STDEV.S, STDEVPA, STDEVP and STDEV.P. Which one should be used? To choose the correct option, one needs to know about the difference between the SD of a population and of a sample, but also about some Excel conventions.

4 How well prepared do interns feel?

The questionnaire results indicated that 45.7% of our sample felt well prepared by vocational laboratory school, but 27.0% thought they were not well prepared. 16.3% wrote they had not used any mathematics or statistics at all over the past weeks or had learned more than necessary. 3.3% noted insufficient preparation for specific topics and 7.7% gave no answer or an answer I could not classify. Some students felt competent in terms of the statistics, but not the mathematics needed; some students from other schools reported the reverse. I speculate this can be explained by when statistics is taught. The former group learned their statistics in their third and fourth years and mathematics in the first and second years (in one school the teacher had been ill). A few also complained that they had forgotten much of what they had learned in the first year by the time they did their internship. Some mentioned that they had not learned to use Excel spreadsheets at school, although these were often used in workplaces. This points to the importance of developing technology-mediated statistical knowledge (acknowledged in most schools). By and large our impression is that laboratory schools prepare students well, considering the limited amount of instruction time, but the fact that 30.3% (27 + 3.3) did not feel well prepared is worrying. Not surprisingly, all teachers complained that their hours for these disciplines had diminished due to longer internships and the tendency to work in a competence-based manner – a trend that several teachers considered to be a matter of economizing on costs. In practice this meant more time on projects and learning on demand, and less on general subjects such as languages and mathematics.

Conclusion and discussion

The answers to the four questions form the basis of a tentative answer to the main question of what student lab technicians need to learn about statistics given the computerisation trend. If I only look at the topics taught in vocational laboratory schools and encountered in laboratories, there seems to be a good match. However, the computerisation trend might require a different approach to these topics. For example, knowing how a SD is computed seems to have become less important, at least at this vocational level, but knowing how to use Excel and dedicated laboratory software has become more important. This is not necessarily easier, for several reasons. First, because computations are black-boxed (Williams and Wake 2007), it can be hard to detect what has gone wrong if the outcome is odd. Secondly, software often offers many choices, as the STDEV example illustrated, and might do things automatically that the user is not aware of (e.g., removing an outlier from a computation, reported in Hoyles et al. 2007). Students therefore need to learn more about the software packages themselves. Thirdly, I see a need for learning about the use of statistics as a consumer rather than a producer: Is it OK that a t test was used? Is the outcome reasonable? If the range of the data set is 10, is it reasonable that the SD is 1.8? In fact, there is a similarity with the calculator discussion in general education, in which estimation has become more important so that pupils can check the outcomes (Ruthven 2006).

With 30.3% feeling underprepared and about 70% of the MBO lab technicians at level 4 deciding to move on to a BA degree or certificates in higher professional education, where a higher level of abstraction is expected, it seems wise to spend more time on mathematics and statistics in laboratory schools. Moreover, this 30.3% might well be an overestimation: As the interviews indicated, supervisors generally adapted workplace tasks to the level of their particular interns. Difficult and important tasks are carried out by higher-level or more experienced lab technicians. Thus students might feel sufficiently educated whereas the workplace system is serving as an ecology adapting to particular gaps or weaknesses in interns' knowledge. Such adaptivity and division of labour also has a complementary side: I was told about lab technicians with an affinity for statistics who were given the opportunity to develop their statistical knowledge and become the team's statistics expert.

One of the main problems I observed is the discrepancy between how statistical measures and techniques are typically represented at school and workplace courses on the one hand, and how they are used in practice on the other. In course materials standard deviation and the t test are typically represented in a symbolic language with $\sum$-signs – a language that is inaccessible to most vocational students. Our impression from observations and previous research (Bakker et al. 2009) is that many teachers and trainers think the essence of, say, a t test is captured by its formula, just like the mean by its calculation, and that they see little opportunity to represent such concepts alternatively, or emphasize their meaning in usage. However, what intermediate-level employees need to know about such techniques is what their purpose is and how they should be interpreted when produced by a computer system, and some conditions of usage. To us it seems sufficient for student lab technicians who do not plan to attend higher professional education to know that a t test is useful for comparing means of data sets (e.g., to check if a new instrument is as accurate as the standard), and what it means when there is a significant difference, and more concretely that there is a small chance of two types of error. The little time attributed to teaching the t test (typically one lesson) is perhaps better spent on such insights, including how to perform a t test in Excel, than on explaining and applying the formula. This example shows that the what-question cannot completely separated from the how-question: What should be taught about the t test is closely related to how it is taught, in particular how concepts and techniques are represented.

The problem of representation of statistical concepts and tests is also reported by Bakker et al. (2009) in the context of process improvement in a car factory. To avoid the symbolic language about process capability indices, they designed relatively simple, visual computer tools with which employees could get a sense of what these indices conveyed, and how their indices could be manipulated by changing mean, control limits or specification limits. These tools proved to facilitate communication between employees with diverse educational backgrounds. I therefore expect that it is in principle possible to convey the practical usage and implications of many statistical concepts and techniques in the context of work without anxiety-evoking formulae. In mathematics education there is a tradition of democratizing powerful ideas (Hegedus and Lesh 2008) through carefully designed computer tools, so that more people can gain from some practical understanding of them. Statistics education research should in my view continue this line of research, in several settings, so as to complete the jigsaw puzzle on impact of technology on what people need to know and hence to learn. Although this chapter focused on vocational education, the key question is also pressing in other areas of education.

Acknowledgments   The research was funded by the Netherlands Organisation for Scientific Research (PROO), grant number 411-06-205. I would like to thank my colleagues Sanne Akkerman, Koeno Gravemeijer and Monica Wijers for the pleasant collaboration in this project.

References

Akkerman, S. F., & Bakker, A. (2012). Crossing boundaries between school and work during apprenticeships. *Vocations and Learning, 5*(2), 153-173. doi: 10.1007/s12186-011-9073-6.

Bakker, A., Kent, P., Derry, J., Noss, R., & Hoyles, C. (2008). Statistical inference at work: Statistical process control as an example. *Statistics Education Research Journal, 7*(2), 131-146.

Bakker, A., Kent, P., Noss, R., & Hoyles, C. (2009). Re-presenting statistical measures in computer tools to promote communication between employees in automotive manufacturing. *Technology Innovations in Statistics Education, 3*(2). Retrieved from http://www.escholarship.org/uc/item/53b9122r.

Bakker, A., Wijers, M., & Akkerman, S. F. (2011). The challenges of teaching statistics in secondary vocational education. In M. Pytlak, T. Rowland, & E. Swoboda (Eds.), *Proceedings of the Seventh Congress of the European Society for Research in Mathematics Education* (pp. 735-744), 9 - 13 Feb 2011. Rzeszow, Poland: University of Rzeszow. Retrieved from http://www.cerme7.univ.rzeszow.pl/WG/5/CERME_Bakker-Wijers-Akkerman.pdf.

Biehler, R. (1997). Software for learning and for doing statistics. *International Statistical Review, 65*(2), 167-189. Retrieved from http://www.stat.auckland.ac.nz/~iase/publications/isr/97.Biehler.pdf.

Biehler, R., Ben-Zvi, D., Bakker, A., & Makar, K. (2013). Technological advances in developing statistical reasoning at the school level. In A. Bishop, K. Clement, C. Keitel, J. Kilpatrick, & A. Y. L. Leung (Eds.), *Third International Handbook on Mathematics Education* (pp. 643-689). New York: Springer. doi: 10.1007/978-1-4614-4684-2_21.

Felstead, A., Gallie, D., Green, F., & Zhou, Y. (2007). *Skills at work, 1986 to 2006.* Oxford: ESRC Centre on Skills, Knowledge and Organisational Performance.

Hoyles, C., Bakker, A., Kent, P., & Noss, R. (2007). Attributing meanings to representations of data: The case of statistical process control. *Mathematical Thinking and Learning, 9*(4), 331-360.

Hoyles, C., Noss, R., Kent, P., & Bakker, A. (in press). *Improving mathematics at work: The need for techno-mathematical literacies.* London: Routledge/Taylor & Francis.

Hoyles, C., Wolf, A., Molyneux-Hodgson, S., & Kent, P. (2002). *Mathematical skills in the workplace.* London: The Science, Technology and Mathematics Council. Retrieved 30 October 2009 from http://www.lkl.ac.uk/research/technomaths/skills2002/.

Klaessens, J. W. A. (2009). *Statistiek in het laboratorium met Excel* [Statistics in the laboratory with Excel]. Oosterbeek: Syntax Media.

Lesh, R., & Hegedus, S. (2008). Democratizing access to mathematics through technology: Issues of design, theory and implementation - In memory of Jim Kaput's work. *Educational Studies in Mathematics, 68* (Special Issue).

Raadschelders, H. M., & den Rooyen, M. F. M. (2005). *Kwaliteitszorg en statistiek in het laboratorium* [Quality assurance and statistics in the laboratory]. Oosterbeek: Syntax Media.

Ruthven, K. (2009). Towards a calculator-aware mathematics curriculum. *Mediterranean Journal for Research in Mathematics Education, 8*(1), 111-124.

Van den Berg, N., & De Bruijn, E. (2009). *Het glas vult zich. Kennis over vormgeving en effecten van competentiegericht beroepsonderwijs; een review.* [The glass is filling up. Knowledge about the design and effects of competence-based vocational education; A review]. Amsterdam/Den Bosch: Expertisecentrum Beroepsonderwijs.

Van den Heuvel-Panhuizen, M. (2005). Can scientific research answer the 'what' question of mathematics education? *Cambridge Journal of Education, 35*(1), 35–53.

Williams, J. S., & Wake, G. D. (2007). Black boxes in workplace mathematics. *Educational Studies in Mathematics, 64*(3), 317–343.

Chapter 12
Tools for Learning Statistics: Fundamental Ideas in Statistics and the Role of Technology

Gail Burrill

Michigan State University

Abstract As the focus of statistics shifted from procedures and formulas towards a more data centric perspective emphasizing understanding big ideas, the role of technology became increasingly important. In the context of this shift, this chapter examines how technology can be used to support the understanding of seven fundamental ideas in teaching statistics, ideas developed in an earlier paper. The discussion briefly considers types of technology ranging from that used in the 1970s to innovative technology that seems promising in 2012.[1]

"The technology revolution has had a great impact on the teaching of statistics, perhaps more so than many other disciplines (Chance et al., 2007)."

1 Introduction

Typically students taking statistics in the 1970s and early 1980s had limited access to computers or even computing packages such as Minitab. The primary tool for most students and teachers in that era was a scientific calculator - with individual keys for $\Sigma(xy)$; $[\Sigma(x)]^2$ or $(\Sigma(x))^2$ to calculate pieces of formulas for statistics such as standard deviation or correlation. The student was expected to assemble the intermediate calculations and perform the final calculation to obtain the value.

In the early 1980s Exploratory Data Analysis (EDA) (Tukey 1977) began to appear in school statistics, engaging students in searching for the "story" in data. This meant looking for ways to maximize insight into a data set, uncover underlying patterns and structure, detect outliers and test assumptions. The process relied heavily on graphical representations - the "inter-ocular impact" (Tukey 1984). The use of graphical representations to explore statistical concepts is probably the most important lesson from EDA (Biehler 1982; 1993). But initially materials for teaching EDA such as the Quantitative Literacy Project (Scheaffer 1986) were developed without access to technology and provided students with premade graphs,

---

[1] Detailed descriptions of the use of technology in teaching and learning statistics can be found in Ben-Zvi (2000) and Chance et al. (2007).

small data sets, and tools such as coins, dice and random number tables to generate simulations.

2 Teaching statistics with technology

In tandem with the introduction of EDA into schools, the introduction of modern computing technology transformed practicing statisticians' work and contributed to the range, nature and complexity of theoretical statisticians' research. However, little, if any, software was designed to engage students in exploring data (Biehler 1993). In 1990, responding to requests from high school statistics teachers in the United States, Texas Instruments released the TI-81 graphing calculator. Although designed for upper level algebra and calculus, the device enabled students to compute basic statistics for a set of data, make box plots and histograms, plot bivariate data, and generate random numbers. The TI-81 was also programmable, which allowed teachers and students to develop programs related to simulation and least squares regressions. Such graphing calculators, though a great step forward from scientific calculators, were still limited and lacked flexibility in what they enabled students to do. Over the next several years, a variety of improvements were made in graphing calculators, but the inability to enter large amounts of data and limits on graphical representations imposed by the small screen inhibited their use in real data settings, although they did provide relatively good simulation capabilities.

As EDA became more and more ingrained as an essential component of teaching statistics, educators such as Cobb (1992) and Wild and Pfannkuch (1999) laid out principles and frameworks for developing statistical thinking.

3 Fundamental ideas in statistics and the role of technology

Building on this work and the Guidelines for Assessment and Instruction in Statistics Education (GAISE) Report (Franklin et al. 2007), Burrill and Biehler (2011, p. 62-63) proposed the following seven ideas as fundamental to developing understanding of the statistical process.

1. Data – including types of data, ways of collecting data, measurement; recognizing that data are numbers with a context;
2. Variation – identifying and measuring variability to predict, explain or control, where "variability" indicates the general phenomenon of change and "variation" the total effect of the change;
3. Distribution – including notions of tendencies and spread that are foundational for reasoning about statistical variables from empirical distributions, random variables from theoretical distributions; and summaries in sampling distributions;

4. Representation – graphical or other representations that reveal stories in the data including the notion of transnumeration;
5. Association and modeling relations between two variables – nature of the relationships among statistical variables for categorical and numerical data including regression for modeling statistical associations;
6. Probability models for data generating processes – modeling hypothetical structural relationships generated from theory, simulations or large data set approximations, quantifying the variability in data including long term stability;
7. Sampling and inference – the relation between samples and the population; deciding what to believe from how data are collected and drawing conclusions with some degree of certainty.

What is the link between these ideas and technology? How does the use of technology enable the learning of the fundamental ideas?

1. Data: Technology allows the use of real data with large data sets rather than small, contrived ones and enables students to actually solve real problems. Such exercises engage students in the practice of statistics and, as a consequence, contribute to their understanding of what statistics is about (Ben-Zvi 2004).
2. Variation: Technology allows students to build intuition about variability through unlimited repetitions of a simulation or process quickly generating thousands of samples; for example, drawing random samples and representing them by a collection of box plots (Biehler 1989; 1999). The variability in repeated samples can be illustrated through graphical representations (Biehler 1997a).
3. Distribution: Technology displays graphs of frequency distributions that can be used to investigate patterns of variation, easily converts frequency distributions into relative frequency distributions to compare data from different sized subgroups within a population, and allows students to compare visual representations of theoretical distributions to empirical distributions, all important for understanding distribution as a "fundamental given of statistical reasoning" (Wild 2005).
4. Representation: Graphical representations can be quickly generated allowing students to visualize distributions, look for trends, patterns and interesting aspects of the data that might lead to better understanding. EDA as characterized by Velleman and Hoaglin (1992) is an "iterative process of describing patterns, subtracting them, and searching anew for pattern in the residuals [that] continues until the data analyst decides to stop." Technology makes this iterative process easy to carry out.
5. Association and modeling: Technology enables students to investigate the existence of dependence between variables and create and revise models to describe relationships. Students can build their own models and see real world phenomena through a mathematical model (Pratt et al. 2011).
6. Probability models: Simulated models can build probabilistic intuitions and help students construct meaning of concepts such as probability, random variable, or expected value (Maxara and Biehler 2006). Technology can also be used

to improve students' understanding of probability by allowing them to explore and represent statistical models, vary assumptions or parameters, and analyze data generated by using the models (Biehler 1991; Jones et al. 2007) as well as explore models of real situations (Biehler 1994).

7. Sampling and inference: Simulations and graphic representations of the results can offer convincing arguments of the truth of important statistical concepts such as the central limit theorem (Moore 1997). Students use simulations - carry out repetitions, change parameters (i.e., sample size, number of repetitions), and describe and explain the observed behavior to make abstract concepts such as sampling distribution, p-values and confidence intervals become more concrete (delMas et al. 1999).

Envisioning how technology could support teaching and learning statistics led to the development of technology to make the vision a reality.

4 The Evolution of technology for learning statistics

4.1 Technology created environments for learning

Software designed for doing statistics is not necessarily appropriate for learning statistics. Such packages are often difficult to learn, limited in what they can do, and not easily adaptable (Biehler 1997b). Ben-Zvi (2000) argues computer-based learning environments should allow students to: (a) practice data analysis and combine exploratory and inferential methods, graphical and numerical methods; (b) use dynamic linked representations and simulations to construct meanings for statistical concepts; and (c) build models for statistical experiments, and use computer simulation to study them.

To support the learner of statistics rather than the practitioner, software packages such as MEDASS Light (Biehler 1995; 1997b) and Fathom (Finzer 2001; Biehler 2003) were designed as tools for learning statistics for secondary students in a first statistics course. The interactive dynamic software allowed students to interact on a direct level with statistical concepts and observe the results of that interaction. Although such software was intended to develop understanding, some still presented a learning hurdle for both teachers and students, and the process of how they use and learn to use the tools in some instances had to be supported by designed interventions (Maxara and Biehler 2006).

4.2 Applets

Another approach to using technology to develop understanding is the use of applets, typically a short visual, interactive dynamic file that enables students to explore a concept or work through a problem in a particular context (see CAUSE, www.causeweb.org, for an annotated, peer reviewed list of sources). Most applets, usually available online, do not come with detailed instructions for teachers on how they might be used.

Action/consequence documents (Dick and Burrill 2008) are related to the applet-family concept. They are pre-constructed interactive files focused on the development of a core mathematical/statistical concept; the documents provide "environments" in which students can play with a statistical idea in a variety of ways but where the opportunity to go astray is limited. (See for example, http://education.ti.com/calculators/timathnspired/US/Activities/Subject?sa=5026). These documents are typically targeted at concepts identified as "tough to learn" by the research literature and student achievement results. They allow students to take purposeful and deliberate actions, observe the mathematical or statistical consequence and reflect on the results (Figures 12.1, 12.2, and 12.3). Materials that accompany these documents provide support for teachers in the form of suggested student activities and possible questions for students with teacher notes that might highlight the importance of the concept as well as provide guidance for implementing the activity.

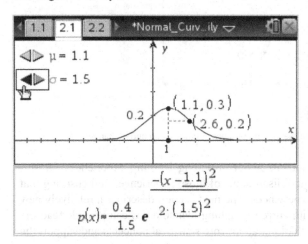

Fig. 12.1 Connecting parameters in the equation of a normal curve to the graph

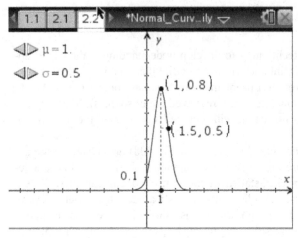

Fig. 12.2 Effect of changing μ and σ on a normal curve

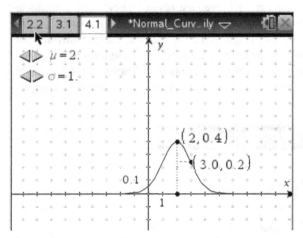

Fig. 12.3 Using a grid to estimate the area under normal curves as μ and σ change

Teaching, however, is not just about providing learning experiences for students but also about managing discussions of those experiences and ensuring that students learn from those experiences. The next section describes a relatively new technology that together with current graphing technologies can provide teachers with ways to probe student thinking about core statistical concepts and support the development of understanding.

5 Technology to support the work of teaching

The teacher is the mediator between students, content, and the use of technology. Teachers shape students' thinking through the tasks they provide, the questions they ask, the norms they establish in their classrooms, the feedback they provide, and their own beliefs about the nature of the subject (Gal and Ginsburg 1994). Classroom connectivity technology (CCT), a tool that can support teachers in carrying out these tasks, consists of a network of devices that connect a teacher's computer to students' handheld devices or personal computers. The network provides a channel for interaction among teacher and students that (a) allows the exchange of information (words, numbers, symbolic expressions, graphs, pictures) between them, (b) enables the teacher in real time to monitor student work, and (c) has a screen capture feature that displays simultaneously all of the student screens at any given instant. Teachers can pose questions; the network instantly collects and displays student responses as well as a summary of those responses. In the other direction, teachers can aggregate data collected from students and display the results, anonymously or by student. The ability to simultaneously display student work provides opportunities for students to share their own reasoning and critique the reasoning of others, consider alternate solution strategies and weigh the merits of each, observe patterns in phenomena such as random behavior and compare how changes in conditions such as sample size affect a distribution.

CCT supports the seven fundamental ideas in statistics from multiple perspectives. The network provides easy transmission of data between student and teacher in addition to providing data for teachers related to student understanding. Mindful of Hawkin's (1997) warning that students do not always take from interaction with technology what we think they do, teachers can use CCT to send students a quick poll asking them to describe in three words how changing the standard deviation will affect the shape of a normal distribution or to send the equation of a linear regression for a set of data. The responses can provide the teacher with opportunities to probe student thinking about their responses, compare responses, or consider whether different expressions (verbal or symbolic) are equivalent and why.

Collecting information from students can motivate the development of statistical concepts. Figure 12.4 shows responses to the question: Give five temperatures collected over the course of a year that would produce a mean temperature of 76° for the year. The responses (with ranges from 110° to 0°) prompted a discussion of how the mean alone is not very informative about a distribution, leading to the need for some measure of spread.

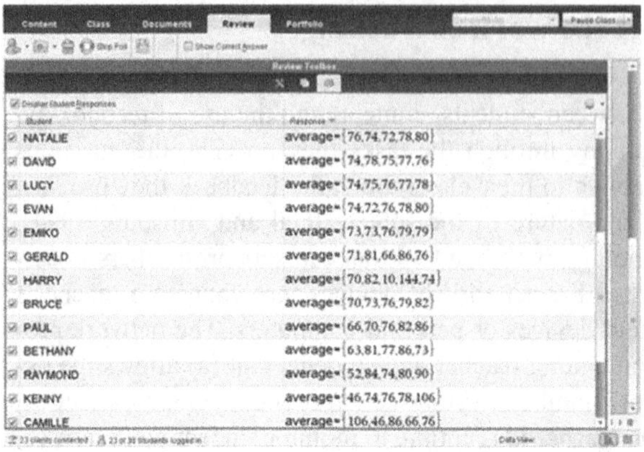

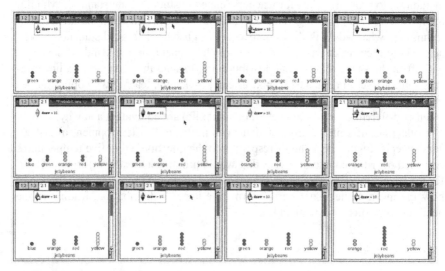

Fig. 12.4 Data collected from students: five temperatures with mean of 76°

Simultaneously capturing student screens can provide a window into variability and distributions of data. While an individual student can create multiple simulations one at a time, being able to see the results across a class of students at the same time allows students to look for similarities and differences and discuss these as a class and enables the teacher to raise important statistical questions. In real time, a student can become a live presenter and demonstrate how they generated their distribution. The figures 12.5 and 12.6 provide examples of the opportunity to discuss the role of sample size (Figure 12.5) and the shape and characteristics of a sampling distribution of sample means (Figure 12.6).

Fig. 12.5 Random samples of 10 jellybeans

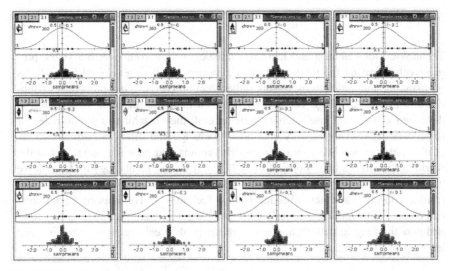

Fig. 12.6 Sampling distributions of 300 sample means

Screen captures show different approaches students use to represent a data set; discussion can highlight how the representations differ and what can be learned from each. While this is possible without CCT, visualizing a variety of representations simultaneously adds a new dimension. The same argument can be made about association and modeling; having students simultaneously display different scatterplots of bivariate data with nearly the same correlation can make visible the limitations of a correlation coefficient. Theoretical probability models can be connected to empirical data. Giving students a population and asking one group of students to display a plot of the population, another a random sample of a given size from that population, and a third group a sampling distribution of a sample statistic from that population can reveal whether students understand the difference in the distributions; the individual screens can be grouped according to the categories, and the discussion can focus on the differences and why these exist. Simultaneously examining simulated sampling distributions to see whether an observed outcome is likely to have occurred by chance can make visible the meaning of a *p*-value or reinforce the concept of confidence levels.

Evidence is emerging that CCT can make a difference in student achievement (Pape et al. 2012; Clark-Wilson 2010; Slovin and Olson 2010). While few if any studies have focused on the use of CCT in teaching statistics, research does suggest that formative assessment makes a difference (Black and Wiliam 1998; Leahy et al. 2005). CCT provides real time information on student understanding that teachers can use to shape their practice.

6 Conclusion

Technology for teaching and learning statistics has evolved over the years, progressively allowing the work to shift to a higher cognitive level enabling a focus on planning and anticipating results rather than on carrying out procedures (Ben-Zvi and Friedland 1997). Technology can also facilitate the work of teaching as described above, enabling teachers to better craft their teaching to promote deeper understanding. But technology continues to change and offers opportunity to rethink what and how we operate in our classrooms and how that is related to the world outside of the classrooms (Gould 2011). If we use technology to do what we have been doing, we will get the same results (Ehrmann 1995). "To get different results, you must add new thinking to new technology. The reason: Technology empowers. But thinking enables" (Moore 1998). The challenge is to continue to add new thinking to new technology with the goal of enhancing the teaching and learning of statistics.

References

Ben-Zvi, D., & Friedlander, A. (1997). Statistical thinking in a technological environment. In J. B. Garfield, & G. Burrill (Eds.), *Research on the role of technology in teaching and learning statistics* (pp. 45-55). Voorburg, The Netherlands: International Statistical Institute.

Ben-Zvi, D. (2000). Towards understanding the role of technological tools in statistical learning. *Mathematical Thinking and Learning, 2*(1&2), 127–155.

Ben-Zvi, D. (2004). Reasoning about variability in comparing distributions. *Statistics Education Research Journal, 3*(2), 42-63.

Biehler, R. (1982). *Explorative Datenanalyse - Eine Untersuchung aus der Perspektive einer deskriptiv - empirischen Wissenschaftstheorie.* IDM Materialien und Studien 24 (Exploratory data analysis - an analysis from the perspective of a descriptive-empirical epistemology of science). Bielefeld: Universität Bielefeld, Institut für Didaktik der Mathematik.

Biehler, R. (1989). Educational perspectives on exploratory data analysis. In R. Morris (Ed.), *Studies in mathematics education: Teaching of statistics* 7 (pp. 185-201). Paris: United Nations Education, Scientific and Cultural Organization (UNESCO).

Biehler, R. (1991). Computers in probability education. In R. Kapadia, & M. Borovcnik (Eds.), *Chance encounters: Probability in education* (pp.169–211). Amsterdam: Kluwer.

Biehler, R. (1993). Software tools and mathematics education: The case of statistics. In C. Keitel, & K. Ruthven (Eds.), *Learning from computers: Mathematics education and technology* (pp. 68-100). Berlin: Springer.

Biehler, R. (1994). Probabilistic thinking, statistical reasoning, and the search for causes: Do we need a probabilistic revolution after we have taught data analysis? In J. B. Garfield (Ed.), *Research papers from the Fourth International Conference on Teaching Statistics* (pp. 20-37). Minneapolis MN: The International Study Group for Research on Learning Probability and Statistics.

Biehler, R. (1995). *Toward requirements for more adequate software tools that support both: Learning and doing statistics.* Revised paper presented at Fourth International Conference on Teaching Statistics, Marrakech, Morocco, 1994. IDM Occasional Paper 157. Bielefeld: Universität Bielefeld, Institut für Didaktik der Mathematik.

Biehler, R. (1997a). Students' difficulties in practicing computer-supported data analysis: Some hypothetical generalizations from results of two exploratory studies. In J. B. Garfield, & G. Burrill (Eds.), *Research on the role of technology in teaching and learning statistics* (pp. 176-197). Voorburg, The Netherlands: International Statistical Institute.

Biehler, R. (1997b). Software for learning and for doing statistics. *International Statistical Review, 65*(2), 167-189.

Biehler, R. (1999). Discussion: Learning to think statistically and to cope with variation. *International Statistical Review, 67*(3), 259–262.

Biehler, R. (2003). Interrelated learning and working environments for supporting the use of computer tools in introductory courses. In J. Engel (Ed.), *Proceedings of International Association for Statistics Education (IASE) Satellite Conference on Teaching Statistics and the Internet* [CD-ROM]. Berlin: Max-Planck-Institute for Human Development.

Black, P., & Wiliam, D. (1998). Inside the black box: Raising standards through classroom assessment. *Phi Delta Kappan, 80*(2), 139-148.

Burrill, G., & Biehler, R. (2011). Fundamental statistical ideas in the school curriculum and in training teachers. In C. Batanero, G. Burrill, & C. Reading (Eds.), *Teaching statistics in school mathematics - Challenges for teaching and teacher education: A joint ICMI/IASE Study* (pp. 57-69). New York: Springer.

Chance, B., Ben-Zvi, D., Garfield, J., & Medina, E. (2007). The role of technology in improving student learning of statistics. *Technology Innovations in Statistics Education Journal 1*(1). Retrieved from http://www.escholarship.org/uc/item/8sd2t4rr.

Clark-Wilson, A. (2010). Emergent pedagogies and the changing role of the teacher in the TI-Nspire Navigator-networked mathematics classroom. *ZDM, 42*(7), 747-761.

Cobb, G. (1992). Teaching statistics. In L. A. Steen (Ed.), *Heeding the call for change: Suggestions for curricular action* (MAA Notes No. 22, pp. 3-43). Washington, DC: Mathematical Association of America.

delMas, R., Garfield, J., & Chance, B. (1999). A model of classroom research in action: Developing simulation activities to improve students' statistical reasoning. *Journal of Statistics Education, 7*(3). Retrieved from http://www.amstat.org/publications/jse/secure/v7n3/delmas.cfm.

Dick, T., & Burrill, G. (2008). Technology and instruction: A focus on inquiry-based learning. Paper presented at the annual meeting of the Association of Mathematics Teacher Educators. Tulsa, OK.

Ehrmann, S. C. (1995), Asking the right questions: What does research tell us about technology and higher learning? *Change, 27*(2), 20 –27.

Franklin, C., Kader, G., Mewborn, D., Moreno, J., Peck, R., Perry, M., & Scheaffer, R. (2007). *Guidelines for assessment and instruction in statistics education (GAISE) report: A preK–12 curriculum framework.* Alexandria, VA: American Statistical Association. Retrieved from http://www.amstat.org.

Finzer, W. (2001). *Fathom dynamic statistics^{TM} software.* Key Curriculum Press.

Gal, I. & Ginsburg, L. (1994). The role of beliefs and attitudes in learning statistics: Towards an assessment framework. *Journal of Statistics Education, 2*(2), 1-15.

Gould, R. (2011). Statistics and the modern student. Department of statistics papers. Department of Statistics, University of California Los Angeles,

Hawkins, A. (1997). Myth-conceptions! In J. Garfield, & G. Burrill (Eds.), *Research on the role of technology in teaching and learning statistics. Proceedings of the 1996 IASE Round Table Conference* (pp. 1-14). Voorburg, The Netherlands: International Statistical Institute.

Jones, G., Langrall C., & Mooney E. (2007). Research in probability: Responding to classroom realities. In F. Lester (Ed.), *The second handbook of research on mathematics* (pp. 909–956). Reston, VA: National Council of Teachers of Mathematics.

Leahy, S., Lyon, C., Thompson, M., & Wiliam, D. (2005). Classroom assessment: Minute by minute, day by day. *Educational Leadership, 63*(3), 19-24.

Maxara, C., & Biehler, R. (2006). Students' probabilistic simulation and modeling competence after a computer-intensive elementary course in statistics and probability. In A. Rossman, &

B. Chance (Eds.), *Working Cooperatively in Statistics Education. 7th International Conference on Teaching Statistics.* Salvador da Bahia, Brazil, 2006 (pp. 1–6). Retrieved from www.stat.auckland.ac.nz/~iase/publications/17/7C1_MAXA.pdf.

Moore, D. (1997). New pedagogy and new content: The case of statistics. *International Statistical Review, 65*(2), 123–165.

Moore, D. (1998). Statistics among the liberal arts. *Journal of the American Statistical Association, 93*(444), 1253-1259.

Pape, S., Irving, K., Abrahamson, A., Owens, D., Silver, D., & Sanalon, V. (2012). The impact of classroom connectivity in promoting Algebra I achievement: Results of a randomized control trial. Manuscript submitted for publication.

Pratt, D., Davies, N., & Connor, D. (2011). The role of technology in teaching and learning statistics. In C. Batanero, G. Burrill, & C. Reading (Eds.), *Teaching statistics in school mathematics - Challenges for teaching and teacher education: A joint ICMI/IASE Study* (pp. 57-69). New York: Springer.

Scheaffer, R. (1986). The Quantitative Literacy Project. *Teaching Statistics, 8*(2), 34-38.

Slovin, H., & Olson, M. (2010). Changing the mathematics teaching and learning environment through the use of networked technology. Paper presented at Hawai'i International Conference on Education. Honolulu, HI.

Tukey, J. W. (1977). *Exploratory data analysis.* Addison-Wesley, Reading, MA.

Tukey, J. (1984, July). Paper presented at Woodrow Wilson Leadership Program for Teachers. Princeton, NJ.

Velleman, P., & Hoaglin, D. (1992). Data analysis. In D. Hoaglin, & D. Moore (Eds.), *Perspectives on contemporary statistics* (pp. 19–39). Washington, D.C.: Mathematical Association of America.

Wild, C., & Pfannkuch, M. (1999). Statistical thinking in empirical enquiry. *International Statistical Review, 67*(3), 223-265.

Wild, C. (2005). A statistician's view on the concept of distribution. In K. Makar (Ed.), *Reasoning about distribution: A collection of current research studies. Proceedings of the Fourth International Research Forum on Statistical Reasoning, Thinking and Literacy (SRTL-4)* [CD-ROM]. Brisbane, Australia: University of Queensland.

Chapter 13
Chance *Re*-encounters: 'Computers in Probability Education' Revisited

Dave Pratt

Institute of Education, University of London

Janet Ainley

School of Education, University of Leicester

Abstract In recognition of Rolf Biehler's contribution to probability and statistics education, we chose to re-visit his chapter in the edited book, Chance Encounters: Probability in Education. In particular, we examine three themes concerning the concept-tool gap, levels of access to concepts, which can be concealed by technology, and issues around demonstration and proof. We found many insights that resonate with current practice two decades later. Nevertheless, we argue that during that time, some progress has been made in how we can conceptualise the issues. For example, we discuss: (i) how it is now possible to unpick the metaphorical understanding that could emerge from the use of black boxes by reference to utility-based understanding; (ii) four principles that could inform how black boxes might be designed to support utility-based understanding; (iii) how the importance of explanation may overshadow a more traditional emphasis on proof.

Introduction

In 1991, Kapadia and Borovcnik invited leading researchers at the time to discuss issues on probability education. As a result, the edited book, *Chance Encounters: Probability in Education,* set out a critical review of the state of research in this field in 1991. In a sense the book demarcates an era dominated by the seminal research of psychologists such as Piaget, Fischbein and Kahneman and Tversky from a modern era in which educationalists in mathematics and statistics education have sought to resolve the many issues raised in the earlier work by conducting experimental studies to further theoretical development on how children learn to deal with uncertainty and how tools might be designed to support that learning.

In this chapter, we focus on Chapter 6, *Computers in Probability Education*, in the original book (Biehler 1991). Rolf Biehler reviewed the opportunities and constraints being offered by what was then a newly emerging technology. At the time of writing the book, Rolf was in his late 30s and now two decades later, it seems timely to revisit his text to consider whether the issues he raised have been resolved or at least come to be better understood, whether they have simply disappeared as the technology itself has improved, and indeed whether new opportunities and constraints have since emerged.

It is worth pausing to remind the reader about the context in 1991. We were early career researchers, studying the potential for using the latest digital tool, the laptop, in primary schools. Secondary schools in the UK possessed at best a room of computers, based on idiosyncratic operating systems, being increasingly used by departments other than mathematics for subjects like Computer Studies. Primary schools might have a single machine in each classroom, but with only a small monitor for use by individuals or pairs. Other than in unusual projects like our own, there was no handheld technology in schools apart from calculators; mobile phones were a distant dream (or nightmare for some of us). The internet was just beginning to emerge. It was not in use in homes or schools but university departments would have individual machines that could access 'JANET', the joint academic network.

This background context is very important as it would be easy to imagine that technological development since then has been so fundamental that little that was said about computers in probability education in 1991 could possibly be relevant today. Such a view would be far from the truth. As we will see, there remains a strong resonance between Biehler's analysis and the current situation, though our re-analysis perhaps also serves to emphasise where progress has been made.

Our approach to this re-analysis will not be exhaustive. We do not intend to work systematically through the many aspects of the original text, space would not allow such a method, but we do invite the reader to re-visit for themselves the original chapter. Instead, we plan to consider three themes, which emerged for us in our reading of the text. First, we will focus more explicitly on design by discussing the concept-tool gap, a construct that pervades the original Biehler chapter. Second, we will debate the extent to which technology hides or exposes underpinning statistical ideas by considering how design directs levels of access. It will be evident that the nature of the tools and indeed the tasks to which those tools are deployed provides a running theme throughout this chapter. Finally we will consider how demonstration, explanation and proof are played out when technology is used to support probabilistic learning. Inevitably such a discussion must be nuanced by considerations of design.

Design and the concept-tool gap

Biehler refers to the 'concept-tool' gap to articulate a concern that the meanings for uncertainty held by students are not well integrated with the performance demands made in curricula and examination syllabi. For young students, probabilistic algebra is still most often learned as a set of rules to be followed in order to get the right answer, leaving the concepts themselves distant and meaningless. The gap to which Biehler refers is, in this instance, between the concepts, such as independence, sample space and the Law of Large Numbers, and their use as tools to solve set problems. Later, students learn about hypothesis testing and confidence intervals but in most cases the learning is restricted to the application of little understood algorithms. Biehler raised this issue in 1991 but unfortunately the concept-tool gap continues, live and kicking, today. Although it is certainly not an issue only for probability education (consider, for example, the teaching and learning of algebra), Biehler was correct to identify this as a problem with particular relevance to probability education, when so often students and teachers conspire to answer examination questions in probability using methods that are understood only in instrumental ways.

Even in those early days, when computers were scarcely evident in schools, Biehler saw the special affordances that might mean the new technology could offer the potential to close the concept-tool gap. In particular, he listed three (p. 188):

1. the number of repetitions is easily increased so that uncertainty and variation in the results can be reduced; new kinds of patterns become detectable,
2. an extensive exploration is possible by changing the assumptions of the model, making further experiments, changing the way generated data are analysed etc.,
3. new and more flexible representations are available to express models and stochastic processes and display data with graphical facilities.

This list stands up to criticism even today. We might, for example, evaluate this list against the design of perhaps the most recent and innovative of computer applications in the field of probability education, *Tinkerplots 2*. In the earlier version of this toolkit, students were able to organize and graph data in largely intuitive ways. Tinkerplots 2 allows students additionally to model uncertain phenomena using probabilities represented as samplers, which can range from simple urn models to histograms and even to hand-drawn probability density functions. When a model has been created by the student, it can be used to generate data and analysed using the original Tinkerplots tools.

The example in Figure 13.1, taken from Konold, Harradine and Kazak (2007), uses urns and spinners to create a machine that could generate the varied attributes of cats. The likelihood of the two genders is set to be equal as are the probabilities of the three eye colours envisaged by the children. The various lengths of cats were represented as balls in an urn so that more common lengths could easily be modelled by the addition of extra balls with that length; older students might have used a sampler that would allow a continuous variable. Once the model has been created, the children are able to generate many cats with varying attributes.

The toolkit makes full use of affordance (1) above, in that it is very easy for students to generate as much data from their model as they like. Thus, small and large data sets can be compared. As imagined by Biehler, students can compensate for the lack of experience in the material world by generating extensive data in the virtual world of Tinkerplots 2. Furthermore, the use of different samplers whose parameters can easily be changed and methods of analysis whether based on collections of individual cases or grouped data exploits affordance (2) above. In response to affordance (3), Cliff Konold who leads the Tinkerplots development team not only allows students to use conventional representations of data, such as box, dot and scatter plots but also introduced novel representations in the light of research which suggests younger students might need transitional tools for organising and representing data prior to being able to intuitively manage the conventional approaches.

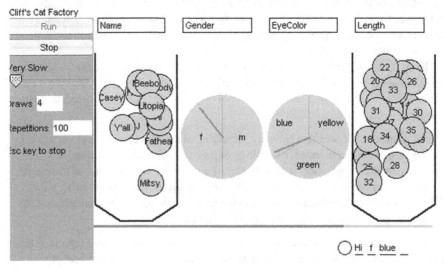

Fig. 13.1 A machine designed for making cats with various attributes

Biehler imagined that simulations of this type might enable students to close the concept-tool gap by making theoretical objects experiential. In a recent inaugural lecture, Pratt (2012) referred to 'making mathematics phenomenal' through the creation of on-screen objects and tools that can be explored and manipulated, showing the currency even now of Biehler's early insight. Indeed, both Biehler's and Pratt's ideas resonate with the constructionist movement (Harel and Papert 1991), which advocates the building of public entities by students to facilitate mathematizing. Papert's *power principle* (1996) argues that such an approach allows students to learn mathematics in a more natural way since they are able to draw on experience to construct mathematical meaning, which he sees as a parallel process to how humans most often learn outside of school mathematics. The constructionist vision is seen by Pratt (2012) as a challenge to design environments that facilitate making mathematics phenomenal and many of his examples come from probability education. For example, in building ChanceMaker, a domain of stochastic abstraction designed to research young children's meanings for chance, distribution and the Law of Large Numbers (Pratt and Noss 2002; 2010), he challenged 10-11 year olds to identify which of a range of virtual random generators were 'working properly'. The students were able to mend any so-called *gadget*, which in their view was not working properly, in particular by editing its workings box, an unconventional urn-type representation of the probability distribution. By editing the gadget, and reviewing the feedback in the form of the animation of the gadget, lists of previous results and charts aggregating those results, the students were able to learn *through use* about the nature of short-term and long-term randomness and come to situated understandings of the Law of Large Numbers and distribution.

Although, perhaps surprisingly, there is no mention of Papert's ideas in the 1991 text, there is direct reference to diSessa's (1986) work in mechanics. Biehler points out diSessa's defence against criticism that his Dynaturtle was not providing a 'real' experience of Newton's Laws. diSessa's counter-argument was that Newton's Laws are themselves a human construction; physics is not a direct perception but an intellectual abstraction, achieved over prolonged periods. The Dynaturtle offered an opportunity to construct Newton's Laws exactly because of its difference from the material world. It is often argued as criticism against virtual environments such as ChanceMaker that randomness on a computer is not the same as randomness as experienced through material objects like coins, spinners and dice. This is true but, because of the affordances identified by Biehler, virtual environments can offer experiences that facilitate the construction of powerful ideas like randomness more effectively than experience in the material world alone can afford.

In more recent years, Biehler and co-workers have designed integrated programmes of learning, built around simulation, in an attempt to bridge the gap (Batanero et al. 2005; Biehler and Hoffman 2011). Ainley, Pratt and Hansen (2006) have also looked at the concept-tool gap from the point of view of task design. In a sense, computer environments are only as good as the tasks to which they are deployed and so many of the design arguments that apply to the provision of tools also apply to how teachers and researchers choose and design tasks. They place some emphasis on the importance of designing the task so that it is seen as purposeful by the students. Many of the examples above would fit this design criterion but sadly many other tasks to which computer tools are deployed are no more engaging than their equivalents in traditional textbooks.

Ainley et al. would see the consideration of purpose as only part of the task design: the task must be directed towards the learning of mathematical or statistical concepts. Tasks that are purposeful may misdirect attention and can easily distract students from mathematics. They coined the term *utility* to describe the important element of task design that aims to focus the students' attention on the power of mathematical ideas to get stuff done. This type of learning is importantly different from tasks that focus on techniques, such as the instrumental learning of the Laws of Probability. It is also different from the type of relational understanding that enables students to appreciate the logic of why the mathematics works. To use a metaphor about car engines, utility would not relate to knowing precisely which controls to use, and in which order, to start the engine; neither would utility relate to understanding the engineering science of how an engine works. Utility would consist in an appreciation how useful an engine is in moving a car, which has obvious value in living one's everyday and professional life.

In some respects this example from everyday life appears ridiculous. The utility of an engine and the car it powers seems patently obvious: no one would embark on learning to drive without appreciating this utility. Ainley et al. argue that it is precisely the contrast between a real life example and the situation in mathematics and statistics classrooms, where students are generally required to learn to operate new tools before having any sense of their utility, which makes utility such an important idea. If mathematical and statistical ideas were experienced through purposeful tasks that made the utility transparent, their power might become rather more evident to students than is currently the case, and the concept-tool gap, if not closed, might be bridged in a different way.

However, this leads us in our re-analysis to a point of departure from the original text. Biehler imagined that simulation could lead to a closing of the concept-tool gap. We would now argue that this is over-simplistic. Simulations can provide purposeful tasks but students using tools such as Konold's Tinkerplots, diSessa's Dynaturtle or Pratt's ChanceMaker will learn about the scope and power of the statistical, mechanical or probabilistic concepts. Such knowledge is very important but is fundamentally different from the demands made by curricula and syllabi, which, we might argue, tend to heighten the concept-tool gap. It is a research question whether students with an appreciation of the power of probability will be better prepared to learn the formal knowledge and then grapple more effectively with the problems set in examination based on these curricula and syllabi. We would conjecture that this would indeed be the case but meanwhile the concept-tool gap remains largely inviolate.

Levels of access

Biehler (1991) in fact suggested that simulations could address the concept-tool gap by conceptualizing the simulation as a black box; "It may be possible that a metaphorical understanding of a 'black box' can guide a reasonable application" (p. 180). He made this argument in response to common criticism that concealing the computational algorithms relies on students accepting the outputs without knowing the underlying formulas, particularly common when people use statistical packages without understanding.

Indeed, the tendency for technology to hide the underlying mechanism is now common practice outside the classroom. Noss (1997) used his inaugural lecture to emphasise that technology often makes the mathematics that drives so much of everyday culture invisible, and that this trend demands the invention of new numeracies. A simple example takes place every minute of the day in the supermarket, when the customer goes through the checkout without either the shopper or the assistant needing to do any calculation. Not only does the computer total the cost of the goods purchased (and keep a stock check) but also the shopper hands over a credit card and no material exchange of notes or coins occurs. It is common for mathematics teachers to claim enthusiastically that mathematics is everywhere but, although more and more mathematics underpins the functions of everyday technology such as intelligent tills, never has this been less obvious to younger students, and indeed to the general public.

It is therefore an argument worthy of consideration that students cannot possibly come to appreciate the power of mathematics if the mathematics classroom mirrors the tendency to hide the algorithms, first through the use of calculators and second by wrapping the mathematics up in impenetrable black boxes. Nevertheless, in 1991, Biehler envisaged that through the use of, for example, a program that calculates Binomial probabilities, students might come to apply a metaphorical understanding without detailed knowledge of the formula.

What did Biehler mean by a 'metaphorical understanding'? This is not clear but we would argue that students who were using the imagined Binomial program in this way might gain a sense of the scope and limitations of the Binomial Distribution. By playing with the program, they might never learn the formula, since that is concealed within the black box, but they might come to appreciate when the Binomial distribution is a useful tool and when it is not. We see a parallel here with the ideas of Meira (1998) who uses the idea of *transparency* (developed from the work of Lave and Wenger, 1991) as 'an index of access to knowledge and activities'. Meira contrasts a view of transparency, or lack of transparency, being inherent in a tool or device, with one of transparency emerging through use:

> "... the transparency of devices follows from the very process of using them. That is, the transparency of the device emerges anew in every specific context and is created during the activity through specific forms of using the device." (Meira 1998, p. 138)

Biehler's 'metaphorical understanding' seems to be close to this notion of transparency. This is precisely the sort of knowledge and learning that Ainley et al. (2006) intended when they referred to utility-based understanding. Utility-based understanding will not emerge automatically and a good deal of careful design is needed to ensure that the tools and tasks lead to such an appreciation. Biehler of course referred to the Binomial program as an instance of a set of black boxes and indeed we also are not making a specific claim about the Binomial Distribution; on the contrary, we make a quite generic connection with utility-based understanding.

When considering the design of tools, one might ask what design decisions might be taken to avoid concealing the algorithms and instead foster utility-based understanding of probability? This question brings the reader explicitly to the issue of levels of access, headlining this subsection. We identify four elements of design, though implementation of any one in isolation is unlikely to be effective:

1. The use of evocative imagery that iconically carries a key aspect of the probabilistic concept;
2. The possibility of manipulating parameters of the simulation and receiving feedback that facilitates understanding of the behaviour of the probabilistic concept as instantiated on-screen;

3. Semantic layers that can be opened by the student who wishes to dig out a deeper meaning;
4. A blurring of control and representation.

An example of (1) above can be found in the innovative developments by Chris Wild's team to support the understanding of core concepts in statistical inference (http://www.stat.auckland.ac.nz/~wild/VIT/index.html). Their visual inference tools focus on how animation can be deployed to give a sense of variation. For example, a statistic such as sample mean can be plotted as a short vertical segment on a horizontal axis during continuous re-sampling. When the trace of the sample mean is captured, the effect is that gradually a box appears as the sample mean varies to the left and right, taking smaller and larger values. Some values of the sample mean occur more often and so there is increased intensity towards the centre. The vibrations of the box give an ongoing sense of the variation and yet the signal and noise apparent in the sampling distribution of the sample mean is visible.

In fact one can imagine a student altering the size of the samples being taken and re-taken and beginning to notice a relationship between the sample size and the width of the animated box that emerges. This is an example of (2) above. The facility to be able to manipulate the parameters in the black box is a minimal requirement if the level of access to the utility of the probabilistic concept is not to remain superficial. Above, Pratt's work with the ChanceMaker tool demonstrated how the students could change parameters such as the number of throws of the gadget and begin to appreciate from the graphical feedback that 'the more times you throw the die, the more even is its pie chart' (assuming of course that the probabilities of each outcome are in fact equal).

ChanceMaker allows the student to open up the gadget and change the workings (in other words the probability distribution). In one sense this is an extended version of (2) since it is possible to think of the workings box as a parameter. However, editing the workings box is a significantly more substantial act than simply changing a parameter. In that sense, deeper meanings may become apparent and so perhaps this is a simple example of (3) above. The design challenge here is to balance access with expressability. Wild's work with VIT is easy to use (though the underpinning ideas are far from simple) but students may be hindered as they have few opportunities to express their own ideas, a key design feature from the Constructionist perspective. ChanceMaker perhaps offers more opportunity for expressing personal conjectures about how the gadget should work through the challenge to mend the gadget, in particular by editing the working box.

A far more ambitious project to make available access to layer upon layer of meaning was diSessa's Boxer project (http://soe.berkeley.edu/boxer/). Boxer is a programming language, related to Logo but as part of a project aimed at developing an all-inclusive medium for computational literacy. The relevant feature here is how Boxer provides closets that by default are closed. Within these closets can be objects or pieces of code that may be best hidden from the naïve learner at the outset but may be of interest later. Since closets can themselves contain closets, there is potential for digging more and more deeply into the semantic layers of the concept. Although Boxer is not used widely, it is perhaps the tool which has taken most seriously the notion that the black box could be opened at the discretion of the user to uncover new layers of meaning. Biehler originally hinted that black boxes could, despite hiding underlying algorithms, prove useful to students who need to capture a metaphorical understanding. In trying to unpick that metaphor, we posit as one of the four design elements the notion that black boxes do not need to be so black but, aside from Boxer, few developments in this direction have yet materialised.

In fact, the Boxer closets could be opened to reveal the underpinning algorithms, offering a possibility of closing the concept-tool gap by, not only supporting utility-based understanding but beginning to appreciate the formulas themselves. Often principles in the design of tools can also inform task design. For example, a teacher might design a task involving measuring the time of flight of a sycamore seed dropped from different heights as part of an exploratory investigation. Several measurements of time might be taken for the same height and inserted into a spreadsheet. It would then be natural perhaps for the teacher to introduce the child to the 'average' function as a black box so that a better estimate could be calculated from a number of measurements. Inquisitive children might ask the teacher what the function does. A well-prepared teacher may have anticipated such a scenario and may have materials on hand that explain the mean average, and so open up the black box, either by exploring more systematically how average behaves for different number sets or by introducing methods of calculation.

When diSessa (1988) refers to integrating the formal and the informal, he has in mind, for example, that, in a computational medium such as Boxer, it is possible to execute mathematical formalisms within a generally expressive and creative environment. Two limited examples are: (i) the word processor, where it is possible to click a url embedded in the text and gain immediate on-line access to the internet; (ii) in Mathematica (http://www.wolfram.com/mathematica/), algebraic text can be executed to create simplifications and computations of that text. More generally, expressive environments that allow textual and graphical creation can also integrate mathematical and statistical formalisms so that the informal creative process, expressed through text and graphics, can be supported by the execution of formal algorithms. When such expressive environments contain tools and structures that behave in ways that reflect underlying mathematical or statistical principles, the tools are described by Noss and Hoyles (1996) as *autoexpressive.*

In (4) above, we take the notion of autoexpressive tools one stage further by operationalising this with the designer in mind. We suggest that a particularly felicitous way of integrating the formal and the informal is to blur control and representation. For example, in Logo, the young child controls the turtle by issuing commands such as fd 50. The command is both a means of control and a representation of how the turtle will move. In ChanceMaker, the workings box controls the behaviour of the gadget but, insofar as students learn to predict what will happen by inspecting the workings box, it becomes an unconventional representation of probability distribution. In both these examples, a key mathematical idea (linear distance; probability distribution) is placed exactly at the point of control so that the student can scarcely avoid engaging with the representation and gradually become aware of its meaning through use. These tools are indeed autoexpressive but to implement that the designer can privilege key representations by making them points of control.

By bridging the formal and informal in this way, the black box becomes rather more meaningful, especially in relation to utility-based understanding of the concept. These four design elements, especially if designed to work together, might begin to describe how the design of tools can create black boxes whose use could support metaphorical, or utility-based understanding, of probabilistic (and other) concepts.

Demonstration, explanation and proof

Biehler describes computers as supporting a new type of scientific method, which is characterised by an experimental style of working with models and data, and suggests that in such a method proofs may be valued to the extent that they offer explanations (p. 172). We understand this to mean that just as formulae may become black boxes when used instrumentally, so too may proofs, however rigorous, which do not support ways of understanding phenomena. In statistics, proof is evidence-based in contrast to mathematics where theorems are established logically from previously proven theorems. Nevertheless, a hypothesis test might be taken as a proof in its everyday sense but the test scarcely explains when the notion of a hypothesis test is obscure, the mathematical basis for the specific test used is not understood and when the procedure has been applied instrumentally. On the other hand, EDA techniques might enable re-presentation of the data that explains the basis of an inference.

We see Biehler's claim about an experimental style of working as closely linked to the ways in which computers support working with graphs and other visual images. Biehler himself provides a beautiful example of a visual simulation to solve the 'Abel and Kain' problem, which not only offers a way to calculate the relative probabilities of the two outcomes (1111, 0011) but also an explanation of why one is more likely than the other (p. 184).

The importance of explaining phenomena, rather than relying of the application of 'black box' formulae or proofs, is particularly pertinent in probability where intuitions play a significant role in thinking. Indeed 'Abel and Kain' only works as a problem because intuition might suggest that the two outcomes are, in fact, equally likely. An approach based on the application of formulae may produce an answer, but unless this solution explains the situation, intuitions are likely to be left untouched.

In the real world, the consequences of relying on intuitive understanding of probabilities may be problematic. Unfortunately Biehler's idea that computers may support an increased focus on subjectivist aspects of probability has not been widely realised, and school curricula still focus largely on coins and dice. An exception is the work of Pratt et al. (2011) on the design of computer environments to support learning and teaching about risk. They developed a tool to research mathematics and science teachers' knowledge about risk. They propose a scenario in which a fictitious young woman, Deborah, has a back condition, which might be cured through an operation. Deborah's dilemma lies in the fact that the operation could result in side-effects, some of which are severe. She needs to balance in some way the positive and negative outcomes with their likelihoods. The probabilities need to be estimated subjectively because the data that are provided give some indication of the likelihoods of the various outcomes but there are contradictions and discrepancies in the information. The teachers are encouraged to model what might happen if Deborah were to have the operation and what might happen if she did not. By running the model, the teachers are able to witness her possible futures and hence gain feedback on their subjective estimates and evidence about how the dilemma might be resolved.

In the years since Biehler's chapter was written, EDA has become a much more widespread pedagogical approach, supported by the substantial development in appropriate software, such as Tinkerplots and Fathom. The EDA approach foregrounds the use of visual methods to explore and identify patterns in data, though it is less clear the extent to which this encourages a search for explanation, rather than focussing on observing relationships. An issue that is becoming more widely recognised in statistics education (Pratt et al. 2008) is the potential for confusion between exploration of a set of data, which is *the whole population*, and exploration of a set of data, which is *a sample from which information about the population can be inferred*. Whilst it may be clear to the teacher that a task involves the second of these situations, careful design is required to produce tasks in which it is very clear to pupils that this is the case. The nature of what might count as proof, or convincing explanation will differ considerably in the two situations: if I am exploring the whole population I can make claims with a level of certainty which is inappropriate if I am working with a sample. Since the tools, both conceptual and technical, that are available can be applied in the same way to either situation, task design is crucial to enabling students to understand the importance of inference.

Conclusion

In re-visiting this chapter we have been struck many times by Rolf Biehler's vision in anticipating future possibilities offered by technology in the field of probability, and indeed in statistics and mathematics education more generally. For example, Biehler's list of affordances that position technology as being especially felicitous in closing the concept-tool gap is identifiable even in the most recent of statistics educational software developments. In particular, the notions that technology can make theoretical objects experiential and that virtual experience can enhance everyday experience still seem very current. Progress has been made on these notions so that the constructs of purpose and utility describe how such an emphasis generates a different sort of understanding from that typically demanded by curricula and syllabi. As a result the gap remains apparent and it may need further work on how to design levels of access, perhaps making use of how to blur control and representation, before those demands can be met in a pedagogically sound way. Some areas anticipated in Biehler's original text have not yet been developed so that, for example, there remains little evidence of the subjective use of probabilities being taught in schools.

There is also a realisation of how uneven and inconsistent the exploitation of that potential has been. In parallel to the development of flexible and creative applications such as Tinkerplots, much of the commercially produced software available to schools embodies approaches that prioritise technique and right answers above deep understanding. Technology which is hugely more sophisticated than that available in 1991 is often used for display and demonstration, rather than for developing the experimental style of working that Biehler envisaged. We would argue that in this context issues of design, of both environments and tasks, are as important as ever.

References

Ainley, J., Pratt, D., & Hansen, A. (2006). Connecting engagement and focus in pedagogic task design. *British Educational Research Journal. 32*(1), 23-38.

Batanero C., Biehler R., Maxara C., Engel J., & Vogel M. (2005). Using simulation to bridge teachers' content and pedagogical knowledge in probability. In *15th ICMI Study Conference: The Professional education and development of teachers of mathematics*. Retrieved from http://stwww.weizmann.ac.il/G-math/ICMI/strand1.html.

Biehler, R. (1991). Computers in probability education. In R. Kapadia, & M. Borovcnik (Eds.), *Chance encounters: Probability in education* (pp. 169-211). Dordrecht, The Netherlands: Kluwer Academic Publishers.

Biehler, R., & Hofmann, T. (2011, August). *Designing and evaluating an e-learning environment for supporting students' problem-oriented use of statistical tool software*. Paper presented at the 58th ISI Session, Dublin, Ireland. Retrieved from http://isi2011.congressplanner.eu/pdfs/450437.pdf.

diSessa, A. (1986). Artificial worlds and real experience. *Instructional Science, 14*(3-4), 207-227.

diSessa, A. (1988). Knowledge in pieces. In G. Forman & P. Pufall (Eds.), *Constructivism in the computer age* (pp. 49-70). Hillsdale, New Jersey: Lawrence Erlbaum Associates.

Harel, I., & Papert, S. (1991). *Constructionism*. Norwood, New Jersey: Ablex.

Konold, C., Harradine, A., & Kazak, S. (2007). Understanding distributions by modeling them. *International Journal of Computers for Mathematical Learning, 12*(3), 217-230.

Kapadia, R. & Borovcnik, M. (Eds). (1991). *Chance encounters: Probability in education*. Dordrecht, The Netherlands: Kluwer Academic Publishers.

Lave, J., & Wenger, E. (1991). *Situated learning: Legitimate peripheral participation*. Cambridge: Cambridge University Press.

Meira, L. (1998). Making sense of instructional devices: The emergence of transparency in mathematical activity. *Journal for Research in Mathematics Education, 29*(2), 121-142.

Noss, R. (1997). *New cultures, new numeracies,* London: Institute of Education, University of London.

Noss, R., & Hoyles, C. (1996). *Windows on mathematical meanings: Learning cultures and computers.* London: Kluwer Academic Publishers.

Papert, S. (1996). An exploration in the space of mathematics educations. *International Journal of Computers for Mathematical Learning, 1*(1), 95-123.

Pratt, D. (2012). *Making mathematics phenomenal,* London: Institute of Education, University of London.

Pratt, D., Ainley, J., Kent, P., Levinson, R., Yogui, C., & Kapadia, R. (2011). Role of context in risk-based reasoning. *Mathematical Thinking and Learning, 13*(4), 322-345.

Pratt, D., Johnston-Wilder, P., Ainley, J., & Mason, J. (2008). Local and global thinking in statistical inference. *Statistics Education Research Journal, 7*(2), 107-129.

Pratt, D., & Noss, R. (2002). The micro-evolution of mathematical knowledge: The case of randomness. *Journal of the Learning Sciences, 11*(4), 453-488.

Pratt, D., & Noss, R. (2010). Designing for mathematical abstraction. *International Journal of Computers for Mathematical Learning, 15*(2), 81-97.

Chapter 14
Multiple representations as tools for discovering pattern and variability – Insights into the dynamics of learning processes

Susanne Schnell, Susanne Prediger

Institute for Development and Research in Mathematics Education, TU Dortmund University

Abstract Dealing with multiple representations is an important learning goal, but also a tool for fostering teaching and learning. It has mostly been explored for algebra and statistics. In this article, we explore how relating different representations can also be used as a tool to foster learning about phenomena of randomness, here concretely for discovering pattern and variability in the context of the empirical law of large numbers. Our case study shows that the process of connecting different representations is quite complex for students and does not establish itself, but helps to foster deep conceptual insights into the duality of pattern and variability.

1 Dealing with multiple representations – learning goal and means

1.1 Multiple representations as a learning goal, means and challenge for learning statistics and other mathematical topics

Multiple representations have played a prominent role in statistics education research, because dealing with different representations (like diagrams, tables etc.) is an important *learning goal* (e.g., Wild and Pfannkuch 1999; Shaughnessy 2007). Biehler (1982) emphasized the importance of multiple representations for explorative data analysis because of their dual function: "On the one hand, they represent data, they have simulative function. On the other hand, they are tools for the activity of exploring data, they have an explorative function." (Biehler 1982, p. 13, translated by authors).

The fundamental ability to relate representations (being referred to as 'transnumeration' for statistics by Wild and Pfannkuch 1999), is also widely discussed for algebra and functions, here called 'representational fluency', defined as "ability to move between and among representations, carrying the meaning of an entity from one representation to another and accruing additional information about the entity from the second representation" (Zbiek et al. 2007, p. 1192).

Even when representational fluency is not a learning goal in itself, the transition between different representations has widely been used as a *means for facilitating learning* of mathematical ideas and concepts since Bruner (1967). Duval (2006) provided a theoretical semiotic foundation that explains its importance by the abstractness of mathematical objects.

As a consequence, many computer tools have been developed that can support the easy switch between representations, varying from graphic plotters up to multi-representation systems like Geogebra. For algebra and calculus, empirical studies have shown particularly that the use of computer-aided graphical representations – in juxtaposition with other representations – can enhance individual and social processes of developing meanings and conceptions (cf. meta-analysis by Fey et al. 2003; Clements et al. 2008). However, distinct literature surveys on technology use (like Zbiek et al. 2007) seem to focus mostly on algebra, functions and geometry whereas stochastic education is rarely considered.

Nevertheless, different design projects and empirical studies in statistics education have produced comparable design and research results: For example, Ben-Zvi (2004) shows that grade 7 students who acquired representational fluency in EDA, make use of tables and graphs also when they investigate data in order to find generalities (global view) and outliers (local view). For many years, software tools like Fathom have provided multiple representations as a way of acquiring statistical conceptions (Biehler 1993).

Although the importance of multiple representations for facilitating learning is widely accepted, many empirical studies show that students do not make use of it automatically and easily. Biehler (1997, p. 174) reports that students tend to think that using one representation is enough. Pfannkuch and Rubick (2002) highlight that different representations produce serious issues of transfer: "[V]ariation thinking with raw data is different from variation thinking with graphs. [..] Unless prompted, the thinking did not change for some of the students and continued to be the same for the graph as it was for the raw data" (p. 17 f.). Hence Biehler warned not to underestimate these *challenges*: "We must be more careful in developing a language for this purpose and becoming aware of the difficulties inherent in relating different systems of representation" (Biehler 1997, p. 176). The intention of this paper is to contribute to the development of a theory of students' use of multiple representations: For this, we provide empirical insights into the learning processes of students and interpret them through the theoretical lens of context-specific constructs (Schnell and Prediger 2012).

1.2 Multiple representations - a means for learning probability?

The importance attributed to multiple representations as a means for supporting learning in algebra and statistics suggests exploring their role also for probability.

Although multiple representations play a significant role in designs for probability education, for example when applying an experimental approach to probability by EDA (Biehler 1994) or when a simulation shows results graphically and numerically, little is known about the dynamics of learning processes when students gradually make sense of these random phenomena by comparing and combining different representations. The issue of transfer between different representations appears implicitly in some empirical studies (e.g., Pratt 1998), but more systematic investigation is needed to give an account of the effects of multiple representations while conducting and interpreting random experiments with computerised tools.

In the following section, we report on a case study focussing on the effects of multiple representations for the probabilistic topic of pattern and variability for chance experiments with small and large total numbers of throws with a die. We roughly present the learning arrangement (Section 2), the research design (Section 3) and some empirical findings (Section 4).

2 Pattern and variability
in the learning arrangement "Betting King"

The learning arrangement (published in Prediger and Hußmann 2014) offers opportunities to make systematic experiences with pattern and variability in chance experiments. The main activity consists of guessing outcomes of a chance experiment (as suggested by Fischbein 1982). The arrangement includes a set of tasks, work sheets and questions guiding the students' investigations.

In the game "Betting King", the players bet on the winner of a race with four coloured animals, being powered by throws of a coloured 20-sided die with asymmetric colour distribution (red ant: 7/20, green frog: 5/20, yellow snail: 5/20, blue hedgehog: 3/20). The game is first played on a board (see Fig. 14.1 left), then in a computer simulation that always shows results in numerical and graphical representation (Fig. 14.1 right). In the latter, the maximum value of diagram's Y axis is changed in relation to the highest frequency of wins (here the red ant's 7 wins), thus providing a pseudo-relative view on the results.

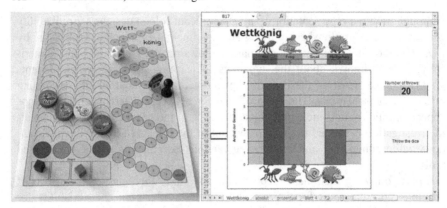

Fig. 14.1 Left: The game 'Betting King' **Right:** Computer simulation; here positions according to the expected values for 20 throws

The game is played in two variations: In the first game "Betting on Winning", students bet on the animal that will go the farthest. Usually, students discover quite quickly that the red ant has better chances to win because of the colour distribution of the die. In the second game "Betting on Positions", students have to quantify their ordinal estimations when asked to predict the *positions* on which each animal will land on the board, in relation to a pre-set total of throws i.e. on the absolute frequencies of each animal. The sum of the positions reached by all four animals therefore equals the total number of throws (i.e. 20 in the example).

The need to *find good betting strategies* motivates the exploration of patterns and variability of absolute and relative frequencies. *The possibility to choose* the total die throw count (between 1 and 40 for the board game and 1 and 10,000 for the computer game) motivates the distinction between short term and long term contexts for deciding when betting is easier (Schnell and Prediger 2012). In order to focus on *relative* deviations instead of absolute ones, the bets of two players are compared, and the bet closer to the actual finishing position of each animal gets a point (or both win if it is a tie).

Table 14.1 shows typical representations and the intended insights (noted as so-called 'constructs', cf. Section 3) in the first sequence of the second type of game – i.e. 'Betting on Positions' - with respect to pattern and variability. Although theoretically, all important discoveries can be made in both representations, students usually first discover patterns (i.e. regularities such as seeing repeatedly the same heights of bars for the same high number of throws) in graphical representations, and effects of variability in numerical representations (i.e. different positions of the animals for every game). We will present an exception, a case of two students who focus on pattern in the numerical and on variability in the graphical representation.

Table 14.1 Intended constructs on patterns and variability in two main representations

	Numerical representation: Tables of absolute frequencies	Graphical representation: Pseudo-relative bar-charts of absolute frequencies
Visualisations in the learning arrangement	Exemplary visualisation: (white rows: students' predictions, grey rows: results of simulations)	Exemplary visualisation: (first two games with a total number of 200 from the table on the left)
Pattern	<In a series of games with high throw totals, most absolute frequencies seem to be in a similar range>	<Even for varying high throw totals, the heights of pseudo-relative bars seem quite stable, but not for small throw totals>
Variability	<Absolute frequencies vary from game to game>	<Heights of pseudo-relative bars vary from game to game>
Pattern in Variability	<In a series of games, absolute frequencies vary within a certain interval which is absolutely larger for higher total numbers of throws>	<In a series of games, heights of pseudo-relative bars vary within a certain range which is relatively smaller for higher throw totals>

In addition to the here presented inquiries, students were also asked to investigate percentages in games of gradually growing total numbers of throws as well as some reflection and transfer tasks; these will not be addressed in this article (see Schnell 2013 for more details).

3 Design of the Study and Research Questions

Following the paradigm of *design research in the learning process perspective* (Gravemeijer and Cobb 2006), the learning arrangement was tested and improved cyclically over three courses of evaluation in six classes (grade 5 and 6, students aged 11 to 13). While classroom experiment provided a positive evaluation on the macro-level, complementary *design experiments in a laboratory setting* were necessary to investigate the processes on the micro-level. For this, the first author of this article conducted design experiments based on the presented learning situation with nine pairs of students (age 11 to 13), with at least four sessions, each of about 90 minutes (40 sessions in total). The design experiments were semi-structured by an experiment manual, determining the sequence of tasks, the use of materials and some reactions of the interviewer in specific situations. The following empirical findings will focus on the second session with one pair of students.

For *data gathering*, each session was videotaped and computer screens recorded. For two pairs, all four sessions were completely transcribed (a detailed analysis is offered in Schnell 2013). For the other seven pairs (including Elisa and Jacob who will be presented here), complete episode plans (capturing the sequence of all questions, activities and crucial statements) served as survey and basis for selecting sequences for more detailed transcription.

The *in-depth data analysis* is guided by an interpretative approach, identifying scenes in which new discoveries are made and integrated into the individual conceptions, here conceptualized as network of constructs (cf. Schnell and Prediger, 2012 for the theoretical background of the analytical model). In these scenes, students' individual *constructs* are reconstructed with respect to four elements: the emerging insights (summarised by the *proposition, which is noted in angle brackets <>*), the *stochastic context* in which an insight occurs (e.g., small/high throw totals), the *representation* to which it refers (among other elements of the situational context which will not be used here) and which *function* the construct has in the specific situation (here reduced to differentiation between describing patterns or describing variability). We refer to the constructs by a short name like ORDINAL-ESTIMATIONS, noted in caps to emphasize that this is referring to a broader insight. In this article, we only give insights into the analysis regarding paradigmatic individual constructs that contribute to the following focused research question:

What constructs do students build for patterns and variability from within the numerical and graphical representations and how do they relate these constructs to each other?

4 Empirical findings: The Case of Elisa and Jacob

Elisa and Jacob (both 12 years old) cooperate intensively in all four experiment sessions. Having played the first game "Betting on Winning" for five times, they soon discover the unequal colour distribution on the die and use it to explain the red ant's empirically observed benefit in an ordinal sense: The more often the colour appears on the die, the more often it appears when the die is thrown

(ORDINAL-PREDICTION). They reach the learning goals of the first game when they draw a connection between the chance of winning with the red ant and the size of the total number of throws: "the ant is better for games with a total number of throws of 100 and 1000 as it always won there".

In this section, we analyse their emerging constructs for the second game "Betting on Positions" in Session 2, as it is an interesting case for analysing the role of multiple representations. Although the computer simulation always simultaneously provides the numerical and graphical representation (tables of absolute frequencies and pseudo-relative bar-charts), both students consider only the numerical representation until they are explicitly prompted to look at the graphical representation in Scene 3. Before analysing this interesting transition, we report on their complex discoveries in tables in Scene 1 and 2.

4.1 Preliminary discoveries in tables of absolute frequencies

Table 14.2 Construct identified in scene 1

Short-title	Proposition	Stochastic context	Function
ORDINAL-ESTIMATIONS	\<Red ant gets the highest position, second highest for green frog and yellow snail (though not necessarily on the same position) and blue hedgehog gets the lowest position\>	Single games of 9 to 40 throws	Description of pattern and prediction of positions

In Session 2, Elisa and Jacob start playing the game "Betting on Positions" with the computer simulation. In the first five games, they choose the total numbers of throws of 9, 11, 20, 30 and 40. They build with their ordinal constructs from the first variant of game ('Betting on Winning') and develop the strategy we are going to call ORDINAL-ESTIMATIONS for predicting finishing positions: the highest position for the red ant, second highest for green frog and yellow snail (though not necessarily on the same position) and blue hedgehog on the lowest position. Additionally, Jacob points out that the sum of the positions has to equal the total number of throws.

Scene 1. Patterns and predictions in tables with absolute numbers

After this introduction, a new record sheet is presented with tables of predefined total numbers of throws: 1 x 20, 4 x 200 and 4 x 2000. The analysis of transcripts shows that in Scene 1 (Session 2; 25:11 – 40:19), they develop the following new construct:

Scene 1 (Session 2, from 25:11)
5   Iv: Ok, now we play with 20 again… which positions?
6   *The students move laptop and mouse so that Jacob can operate them.*
7   E: *(takes pen and record sheet)* Ok, how many?
8   J: What do we have here? *(points to the following old record sheet)*
           Do we have 20 here, too? Then we take these again […]

	Bets and results:	Red ant	Green frog	Yellow snail	Blue hedgehog	Points
Total number of throws: 20	Students' bet	7	5	8̶ 6	2	2
	Interviewer's bet	9	4	4	3	2
	Thrown positions	9	4	6	1	

11ff   *E and J write "9,4,6,2" as prediction into the new record sheet, then correct to 1.*
       The interviewer also makes a prediction, then the game is simulated and the results
       compared.

	Bets and results:	Red ant	Green frog	Yellow snail	Blue hedgehog	Points
Total number of throws: 20	Students' bet	9	4	6	2̶1	4
	Interviewer's bet	7	5	4	4	1
	Thrown positions	8	4	6	2	

27   E: We have got [a point for] the hedgehog, too. Four points *(laughs)*.
28   Iv: I have only one overall.
29   J: You can really win with this strategy, when you look at this *(points to old record sheet)*.

Jacob consciously makes use of empirical positions in the old record sheet in order to predict positions for the total number of throws of 20 (line 6-8). By repeatedly applying the strategy of using the previous results, Elisa and Jacob emphasize the (perceived) stability of the positions and make use of them as a *pattern* that helps predict future positions. Although the numerical representations were expected to prioritize the discovery of variability, the students focus on pattern and evaluate their strategy as very useful when they get four out of four points in the game of 20 (line 29).

For the next game, which is the first with a throw total count of 200, the students apply the strategy of ORDINAL-ESTIMATIONS and win two points. After applying the betting strategy PREVIOUS-RESULTS for the second game of 200, Jacob evaluates its success:

Scene 1b (Session 2, from 29:50)
73   E: Three points.
74   J: But it seems the rough estimation is always good.

The use of the term "rough" might refer to the fact that the students got only 3 out of 4 points, or it might indicate an awareness of the variability of the positions: the animals reach only similar, but not exactly the same positions. In both interpretations, Jacob seems to raise a first idea of variability which he does not make further explicit or come back to it.

Scene 2. Ranges for absolute frequencies

After filling in the record sheet and writing down their strategy (i.e. "we always use the positions from the game before"), Scene 2 begins (Session 2; 40:20 – 66:14), in which the Elisa and Jacob develop the following constructs:

Table 14.3 Constructs identified in scene 2

Short-title	Proposition	Stochastic context	Function
RANGES-ABSOLUTE	\<For series of games with the same total number of throws, the animals stay in certain ranges\>	First verbalised for series of games of 200; Later used for series of games of 100 and 2000	Description of pattern
OUTLIERS	\<Each time, one position is outside the ranges\>	Single games of 200 and 100	Description of variability

Elisa and Jacob are introduced to a new record sheet, which is supposed to help focussing on the intervals of the game results ("ranges") for the green frog in a series of eleven games of 200 (see Table 14.1 for their record sheet). They occasionally use PREVIOUS-RESULTS for the green frog and start experimenting with other numbers.

After ten games, the students claim the range is "so far 30 to 55", which is then also true for the eleventh game. This range applies for many more trials, with only one exception. After writing down their range, the students are asked to consider a new total number of throws (they pick 100) for which they have no trouble first predicting and then empirically identifying a range of positions. They evaluate their range of positions in a series of game, where again only one is not within the range. These outliers are mentioned by Elisa after the second series of games: "It is like before, like here *(points to betting range for 200)*, one exception so far to what we said".

Like before, the students only focus on the stability they perceive, i.e. the pattern that the positions stay in certain ranges. They choose their range of prediction by using the highest and the lowest score in their experiments. The outlier, which shows up for 200 as well as for 100, is mentioned by Elisa in a way whereby one could think she sees the existence of exactly one outlier as a new pattern. It is not further addressed and does not lead to changing the previously stated range of positions.

The tables of absolute frequencies were designed to focus on variability between single games, but a more or less stable interval of positions across the series of games. Elisa and Jacob focus mainly on the stability of their empirically based predictions even in transferring from the results of one game to predictions of another. The idea of ranges seems thus very natural for them and they do not pay attention to the two outliers. The idea of variability as creating an uncertainty in predicting from single outcomes or short series of games seems so far not be established for the students.

4.2 Transition from numerical to graphical representation

Scene 3. Patterns and variability in quasi-relative bar-charts

In Scene 3 (Session 2; 66:15-76:30), Elisa and Jacob are explicitly prompted to investigate the bar-charts. When the computer simulation is introduced (for Elisa and Jacob in the first design experiment session), the bar-charts are only explained by the interviewer as a visualisation of the game results and the students are asked to relate the outcome of one game to the respective situation on the board. All further insights are discovered by the students. Although Elisa and Jacob first do not connect the graphical representation to previous observations in the numerical representation, they later relate both and successively develop the following constructs:

Table 14.4 Constructs identified in scene 3

Short-title	Proposition	Stochastic context	Function
TWO-UP	\<Two animals' bars always go up and two down\>	Series of games of 999	Description of variability
DIGIT-PATTERN	\<The number of animals changing is related to the number of digits of the total number of throws\>	Short series of games of 1,528	Description of variability
ANT-LIMIT	\<The ant will always stay above 1,000 in the diagram\>	Series of games of 3,456	Description of variability and pattern
BARS-RANGES	\<The animals' bars will always stay in certain ranges of absolute numbers\>	Series of games of 3,456, later series of games of 10,000	Description of variability and pattern
SMALL-HIGH-BARS-MOVEMENT	\<For small throw totals, all animals' bars change a lot, for high throw totals they change just a bit\>	Series of games of 12 and series of games of 3,456	Description of variability and pattern

Elisa first states "I don't understand that, it is clear what you can see: who won" and Jacob agrees, referring to other external elements of the image such as "the colours of the animals". These statements seem to illustrate that the transfer from one representation to the next one is not evident for students. To facilitate deeper investigations of the changes of the bars when producing series of games, the interviewer then suggests producing many games with the same total number of throws and comparing the diagrams. This means, whenever a new game is simulated (by clicking on the "Throw the die"-button in Fig. 14.1 right), a new barchart overlays the old one and thus creates an impression of "up and down moving" bars. The students choose 999 as the total of throws for 13 games. In minute 68:15, Elisa explicates a first pattern TWO-UP: "Two always go up clearly, but they don't have to be on the same position, only go up. And two go down, always mutually", without referring to the discoveries made in the numerical representation.

It is mainly the interviewer who suggests looking at new (low and high) total numbers of throws, so that the students investigate games of throw totals of 999, 555, 12, 1,528, 4, 3,456 and 10,000. Elisa seems to first realize the variability of the bars and spends some time investigating the number of "changing" bars. Then she suggests a dependency between numbers of digits in the total of throws and numbers of changing bars:

Scene 3a (Session 2; from 70:15)
34   E: *(generates the first two games with a total number of 1,528)* I think, all of them
            change, here. I think, when you have four numbers [i.e. digits of the total
            number of throws], all four change *(points to bars with cursor)*, when you have
            three numbers, only three change and when you have two numbers, maybe only
            two change.
35ff   *For contradiction, Jacob generates games with a throw total of 4, in which up to
       all animals change. Thus, the students refute Elisa's suggestion.*

Elisa's construct DIGIT-PATTERN is a paradigmatic example for idiosyncratic discoveries. Students do not necessarily see the intended structures in graphical representations but find individual ones: here she searches for a deterministic functional connection between the number of digits of the total number of throws and the number of moving bars instead of considering the height of the bars. Thus, for a total number of 1,528 (four digits), Elisa expects the bars of all four animals to change in height.

The interviewer now suggests using a "very high number", for which Jacob picks 3,456. After about ten games, Jacob states that the hedgehog "almost never seems to change". Elisa again builds with TWO-UP, but then addresses another feature of the diagram to describe the movement:

Scene 3b (Session 2; from 73:00)
65   E: For the ant, it always goes to this line and then up again.
66   J: Stays over 1,000.
67   E: It always stays there. The frog, he always moves just a tiny bit.
68   J: Yes, and he stays in the 800s range or until the 1,000s range.
69   E: And the snail also always stays with 800 to 1,000, but
            |it then goes up a bit.
70   J: |it always tries to get a tie with the frog.
71   E: Yeah.
72   Iv: Right
73   E: Yes, the hedgehog stays roughly in the 400s to 600s range. It always goes up.

In lines 65-67, the students co-construct ANT-LIMIT: Elisa uses the line in the background of the diagram (see Table 14.1, probably in this case the mark for 1,200) to describe a limit for the red bar's movement. This is the first time she focusses not on single bars as objects, but on their heights in a series of games with throw total 3,456. While Elisa uses the descriptive term "line" in line 65, Jacob uses an absolute number to describe how the bar is always higher than the 1,000 mark in line 66.

In line 68, Jacob first transfers a construct from the numerical to the graphical representation: He uses the expression "range" that was established for the numerical representation (in Scene 2) to describe how the position of the green frog is stable in a way. Elisa adopts the term "range" in line 73, but adds in both statements how the animals "go up". It seems that the use of the construct RANGES that was transferred from numerical representations helps the students to cope with the variability of the bars, which they observed the whole time. While using ranges as a seemingly unproblematic instrument to describe their observations in the numerical representation of tables of absolute frequencies, for the bar-charts especially Elisa struggles with a clear focus and finally transfers the construct RANGES to overcome this.

The transfer from one representation to the next one helps both students to develop deeper insights into the duality of pattern and variability for short and long term contexts. This becomes visible in the next scene when the interviewer asks the students to compare these findings with a low total number of throws for which the students choose twelve. Students evaluate their idea for the total number of 10,000 and both students agree that it seem to be true. Then, the interviewer asks to explain the differences:

Scene 3c (Session 2; from 75:30)
85   Iv: So, what would you say, how do the bars change for small throw totals,
        for 12, and how do they change for large throw totals?
86   E: For the small ones, I think, it changes big times. They sometimes go down a lot,
        as the three [bars] now, and then sometimes really high. And for bigger total
        numbers of throws, I think, it changes just a tiny bit.

So finally, the students establish the intended construct SMALL-HIGH-BARS-MOVEMENT and integrate their constructs for graphical and numerical representations while distinguishing short and long term contexts.

5 Conclusion

Pattern and variability are two dual perspectives on one phenomenon that have to be balanced. In our learning arrangement, connecting different representations is intended to foster students' access to this duality. The summary of Elisa's and Jacob's constructs in Table 14.5 shows that although this intended learning goal was finally reached, the process needed the support of the interviews because forming connections between representations does not happen automatically.

Table 14.5 Some of Elisa's and Jacob's constructs in the process (partly revised)

	Numerical representation: Tables of absolute frequencies	Graphical representation: Quasi-relative bar-charts of absolute frequencies
Pattern	PREVIOUS-RESULTS <Observed positions are a good rough estimation for the positions in a game with the same total number of throws>	DIGIT-PATTERN <The number of animals changing is related to the number of digits of the total number of throws>
Variability		TWO-UP <Two animals' bars always go up and two down>
Pattern in Variability	OUTLIERS <Each time, one position is outside the ranges>	ANT-LIMIT <The ant will always stay above 1000 in the diagram>
	RANGES-ABSOLUTE <For series of games with the same total number of throws, the animals stay in certain ranges>	BARS-RANGES <The animals' bars will always stay in certain ranges of absolute numbers>
		SMALL-LARGE-BARS-MOVEMENT <For small throw totals, all animals' bars change a lot, for large throw totals they change just a bit>

While the case of Elisa's and Jacob's process is specific in some aspects (e.g. their exclusive focus on patterns in the numerical representation), the transition to the graphical representation shows typical phenomena that we could reconstruct in different case studies with the same learning arrangement:

1. Even if multiple representations are offered by a (teaching-)learning arrangement, students can prioritize one without considering the others.
2. The explicit prompt to switch the representation helps students to develop new perspectives.
3. Graphical representations are not self-explaining; students can observe very idiosyncratic structures.
4. It is not evident for students to transfer discoveries from one representation to the next one automatically, but when it happens, it helps to deepen students' conceptions.
5. Especially for the duality of patterns and variability, the transition between representations can help to gain complex insights into core phenomena of random since it allows students to draw connections.

References

Ben-Zvi, D. (2004). Reasoning about variability in comparing distributions. *Statistics Education Research Journal, 3*(2), 42-63.

Biehler, R. (1982). *Explorative Datenanalyse - Eine Untersuchung aus der Perspektive einer deskriptiv - empirischen Wissenschaftstheorie*. IDM Materialien und Studien 24. Bielefeld: Universität Bielefeld, Institut für Didaktik der Mathematik.

Biehler, R. (1993). Software tools and mathematics education: the case of statistics. In C. Keitel, & K. Ruthven (Eds.), *Learning From Computers: Mathematics Education and Technology* (pp. 68-100). Berlin: Springer.

Biehler, R. (1994). Probabilistic thinking, statistical reasoning, and the search for causes - Do we need a probabilistic revolution after we have taught data analysis? In J. Garfield (Ed.), *Research Papers from ICOTS 4*, Marrakech 1994. Minneapolis: University of Minnesota.

Biehler, R. (1997). Students' difficulties in practising computer supported data analysis - Some hypothetical generalizations from results of two exploratory studies. In J. Garfield, & G. Burrill (Eds.), *Research on the Role of Technology in Teaching and Learning Statistics* (pp. 169-190). Voorburg: International Statistical Institute.

Bruner, J. (1967). *Toward a theory of instruction*. Cambridge, MA: Harvard University Press.

Clements, D., Sarama, J., Yelland, N., & Glass, B. (2008). Learning and Teaching Geometry with Computers in the Elementary and Middle School. In M. K. Heid, & G.W. Blume (Eds.), *Research on Technology and the Teaching and Learning of Mathematics* (pp. 109-154). Charlotte: NCTM.

Duval, R. (2006). A cognitive analysis of problems of comprehension in a learning of mathematics. *Educational Studies in Mathematics, 61*(1-2), 103-131.

Fey, J.T. (2003). CAS and Assessment of Mathematical Understanding and Skill: Introduction. In J.T. Fey, A. Cuoco, C. Kieran, L. McMullin, & R. M. Zbiek (Eds.), *Computer algebra systems in secondary school mathematics education* (pp. 287-288). Reston, VA: NCTM.

Fischbein, E. (1982). Intuition and proof. *For the Learning of Mathematics, 3*(2), 9-19.

Gravemeijer, K. & Cobb, P. (2006). Design research from the learning design perspective. In J. Van den Akker, K. Gravemeijer, S. McKenney, & N. Nieveen (Eds.), *Educational design research*. (pp. 45–85). London: Routledge.

Pfannkuch, M., & Rubick, A. (2002). An exploration of students ' statistical thinking with given data. *Statistics Education Research Journal, 1*(2), 4-21.

Pratt, D. (1998). *The construction of meanings in and for a stochastic domain of abstraction* (Doctoral Thesis). University of London.

Prediger, S., & Hußmann, S. (2014, in press): Spielen, Wetten, Voraussagen - Dem Zufall auf die Spur kommen. In T. Leuders, S. Prediger, B. Barzel, & S. Hußmann (Eds.), *Mathewerkstatt 7*. Berlin: Cornelsen.

Schnell, S. (2013, in press). *Muster und Variabilität erkunden – Empirische Analysen von Konstruktionsprozessen kontextspezifischer Vorstellungen* (PhD Thesis). Wiesbaden: Springer Spektrum.

Schnell, S., & Prediger, S. (2012). From "everything chances" to "for high numbers, it changes just a bit". Theoretical notions for a microanalysis of conceptual change processes in stochastic contexts. *ZDM - The International Journal on Mathematics Education, 44*(7), 825-840.

Shaughnessy, J. M. (2007). Research on statistics learning and reasoning. In F. Lester (Ed.), *Second Handbook of Research on Mathematics Teaching and Learning* (pp. 957-1009). Greenwich: Information Age.

Wild, C. J., & Pfannkuch, M. (1999). Statistical Thinking in Empirical Enquiry. *International Statistical Review, 67*(3), 223-265.

Zbiek, R., Heid, K., Blume, G., & Dick, Th. (2007). Research on Technology in Mathematics Education. In F. Lester (Ed.), *Second handbook of research on mathematics teaching and learning* (pp. 1169-1207). Charlotte: Information Age.

Chapter 15
EDA Instrumented Learning with TinkerPlots

Dani Ben-Zvi, Tali Ben-Arush

The University of Haifa, Israel

Abstract   Recent investigations of technology-supported learning conducted from an *instrumental* perspective provide a powerful framework for analyzing the process through which artifacts become conceptual tools and for characterizing the ways students come to understand and implement a tool in solving a task. In this chapter, we focus on *instrumentation* – the process of transforming an artifact (component/s in the tool) into an instrument that is meaningful and useful to the learners – in the context of statistics education. Our goal is to characterize children's instrumentation in solving Exploratory Data Analysis (EDA) tasks. To illustrate this process, we bring short episodes from a case study of two fifth graders studying EDA with *TinkerPlots* in the 2012 *Connections* project. We suggest three types of instrumentation: *unsystematic*, *systematic*, and *expanding*. We also note that expanding instrumentation is hindered sometimes by *instrumented fixation*. We conclude by presenting several challenges stemming from the implementation of instrumental theory in the context of learning statistics.

1 Introduction

This chapter is written in tribute to Rolf Biehler's sixtieth birthday and with deep appreciation of his profound contributions to the understanding of technology's role in statistics education (e.g., Biehler 1989; 1993; 1995; 1997; Biehler and Hofmann 2011). We follow Biehler's legacy to study the nature and added value of technology in helping students to better understand statistics. Although the integration of learning technologies in statistics education is a common practice nowadays (Biehler et al. 2013), it is a challenging endeavor (Salomon and Ben-Zvi 2006). In response, researchers study the potentials and limitations of these tools in transforming statistics education and how they can play a critical role in helping students become statistically literate (Ben-Zvi 2000; Chance et al. 2007). In this chapter, we focus on *instrumental* aspects of this challenge.

Instrumental Genesis (IG, Verillon and Rabardel 1995) provides a powerful framework for analyzing learning processes that involve (computerized) tools: How designers and teachers construct affordances for learning tools, and how students acquire techniques for using those tools efficiently (Artigue 2002). IG was used in mathematics education (e.g., Guin and Trouche 1999; Haspekian 2005;

White 2008) but has been rarely discussed in statistics education (Pratt et al. 2006; Pratt and Noss 2002; 2010).

We discuss IG in the context of learning Exploratory Data Analysis (EDA) with *TinkerPlots* (TP) – dynamic interactive statistics software (Konold and Miller 2011). Our goal is to describe *how children begin to learn and appropriate a technological artifact to transform it into a meaningful and useful instrument in their data explorations.* To respond to this question, we first explain IG and EDA. To illustrate the use of IG in the analysis of children's learning we present empirical data from a case study that involves a pair of fifth graders using "Bins" and "Separate" – TP artifacts – to make sense of authentic data. We refer also to the circumstances in which this instrumentation process took place: the TP design (Konold 2006) and the nature of the tasks that were given to the students (Ben-Zvi et al. 2007). We then suggest an initial characterization of students' *instrumentation* based on our analysis of the evolution in the course of children's interaction with artifacts. We conclude with several challenges learnt from the application of the IG lens to students' EDA learning.

2 Theoretical background

2.1 The Instrumental Approach

The introduction of technological tools in education has challenged our understanding of epistemic processes of artifactual knowledge construction, raising new questions regarding the relationships between the learner, artifact and task. There was a need for new conceptualizations of the interaction with artifacts that intervene as mediators between the learner and the object of his action. One such conceptualization is *instrumental theory* that provides an analytical framework for learning in instrumented situations on the basis of the idea of *instrumental genesis* comprising both *instrumentation* and *instrumentalization*.

2.1.1 Instrumental theory

Instrumental theory has roots in anthropological and psychological traditions that focused on the subject–object–instrument relationships to explain the interactions with objects and the subject's cognitive development (e.g., Leontiev 2009; Vygotsky 1978). It captures the role of tools in human activities from cognitive-psychological and socio-cultural perspectives (Verillon and Rabardel 1995) and applies to any human activity with instruments. We restrict our discussion to instrumented learning activities and the appropriation of computerized tools in solving instructional tasks. According to this theory, artifacts (one or more components in a tool) become an instrument (with a new significance to the learner) through the implementation of a specific task (instrumented activity) (White 2008).

What role do instruments play in cognition? And how do they afford cognitive development? Instrumental theory elaborates the underlying microscopic cognitive processes that take place in any instrumented context. These processes, modeled in the IAS Model (Fig. 15.1, Verillon and Rabardel 1995), have an effect not only on the Subject (learner)–Instrument interaction, but also change the Subject–Object (educational task) interaction and the nature of the task itself. Instrumented activity can provide for the learner's meaning-making of the task, development of conceptual understanding of the related domain, and improvement of problem solving performance.

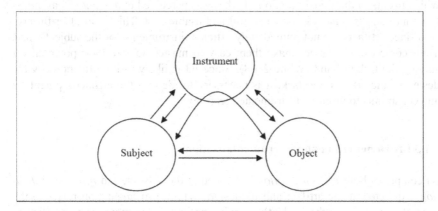

Fig. 15.1 The IAS model: The triad characteristic of Instrumented Activity Situations (Verillon and Rabardel 1995, p. 85. Printed with permission, the European Journal of Psychology in Education).

Instrumental theory includes two fundamental concepts: *instrumental orchestration* and *IG*. While instrumental orchestration refers to the design, learning and teaching processes and structures of a whole class (e.g., Drijvers et al. 2010) IG focuses on the learning processes of the individual learner (or sometimes groups of learners). We refer only to the latter in this chapter.

2.1.2 Instrumental genesis

IG comprises several notions: subject, object, artifact, utilization scheme, and instrument. *Subject* refers to a learner who solve a specific task with the assistance of an instrument. *Object* refers to the given task and its setting, aims and structure. *Artifact* is a man-made physical or virtual device (such as software) that has no meaning for the learner in isolation. *Utilization scheme* is the cognitive scheme that occurs when a learner is trying to solve a task while using one or multiple artifacts together. It includes not only previous knowledge (such as heuristics and school content), but also knowledge acquired while performing the task with the artifact (Verillon and Rabardel 1995).

An *instrument* is the output of the utilization scheme where the input is the artifact. In other words, an instrument can be defined as such, only after the subject has achieved a fair understanding of an artifact, or each of the artifacts and their utility and constraints. Thus, "an instrument is a mixed entity, part artifact, part cognitive scheme which make it an instrument" (Artigue 2002, p. 250). At first, the artifact does not have any instrumental value for the learner. It becomes an instrument through a process called IG.

IG is the process by which the *subject* uses the *artifact* while building and appropriating personal (or sometimes new or pre-existing social) utilization schemes, which results in the transformation of what was perceived by the *subject* as an *artifact* into an *instrument* – a meaningful and usable tool. This is an effortful process since artifacts are not immediately efficient instruments for the subject's use. Their complexity does not make them easy to master nor see their potential for solving the task at hand (White 2008). Since IG is likely to be influenced by the design of the artifact and task, it is necessary to identify the artifact's potentials and constraints to promote instrumented learning.

2.1.3 Instrumentation and instrumentalization

IG comprises both *instrumentation* – directed at the subject, and *instrumentalization* – directed at the artifact and its design. While instrumentation refers to the ways the tool shapes the student's learning process, instrumentalization refers to how the student's knowledge guides the way the tool is used (Drijvers et al. 2010).

Instrumentation is the process by which "the artifact prints his mark on the subject. One might say, for example, that the scalpel instruments a surgeon" (Trouche 2004, p. 290). The affordances and constraints of the tool influence the way the learner carries out a task and the emergence of the corresponding conceptions. During the instrumentation process the learner progressively develops or appropriates pre-existing *utilization schemes*, techniques and application knowledge of the instrument that enable an effective response to given tasks (Artigue 2002).

The *instrumentalization* of the artifact is a process which gradually loads the artifact with potentials, meanings and constraints that eventually transform it into a significant instrument for specific uses. The learner's knowledge guides the way the tool is used and in a sense shapes the tool.

IG provides an opportunity for researchers and designers to examine student learning and teacher's role and to refine the design of software, learning trajectory and tasks. Investigations of technology-enhanced mathematics learning conducted from an instrumental perspective (e.g., Artigue 2002; Guin and Trouche 1999; Haspekian 2005; White 2008) provided an insightful analysis of the process through which artifacts become meaningful conceptual tools. Furthermore, they characterized the ways learners come to understand and implement instruments. In this chapter, we situate this discussion in the context of EDA instrumented tasks.

2.2 EDA

Exploratory Data Analysis (EDA) is the art of making sense of data by organizing, describing, representing, and analyzing data, with a heavy reliance on informal analysis methods, visual displays and, in many cases, technology. Data analysis is typically viewed as a four-stage process: a) specify a problem, pose a question and formulate a hypothesis; b) plan, collect and produce data from a variety of sources (survey, experiments); c) process, analyze and represent data; and d) interpret the results, discuss and communicate conclusions. In reality however, researchers do not proceed linearly in this process but rather iteratively, moving forward and backward, considering and selecting possible assumptions and enquiry paths (Konold and Higgins 2003).

In the 1970s, the reinterpretation of statistics into separate practices comprising EDA and confirmatory data analysis (CDA, inferential statistics) (Biehler 1984; Tukey 1977) freed certain kinds of data analysis from ties to probability-based models, such that the analysis of data began to acquire status as an independent intellectual activity. The introduction of simple data tools, such as stem-and-leaf and boxplots, paved the way for students at all levels to analyze real data interactively without having to spend hours on the underlying theory, calculations and complicated procedures. Computers and new pedagogies would later complete the "data revolution" in statistics education (Ben-Zvi and Garfield, 2008).

Now that we have introduced the IG theory and EDA, we turn to an illustration in an instrumented EDA learning environment.

3 Illustration: Instrumental genesis in learning EDA

In this section we illustrate an implementation of instrumental theory in response to the key question we presented above – how do children begin to learn and appropriate a technological artifact to transform it into a meaningful and useful instrument in their data explorations? We suggest a preliminary characterization of students' instrumentation as they implement the "Bins" and "Separate" TP artifacts.

3.1 Setting

We draw on case study data from the 2012 *Connections* project (Ben-Zvi et al. 2007; Makar et al. 2011). Primary school students in Grade 5 engaged in five extended data investigations (each lasting 2-3 lessons of 90 minutes) using data from a student-administered survey across several grades in their school. The survey, designed with students' participation, collected information about students' body part dimensions, free time activities, pets, etc. (29 variables, n=331), creating a rich, authentic and interesting database for a series of extended investigations. In each investigation, students posed a research question, organized their sample data using TP, and made sense of it to draw informal inferences. We first provide a brief account of the three elements of the IG model (Fig. 15.1):

Subjects. We focus on two communicative students, who happened to be academically successful boys from the same class: Avi and Eyal (aged 11). They had no prior experience with TP, except for a short demonstration by the teacher.

Object. Each of the five data investigations was comprised of a whole class preparation discussion about the investigated topic, followed by an open-ended instrumented inquiry of the data in pairs, using handouts and occasional researcher's interventions. The students later presented their investigations in front of the class to provoke further discussion. The growing sample idea guided the design of these investigations (Ben-Zvi et al. 2012). The data were familiar and interesting to the students since they chose the survey questions and participated in the data collection.

Our examples come from the students' first independent data investigation with TP on issues related to free time (e.g., what students do in their free time, preferred communication method). They received open-ended guiding questions, but no explicit instructions on how to use TP. Each group was given a different sample of 8–10 students from their class (including the members of the group themselves) with 16 variables to analyze and make informal inferences beyond the data at hand.

Instrument. TP is innovative data analysis software designed to support students' development of statistical reasoning. With TP's dynamic graphing tool, children can invent their own elementary data plots (Biehler et al. 2013).

3.2 The "Bins" artifacts

We focus on "Bins", a set of basic but statistically important TP artifacts, which are divisions in data plots that contain all cases of a certain type. Bins can display the distribution of a category attribute (Fig. 15.2) or a numeric attribute, and can provide the count and the percent of individuals who fall under each category or class (Fig. 15.3).

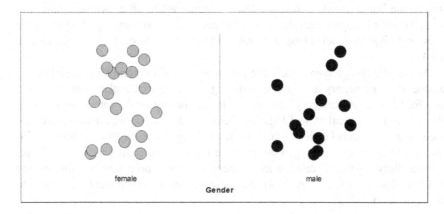

Fig. 15.2 Separation to bins of category attribute in TP.

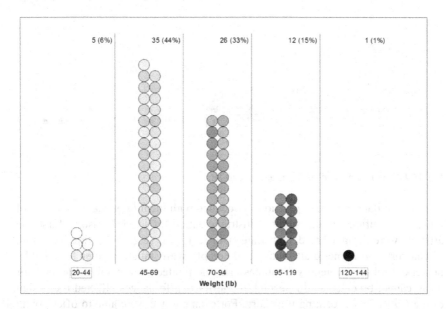

Fig. 15.3 Separation to bins of numeric attribute (stacked, with counts) in TP.

There are three basic ways to graph attributes in TP by *separating* cases into bins: (a) directly drag a case icon in a plot; (b) click on the "Separate" button in the plot toolbar; and (c) drag an attribute name from a "data card"[1] to a plot axis. Each of these ways comes with a set of operations. For example, in method (a), a case icon can be dragged either to the right or up – to make two (or more) groups,

---

[1] Data cards display the data set in a case by case view. Each card presents all the attributes of a case and their value, unit, type (category or numeric), etc. Data cards are used to add or change data, drag attributes into plots, and change the color scheme of attributes.

and to the left or down – to make fewer groups. When cases are separated into bins, TP adds lines between them, puts the name of the separating attribute on the axis, and displays labels along the axis to show what's in each bin (Figs. 15.2 & 15.3).

Not all the Bins features are same for category and numeric attributes. For example, with a category attribute, a first drag of a case icon makes only two bins, one for the dragged category and "other" for all the rest of the categories (such as the bins in the vertical axis of Fig. 15.6 below). To create more bins, another case icon from the group labeled "other" must be dragged out. With a numeric attribute, a case icon can be dragged out to make from two to eight bins, before the attribute "fully separates" and the axis appear as a continuous number line and the bin lines disappear (e.g., Fig. 15.4). All these actions are reversible by dragging to the opposite direction.

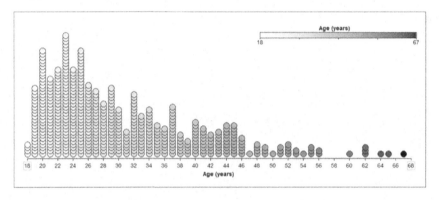

Fig. 15.4 A "fully separated" numeric attribute in TP.

These artifacts and the operations associated with them (e.g. change number of bins) are significant to our study of instrumentation for several reasons. First, the artifacts were designed to demonstrate statistical ideas (e.g. the distinctive nature of category and numeric attributes) and comply with statistical rules (e.g. unlike numeric attributes, category attributes can't be plotted on a number line or their mean cannot be calculated). Second, the artifacts affordances affected the design of the *Connections* learning trajectory. For example, we were able to offer young students to investigate covariation, exceptionally early in the curriculum, without using the conventional scatter plot, but rather a simple plot of one axis for the first attribute, while the second is presented by the varying colors of its case icons. Third, TP is supportive of our inquiry-based pedagogical approach by scaffolding students' free experimentation with data representations in a trial-and-error style. Thus, observing students' emergent appropriation and use of TP artifacts – while they construct meanings for key EDA ideas (e.g., data, distribution) – can inform both their cognitive and instrumental development and the relationship between them.

3.3 Data analysis

The pair's investigations were videotaped along with the use of Camtasia to capture both their computer screen and interactions as they worked. Videos were observed, transcribed, translated from Hebrew to English and annotated for further analysis of the students' instrumentation when learning EDA. Interpretations were discussed until consensus was reached. Differences between Hebrew and English connotations of words were discussed extensively among the authors. A few episodes from the first activity were selected to illustrate the students' instrumentation of TP. These episodes traced students' first encounters with the artifactual power of TP and were found productive in analyzing the emergence of students' instrumentation and cognitive development.

4 Students' instrumentation of TinkerPlots

We demonstrate students' instrumentation of TP during their negotiations with authentic data. We identify three major types of processes, which we label: unsystematic, systematic, and expanding instrumentation. We characterize each type and illustrate it with an episode from the students' artifactual activity.

a) Unsystematic instrumentation
The students' instrumentation process at the beginning of the data investigation can be characterized as unsystematic. They do not make intentional and systematic effort to study the TP artifacts, but rather use them superficially in an unorganized and trial-and-error manner (Makar and Confrey 2013, a chapter in this book). Their key goal at this initial stage is to represent the data satisfactorily based on their informal hypothesis about the question of interest. To support this goal, they therefore make sense and gain control of just a few basic artifacts and the artifactual actions associated with them.

Avi and Eyal started the investigation by getting acquainted informally with their nine-case sample, trying different data organizations in the TP plot window, and verifying they were part of the sample. In response to the first task, "formulate a question which you are curious to explore," they chose to focus on the number of children who prefer sports in their free time and hypothesized that these were mostly boys. By separating the free time activity attribute to bins (Fig. 15.5), they observed straightaway that 4 out of the 9 children prefer sports, and by checking the "data cards" associated with these four icons, they also realized that all of them were boys as they expected.

Note that the data in this plot (Fig. 15.5) is separated twice to bins: horizontal separation by free time activity and vertical separation by number of activities per week. The additional vertical separation is irrelevant to Avi and Eyal's concentration on the free time activity attribute and was unintentionally generated by them when they previously acquainted themselves with the data and some basic TP artifacts. The point is that at this stage, they had neither little control nor understanding of the Bins and Separation artifacts and did not seem to be bothered by the redundant vertical separation in this plot.

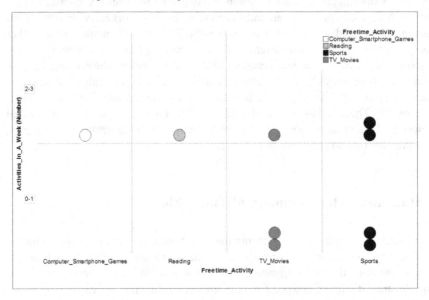

Fig. 15.5 Avi and Eyal's TP plot of the free time activity data. The data are unintentionally separated vertically by the number of activities per week.

In unsystematic instrumentation, students are typically focused on the task at hand rather than the tool, and are satisfied with a data representation (such as Fig. 15.5) that provides "good enough" information to solve the task. Students are naturally not yet fully familiar with the artifactual features of the software. Their resources and attention have to be distributed among the challenges presented by the task and the new domain of EDA (e.g., language, concepts, norms, tools).

b) Systematic instrumentation

Systematic instrumentation is an intentional and somewhat organized exploration of an artifact and the actions associated with it. This process takes place after a basic level of familiarization with the artifacts has been achieved. During systematic instrumentation the learner is focused more on the tool, rather than the given task, to study how artifacts function and co-function and identify their potentials and constraints, and acquire a higher level of control of and confidence with them. To achieve these goals, students typically experiment repeatedly with both familiar and new artifacts and evaluate the results of various artifactual actions.

The first demonstrated systematic instrumentation occurred 7.5 minutes into the instrumented activity, after Avi and Eyal had formulated their first informal inference about the free time activities. Only Avi became involved in a one-minute intensive session of systematic instrumentation, whereas Eyal was engaged in writing down their conclusion. Now that he was familiar to some extent with the

Bins and Separation artifacts of TP, Avi experimented with them, and especially with two of the three methods to separate cases into bins: directly dragging a case icon in a plot, and clicking on the "Separate" button in the toolbar. He swiftly and recurrently dragged icons right and left, up and down, and clicked the "Separate," "Stack" and "Order" menu buttons to understand their potentials. After this series of actions, Avi produced a new plot (Fig. 15.6) and excitedly shared the results with Eyal.

> Avi   Look, I found a method [to best represent these data[2]]! Have a look! This [Fig. 15.5] is the best way to see our question. First, here [in the vertical axis] we see Sports and "other" [a bin that includes all the rest of the categories]. But, we also see [in the horizontal axis] the distribution[3] of the "other." How the other [is distributed]… what is what. This is the best way to see the answer to our question. Do you see that?

The resulting data plot (Fig. 15.6) is simple, but innovative. Avi used both his old and new artifactual knowledge – acquired in the unsystematic and systematic instrumentation phases – to best represent the complex message he wanted to express. On the one hand – Sports is the most popular free time activity, and to emphasize this, he dragged the Sports icons to the far upper right corner of the plot. On the other hand – he presented the distribution of all the categories of the free time activity by "fully separating" the attribute horizontally. By proclaiming, "I found a method," Avi recognized that his invented plot is a useful and powerful expressive instrument in EDA. Furthermore, in response to Eyal's request to explain this unusual plot, Avi was fluent and quick in re-configuring the plot from scratch and explaining it intelligibly. The students' systematic instrumental development was coupled with an evident conceptual development, namely, the emergence of understanding of a key statistical concept – distribution (Ben-Zvi and Ben-Arush under review).

---

[2] These insertions in square brackets are our best guesses of what students mean given the context and history of their work.

[3] Avi used the Hebrew word "pilug" (separation) rather than "hitpalgut" (distribution). Though they both come from the same three-consonant root word (meaning divide, separate), the first is more known to children than the latter, which is a professional term. In any case, it is an unexpected term from a fifth grader.

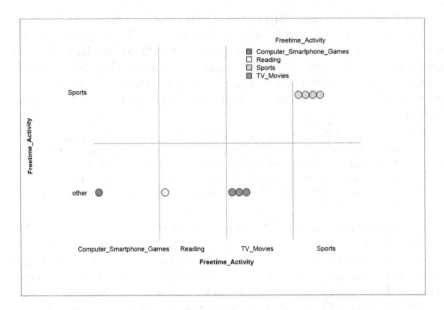

Fig. 15.6 Avi's dual intentional separation of the sample data in TP.

c) Expanding instrumentation and instrumented fixation

This third type of instrumentation refers to artifactual learning that expands the emerging instrumentalization of an artifact, namely its scope and fluency of uses, during the transfer of its application to new contexts and situations. The outcome of this meaningful process is that the set of artifacts and the associated actions on these artifacts transform (partially or fully) into a more usable and powerful instrument together with its utilization schemes for the student.

In our case, once the students gained some confidence and familiarity with plots of category attributes (dragging case icons, separating them to bins, etc.), they tried to apply the same set of artifacts and actions (instruments with utilizations schemes) on a numeric attribute in their sample, the number of free time activities per week. To their great surprise, the artifactual transfer was not smooth. For example, the bins' endpoints were sometimes fractions although the attribute values were only whole numbers. This conflict pushed them to think about the limitations of their previous knowledge, study new related artifacts (e.g., change the endpoints of numeric axis and bin widths), and acknowledge the artifactual and conceptual differences between numeric and category attributes.

The following repetitive experimentation with various attributes brought about – after a great deal of struggle, making errors, and some guidance by the researchers – the students' expansion of the instrumentalization of the Bins and Separation artifacts. This expansion included the development of utilization schemes, such as, when and how to implement the instrument in different circumstances, when to avoid its application, etc. Learning the artifact's potentials, constrains and utilization is the core of the expanding instrumentation stage and the student's development of fluency with technological tools.

Finally, we wish to draw attention to a phenomenon we labeled *instrumental fixation*. In many cases, once the students managed to transform a set of artifacts into a significant instrument, they tended to "fix" its features and utilization schemes, and did not look for new instrumental potentials or constraints. This habitual behavior is obviously problematic, resulted in students' limited use of the powerful TP artifacts for data analysis and representation. Cognitive psychologists similarly identify a fixation effect in instances in which previous experience or familiarity with certain tasks can make problem solving more difficult (Sternberg and Sternberg 2011). This is the case whenever habitual directions get in the way of finding new directions. Hence, fixation seems to be an important hindering factor in the *expanding instrumentation* phase by preventing efficient and full utilization of the instrument. It also presents a challenge for designers, educators and researchers who are interested in students' instrumental and conceptual development.

5 Discussion

Two themes are simultaneously present and closely intertwined in recent investigations of technology-enhanced statistics learning: (a) students' development of statistical reasoning as it occurs in the social context of the instrumented activity and (b) students' instrumental genesis of innovative tools that can offer them new forms of understanding, learning and communication. This chapter places the second theme in the forefront of the discussion, while temporarily pushing the first theme to the background. Thus, we concentrated on the learning processes of the educational tool by employing instrumental theory in a case study on children's learning EDA with TinkerPlots. The analysis of students' instrumented activities resulted in a preliminary characterization of three types of instrumentation: Unsystematic, systematic and expanding, which sometimes is hindered by instrumented fixation.

At first, students approached the artifacts in an *unsystematic* manner. They experimented and partially exposed some of its features, while focusing primarily on the task at hand rather than the tool. As the artifacts of TP began to emerge as a useful instrument together with its utilization schemes, a *systematic instrumentation* process took place. Students concentrated on studying various artifactual features and functions. In our case study, this short but productive process resulted with the creation of an innovative plot of students' free time activities (Fig. 15.6), which was considered by Avi as the optimal representation of their hypothesis. As the emerging instrumentalization was utilized in different contexts and situations, an *expanding instrumentation* occurred, in which additional potentials and constraints of the instrument were revealed together with utilization schemes. This was a productive process that brought about both students' instrumental and cognitive development, but was sometimes hindered by the effect of *instrumented fixation*.

The implications of this characterization of students' instrumentation are limited due to the idiosyncratic nature of our case study. While a fuller empirical account will be provided elsewhere (Ben-Zvi and Ben-Arush forthcoming), we conclude this chapter by suggesting a few significant methodological, empirical and pedagogical challenges when we consider instrumental processes in statistics education. We plan to include some of the following challenges in our future research agenda and call other researchers to follow.

Empirical challenges

- Are all the three identified types of instrumentation required for a full IG? And, do they always come in the same order?
- Our initial observations indicate the value of group work to IG and the existence of individual instrumentation patterns: What is the role of collaborative instrumentation? How does instrumentation develop among different students?
- What are the particular features of TP that support or hinder IG?

Methodological challenges

- What methodological approaches are appropriate in studying students' instrumentation? Are the methods we implement in exploring the development of students' statistical reasoning applicable for the study of instrumental development?
- What are the units of analysis in studying IG? Some options that can be considered: Student's manipulation of an artifact, the products of these manipulations, and the discourse associated with the artifactual activity.
- How can we represent students' instrumentation processes to both analyze and communicate them? Ben-Zvi and Aridor (2012) recently proposed a visual tool ("CDR Maps") to represent students' reasoning about data and context. Is this, or a similar, tool can be compatible for IG research?

Pedagogical challenges

- What are the design considerations and principles of tasks and learning trajectories that support students' IG?
- How can we help students overcome instrumented fixation without changing the pedagogical approach from "guide on the side" to "sage on the stage"?

Studies of these challenging directions can inform educators and educational designers in their quest to support students' instrumented journeys to better understand and appreciate statistical ideas. Our duty is to construct better affordances for learning tools and better learning environments for students to acquire techniques for using them efficiently. These efforts are particularly important in statistics education nowadays when statistical reasoning is increasingly ubiquitous in today's digital world, children are more and more technologically fluent, and innovative yet simple technology is ever more available.

References

Artigue, M. (2002). Learning mathematics in a CAS environment: The genesis of a reflection about instrumentation and the dialectics between technical and conceptual work. *International Journal of Computers for Mathematical Learning, 7*(3), 245-274.

Ben-Zvi, D. (2000). Toward understanding the role of technological tools in statistical learning. *Mathematical Thinking and Learning, 2*(1&2), 127-155.

Ben-Zvi, D., & Aridor, K. (2012). Children's wonder how to wander between data and context. In D. Ben-Zvi, & K. Makar (Eds.), *Teaching and Learning of Statistics* (pp. 307-316). The University of Haifa: Statistics Education Center Press.

Ben-Zvi, D., Aridor, K., Makar, K., & Bakker, A. (2012). Students' emergent articulations of uncertainty while making informal statistical inferences. *ZDM – The International Journal on Mathematics Education, 44*(7), 913-925.

Ben-Zvi, D., & Ben-Arush, T. (under review). Students' Instrumental Genesis: The Case of EDA Instrumented Learning with TinkerPlots. *Technology Knowledge and Learning.*

Ben-Zvi, D., Gil, E., & Apel, N. (2007). What is hidden beyond the data? Helping young students to reason and argue about some wider universe. In D. Pratt, & J. Ainley (Eds.), *Reasoning about Informal Inferential Statistical Reasoning: A collection of current research studies.* Proceedings of the Fifth International Research Forum on Statistical Reasoning, Thinking and Literacy (SRTL-5). Warwick, UK: University of Warwick.

Biehler, R. (1984). Exploratory data analysis - A new field of applied mathematics. In J. S. Berry, D. N. Burghes, I. D. Huntley, D. J. G. James, & A. O. Moscadini (Eds.), *Teaching and Applying Mathematical Modelling* (pp. 117-130). Chichester: Ellis Howard.

Biehler, R. (1989). Enhancing probability education by computer supported data analysis. In W. Blum, I. Huntley, & M. Niss (Eds.), *Modelling, Applications, and Applied Problem Solving: Teaching Mathematics in a Real Context* (pp. 123-130). Chichester: Ellis Howard.

Biehler, R. (1993). Software tools and mathematics education: the case of statistics. In C. Keitel, & K. Ruthven (Eds.), *Learning from Computers: Mathematics Education and Technology* (pp. 68-100). Berlin: Springer.

Biehler, R. (1995). *Towards requirements for more adequate software tools that support both learning and doing statistics.* IDM Occasional Paper 157. Bielefeld: Institut für Didaktik der Mathematik, Universität Bielefeld.

Biehler, R. (1997). Software for learning and for doing statistics. *International Statistical Review, 65*(2), 167-189.

Biehler, R., & Hofmann, T. (2011, August). *Designing and evaluating an e-learning environment for supporting students' problem-oriented use of statistical tool software.* Paper presented at the 58th ISI Session, Dublin, Ireland. Retrieved from http://isi2011.congressplanner.eu/pdfs/450437.pdf.

Biehler, R., Ben-Zvi, D., Bakker, A., & Makar, K. (2013). Technology for enhancing statistical reasoning at the school level. In M. A. Clement, A. J. Bishop, C. Keitel, J. Kilpatrick, J. & A. Y. L. Leung, (Eds.), *Third International Handbook on Mathematics Education* (pp. 643-689). New York: Springer.

Chance, B., Ben-Zvi, D., Garfield, J., & Medina, E. (2007). The role of technology in improving student learning of statistics. *Technology Innovations in Statistics Education, 1*(1), 1-26.

Cobb, P., Confrey, J., diSessa, A., Lehrer, R., & Schauble, L. (2003). Design experiments in educational research. *Educational Researcher, 32*(1), 9-13.

Drijvers, P., Doorman, M., Boon, P., Reed, H. & Gravemeijer, K. (2010). The teacher and the tool: Instrumental orchestrations in the technology-rich mathematics classroom. *Educational Studies in Mathematics, 75*(2), 213-234.

Garfield, J., & Ben-Zvi, D. (2008). *Developing students' statistical reasoning: connecting research and teaching practice.* Springer.

Guin, D., & Trouche, L. (1999). The complex process of converting tools into mathematical instruments: The case of calculators. *International Journal of Computers for Mathematical Learning, 3*(3), 195-227.

Haspekian, M. (2005). An "Instrumental Approach" to study the integration of a computer tool into mathematics teaching: The case of spreadsheets. *International Journal of Computer for Mathematical Learning, 10*(2), 109-141.

Konold, C. (2006). Design a data analysis tool for learns. In M. Lovett, & P. Shah (Eds.), *Thinking with data: The 33rd Annual Carnegie Symposium on Cognition* (pp. 1-32). Hillside, NJ: Lawrence Erlbaum Associates.

Konold, C., & Higgins, T. (2003). Reasoning about data. In J. Kilpatrick, W. G. Martin, & D. E. Schifter (Eds.), *A research companion to principles and standards for school mathematics* (pp.193-215). Reston, VA: National Council of Teachers of Mathematics (NCTM).

Konold, C., & Miller, C. D. (2011). *TinkerPlots: Dynamic data exploration* [Computer software, Version 2]. Emeryville, CA: Key Curriculum Press.

Leontyev, A. N. (2009). *The development of mind* (selected works). Marxists Internet Archive Publications. Retrieved from http://www.marxists.org/archive/leontev/works/development-mind.pdf.

Makar, K., Bakker, A., & Ben-Zvi, D. (2011). The reasoning behind informal statistical inference. *Mathematical Thinking and Learning, 13*(1), 152-173.

Makar, K., & Confrey, J. (2013). Wondering, Wandering or Unwavering? Learners' Statistical Investigations with Fathom. In T. Wassong, D. Frischemeier, P.R. Fischer, R. Hochmuth, & P. Bender (Eds), *Mit Werkzeugen Mathematik und Stochastik lernen – Using Tools for Learning Mathematics and Statistics* (pp. 351–362). Wiesbaden: Springer Spektrum.

Pratt, D., & Noss, R. (2002). The micro-evolution of mathematical knowledge: The case of randomness. *Journal of the Learning Sciences, 11*(4), 453-488.

Pratt, D., & Noss, R. (2010). Designing for mathematical abstraction. *International Journal of Computers for Mathematical Learning, 15*(2), 81-97.

Pratt, D., Jones, I., & Prodromou, T. (2006). An elaboration of the design construct of phenomenalisation. In A. Rossman, & B. Chance (Eds.), *Proceedings of the Seventh International Conference on Teaching Statistics* [CD-ROM]. Salvador, Bahia, Brazil. Voorburg, The Netherlands: International Statistical Institute.

Salomon, G., & Ben-Zvi, D. (2006). The difficult marriage between education and technology: Is the marriage doomed? In L. Verschaffel, F. Dochy, M. Boekaerts, & S. Vosniadou (Eds.), *Instructional psychology: Past, present and future trends (Essays in honor of Erik De Corte)* (pp. 209-222). Elsevier.

Sternberg, R. J., & Sternberg, K. (2011). *Cognitive psychology* (6th edition). Belmont CA: Wadsworth Publishing.

Trouche, L. (2004). Managing the complexity of human/machine interactions in computerized learning environments: Guiding students' command process through instrumental orchestrations. *International Journal of Computers for Mathematical Learning, 9*(3), 281-307.

Tukey, J. (1977). *Exploratory data analysis.* Reading, MA: Addison-Wesley.

Verillon, P., & Rabardel, P. (1995). Cognition and artifacts: A contribution to the study of though in relation to instrumented activity. *European Journal of Psychology of Education, 10*(1), 77-101.

Vygotsky, L. S. (1978). *Mind in society: The development of higher psychological processes.* Cambridge: Harvard University.

White, T. (2008). Debugging an artifact, instrumenting a bug: Dialects of instrumentation and design in technology-rich learning environments. *International Journal of Computers for Mathematical Learning, 13*(1), 1-26.

Kapitel 16
Zur Erfassung sprachlicher Einflüsse beim stochastischen Denken

Sebastian Kollhoff

Universität Bielefeld

Franco Caluori

Pädagogische Hochschule Nordwestschweiz

Andrea Peter-Koop

Universität Bielefeld

Abstract Dieser Beitrag steht im Kontext der Entwicklung und Erprobung eines diagnostischen Interviews zur Stochastik. Das stochastische Denken von Kindern ist zu einem hohen Grad von Intuitivität und Individualität geprägt, welche sich insbesondere im Begriffsverständnis und den sprachlichen Nutzungskonzepten von Begriffen der Stochastik wiederfinden. Aus diesem Grund wurden in einer Vorstudie zehn Interviews mit Schülerinnen und Schülern am Ende der sechsten Klasse durchgeführt, in denen Aufgabenentwürfe zur Erhebung des passiven Sprachgebrauchs in stochastischen Situationen erprobt und reflektiert wurden. Die Interviews lassen vermuten, dass die Nutzungskonzepte der Schülerinnen und Schüler in Bezug auf die Begriffe „zufällig", „wahrscheinlich", „sicher", „möglich" und „unmöglich" am Ende der sechsten Klasse noch stark der Alltagssprache verhaftet sind. Zudem ist anzunehmen, dass sprachlich-begriffliche Nutzungskonzepte nicht unabhängig von stochastisch-begrifflichen Vorstellungen zu erheben sind und die kontextspezifischen Verwendungen stark von der Situation abhängen. Auf dieser Grundlage werden Perspektiven für weiterführende Design-Varianten diskutiert.

Einleitung

Bereits in der frühen Kindheit machen Kinder Erfahrungen mit deterministischen und nichtdeterministischen Ereignissen. Sie spielen Spiele, die auf einem Zufallsprinzip beruhen, treffen Entscheidungen und benutzen umgangssprachliche Begriffe der Stochastik, mit denen sie sowohl im eigenen alltäglichen Umfeld sowie auch zunehmend medial in Kontakt kommen. Besonders im Spiel werden Kinder schon früh mit fairen oder unfairen Situationen konfrontiert, die sie mit ihren Spielkameraden aushandeln müssen. Dabei entwickeln sie nicht nur einen höchst individuellen Zugang zur Bewertung zufälliger Vorgänge, sondern sind bereits früh gefordert, sich sprachlich mit derartigen Ereignissen auseinander zu setzen. Das benötigte Vokabular entnehmen sie zumeist ihrer alltäglichen Umwelt, wobei jedoch das alltägliche Verständnis von Begriffen der Wahrscheinlichkeit nicht immer deckungsgleich mit der fachlich fokussierten Verwendung bzw. einem fachlichen mathematischen Verständnis ist. So werden beispielsweise die Begriffe „Zufall" und „Wahrscheinlichkeit" vermehrt synonym gebraucht (vgl. Jones et al. 2007) oder der Terminus „wahrscheinlich" häufig als Beschreibung für ein nicht sicheres jedoch erwartetes Ereignis gebraucht (Hasemann et al. 2008, S. 150). Ferner prägen vor allem individuelle Erfahrungen und gar animistische Vorstellungen (Wollring 1994) der Kinder ihr stochastisches Denken, womit auch die im mathematischen Sinne wertneutralen Begriffe der Wahrscheinlichkeit eine individuelle Konnotation erhalten.

Ein zentraler Befund der mathematikdidaktischen Forschung zum stochastischen Denken bei Kindern ist die Resilienz intuitiver Fehlvorstellungen, die trotz Unterricht Bestand hat (vgl. Fischbein und Gazit 1984; Garfield und Ahlgren 1988; Rasfeld 2004). Durch die Notwendigkeit etwas ausdrücken zu müssen, was bisher nicht bezeichnet wurde, entsteht eine Spannung zwischen der Sprache und den intuitiven Vorstellungen der Kinder, welche damit ein wesentlicher Faktor für die Entstehung von Fehlvorstellungen ist (Gal et al. 1992).

Der hohe Grad an Intuitivität und Individualität, der das stochastische Denken von Kindern prägt, ist eine wesentliche Schwierigkeit bei der empirischen und diagnostischen Erhebung. Der Sprache kommt bei der Entwicklung von Erhebungsinstrumenten eine zentrale Rolle zu. Sie ist einerseits das Hauptmedium der Kommunikation von Vorstellungen, kann aber andererseits auch ein Hindernis für das Verständnis sein. Das ist nicht zuletzt auch der Grund dafür, dass die Pionierarbeiten auf diesem Gebiet auf der „decision making technique" (vgl. Davies 1965) beruhen, bei der Entscheidungen in bestimmten Situationen beobachtet werden und somit nur eine Kontextsprache zum Versuchsaufbau benötigt wird.

In diesem Beitrag wird über die Erfahrungen im Rahmen der Entwicklung des Elementar Mathematische Basisinterview (EMBI) 3-6 Stochastik, eines handlungsleitenden (vgl. Wollring 2006) diagnostischen Interviews zur Stochastik, berichtet. Im Anschluss an das EMBI Arithmetik (Peter-Koop et al. 2007), das für die Klassen 0-2 ausgelegt ist, ist es Ziel dieses Entwicklungsforschungsprojektes ein diagnostisches Interview zu konzipieren und zu erproben, das für den Einsatz in den Klassen 3-6 geeignet ist. Das Interview soll zum einen differenziert den Leistungsstand von Schülerinnen und Schülern erfassen und zum anderen qualitative Einsichten in ihre Vorstellungen zur Identifizierung von Fehlvorstellungen erlauben. Lehrerinnen und Lehrer sollen so gezielte Impulse für eine weitere unterrichtliche Behandlung stochastischer Inhalte und zusätzlich Informationen über individuelle Leistungsdefizite für die Förderung erhalten.

Einen ersten zentralen Aspekt in der Entwicklung dieses Interviews stellt die Erfassung der sprachlichen Kompetenzen und des Begriffsverständnisses der Schülerinnen und Schüler dar. Im Rahmen der Entwicklung von Items wurden zehn Interviews mit Kindern am Ende der sechsten Klasse durchgeführt, die zum einen die Erfassung des stochastischen Wortschatzes und zum anderen die Evaluation von Item- bzw. Aufgabenformaten zur Feststellung sprachlicher Voraussetzungen zum Ziel hatten.

Curriculare Vorgaben zur Sprache unter der Leitidee Daten, Häufigkeit und Wahrscheinlichkeit

In den Bildungsstandards der Grundschule (Kultusministerkonferenz [KMK] 2004) wird unter der Leitidee „Daten, Häufigkeit und Wahrscheinlichkeit" für die Schülerinnen und Schüler am Ende der vierten Klasse das Ziel formuliert, dass sie „Wahrscheinlichkeiten von Ereignissen in Zufallsexperimenten vergleichen" und die dafür notwendigen „Grundbegriffe kennen (z.B. sicher, unmöglich, wahrscheinlich)" (KMK 2004, S. 11) und nutzen sollen. Hasemann, Mirwald und Hoffmann (2008) konkretisieren weiter, dass die Schülerinnen und Schüler lernen sollen, die Wahrscheinlichkeiten von Ereignissen auf einer Skala von „unmöglich" bis „sicher" einzuschätzen. Dabei sollen sie auch Formulierungen wie „wahrscheinlicher als" und „gleich wahrscheinlich" benutzen. Diese Ziele spiegeln sich in den Lehrplänen der Länder zum Teil in erweiterter Form wieder. So findet sich zum Beispiel im Kernlehrplan des Landes Nordrhein-Westfalen für die Grundschule (2008) die Formulierung: „Die Schülerinnen und Schüler beschreiben die Wahrscheinlichkeit von einfachen Ereignissen (sicher, wahrscheinlich, unmöglich, immer, häufig, selten, nie)" (S. 66).

In den Bildungsstandards und Lehrplänen für die Sekundarstufe I findet der sprachliche Umgang größtenteils keine weitere Erwähnung. Eine Ausnahme stellt auch hier der Kernlehrplan für die Hauptschule in Nordrhein-Westfalen dar, in dem Hinweise für einen sprachsensiblen Unterricht formuliert werden. Es wird die Rolle der Sprache beim Mathematiklernen allgemein hervorgehoben: „Fachliches Lernen und sprachliches Lernen sind unmittelbar miteinander verbunden. Sprache besitzt dabei eine besondere Bedeutung – zum einen für die fachliche Kommunikation, zum anderen aber auch für die fachlichen Verstehensprozesse und die begriffliche Erfassung von Welt" (Ministerium für Schule und Weiterbildung des Landes Nordrhein-Westfalen 2011, S. 30). Auf der Wortebene wird zwischen „Umgangs- und Standardsprache" wie auch zwischen einem „sachbezogene[n] und fachsprachliche[n] Wortschatz" (Ministerium für Schule und Weiterbildung des Landes Nordrhein-Westfalen 2011, S. 31) unterschieden.

Forschung zu Begriffen der Wahrscheinlichkeit

Forschungsergebnisse zur Begriffsentwicklung und besonders zum Begriffsgebrauch in der Stochastik finden sich zu großen Teilen im englischsprachigen Raum (Übersichten finden sich bei Shaughnessy 1992; Jones et al. 2007). Im Folgenden werden einzelne ausgewählte Studien und Befunde unter Berücksichtigung der Methodik kurz vorgestellt.

Ein bewährter und häufig genutzter Zugang zum Sprachgebrauch der Schülerinnen und Schüler ist, sie Situationen reflektieren und interpretieren zu lassen. Fischbein, Nello und Marino (1991) identifizieren auf diese Weise den linguistischen Faktor Sprache als ein Hindernis in der Entwicklung eines umfassenden stochastischen Verständnisses. In ihrer quantitativen Studie mit etwa 600 italienischen Schülerinnen und Schülern im Alter von 9-14 Jahren stellten sie insbesondere fest, dass die Schülerinnen und Schüler einerseits über keine eindeutige Definition der Begriffe „möglich", „unmöglich" und „sicher" verfügten und damit Schwierigkeiten bei der Interpretation und Einschätzung stochastischer Situationen hatten. Die Schülerinnen und Schüler sollten schriftlich vorgegebene Ausgänge von Zufallsereignissen mit den oben genannten Begriffen einschätzen. Zum einen wurde ein Würfel als Zufallsgenerator vorgegeben und zum anderen ein Generator benutzt, der Zahlen von 0 bis 90 ausgab. Besonders auffällig war der nahezu synonyme Gebrauch der Begriffe „selten" und „unmöglich" sowie der Befund, dass es den Schülerinnen und Schülern größere Schwierigkeiten bereitete, sichere Ereignisse zu erfassen und als solche zu verstehen, im Vergleich zu möglichen Ereignissen. An dieser Stelle sei angemerkt, dass dieser Befund sich nicht auf die Kontrastierung zu unsicheren Ereignissen bezieht, sondern auf der Identifizierung von Ereignissen aus einem Ereignispool basiert. Es ist daher nicht auszuschließen, dass im Rahmen einer Kontrastierung eine stärkere Differenzierung zu beobachten gewesen wäre.

Green (1983) berichtet ebenfalls von einem nicht normativen Verständnis stochastischer Begriffe und beschreibt in einer Studie mit ähnlichem Vorgehen, dass einige englische Schülerinnen und Schüler im Alter von 11-16 Jahren Begriffe bezüglich ihrer Bewertung von Wahrscheinlichkeiten überinterpretieren. So werden etwa die Begriffe wie „wahrscheinlich" („likely") mit „geschieht immer" („always happen") und „unwahrscheinlich" („unlikely") mit „kann nicht geschehen" („cannot happen") gleichgesetzt. Die Schülerinnen und Schüler bringen folglich individuelle Konnotationen der Begriffe und unterschiedliche Nutzungskonzepte in ihre Beschreibung und Wertungen mit ein.

Watson und Moritz (2003) berichten von einer Mitte der 1990er Jahre in Australien durchgeführten längsschnittlichen Untersuchung, in der Schülerinnen und Schüler der Klassen 6-11 (mit Ausnahme von Klasse 7) in Anlehnung an Garfield und Gal (1999) acht Begriffe auf einer Zahlengeraden von 0 bis 1 gemäß der von ihnen implizierten Wahrscheinlichkeiten eintragen sollten. Bei den Begriffen handelt es sich zu einem Teil um umgangssprachliche Ausdrücke bzw. Begriffe aus dem Alltag und zum anderen Teil um fachliche Begriffe: „58 per cent", „impossible", „sure thing" „looking good", „unlikely", „no worries",in doubt" und „50-50-chance". Für die Auswertung wurde die Zahlengerade in elf Intervalle (0-4%, 4-15%, 16-25%, ..., 96-100%) geteilt, um die Eintragungen der Schülerinnen und Schüler zu quantifizieren. Sie stellten dabei insbesondere fest, dass etwa ein Drittel der Schülerinnen und Schüler in der sechsten Klasse den Begriff „unmöglich" („impossible") nicht im Intervall 0-4% positionierten und die Angaben der Schülerinnen und Schüler generell in allen Klassenstufen eine hohe Streuung aufwiesen, was darauf hindeutet, dass die Schülerinnen und Schüler einer Klasse über zum Teil deutlich abweichende Begriffsvorstellungen verfügen. Die Ergebnisse des längsschnittlichen Vergleichs stellen zudem deutlich heraus, dass sich die Begriffsanordnungen mit zunehmender Klassenstufe einander annähern und die Streuung abnimmt, wenn gleich auch in der elften Klasse noch einige deutliche Abweichungen erkennbar sind. Das heißt, dass die Schülerinnen und Schüler zunehmend gleiche Interpretationen der Begriffe entwickeln, wenn auch einzelne Schülerinnen und Schüler ihren begrifflichen Fehlvorstellungen verhaftet bleiben. Individuelle Unterschiede sind in dieser Studie nicht überraschend. Die Autoren geben keinerlei Informationen, inwiefern die Skala im Unterricht eingeführt wurde, was zur Beurteilung der Ergebnisse entscheidend ist. Somit ist nicht geklärt, welches nicht an Worte gebundene Wahrscheinlichkeitsverständnis die Skala als eigenes Bezeichnungssystem für die Schülerinnen und Schüler darstellt.

Bei einer Untersuchung in England haben Amir und Williams (1999) die Konzepte von Schülerinnen und Schülern im Alter von 11-12 Jahren zum Begriff des Zufalls („chance") in den Mittelpunkt gestellt. In offen gestellten Fragen stellten sie dabei fest, dass die Mehrheit der Schülerinnen und Schüler diesem Begriff mehrere und mannigfaltige Bedeutung zuweisen. Zumeist verbinden sie mit dem Begriff „Zufall" Konstrukte, wie „etwas, das einfach passiert" („something that just happens"), oder das unerwartete, ungewünschte und ungewöhnliche Eintreten eines Ereignisses („just happens by accident"). Die Autoren wiesen dabei speziell darauf hin, dass die Schülerinnen und Schüler ihr Verständnis von Zufall und Glück nicht erst in der Schule erwerben, sondern ihre Vorstellungen auf alltägliche Erfahrungen zurückführen, wodurch diese besonders durch den alltäglichen Sprachgebrauch gekennzeichnet sind, was insbesondere im Unterricht zu berücksichtigen ist (Amir und Williams 1999, S. 103).

Tarr (2002) thematisiert ebenso wie Amir und Williams (1999) und Watson (2005) eine sprachliche Verwirrung, bei welcher der Ausdruck „50-50-chance" eher eine Unsicherheit, als eine spezifische Angabe einer Wahrscheinlichkeit darstellt. Häufig wird dieser Ausdruck für das Eintreten weniger wahrscheinlicher Ereignisse genutzt.

Die vorgestellten Studien weisen auf eine Vielzahl sprachlicher Schwierigkeiten beim Umgang mit Wahrscheinlichkeiten hin. Es wird deutlich, dass die Begriffsbedeutungen von Person zu Person unterschiedlich sind und individuelle Vorerfahrungen die Begriffe und vor allem den Begriffsgebrauch der Schülerinnen und Schüler beeinflussen. Für die Entwicklung empirischer und diagnostischer Instrumente bedeutet dies, dass zunächst der Sprachgebrauch der Schülerinnen und Schüler individuell erfasst werden sollte, um ihre Erklärungen und

Vorstellungen angemessen verstehen und deuten zu können. Gleichzeitig scheint die Form der Befragung von Bedeutung, da sie den Rahmen der Kommunikation absteckt.

Methodik und Anlage der Erprobung

Als Vordesign wurde ein Interviewleitfaden entwickelt, der aus drei Itemgruppen besteht, die in Tabelle 16.1 aufgeführten Begriffe fokussiert und zugleich verschiedene Frage-/Itemtypen enthält. Auf diese Weise sollte in einer Erprobung folgenden Fragestellungen nachgegangen werden:

1. Über welches Begriffsverständnis verfügen die Schülerinnen und Schüler zu den Begriffen: Zufall/zufällig, möglich, unmöglich, sicher?
2. Welche Fragetypen eigenen sich für die Erfassung dieses Begriffsverständnis besonders? Welche Vor- und Nachteile haben die verschiedenen Fragetypen? Wie müssen die Aufgaben konzipiert sein, um die Nutzungskonzepte und Begriffsvorstellungen der Schülerinnen und Schüler erfassen zu können?

Es wurden drei verschiedene Aufgabentypen verwendet. Tabelle 16.1 gibt eine Übersicht über die Zuordnung von Begriffen und Aufgabentypen. Die Fragen beruhten ausschließlich auf einem theoretischen (a priori) Zugang zum Wahrscheinlichkeitsverständnis.

Bei der rein sprachlichen Bewertung von Ereignissen wurden den Schülerinnen und Schülern mündlich Ereignisse vorgetragen, die sie bewerten sollten. Sie sollten dabei zunächst einschätzen, ob ein vorgetragenes Ereignis ein zufälliges Ereignis ist und dabei begründen, wie sie zu ihrer Einschätzung kommen. In einem zweiten Schritt sollten die Schülerinnen und Schüler die Ereignisse mit den Begriffen „möglich", „unmöglich" oder „sicher" bewerten und ebenfalls ihre Antworten erläutern, so gut sie konnten.

Tab. 16.1 Übersicht der Aufgaben und Aufgabentypen

Aufgabe	Begriffe	Aufgabentyp
1	Zufällig, möglich, unmöglich, sicher	Rein sprachliche Bewertung von Ereignissen
2	Sicher, unmöglich	Bewertung mit Sicht auf Material
3	Sicher, unmöglich	Handlungen am Material

Bei der Bewertung von Situationen mit Sicht auf Material wurden den Schülerinnen und Schülern Bilder von Aquarien vorgelegt, die zunächst mit roten und grünen, später zusätzlich mit gelben Fischen gefüllt waren. Zu diesen Aquarienbildern wurden sie gebeten, das Eintreten vorgetragener Ereignisse zu bewerten, in dem sie zu dem Ereignis passende Bilder bestimmten. In einem weiteren Schritt wurden die Schülerinnen und Schüler gebeten, Aquarienbilder nach bestimmten Vorgaben zu bestimmen (z.B. „In welchem Aquarium ist es sicher, einen grünen

Fisch zu angeln?"). Zuletzt wurden den Schülerinnen und Schülern Bilder von leeren Aquarien vorgelegt, die sie nach einer Vorgabe mit roten, gelben und grünen Fischen füllen sollten. Auch dabei waren sie stets aufgefordert ihre Handlungen zu erläutern.

Zur Erprobung wurden zehn Interviews mit je fünf Schülerinnen und Schülern zweier Realschulen am Ende der sechsten Klasse durchgeführt. Es wurde versichert, dass im Unterricht zuvor keine differenzierte Thematisierung des Begriffsgebrauchs in der Stochastik erfolgt ist. Die Interviews dauerten je 30 Minuten und wurden videografiert und transkribiert. Die Interviewtranskripte wurden im Anschluss inhaltsanalytisch vergleichend ausgewertet, um herauszustellen, inwiefern sich die Antworten der Schülerinnen und Schüler gleichen und in welchen Aspekten sie sich unterscheiden. Im Folgenden werden einzelne ausgewählte Ergebnisse der Erprobung kurz vorgestellt.

Ergebnisse der Erprobung

Rein sprachliche Bewertung von Ereignissen

Im ersten Aufgabenformat wurden die Schülerinnen und Schüler zunächst gebeten, einige mündlich vorgetragene Ereignisse dahingehend zu bewerten, ob ihr Eintreten zufällig oder nicht zufällig ist. Es stellte sich heraus, dass die befragten Schülerinnen und Schüler über weitestgehend zutreffende Nutzungskonzepte vom Zufallsbegriff verfügen, wie es im Folgenden anhand eines Beispiels kurz skizziert wird.

Während Schülerin K Zufall als etwas Ungewisses, ein Ereignis, das man nicht vorhersagen kann, beschreibt, steht in den Beschreibungen von Schülerin I der Aspekt im Vordergrund, dass der Zufall Ereignisse umfasst, die keiner Regelmäßigkeit unterliegen. Aus den Antworten von Schüler F hingegen geht eine weitere Dimension des Zufalls hervor, nämlich, dass Wahrscheinlichkeiten eine Rolle spielen. Es lässt sich beobachten, dass die Erklärungsansätze bei nahezu allen Ereignissen denselben Ansatz verfolgen, jedoch vereinzelt in Abhängigkeit von der Sachsituation angepasst werden. Die Einschätzungen der Schülerinnen und Schüler auf das Ereignis „Thomas würfelt beim Spiel „Mensch ärgere dich nicht" im ersten Wurf eine 5" lauten:

- „Zufällig, das kommt darauf an, wie viel Glück man hat und wie man würfelt." (Schülerin K)
- „Zufällig, er hätte am Anfang auch eine 2 würfeln können." (Schülerin I)
- „Das ist zufällig, man kann ja nicht entscheiden, welche Zahl man würfelt. Da gibt es ja mehr Möglichkeiten." (Schüler F)

Schülerin K beschreibt, dass man den Wurf eines Würfels nicht vorhersagen kann. Gleichzeitig ist sie jedoch der Meinung, dass man den Würfel beeinflussen kann, obgleich der Ausgang des Wurfs dadurch noch nicht mit Sicherheit vorhersagbar ist. Bei einer weiteren Aussage wiederholt sie diese Vorstellung für den Münzwurf und erklärt, dass es auch dabei darauf ankommt, wie man eine Münze wirft, es jedoch trotzdem ein zufälliges Ereignis sei. In den Antworten von Schülerin I steht stets die Unregelmäßigkeit des Eintretens eines zufälligen Ereignisses im Vordergrund. Auch die Antwort von Schüler F folgt einem konsistenten Ansatz, nämlich dass man für zufällige Ereignisse keine Vorhersage treffen kann, den Wurf nicht beeinflussen kann und die Anzahl der möglichen Ausgänge von Bedeutung sind.

Auffallend ist bei diesem Aufgabenformat, dass die Schülerinnen und Schüler ihre individuellen Erfahrungen in ihre Antworten einfließen lassen, wodurch sie zumeist weniger auf die konkrete Situation eingehen und ihre Antworten in Folge dessen nur geringe Einblicke in ihre Begriffskonzepte erlauben. Ferner zeigte sich besonders beim zweiten Aufgabenteil, in dem die Schülerinnen und Schüler Ereignisse dahingehend bewerten sollten, ob sie „möglich", „unmöglich" oder „sicher" sind, dass die Antworten nur wenig Aussagekraft besaßen und insbesondere das Sachwissen der Schüler prägend für ihre Antworten war. Hier zeigen sich die Grenzen des Aufgabenformats. Die Antworten weisen insgesamt nicht auf die Begriffsvorstellungen der Probanden hin, sondern spiegeln ausschließlich ihre Sachkenntnis wider. Verfügen die Schülerinnen und Schüler nicht über ausreichende Sachkenntnis oder Erfahrung, so geben sie Einschätzungen wieder, etwa ob sie ein Ereignis für möglich halten. Die Erfahrungen und das Wissen der Schülerinnen und Schüler erlauben es ihnen nur in hinreichend bekannten Sachverhalten differenzierte Einschätzungen zu geben, so dass dieses Aufgabenformat nur in sehr geringem Maße Einblicke in die Begriffsunterscheidungen „sicher – möglich" sowie „zufällig – nicht zufällig" der Schülerinnen und Schüler ermöglicht.

Bewertung mit Sicht auf Material

In der zweiten Aufgabe wurden den Schülerinnen und Schülern Bilder von vier Aquarien vorgelegt, die mit roten und grünen Fischen gefüllt waren (Aquarium 1: 7 rot, 1 grün; Aquarium 2: 0 rot, 5 grün; Aquarium 3: 3 rot, 2 grün; Aquarium 4: 6 rot, 7 grün).

Die Schülerinnen und Schüler sollten Aquarienbilder mit bestimmten Eigenschaften heraussuchen und ihre Auswahl begründen. Zunächst wurden sie gebeten, die Aquarienbilder zu benennen, in denen es sicher ist, einen grünen Fisch zu angeln. Alle Probanden wählten hier richtigerweise Aquarium 2, jedoch aus zum Teil unterschiedlichen Gründen. Ferner lieferten ihre Erklärungen auch weitere Einsichten, wie es die drei folgenden Beispiele zeigen:

- „In Aquarium 2, weil da viele grüne Fische sind. Bei Aquarium 1 sind mehr rote Fische; da ist es eher sicher, dass mehr rote als grüne Fische geangelt werden, weil da nur einer ist. Bei Aquarium 3 ist es Zufall, wenn man einen grünen fängt, weil da mehr rote als grüne sind. Und bei Aquarium 4 ist es eigentlich auch Zufall, weil alle Fische durcheinander schwimmen." (Schülerin J)
- „In Aquarium 4, weil da die Wahrscheinlichkeit größer ist als in Aquarium 1, weil da ist nur ein roter Fisch. In Aquarium 3 sind es nur drei. Auf jeden Fall ist es in Aquarium 2 sicher, weil da gar kein roter Fisch ist. Aber in Aquarium 4 ist es auch ziemlich sicher, so 75 Prozent. [Wie kommst du auf die 75 Prozent?] Das kann ich nicht erklären." (Schüler D)
- „In Aquarium 2, weil da nur ein grüner Fisch ist. Und in Aquarium 4, weil da mehr grüne Fische an der Oberfläche schwimmen und da gibt es auch mehr als von den roten." (Schülerin K)

In den Aussagen der Schülerin J und des Schülers D zeigt sich, dass der Begriff „sicher" einerseits eindeutig für ein unumgängliches Ereignis steht. Gleichzeitig machen sie jedoch Abstufungen und verstehen unter einem sicheren Ereignis nicht ein Ereignis, das *„immer* eintritt" (Hasemann et al. 2008, S. 153 Hervorhebung i.O.). So erklärt Schülerin J, dass es in Aquarium 1 „eher sicher" ist einen roten Fisch zu fangen und Schüler D erklärt, dass es in Aquarium 4 „auch ziemlich sicher" ist, einen grünen Fisch zu fangen, weil dort mehr grüne als rote Fische vorhanden sind. Diese Erklärungen lassen vermuten, dass die beiden Probanden zwar über ein alltäglich hinreichendes Verständnis des Begriffs „sicher" verfügen, welches jedoch nicht dem eindeutigen fachlichen Ausdruck entspricht, der keine Abstufungen oder Einschränkungen zulässt. Besonders deutlich wird dies in der Erklärung von Schüler D, der bei einem Verhältnis von sechs roten zu sieben grünen Fischen der Meinung ist, dass es annähernd sicher ist, einen grünen Fisch zu angeln. Er versucht weiterhin seine Erklärungen durch Angabe eines Wahrscheinlichkeitsmaßes zu stärken, das er jedoch auch ohne inhaltliche Vorstellungen bemisst. Auch Schülerin K spricht der Verteilung in Aquarium 4 eine Sicherheit zu, einen grünen Fisch zu angeln, weil „mehr grüne Fische an der Oberfläche schwimmen". Diese Erklärung offenbart eine Schwäche, die eher dem Material zuzuschreiben ist. Für die Schülerin spielt nicht nur die Anzahlverteilung der roten und grünen Fische eine Rolle, sondern eben auch die Position, die sie im Aquarium einnehmen. Dieser Faktor mag auch die Erklärung von Schüler D beeinflusst haben und es ist zu vermuten, dass in dieser Aufgabe die Grundannahme, dass alle Fische gleichwahrscheinlich anbeißen, nicht ausreichend geklärt ist.

Die Beobachtung, dass der Begriff „sicher" von der Mehrzahl der Probanden unpräzise gebraucht wurde, ließ sich in ähnlicher Form auch zum Begriff „unmöglich" machen. Sieben von zehn Schülerinnen und Schülern erklären, dass es in einem Aquarium mit sieben roten und einem grünen Fisch unmöglich sei, einen grünen Fisch zu angeln. Als Begründungen geben sie an, dass in diesem Aquarium „sehr viele rote Fische sind" (Schülerin B) oder „weil da nur einer [grüner Fisch] ist" (Schüler C). Schülerin K begründet in gleicher Weise „weil es da mehr rote gibt und man muss eigentlich schon ziemlich Glück haben, da einen grünen zu fangen". Diese Erklärungen zeigen deutlich, dass die Probanden den Begriff „unmöglich" nicht wie von Hasemann, Mirwald und Hoffmann (2008) definiert als ein Ereignis auffassen, „dass in *keinem* Fall eintreten kann" (S. 154, Hervorhebung i.O.), sondern verknüpfen diesen Begriff vielmehr mit einer geringen Wahrscheinlichkeit des Eintretens eines Ereignisses.

Handlungen am Material

Im dritten Aufgabenformat wurden den Probanden leere Aquarien vorgelegt, die sie nach bestimmten Vorgaben mit roten und grünen Fischen füllen sollten.

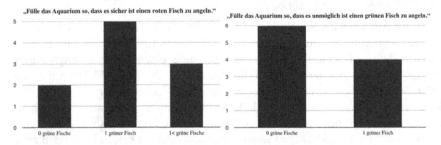

Abb. 16.1 Ergebnisse der Aufgaben 3a (links) und 3b (rechts)

Zunächst wurden die Schülerinnen und Schüler in Aufgabe 3a gebeten, die Aquarien so zu füllen, dass sicher ein roter Fisch geangelt wird. Nur zwei Probanden füllten die Aquarien ausschließlich mit roten Fischen. Die übrigen Probanden fügten mindestens einen grünen Fisch hinzu, drei Probanden sogar mehr als einen und erklärten dennoch, dass es wirklich sicher sei, dass in dem Aquarium ein roter Fisch geangelt wird. Ein ähnliches Ergebnis zeigte sich bei Aufgabe 3b, in der die Schülerinnen und Schüler das Aquarium so füllen sollten, dass es unmöglich ist, einen grünen Fisch zu fangen. Auch hier zeigten sich die Verständnisschwierigkeiten der Probanden und vier von zehn Schülerinnen und Schülern erkannten nicht, dass das Aquarium keine grünen Fische enthalten darf, wenn es unmöglich sein soll, einen grünen Fisch zu angeln.

Auch in dieser Aufgabe konnte beobachtet werden, dass das Material die Antworten der Schülerinnen und Schüler beeinflusst. Schülerin J füllte in Aufgabe 3b das Aquarium mit sechs roten und einem grünen Fisch und erklärte: „Hier ist es jetzt unmöglich einen grünen Fisch zu angeln, weil die roten Fische den Weg zum grünen versperren." Jedoch zeigten die Probanden in ihren Handlungen am deutlichsten, wie sie die Begriffe verstehen.

Diskussion und Ausblick

Die hier beschriebene Untersuchung ist als Vordesign im Rahmen der Entwicklung eines diagnostischen Interviews zur Stochastik zu betrachten. Obgleich es sich bei der ersten explorativen Erprobung zunächst um einen methodischen Zugangsversuch handelt, konnten dennoch einige bestätigende Beobachtungen bezüglich des Begriffsverständnisses der Schülerinnen und Schüler gemacht werden. Am Ende der sechsten Klasse scheint das Begriffsverständnis noch sehr stark der Alltagssprache verhaftet zu sein, obwohl die Begriffserarbeitung laut dem Lehrplan in der Grundschule erfolgt sein sollte. Die Begriffe „sicher" und „unmöglich" werden von den befragten Schülerinnen und Schülern nur in geringem Maße in einer fachlich fokussierten Weise interpretiert und verstanden. Dieses umgangssprachliche Verständnis der Begriffe erschwert die Kommunikation über zufällige Ereignisse und kann den Lernprozess der Schülerinnen und Schüler nachhaltig beeinflussen. Hinsichtlich der Entwicklung eines diagnostischen Instruments ist es folglich notwendig die sprachlichen Voraussetzungen der Schülerinnen und Schüler individuell zu hinterfragen und zu überprüfen, um ihre Äußerungen deuten und interpretieren zu können.

Eine wesentliche Frage, die sich im Anschluss an die erste Erprobung auftut, ist, wie Aufgaben und Testitems konstruiert sein müssen, um einen tieferen Einblick in die sprachlichen Nutzungskonzepte der Schülerinnen und Schüler zu erhalten. Im Wesentlichen sind die Begriffe der Wahrscheinlichkeit alltagssprachliche Vokabeln, die einer Interpretation im Sinne des Wahrscheinlichkeitsbegriffs bedürfen. In diesem Sinne muss auch der alltagssprachliche Umgang in Betracht genommen werden. Im Alltag begegnen Kindern Formulierungen wie etwa „unmögliches Verhalten". In Formulierungen wie dieser werden Begriffe in bestimmten Situationen anders verwendet. Für die Entwicklung von diagnostischen Aufgabenstellungen ergibt sich daraus das Problem, dass sie auf eine fokussierte Verwendung der Sprache ausgerichtet sein müssen. Zudem ist anzunehmen, dass sprachlich-begriffliche Nutzungskonzepte nicht unabhängig von stochastisch-begrifflichen Vorstellungen zu erheben sind und die kontextspezifischen Verwendungen stark von der Situation abhängen. Ein möglicher Ansatz ist es an dieser Stelle, die Schülerinnen und Schüler beispielsweise zu bitten, doch von sich aus Sätze zu bilden, in denen sie die Begriffe „zufällig", „sicher", und „unmöglich" benutzen. Eine derartige Aufgabenstellung würde es ermöglichen, die Begriffsnutzungskonzepte der Schülerinnen und Schüler in eigens konstruierten und damit ihnen vertrauten Situationen zu beobachten.

In der Erprobung hat sich zudem gezeigt, dass das genutzte Material sehr stark die Aufmerksamkeit der Schülerinnen und Schüler lenkt (vgl. Schipper 2009, S. 288 ff.) und ihre Interpretationen und Antworten beeinflusst. Gleichzeitig konnten gute Einblicke in die Vorstellungen und das Begriffsverständnis der Schülerinnen und Schüler anhand von eigenständigen konkreten Handlungen am Material gewonnen werden.

Die Aufgaben in dieser Erprobung beruhen ausschließlich auf einem theoretischen a priori Zugang zum Wahrscheinlichkeitsbegriff. Dieser statische Zugang ist jedoch nicht jeder Schülerin und jedem Schüler gleichermaßen zugänglich. Zugleich ist die theoretische Betrachtung von Ereignissen sowohl auf einer rein sprachlichen Ebene, wie auch anhand der Betrachtung oder der konkreten Handlung am Material mit Schwierigkeiten verbunden, die besonders darin bestehen, dass den Schülerinnen und Schülern zum einen die Situation verständlich und zugänglich sein muss und zum anderen auch eine kontextspezifische Diskussion ermöglicht werden soll.

Es ist deshalb in weiteren Erprobungen angedacht, vor allem frequentistische und simulierende Ansätze zu verfolgen, um die Schülerinnen und Schüler vermehrt in Handlungen einzubinden (vgl. z.B. Davies 1965) und auf diese Weise Einblicke in ihre Vorstellungen und damit ihre Begriffsnutzungskonzepte zu erhalten. Eine Studie von Langer (1975) zeigt auf, dass die persönliche Involviertheit in Zufalls- und Spielsituationen dazu führt, dass die Teilnehmer Situationen anders bewerten als in theoretischen und von ihnen unabhängigen Situationen. Ein weiterführendes Aufgabendesign sollte dementsprechend darauf abzielen, die Schülerinnen und Schüler vor Entscheidungen zu stellen, die es zunächst erfordern, eine Situation zu analysieren und dann im Anschluss eine Entscheidung auszuhandeln und zu erläutern. In diesem Zusammenhang verspricht die Einbindung in Wettkämpfe, Entscheidungs- oder Gewinnsituationen einen gehaltvollen Einblick in das Wahrscheinlichkeitsverständnis der Probanden zu ermöglichen, auf deren Grundlage dann auch die begrifflichen Nutzungskonzepte der Schülerinnen und Schüler in ihren Erklärungen beobachtet und interpretiert werden können.

Literaturverzeichnis

Amir, G. S., & Williams, J. S. (1999). Cultural influences on children's probabilistic thinking. *Journal of Mathematical Behavior, 18*(1), 85-107.

Davies, C. M. (1965). Development of the probability concept in children. *Child Development, 36*(3), 779-788.

Fischbein, E., & Gazit, A. (1984). Does the teaching of probability improve probabilistic intuitions? An explanatory research study. *Educational Studies in Mathematics, 15*(1), 1-24.

Fischbein, E., Nello, M. S., & Marino, M. S. (1991). Factors affecting probabilistic judgements in children and adolescents. *Educational Studies in Mathematics, 22*(6), 523-549.

Gal, I., Mahoney, P., & Moore, S. (1992). Children's usage of statistical terms. In W. Geeslin, & K. Graham (Hrsg.), *Proceedings of the 16th annual meeting of the International Group for Psychology in Mathematics Education* (Vol. 3, S. 160). Durham, New Hampshire, USA.

Garfield, J., & Ahlgren, A. (1988). Difficulties in learning basic concepts in probability and statistics: Implications for research. *Journal for Research in Mathematics Education, 19*, 44-63.

Garfield, J., & Gal, I. (1999). Teaching and assessing statistical reasoning. In L. V. Stiff, & F. R. Curcio (Hrsg.), *Developing mathematical reasoning in grades K-12* (1999 Yearbook, S. 207-219). Reston: VA: National Council of Teachers of Mathematics.

Green, D. R. (1983). A survey of probability concepts in 3000 pupils aged 11-16 years. In D. R. Grey, P. Holmes, V. Barnett, & G. M. Constable (Hrsg.), *Proceedings of the First Interna-*

tional Conference on Teaching Statistics (S. 766-783). Sheffield, GB: Teaching Statistics Trust.

Hasemann, K., Mirwald, E., & Hoffmann, A. (2008). Daten, Häufigkeit, Wahrscheinlichkeit. In G. Walther, M. Heuvel-Panhuizen, & D. Granzer (Hrsg.), *Bildungsstandards für die Grundschule: Mathematik konkret* (S. 141-161). Berlin: Cornelsen Scriptor.

Jones, G. A., Langrall, C. W., & Mooney, E. S. (2007). Research in probability. Responding to classroom realities. In F. K. Lester (Hrsg.), *Second Handbook of Research on Mathematics Teaching and Learning* (S. 909-956). Charlotte, USA: Information Age Publishing.

Kultusministerkonferenz (2004). *Bildungsstandards im Fach Mathematik für den Primarbereich. Beschluss der Kultusministerkonferenz vom 15.10.2004.* Abgerufen von http://www.kmk.org.

Langer, E. J. (1975). The Illusion of Control. In *Journal of Personality and Social Psychology, 32*(2), 311-328.

Ministerium für Schule und Weiterbildung des Landes Nordrhein-Westfalen (2008). *Richtlinien und Lehrpläne für die Grundschule in Nordrhein-Westfalen.* Frechen: Ritterbach Verlag.

Ministerium für Schule und Weiterbildung des Landes Nordrhein-Westfalen (2011). *Kernlehrplan und Richtlinien für die Hauptschule in Nordrhein-Westfalen.* Abgerufen von http://www.standardsicherung.schulministerium.nrw.de/lehrplaene.

Peter-Koop, A., Wollring, B., Spindeler, B., & Grüßing, M. (2007). *Elementar Mathematisches Basisinterview.* Offenbach: Mildenberger Verlag.

Rasfeld, P. (2004).Verbessert der Stochastikunterricht intuitives stochastisches Denken? Ergebnisse aus einer empirischen Studie. *Journal für Mathematik-Didaktik, 25*(1), 33-61.

Schipper, W. (2009). *Handbuch für den Mathematikunterricht an Grundschulen.* Braunschweig: Westermann Schroedel Diesterweg Schöningh Winklers.

Shaughnessy, J. M. (1992). Research in probability and statistics. In D. A. Grouws (Hrsg.), *Handbook of Research on Mathematics Teaching and Learning* (S. 465-494). New York: Macmillan.

Tarr, J. E. (2002).The confounding effects of "50-50-chance" in making conditional probability judgments. *Focus on Learning Problems in Mathematics, 24*(4), 35-53.

Watson, J. M., & Moritz, J. B. (2003).The development of comprehension of chance language: Evaluation and Interpretation. *School Science and Mathematics, 103*(2), 65-80.

Watson, J. M. (2005). The probabilistic reasoning of middle school students. In G. A. Jones (Hrsg.), *Exploring probability in school: Challenges for teaching and learning* (S. 145-168). New York: Springer.

Wollring, B. (1994). Animistische Vorstellungen von Vor- und Grundschulkindern in stochastischen Situationen. *Journal für Mathematik-Didaktik, 15*(1/2), 3-34.

Wollring, B. (2006). Welche Zeit zeigt deine Uhr? - Handlungsleitende Diagnostik für den Mathematikunterricht der Grundschule. *Friedrich-Jahresheft, 24*, 64-67.

Chapter 17
The epistemological character of visual semiotic means used in elementary stochastics learning

Judith Stanja, Heinz Steinbring

Universität Duisburg-Essen, Campus Essen

Abstract The paper has a theoretical focus and outlines the epistemological character of visual semiotic means[1] used in elementary stochastics learning. The role of these semiotic means as knowledge-oriented tools will be discussed with respect to Exploratory Data Analysis (EDA) and its contribution to the construction of new knowledge. One particular characteristic of the visual semiotic means used in primary school is their reliance on gestures and language as we will illustrate with a few examples taken from a qualitative empirical study with 3rd graders.

Introduction

There is a wide-spread discussion about the status of semiotic means in learning mathematics. For stochastics, several studies suggest that visual means play an important role in the development of stochastic thinking or the solving of stochastic problems. However, the status of semiotic means is rarely discussed. A very detailed discussion of the functions of visual semiotic means was given by Biehler (1982) in his epistemological analysis of Exploratory Data Analysis (EDA) as a discipline. We will present key ideas of his study and of following work, and use them to search for parallels and differences between Early Stochastics (primary school level) and EDA. This can be understood as a thought experiment that helps to discuss parts of a theoretical foundation for studying early stochastic thinking. To clarify our theoretical considerations, we will present illustrating examples from an on-going qualitative research project on early stochastic thinking in primary school.

The importance of visual semiotic means in elementary stochastics compared to semiotic means used in arithmetics and algebra is a result of the curricula and the available pre-knowledge of primary school children in these domains. We argue that the particular semiotic means (see Figure 17.1) used in the mentioned

---

[1] The notion of semiotic means must not be confused with the mathematical notion of mean (that is not used in the whole article). We will use semiotic mean as a generic term for communication tools, artefacts, general representations and mathematical signs.

qualitative study, only allow to express stochastic ideas in a first vague way. We show that those ideas could potentially be expressed in a more precise way and further developed with more elaborated semiotic means in secondary school stochastics.

How can ideas be expressed in Primary school? How can one work with the visual semiotic means given in Figure 17.1? How are they related to other types of semiotic means? When studying the epistemic functions of the visual means we see a certain comparability of elementary stochastics with EDA and ask what parallels can be drawn? Where are differences? One parallel to EDA concerns the importance of visual means that are predominantly used though in different ways. Another objective will be the role of the subject (student, data analyst) that is using the means and we draw a connection to constructivism.

Fig. 17.1 List and elementary diagramm

We start with an excursion on EDA where we mainly refer to the work of Biehler that is concerned with an analysis of EDA as a discipline. A next section deals with the description of some possible visual means for Primary school stochastics and their functions as well as the ways of working with them. We will also give some examples from the qualitative empirical study in 3rd grade to illustrate this. The last part presents parallels and differences in the use of semiotic means between EDA and Early Stochastics.

Excursion on Exploratory Data Analysis

Emergence and Classification

From the perspective of mathematics education as a constructivist science with the aim of curriculum development, Biehler conducted an epistemological analysis of EDA. According to his analysis (1982), EDA emerged in the 20th century in the struggle for a more applied statistics both in practice and at the university level. Statistics in the first half of the 20th century was mainly concerned with the testing of hypotheses, and data was analyzed for this purpose only. The outstanding work of Tukey, who is considered to be the founder of EDA, made clear that there are other problems in applications that needed to be considered such as good descriptions of data and construction of hypotheses on the basis of data. EDA tried to develop means and principles to work more effectively on such problems (Biehler 1982, p.166). When classifying EDA, different perspectives could be taken. For instance, Polasek (1988) understood EDA as part of descriptive statistics. According to him descriptive statistics is concerned with the description of the distribution of a characteristic whereas EDA is concerned with identifying what is conspicuous/noteworthy or explaining about the distribution of a characteristic. For others statistics and EDA are seen as separate subject fields (Biehler refers to Cox (1982) as an example). For his analysis of EDA Biehler identified three levels and their relationship: 1. meta-knowledge of statistics; 2. methods, means and techniques, mathematical theory of statistics; and 3. practice of data analysis (Biehler 1982, 168 pp.). Meta-knowledge, that is epistemological and philosophical orientation and self-conception of statistics, influences applications and plays an important role in the development and evaluation of statistical methods. When the relation between methods and practice changes, the meta-knowledge changes as well. With EDA, the role that probability usually played in statistics changed. Tukey particularly avoided terms of probability in his work to strengthen the perspective that real data does not need to have a probabilistic basis. He considered EDA as an empirical science that deals with real data. This claim cannot only be supported by Tukey's references to natural sciences and their experience in working with data but also by the standards for justification of methods and the evaluation of them. Biehler refers to Tukey (1962) who states that standards for deductive reasoning are not sufficient and sometimes not possible or necessary. Tukey also stresses the meaning of empirical-practical components of justification and the dominance of utilitarian considerations and broad applications instead of "security". Methods are evaluated by testing them with real data ("experimental sampling", Biehler 1982, p. 182) and the results obtained by EDA might be revised in the future since their origin is fraught with uncertainty (see Biehler 1982, p. 184).

In its original understanding, EDA is more concerned with knowledge construction in other domains and not in mathematics. The possible contributions of EDA methods to knowledge construction in other knowledge domains was already used as an argument for the relevance of EDA in school mathematics in the eighties. Moreover, it was discussed that the means used in EDA were relatively basic compared to the prerequisites in probabilistic statistics. An elementary way of using data could be the basis for more complicated methods of statistics (see Biehler 1982, p. iii). Nowadays, EDA is as well established as the basis or starting point for inductive statistics or confirmatory data analysis (CDA). As Tukey says: "Exploratory analysis is detective in character. Confirmatory data analysis is judicial or quasi-judicial in character;" and "Exploratory analysis can never be the whole story, but nothing else can serve as the foundation stone – as the first step." (Tukey 1977, p. 3)

Principles of EDA and the role of visual means

EDA deals with data as a system of characteristics – mostly numbers. The techniques concern "handling and looking" of and at "graphical, arithmetic or intermediate" semiotic means (Tukey 1977, p. 3). The representation of data is a crucial problem and necessary for the construction of knowledge. Data as representations are relatively autonomous compared to their meaning as numbers (Biehler 1982, p. 280). They have a dual character: a *simulative* function which indicates they store or represent knowledge about some objects and an *exploratory* function which indicates that they are used as tools to develop the knowledge further that is represented by them (Biehler 1982, p. 13). Biehler and Steinbring characterize the techniques and the functions of visualizations by the following aspects: exploratory and communicative means, storage of values and relational diagram, model character, variation and diversity (1991, p. 9). They describe the working methods as experimental and interactive. The interactivity characteristic can be understood in the following way: The analyzer explores the data by multiple analyses and representations and the obtained results influence the further activity of the analyzer (Biehler 1982, p. 227). EDA usually uses several analysing techniques and representations. As Biehler writes: "EDA is related to a more general movement in statistics towards using graphs as tools in research, as tools for analyzing data and not only as a means of communicating 'the obvious to the ignorant'" (1986, p. 79). EDA tries to represent as much information graphically as possible (Biehler 1982, p. 14). The emphasis on visual means is justified by some assumed advantages like easier information processing. Visual means are preferred to numerical summaries that are supposed to have the tendency to hide every structure for which they have not been constructed for (Biehler 1982, p. 313). Actually this might also be the case for visual means (for example, look at the elementary frequency diagram in Figure 17.1). Biehler emphasizes this too, arguing that graphic representations are not per se easier and more elementary than algebraic semiotic means. Accordingly, it depends on the intended usage how easily they can be understood and used (Biehler 1982, p.11).

The means may differ quite a lot – different relations might be emphasized. This explains the necessity and the efficacy of the variation of representations when exploring. There are several principles that characterize EDA. One is *variation of representations* which includes the activity of looking at data from a variety of points of view (Biehler 1982, p. 12). Another one deals with *visual inspection* which means to look for structures and characteristic features in the representation.

The multiple means help to focus on particular aspects of the data; they don't deal with all relations that could be imagined but emphasize some of them. Operating with data as numbers and signs without reference to the specific objects, is very important. The relevance of discovered structures can only be judged by taking the context into account. If relations are discovered in the data as a system of numbers the question arises whether they have a meaning at all, whether there are relations in the empirical context that correspond to the found relations?

Let us look at two of the representations that were introduced by EDA – the semi-graphic *stem-and-leaf* display and the graphic *box plot*. These representations are "often rendered prominent with regard to the secondary curriculum" (Biehler 1986, p. 80, compare also Bakker et al. 2004). The displays were introduced by Tukey (1977, p. 7; see Biehler 1982 for a comparison to histograms). There are versions of stem-and-leaf displays that store the original data, with ordered data and versions with less detailed information. The display can be stretched by adding stems if the original seems to be too "crowded" and equally can be squeezed by reducing the number of stems if the original seems to be too "spread out" (Tukey 1977, p. 11). Stem-and-leaf displays are used "for storage" and "for looking at" (Tukey 1977, p. 14). With some experience, we can find indications for instance for groups, the center of data or symmetry which might be extended and precised.

This empirical knowledge is relevant in applications of this technique in a two-fold way. On the one hand, to know that these characteristics might be seen in a stem-and-leaf display influences the choice of technique for analyzing the data. On the other hand, the knowledge of these kinds of indicators or patterns regulates to a certain extent what one will actually see and be able to communicate (Biehler 1982, p. 32). For examples of stem-and-leaf displays look at Figure 17.2 that shows two versions of the mother's ages of first graders in a Swedish classroom that were used for descriptions (constructed by the authors according to the originals by Dunkels 1986). (A possible continuation could be the comparison of the data sets of the ages of mothers and fathers in that class. Furthermore, hypothesizing about the distribution in parallel classes could be a starting point for informal inference.)

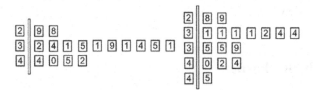

Fig. 17.2 Stem-and-leaf display and stretched stem-and-leaf display showing the mother's ages

Through the handling of the original data list – the building of a stem-and-leaf representation - we get insights about relations among the data. The activity of sorting and counting yields new properties such as the center (→ median) and extreme values (→ outliers) and so on. The new properties might then be made more precise and explicit in another representation, such as a box plot, which then serves as the basis for new insights or the discovery of more complex relations. This process of creating a visual means that "shows" the properties and relations discovered is closely related to what Sfard (2000) calls *objectification*. According to Sfard, introducing a new symbol is a crucial step to the generation of a concept and symbol and meaning constitute each other.

The box plot is the first plot that Tukey introduces for showing graphically the summary of some characteristics such as extreme values and middle values of data. Figure 17.3 shows a box plot based on the ages represented in Figure 17.2. The box plot is particularly useful for large data sets and the comparison of various data sets. As Biehler (1982) writes: "They provide a compact view of where the data are centered and how they are distributed over the range of the variable. They also provide easy ways to compare parts of distributions to see, for example, how the data in the top quartile compares in two groups" (p. 164). In box plots, the frequencies of the characteristics are not represented any more, but the densities are. So, stem-and-leaf and box plots differ in what they make explicit and therefore may serve different purposes in knowledge construction (compare this to the ideas of Bertin 1974 or Duval 2011).

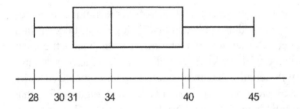

Fig. 17.3 Box plot

Because the analyzer has to "look at data", interpret them and make decisions such as which representations he uses and when to stop the analysis, Biehler speaks of a personalistic attitude, which means that results of data analyses entail subjective and personal elements that cannot be eliminated. This perspective corresponds to a constructivist view.

Visual Means in Early Stochastics

This section provides a description of visual semiotic means that were used in the qualitative research project on early stochastic thinking. We will discuss their functions as well as the ways of working with them. We will also give some illustrating examples from our study.

Presenting possible tools for Early Stochastics

The choice of means at primary school is determined by the possible pre-knowledge in arithmetic, geometry and algebra. For instance, the pre-knowledge in arithmetic forces us to work with natural numbers and absolute frequencies. In 2011/12, we conducted a qualitative study in 3rd grade with the purpose to provide children with tools that allow them to enter the stochastic discourse.[2] We chose two types of visual means, lists and elementary diagrams (Figure 17.1), that – as we assumed – allow students to take different perspectives on the situation (experiments with spinners). Activities with these means in the classroom lead to further visual semiotic means such as sorted sets of lists and diagrams that might be understood as pre-forms for conventional representations in more advanced stochastics. The choice of means to be used in the classroom was made by the teachers and not by the children. So, they did not work with self-created means. Also variations of means were requested by the teacher and not done by the child's initiative. One task in the classroom was, for example, to compare the means, to explore the possibilities and limitations of each and to discuss what one is able to see or not. To our knowledge it is not clear whether young children can consciously change visual or other means for the purpose of exploring the data. We suspect that the conscious variation of means presupposes to a certain extent some experience with the representations. The change of types of representations is a crucial step in the learning process as Duval (2011) pointed out. The templates facilitated the construction of the means. The reading of the means can be facilitated by coloured codings (see Figure 17.1 or the examples below).

The visual means allow students to study questions like: What would be the outcome of an experiment? In the case of colours this is a categorical attribute with the characteristics "yellow" and "blue". If we are interested in the outcome of an experiment without sorting according to the characteristics we can use the list. When recording the outcomes as they happen, the list represents a time sequence and allows to study questions about time issues such as: When did blue occur for the first time? How often in a row did blue occur?

The elementary diagram (see Figure 17.1) is a summary display and can be used when we are interested in the outcomes that are sorted according to the characteristics. The diagram is also useful when we ask "How often did the colours appear in an experiment with 20 outcomes?" Here, we are interested in absolute frequencies or in the quantity of the colour which is a numerical ordinal attribute with the 21 characteristics {0, 1, ..., 19, 20}. We can also ask for simple comparisons such as whether there was more blue or yellow.

Up to now we only considered a single representation list and diagram. Other representations can be constructed through activity in the classroom using the lists and diagrams. For instance, when the initial question was how the outcomes of the other children in the class looked like, the *protocols* of the students were gathered together and were sorted according to some characteristic for the purpose of clarity. These sorted sets of diagrams or lists served as new representations that were described by the children. By reference to the experimentation and the spinner, they tried to make sense of the observed results such as "always different looking"

---

[2] The participating classes had no or only little experience in stochastics.

(variability), "mostly more blue than yellow", "most children had about ...times blue", "yellow occurred only ...times in a row".

Functions of visual means

In the described situations and tasks for the young students, the semiotic means function as *protocols* for the conducted experiments and serve as a basis to describe the experiments. Using the means while conducting the experiment for control illustrates the *protocol function*, for example: the length of a trial by checking how many turns with the spinner are left, or the correct recording is checked by comparing pointer positions of the spinner with the corresponding entry, and for the purpose of memorizing the actual outcome of an experiment or a possible outcome when giving a prognosis. Besides this *protocol function*, the means possess also an *epistemic function*. The diagrams and lists resulting from experiments might be used as *knowledge-oriented tools* to see patterns and structures, describe the outcomes, construct connections to the random generator and the conducted experiment (proportions, how was the experiment done) as well as connections between several outcomes (similarities and differences).

When the problem of giving a prognosis for future experiments is introduced, the existing representations serve as the basis to hypothesize, to reflect about the uncertainty of statements about future outcomes and hence to construct new stochastic knowledge. Thus, whether a means has a protocol or epistemic function is determined or framed by the activities such as experimenting, recording, sorting, describing, observing, comparing, hypothesizing, and communicating in general and by the purpose of the activity. These protocol and epistemic functions are not hierarchical in the sense of a one-way road from protocol to knowledge-construction. Resorting lists according to some characteristic after a hypothesis was given, is one example in the opposite direction. So we get an interplay of protocol and epistemic function.

We will now give a few examples from interviews with 3rd graders (age 8-9)[3] to illustrate some *epistemic* and *communicative function(s)* of the means. In the interviews a spinner was used as a random generator, which is divided in 6 parts of equal size, one yellow and five blue parts. Here, we focus on questions about how stochastic ideas can be expressed with the given visual means and how children can work with these.[4]

Let us first look at what the student Mats did. His initial statement about a possible outcome was:

"Hm, I would take so five times yellow and fifteen times blue."

---

[3] The interviews are part of a qualitative research project on early stochastic thinking. The purpose of the interviews is to get some insight in how young children use and interpret semiotic means and how they understand stochastic prognoses.

[4] The interviews are originally in German and the statements were translated.

When being confronted with a template for a list and the task to make explicit what a possible outcome might be, he changed his focus. He used his filled list as a reference to say something about other possibilities for the outcomes and to make clear what he was focusing on. He has filled the list with a possible outcome (see Figure 17.4).

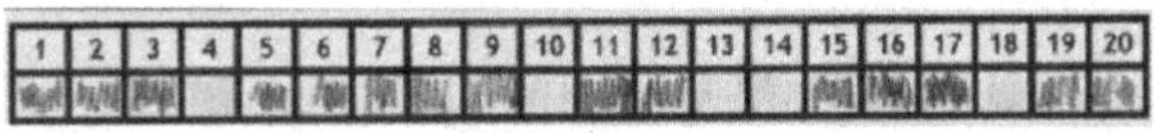

Fig. 17.4 Possible outcome for Mats

The interviewer wanted him to use the list to explain what he was thinking. Mats pulled the list to himself and said:

> "I have thought now, that, well blue is very often (*looks to the spinner*), that it could also come out more often next to one another but there (*points to the 13ᵗʰ and 14ᵗʰ field*) I have thought that sometimes it could be somehow two times or three times but three times I have thought is … not so, [...]"

In this statement we can also reconstruct the relation he sees to the spinner on the basis of the characteristic of more blue than yellow. It becomes also clear that Mats was focusing on the block length for the colors. When Mats was asked how certain he is that the outcome would look that way, he answered "60 percent" and "certain but not so certain". When the interviewer asked if the outcome could be different, he answered:

> "Yes, it could be that three times yellow comes out or so, that can also be but I have done it like that right now, because this … is a little chance [Zufall] if it is three at a time. Well, next to one another."

So, Mats focused on block lengths, here. With respect to this focus, he used the list as one example and as a reference to make explicit how other outcomes could look like and to evaluate the chance of the other outcomes with respect to his focus on the length of blocks. Other foci could be taken using a list. The student Ellen, for example, was looking at the single entries when asked to *compare* her initial idea with the actual outcome after an experiment:

> "Well, the first is already the same. The second and third as well. The fourth not. The fifth as well, the sixth not again, the seventh as well, the eighth not, the ninth not (.) the tenth as well, the eleventh also, the twelfth not, the thirteenth not as well, the fourteenth yes, fifteenth and sixteenth not, seventeenth yes, eighteenth not, nineteenth not, twentieth also not." (*she points at the corresponding fields in both lists, Figure 17.5*)

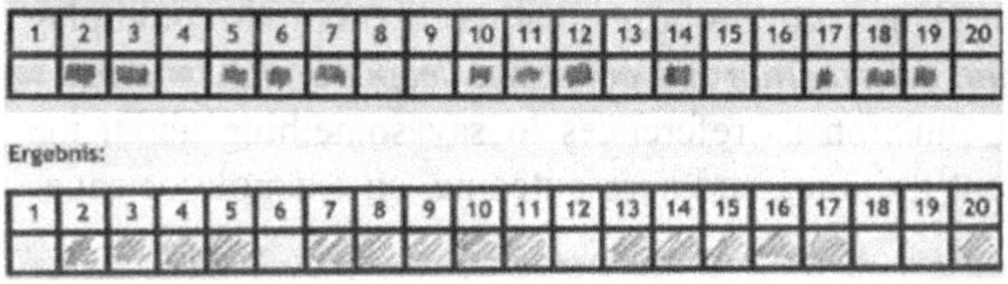

Fig. 17.5 Ellen's initial idea of an outcome (above) and the actual outcome of her experiment

When the interviewer put the focus on frequencies, she first focused on the length of blocks:

"One time yellow, four times blue,[...]"

and later on the absolute frequencies:

"Blue I have in total 15 times."

The preceding two examples show that different foci could be taken using a list: (non-) patterns, absolute frequencies, blocks in random sequences and so on. This is a strength of the list because it offers many possibilities for the stochastic discourse in classroom. At the same time this might be considered as a weakness or difficulty since the list, as part of the expression, relies on language and gestures. The list alone will not tell us what focus is taken.[5]

Let us look at the verbal expressions from another example of the student Lilly. She focuses on the absolute frequency of yellow, and she uses the list from her first experiment partially as a basis for her prognosis (for hypothesizing) about a new experiment:

"I would say for the next time that it is about five times yellow."

and

"Because, ehm, there (*looks at the list showing the outcome from her experiment before*) also six times came out, actually and I thought, I believe that in some cases when I got so many then, that then one less or two less will come out. And with five and fifteen, that has come out two or three times with a similar spinner with Mrs. B. That's why I believe it could look like that."

Later in the interview, she explains:

"I also look at the outcome. That is why I said one less yellow." and "Then this can be similar. Because, if one has done, if this came out, then something similar can come out just like that, because one cannot always get the same outcome with the same spinner. That is actually very very unlikely."

Here, the additional statements make clear that Lilly is aware of the idea of variability which cannot be expressed with the list alone.

The *reliance on gestures and language* (to express pre-forms or aspects of concepts of more advanced school stochastics) is a particular characteristic of visual semiotic means in primary school. Although used in a different context Ainsworth's (2006) ideas about several functions of representations in multiple representation systems could be interesting to get more insight about the interplay of these ways of communication. Ainsworth proposes that representations could play *complementary roles, constrain information* and *construct deeper understanding*.

Diagrams and lists might function as references to say something about the outcomes: they could represent the imagined sure outcome, an example or coun-

---

[5] The same ambiguity, although not as extreme as with the list holds for the use of elementary diagrams (see Figure 17.1). Interestingly, when working with the diagram, both Mats and Ellen focused on the absolute frequencies.

ter-example for outcomes that are possible, more or less likely or impossible, or they might be used to say something about ranges of frequencies or (non-) patterns.

Parallels and differences in relation to EDA

In this part we will focus mainly on two topics: the role of visual means and the role of the subject (student/ data analyst) in knowledge construction.

One purpose for the use of visual means is the construction of new knowledge. In the original understanding of EDA (see Tukey 1977), data analysis aimed at knowledge construction in applications outside of mathematics. Today, the visual means invented by EDA are used also for knowledge construction of confirmatory data analysis (inference) inside of mathematics. In school stochastics we understand the "Graphical means as representation for mathematical notions" (Biehler 1986, p. 80). He writes for example that "these displays, or slight modifications of them, can also enrich those parts of curricula where probability and inference is the main goal. Visualizing chance variability by drawing several random samples and representing them by a collection of box plots may be such an application. Thus, the displays can be used to initiate a *fruitful* comparison, interaction and restructuring of experience had with these diagrams in an exploratory setting, and in the context of ideal chance variability [...]" (p. 80). Biehler also refers to Landwehr, Swift and Watkins (1984), who used box plots to teach confidence intervals.

In our project the data had a probabilistic background and the intentions were to provide children with different elementary tools that allow them to take different perspectives on experiments with random generators and by sorting these visual means in different ways, to observe patterns and variability. The aim was also to support students in communicating their ideas in the classroom by allowing gestures such as pointing and showing ranges. A characteristic of elementary visual means in primary school is a *reliance on language and gestures*. The means allow only to express stochastic ideas in a rather vague way (as we could see in Lilly's case). To further develop these ideas and be able to communicate them in a more precise and structured way in secondary school, more complex visual means such as the box plot or combined means (as dot plot and box plot) or animations[6] might be used as well as other types of semiotic means as advanced arithmetic and algebraic means.

We can summarize from the previous sections that visual means have several functions. There is an analogy between the purposes in EDA and Early Stochastics: visual means are said to have a *communicative*, a storage or *protocol* function and a *knowledge-oriented* or exploratory function.

---

[6] See http://www.censusatschool.org.nz/2008/informal-inference/ for simulation of various samplings (from the same population) with box plots that keep traces of previous samples and allow to get information about sampling variation.

There are two types of learning involved in EDA: the *learning from data* and the *learning to learn from data* (see Biehler 1982, p. 34). This is analogous to the situation in primary school. Getting to know the tools challenges the learners since the use of tools can only be learned by using them (See Biehler and Steinbring 1991). Both in EDA and Early Stochastics, knowledge about possible patterns, indicators and so on is based on experiences. These are necessary for developing stochastic reasoning and the understanding of stochastic predictions. A crucial point is that semiotic means are necessary in stochastics since they are the only way to get access to abstract notions, but these semiotic means must not be identified with the stochastic concepts (see Duval 2000, p. 61) and young children are not familiar with these notions yet.

Children have to learn how a representation works, how and what information can be read from the representation and how the representation is related to the domain it represents. (see Ainsworth 2006) This is in agreement with Bauersfeld (2000), who states that every visualization needs to be learned and meanings have to be actively constructed by the learning children. Learned or spontaneous interpretations of means that occur due to a transfer from other mathematical domains have to be taken into account. For example, areas from geometry need a new interpretation when dealing with box plots. A smaller area means more dense which might be counter intuitive (see also Bakker et al. 2004). Other examples are symbolic formats from arithmetic like fractions and percentages or function graphs from analysis.

We have also seen in the examples that children do use the visual means in different ways. So, considering the role of the subject should be part of reflections about knowledge construction and learning in stochastics from a perspective of a constructivist understanding of learning.

References

Ainsworth, S., Bibby, P., & Wood, D. (2002). Examining the Effects of Different Multiple Representational Systems in Learning Mathematics. *The Journal of the Learning Sciences, 11*(1), 25-61.

Ainsworth, S. (2006). A conceptual framework for considering learning with multiple representations, *Learning and Instruction, 16*, 183-198.

Bakker, A., Biehler, R., & Konold, C. (2004). Should young students learn about box plots? In G. Burrill, & M. Camden (Eds.), *Curricular Development in Statistics Education. Proceedings of the International Association for Statistical Education (IASE) Roundtable* (pp. 163-173). Auckland: IASE.

Bauersfeld, H. (2000). Radikaler Konstruktivismus, Interaktionismus und Mathematikunterricht. In E. Begemann (Ed.), *Lernen verstehen – Verstehen lernen* (pp. 117-146). Frankfurt a.M.: Peter Lang.

Bertin, J. (1974). *Graphische Semiologie. Diagramme Netze Karten.* Berlin: De Gruyter.

Biehler, R. (1982). *Explorative Datenanalyse – Eine Untersuchung aus der Perspektive einer deskriptiv-empirischen Wissenschaftstheorie.* IDM Band 24. Bielefeld: Universität Bielefeld.

Biehler, R., & Steinbring, H. (1991). Entdeckende Statistik, Stengel-und-Blätter, Boxplots: Konzepte, Begründungen und Erfahrungen eines Unterrichtsversuches [Statistics by discovery, stem-and-leaf, box plots: Basic conceptions, pedagogical rationale and experiences from a teaching experiment]. *Der Mathematikunterricht, 37*(6), 5-32.

Dunkels, A. (1986). EDA in the primary classroom – graphing and concept formation combined. In R. Davidson, & J. Swift (Eds.). *The Proceedings of the Second International Conference on teaching statistics* (pp. 79-85). Victoria, B. C.: University of Victoria.

Duval, R. (2011). *Idées directrices pour analyser les problèmes de compréhension dans l'apprentissage des mathématiques.* Recife, Brasil: XIII CIAEM-IACME.

Duval, R. (2000). Basic Issues for Research in Mathematics Education. In T. Nakahara, & M. Koyama (Eds.), *Proceedings of the 24th International Conference for the Psychology of Mathematics Education* (Vol. I, pp. 55-69). Hiroshima, Japan: Nishiki Print Co., Ltd.

Polasek, W. (1988). *EDA - Explorative Datenanalyse.* Heidelberg: Springer.

Sfard, A. (2000). Symbolizing mathematical reality into being: How mathematical discourse and mathematical objects create each other. In P. Cobb, K. E. Yackel, & K. McClain (Eds), *Symbolizing and communicating: perspectives on Mathematical Discourse, Tools, and Instructional Design* (pp. 37-98). Mahwah, NJ: Erlbaum

Tukey, J. W. (1977). *Exploratory Data Analysis.* Reading, MA: Addison-Wesley.

Chapter 18
Contexts for Highlighting Signal and Noise

Clifford Konold

University of Massachusetts Amherst

Anthony Harradine

Prince Alfred College

Abstract   During the past several years, we have conducted a number of instructional interventions with students aged 12 – 14 with the objective of helping students develop a foundation for statistical thinking, including the making of informal inferences from data. Central to this work has been the consideration of how different types of data influence the relative difficulty of viewing data from a statistical perspective. We claim that the data most students encounter in introductions to data analysis—data that come from different individuals—are in fact among the hardest type of data to view from a statistical perspective. In the activities we have been researching, data result from either repeated measurements or a repeatable production process, contexts which we claim make it relatively easier for students to view the data as an aggregate with signal-and-noise components.

1 Introduction

Suppose you wanted to introduce 12-year-old students to basic ideas in statistics such as center and spread. Here are two short descriptions of classroom activities involving the collection and analysis of data which you could use. Which option would you select and why?

Activity 1. Students collect information about themselves including their gender, height, and distance traveled to school. They explore questions such as whether 12-year-old boys are taller than 12-year-old girls.

Activity 2. Each student measures the length of a table using two different measurement instruments. They explore questions such as whether one instrument gives a more accurate measure than the other.

Based on an informal analysis of published data activities, our guess would be that most readers would prefer Activity 1. The overwhelming majority of activities for all age levels are similar to Activity 1, in that students work with data from contexts where the attributes they are dealing with, such as height, result from what we will refer to as "natural" variability. Rarely do we encounter published activities similar to Activity 2 that involve repeated measurements. As to the reason that educators might give for their preference, we imagine many would regard the question posed in Activity 1 as being of interest to students whereas the question in Activity 2 seems rather boring. Would students really care about the length of a table, let alone the characteristics of repeated measurements of it?

In this chapter we argue the pedagogical merits of using contexts such as Activity 2. During the past 8 years, we have conducted several instructional interventions with students aged 12 – 14 as well as with teachers using contexts similar to Activity 2 where our objective has been to establish a solid foundation for statistical thinking. In this article, we describe this objective and explore the affordances of different contexts in making those ideas visible to students and in supporting classroom discourse about important aspects of those contexts.

2 Characteristics of Three Statistical Contexts

Fig. 18.1 Fruit sausages made by three different students. Ideally, sausages would all be the same length and thickness and thus weigh the same.

In collaboration with Rich Lehrer and Leona Schauble, we have been pursuing a speculation put forward in Konold and Pollatsek (2002)—that data from some contexts are considerably easier than others to conceive of statistically as combinations of signal and noise (see Konold and Lehrer 2008, for a review of some of this research). In the context of repeated measurements, we have involved students in measuring various lengths (e.g., their teacher's arm span, a table, the footprint of a crime suspect). More recently we have tested manufacturing contexts including packaging toothpicks, cutting paper fish to a desired length, and producing Play-Doh "fruit sausages" of a specified size (see Figure 18.1).

There are many similarities between the contexts of manufacturing and repeated measurements but also some interesting differences (Konold and Lehrer 2008). These differences have led us to consider whether manufacturing processes might provide a more suitable context in which to involve young students. The main advantage is that it is clear in the manufacturing context why we are producing multiple objects—it is the nature of manufacturing. And we measure samples of them for quality assurance. By contrast, in most repeated-measurements contexts (such as determining the length of a person's arm span), the reason for repeatedly measuring is not as clear. In real life, we measure once or twice, and exercise care if the measurement matters. For this reason, it might be rather challenging to motivate students to repeatedly measure and to sustain their interest (though there appears to be no lack of interest in Lehrer's classrooms). Secondly, the outputs of a production process are physical objects and not just values. Students can look at the manufactured objects and note the variability even before they measure them (see Figure 18.1). Later, when looking at measurement values, students can re-inspect the physical objects, coordinating observations from graphs with features of the real objects. Finally, that the data from the production process are individual objects makes the context closer, conceptually, to the context of natural variability. Because of this similarity, it seems reasonable to expect that students could more easily apply knowledge formed in the study of production processes to situations involving natural variability.

In short, our claim is that in the contexts of repeated measurement and production it is clear that 1) we are using our data to try to infer a single value (a signal) and that 2) the variability in the observed values is a nuisance (noise) obscuring the signal. By contrast, both the existence and nature of signal and noise in contexts of natural variability are difficult to conceive. When, for example, we summarize with a single value the distribution of heights of a sample of adult males, we find it rather hard to explain what we are trying to capture beyond perhaps the population parameter if we are trying to make an inference. You can point to nothing in the real world to which the mean of this sample of heights refers, whereas the mean of a sample of measurements of a table refers to the actual length of that table.

When we claim that it is easier to perceive signal and noise in the contexts of data production and repeated measurement we do not mean to suggest that these components can be directly perceived in data. The development and refinement of these and other statistical constructs and perceptions are goals of our instruction. Rather, our contention is that these contexts provide a more suitable beginning point for developing statistical practices and perspectives in our students.

Our take on the nature of what it is students ideally learn and how they learn it is consistent with the views of Bakker and Derry (2011). Their view is, in turn, grounded in the philosophy of Robert Brandom (as cited in Bakker and Derry 2011) who argues that knowledge in a particular domain is more than a collection of mental representations but rather comprises an interrelated web of ideas, skills, and justifications. Consider, for example, using a mean of several measurements to estimate the true length of a table. To do so, we probably have facility with the algorithm along with knowledge about various characteristics of the mean. But more fundamental to the use of the mean is this context are the reasons we give for computing and using it, the hedges that we offer, our justification for removing particular extreme values from our computation, the alternatives that we considered and our reasons for rejecting them, the explanations we give for our observations. Accordingly, as we describe the contexts and activities below, we pay particular attention to the classroom conversations that typically arise. It is from these conversations about what students are doing, and why, that we believe a more sophisticated "web" of statistical understandings emerge. Thus we can refine our central claim as follows: by using activities involving repeated measurement and production process we can motivate and shape the kinds of conversations among students from which they can learn to perceive data as a combination of signal and noise. Below, we elaborate our claim and analyze episodes from one of our classrooms.

2.1 Repeated Measures and Manufacturing Contexts

When we repeatedly measure a feature of some object, we obtain a distribution of values for that measure. Below, for example, are 19 measurements of the length of a table made by different students using the same metal tape measure.

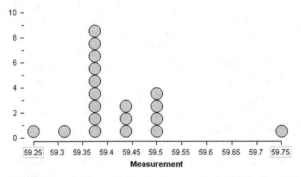

Fig. 18.2 Distribution of 19 measurements of the length of a table

Two of the questions we pose to students once they see the class data are: 1) What is your best guess of the actual length of the table and 2) Why are the measurements not all the same? These questions provoke some interesting responses. Having observed one another measure the table, students can quickly generate several explanations for the variability. Aspects they mention include the placement of the tape's end hook, where at the other end a person sights, how parallel the tape is placed to the table's edge, and how the measurement is rounded. Similarly, for the various manufacturing contexts we have tried, students can provide detailed descriptions of process components that produce variability. They can do this because they have participated in the manufacturing or measurement process and have observed their classmates doing the same and have noted differences in how measurements are made or the product is created. In producing the fruit sausages shown in Figure 18.1, for example, students used a device that extruded a long strand of Play-Doh. But these strands had a different surface texture and were of different width depending on how hard a student pressed down on the extruder's handle. Additionally, students measured the strand to cut it into sections that were supposed to be 5 cm long. However, the lengths of these pieces varied both due to measurement error and also to differences in the cutting procedure. Finally, students weighed the cut pieces and here again they could observe how carefully this was done in different groups.

Thus it is not only the contexts themselves but the students' direct involvement in making measurements or product that we believe allows them to fairly naturally regard the resulting data as having two general components. One of these is the signal, or fixed component. The true length of the table does not vary. Because each measurement is performed on the same table, the length of the table is captured in some sense in each measurement. Similarly, in the manufacturing contexts there is a specific target—how much a fruit sausage should weigh or the number of toothpicks in a package. We do not initially reveal this target number to the students but rather tell them that if they carefully employ the prescribed production procedure they will get very close to the target value. Again, since this target does not change and each fruit sausage, for example, is made using the *same* procedure, the signal from that process is present in the measured weight of both an individual sausage and aggregate characteristics of many sausages.

The other component of data from the measurement and manufacturing contexts is its variability. This component is undesirable (hence the term noise) and is caused by errors of various types. Because the students have themselves made these measurements or manufactured the product, they know a lot about the factors which contribute to that variability. To highlight these features, we often have them do measurements with two different tools or produce a product using two different procedures, one of which tends to produce much more variability than the other.

To summarize, in the repeated measurement and manufacturing contexts students can generate reasonable hypotheses about causes of error, suggest ways to reduce error, anticipate how the distribution of data obtained with a more accurate measuring tool or more controllable process will compare with data obtained using a less accurate tool or process, and expect that the actual measurement or manufacturing target lies somewhere in the middle of the distribution where the observations tend to cluster.

A critical feature of these contexts is that the inference one needs to make from data is clear. We want to know the length of the table or to determine whether our process is creating product according to specification. This clarity of question and purpose is often missing from activities with data and, as a result, the need for making inferences and justifying how they are arrived at is never established (see Makar and Rubin 2009). These conditions often exist when students are exploring data from what we are calling the context of natural variability.

2.2 The Context of Natural Variability

The ability we described above to observe and even control the process from which the data result is typically absent in the world of natural objects. Given data about the heights of boys and girls in their classroom, for example, students might have filled out and observed others filling out a questionnaire from which the data were collected. But they cannot observe firsthand the environmental and genetic factors that influence individual difference in height. Especially for younger students, the causes associated with these individual differences are rather mysterious. Furthermore, why should we consider all the heights of boys together in a single distribution and summarize that distribution with something like a mean? According to Stigler (1999, pp. 73-74) this conceptual difficulty was the reason that it took as long as it did to apply statistical thinking to what we have called natural objects:

> The first conceptual barrier in the application of probability and statistical methods in the physical sciences had been the combination of observations; so it was with the social sciences. Before a set of observations, be they sightings of a star, readings on a pressure gauge, or price ratios, could be combined to produce a single number, they had to be grouped together as homogeneous, or their individual identities could not be submerged in the overall result without loss of information. This proved to be particularly difficult in the social sciences, where each observation brought with it a distinctive case history, an individuality that set it apart in a way that star sightings or pressure readings were not. ... If it were felt necessary to take all (or even many) of these [distinct case histories] into account, the reliability of the combined result collapsed and it became a mere curiosity, carrying no weight in intellectual discourse. Others had combined ... [the individual cases], but they had not succeeded in investing the result with authority.

2.3 Comparing Critical Features of the Contexts

In Table 18.1 we summarize the critical features of data from these three contexts, repeated measurements, manufacturing, and natural objects.

Consider first the types of processes that produce the objects in the three contexts. Both repeated measurement and manufacturing involve processes that, in theory, we can observe; the processes involved in the creation of natural objects are generally unobservable and complex. Consider next the nature of the variability (or noise). In the repeated measurement and manufacturing contexts, variability is undesirable. In these contexts, if it were possible, we would eliminate all variability. Indeed, because we are partially in control of these processes, we can work to minimize variability. In the case of living natural objects, variation is a positive feature, critical to species survival. If we refer to variability in these natural contexts as noise, we do so only in a metaphoric sense. Finally, consider the signal as indicated by the mean, for example. In distributions of repeated measurements and manufactured objects, the mean of a sample is an estimate of a real-world entity—the length of a table or the target of the manufacturing process. To refer to means as signals in the case of natural objects is, again, only metaphoric as they have no specific real-world referents. Quetelet (1842), who was the first to apply statistics like the mean to measurements on people, invented the imagery *l'homme moyen* (average man) to stand in place of a concrete referent.

Table 18.1 Comparison of critical features of data from three different contexts

Context	Process	Noise	Signal
Repeated Measurements	Observable measurement protocol	Measurement error	True value
Manufactured Objects	Observable mechanical process	Process variation	Target
Natural Objects	Unobservable multi-system process	Individual differences	?

3 Making Sausages: A Teaching Experiment

In April 2007 we conducted a weeklong teaching experiment with 15 students, aged 13-14, at Prince Alfred College, an all-boys private school in Adelaide, Australia. Our activity was centered around the production of "fruit sausages," small cylindrical pieces of Play-Doh that ideally were to measure 5 cm in length and 1 cm in diameter. If made according to these dimensions we determined that the sausage would weigh 4.4 grams, but we did not reveal this target weight initially to the students; we hoped that they could eventually infer it from analysis of their data. On the first day of production, each student made five fruit sausages by hand. Working in teams of three they then weighed the sausages using first a triple beam balance and then a digital scale. Analysis of the data showed the weights to be quite variable, which set the stage for introducing a pressing device that students could use to squeeze out a 1 cm diameter length of material. Continuing to work in teams, students made another batch of sausages using this device. Analysis of these data culminated with students developing measures of center and variability to decide which team's product was 1) closest to specification and 2) was least variable. We then asked them to anticipate what the data would look like if we made 1,000 fruit sausages. Finally, students ran and critiqued a model built in *TinkerPlots* (Konold and Miller 2012) to simulate their manufacturing process to test their hypothesis about what this distribution of 1,000 would look like. Below, we analyze excerpts from the classroom conversation, highlighting aspects of the context that we believed helped to support the development of ideas of signal and noise.

3.1 Measurements on the Hand-Made Sausages

Looking at the data of their weight measurements sparked animated conversation about the variability they observed. One student's data attracted particular attention as his sausages were very fat, weighing over twice as much as most other students' product. Students concurred that these fat sausages would be extremely popular with consumers. But representing management we reminded them of the bottom line and the importance of keeping down costs. This analysis and discussion motivated the need for a process that would produce more consistent results closer to the product specifications.

In most of our activities we have students use at least two measurement methods and/or production processes so as to produce distributions of different characteristics. Here, students expected the measurements using the triple beam balance would be more variable. In fact, they were of the opinion that the digital scale would measure nearly perfectly, and because it served our purposes we did not try to disabuse them of this belief. By subtracting the weights obtained from the digital scale from those from the triple beam, we created a new attribute (see Figure 18.3) that the students could consider as the error in the weighings from the triple beam. The color indicates the team from which the measurements came.

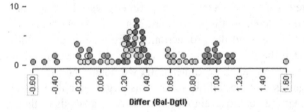

Fig. 18.3 The differences between the weights as determined on the digital scale and triple beam balance. The sausage on the far right weighed 1.6 grams more on the triple-balance beam than on the digital scale. Circles are colored according to the student team who made the measurements.

Looking at this graph, students quickly focused on the cluster of values on the far right. To help the conversation, we separated the distribution into the different teams (Group) and placed a reference line at 0 on the axis (Figure 18.4).

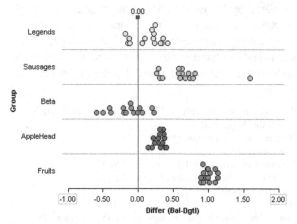

Fig. 18.4 This graph was made by pulling the groups out of the distribution shown in Figure 18.3 and adding a reference line to indicate where the values would be if a sausage weighed the same on both scales.

Below is part of the classroom discussion prompted by the graph in Figure 18.4. Ted was a member of the Fruits group whose data is on the lower right of the graph. CK is Konold.

Ted: We thought our triple-beam balance was out of whack by about 1 gram.
CK: So why do you think that?
Ted: We were over weighing it, so, because all of ours [the differences] were consistently around 1.
CK: ... And how would it be out of whack by a---
Jay: The thing [adjustment mechanism] at the back.
Ted: Yeah, because you can adjust it.
CK: Did you guys not adjust it?
Ted: No. ... And then what we decided is that it was out of whack by 1 because all our ---, we have a big clump around the one mark, and nothing else anywhere else.
CK: And what does it mean that you guys are all close here?
Ted: We're pretty consistent weighers.

From this graph Ted makes an inference about the amount of bias in the weighings from their triple beam balance. His estimate of 1 gram is based on where their difference values are "clumped" in the distribution. At this early stage in the multi-day activity, we were not expecting students to characterize their data using formal averages. Our hope was that, as Ted does here, students would begin to use their informal notions of modal clump (Konold et al. 2002) as an indicator of center and draw inferences from that about the location of an unknown but specific value—a signal. In this case, that signal is the amount of bias introduced by their failure to calibrate the scale before using it. As indicated by his final comment, by this time Ted was interpreting variability in terms of *consistency*.

Students were also beginning to distinguish between consistency (limited noise) and accuracy (being close to the true value) as indicate by the exchange below. This conversation was initiated by Harradine (AH) who directed the class's attention to the data in Figure 18.4 from some of the other teams. These notions of consistency and accuracy are, of course, interpretations of noise and signal. Data from the contexts of production and repeated measurement lend themselves to these interpretations of consistency and accuracy in a way that data from contexts of natural variability usually do not.

AH:	So, I want to know then, who are the Appleheads? So, if what they [Ted] said is true, what's the story with the Appleheads?
CK:	So are they as consistent as measurers as you guys?
Keith:	No, but they were the most accurate.
AH:	So hang on, whoa. So what does it mean that they're closer to 0?
Jamie:	It means they adjusted their triple-beam balance closer to the 0 mark.
AH:	Now these four other groups up here?
?:	They're all spread out.
AH:	So who's the Beta group? You guys. How do we now, how do we explain why you guys were like this?
Jamie:	… Because we adjusted the scales after I'd used it [Jamie's data are on the far left of the Beta group.]
AH:	Ahhh, you adjusted the scale. … So these have shifted, so you reckon if they…
Jamie:	If they, if Ken and Keith hadn't changed it, then they'd be kind of on top of each other, because they wouldn't have been…
AH:	Oh, you reckon they were adjusted in between when you measured them?
Keith:	Yeah, he measured it, and then I adjusted it to 0.
AH:	And then you measured some more. Ah, so where would these be [if the scale had been adjusted first]?
Jamie:	Underneath the others, the big purple clump.
AH:	These ones [on the right]?
Jamie:	Yeah.

Explaining variability is one of the primary objectives of data analysis. Here Keith is imagining their single distribution of weight differences as comprising two groups—those made before and those made after the scale was recalibrated. These students' experience of producing and then measuring the sausages provides the critical source for these explanations. They can relate some of the variability in their data to their own actions and thus offer explanations for the variability. With prompting, Jamie could imagine transforming their data to remove the effects of this manipulation, mentally moving one cluster of data into the other, effectively removing the effect of the recalibration and reducing the overall variability.

3.2 Models of Measurement and Production

In most of our activities involving measurement error or production systems, we conclude by having students build and/or run simulations of the situations they have been investigating. One purpose of this modeling component is to explore statistical aspects of the situations that are not possible with data from the real situation, mainly because with limited time, we cannot collect enough data. The more important reason we include the modeling component is to objectify the signal-noise, statistical view of the situation. By building and working with models of the situations, we hope to foster a more generalized view of signal-noise, one we hope students can eventually apply to a wide variety of contexts, including contexts of natural variability.

Figure 18.5 shows the model we introduced to the students on the final day of the teaching experiment, describing it as a model of the process for making sausages using the extruder according to our specifications. By this point, we had informed students that the ideal sausage would weigh 4.4 grams. This value is entered in the counter object on the left of the sampler. The range and distribution of values in the "variation" spinner are based roughly on the class's experience with making and weighing sausages. To model the making of, for example, 30 sausages, we hit run. The sampler first selects the value 4.4 from the counter object and then selects a value for weighing error by activating the spinner on the right and taking the value from the slice the spinner's arrow randomly lands in. These two values, the target and the error, are sent to the data table on the right where they are summed to give a final modeled weight. For the first simulated sausage, that weight is 4.4.

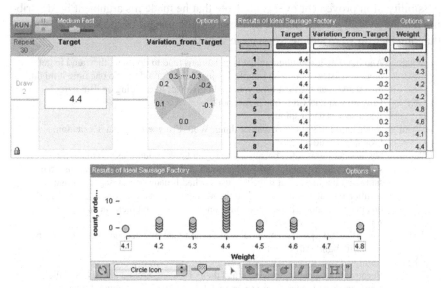

Fig. 18.5 A model of the process of making fruit sausages with the extruder built in TinkerPlots. The sampler in the upper left contains the target weight of 4.4 grams. A random value for error from the target weight is selected from the variation spinner. On each trial, these two values are selected and sent to the table on the right where they are added to get the value of Weight. The shape of the distribution of values shown above is determined by the make up of the variation spinner; its position along the axis is determined by the target value.

After briefly explaining the model and then running it to make four simulated sausages, we asked:

CK: So, what do you think?
?:   Uh, good!
AH: Will it make data like our sausages? Is it a good model do you reckon?
Various: Yeah.
AH: Go on, Vern.
Vern: It's a decent model, but, yeah, but it's not exactly humanlike.
CK: It's not humanlike? What do you mean?
Vern: As in humans usually do something wrong with theirs...
CK: Give me an example of a human kind of error that we wouldn't get here.
Vern: Like, if someone just repeatedly smashed on the squisher [pressed hard on the Play-Doh extruder]. That would make a different weight than, probably, one of those [simulated sausages from the sampler].

Examining a model tends to surface a number of ideas students have about the process. Vern proposed that while the model is "decent" it does not take into account the sort of systematic effects they had observed, for example that pressing hard on the extruder produced thicker sausages that weighed more than those made by pressing more gently. Wyatt added that the model did not allow for the possibility of improvement over time. Note that he made his argument by describing not the overall weight but rather the error component of the weight, a way of thinking that we were aiming to promote with the model.

Wyatt: Another way it's not humanlike is as humans tend to do more, they tend to get more consistent. This [the sampler] is just doing it random. So one time it might do a 0 and then the next time you might do a 0.5, like adding on. But here …
CK: How do you know this one is random?
?:   Because it is just a spinner.
Wyatt: Yeah, cause this is just a spinner thing. Well, I'm guessing that it's random. Probably would be....
Wyatt: No, it's not humanlike, because, um, if you think about it, as you make more sausages, like by hand, you'd get generally more consistent and be around the same number more. But this, like just say had I made one sausage, sausage number 100, at 4.4 in grams, then the next one 4.5, to be consistent. But then on this [sampler model] you might have one that is 4.0 and then one at 4.05 or something, 'cause it's just random.
Ted: Aah, [I've got a] comment about it. It doesn't take into account that each group makes it slightly differently.
CK: Right.
Ted: Because, for example, the Fruits had theirs focusing around about 4, and the Appleheads had theirs focusing around 4.4. And we're assuming the factory always makes them precisely on 4.4.

People have a strong inclination to offer explanations for events or trends even when there is good evidence that nothing but chance is responsible. So at the same time we encourage students to pose theories that account for trends, we also want them to develop the ability and habit to question whether the patterns they notice might have resulted from chance. Taking this idea into consideration is the prime motivator for statistical inference. While it certainly is reasonable to expect that makers of fruit sausages get better with practice, we have students both look at their data for evidence of this and also run the model and see whether they can get similar "trends." Thus we do not typically compute any probabilities, but find that without many trials from the sampler, students develop the sense that some rather stunning patterns can occur just by chance. For this purpose we have used the kind of display seen in Figure 18.6 because it more directly depicts than does the stacked dot plot the idea of a signal with noise scattering data around it randomly. It is especially powerful (and entertaining) to watch it build up in real time.

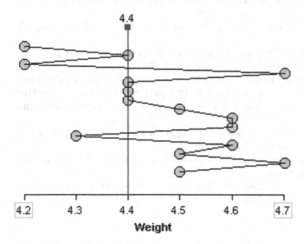

Fig. 18.6 The weights of 15 simulated sausages displayed as a time series

4 Conclusion

We have argued that classroom activities that involve repeated measurements and manufacturing are particularly suitable for introducing students to statistics. This is primarily because in these contexts statistical properties of distributions, including indicators of center and spread, refer to properties of actual objects. In these contexts, it is clear to students that the objective is to infer these properties (such as the length of a table or the target of the fruit-sausage process) from the data they collect. Furthermore, in the activities we have tested and described, students not only collect the data but exercise some amount of influence over them. Their actions and decisions impact both the accuracy and the consistency of their data. Observations they make during the data-collection phase provide a source of explanations of the trends and variability in the data. These characteristics function together to fuel conversation among students about the nature and meaning of their data and the conclusions they can draw from them.

Acknowledgments This work was supported by grants ESI 0454754 and DRK-12 0918653 from the National Science Foundation. The views expressed are our own and do not necessarily represent those of the Foundation.

References

Bakker, A. (2004). Reasoning about shape as a pattern in variability. *Statistics Education Research Journal, 3*(2), 64-83.

Bakker, A., & Derry, J. (2011). Lessons from inferentialism for statistics education. *Mathematical Thinking and Learning, 13*, 5-26.

Konold, C., & Lehrer, R. (2008). Technology and mathematics education: An essay in honor of Jim Kaput. In L. English (Ed.), *Handbook of International Research in Mathematics Education*, (2nd edition) (pp.49-72). New York: Routledge.

Konold, C., & Miller, C. D. (2012). *TinkerPlots 2.1: Dynamic data exploration.* Emeryville, CA: Key Press.

Konold, C., & Pollatsek, A. (2002). Data analysis as the search for signals in noisy processes. *Journal for Research in Mathematics Education, 33(4),* 259-289.

Konold, C., Robinson, A., Khalil, K., Pollatsek, A., Well, A., Wing, R., & Mayr, S. (2002). Students' use of modal clumps to summarize data. In B. Phillips (Ed.), *Proceedings of the Sixth International Conference on the Teaching of Statistics* [CD ROM]. Cape Town: International Statistical Institute.

Makar, K., & Rubin, A. (2009). A framework for thinking about informal statistical inference. *Statistical Education Research Journal, 8(1)*, 82-105.

Quetelet, M. A. (1842). *A treatise on man and the development of his faculties.* Edinburgh, Scotland: William and Robert Chambers.

Stigler, S. M. (1999). *Statistics on the table: The history of statistical concepts and methods.* Cambridge, MA: Harvard University Press.

Kapitel 19
Simulation als Bindeglied zwischen der empirischen Welt der Daten und der theoretischen Welt des Zufalls

Andreas Eichler

Pädagogische Hochschule Freiburg

Abstract Ein wahrscheinlichkeitstheoretisches Modell kann auf der Basis von Symmetrieüberlegungen oder auf der Basis empirischer Daten erzeugt werden und dient der Prognose zukünftiger Daten. Empirische Daten stehen dadurch in einem stetigen Wechselspiel mit den theoretischen Modellen, sie basieren auf den Modellen und dienen gleichzeitig zum Aufstellen, Modifizieren oder Überprüfen dieser Modelle. Eine Schwierigkeit im Stochastikunterricht ist es allerdings dabei, insbesondere den Bezug von den abstrakten Modellen der Wahrscheinlichkeitsanalyse zu empirischen Daten herzustellen, da sich in der Regel keine empirischen Daten zu abstrakten Modellen erzeugen lassen.
In diesem Beitrag soll daher die Simulation als Möglichkeit verdeutlicht werden, den Übergang von den Modellen zu den Daten, in diesem Fall virtuellen Daten, anschaulich zu machen. Dazu wird vorbereitend zunächst das Wechselspiel von Theorie und Empirie in der Stochastik erläutert sowie der Einstieg in rechnergestützte Simulationen über händische Simulationen diskutiert. Anschließend werden verschiedene, potentiell bekannte Beispiele beider Sekundarstufen vorgestellt, die die Möglichkeiten rechnergestützter Simulationen illustrieren können. Abschließend werden die mit der rechnergestützten Simulation verbundenen Vorteile für das schulische Lernen zusammengefasst.

1 Einleitung

Die Sammlung von Daten in Erhebungen, die Schülerinnen und Schüler planen und durchführen, sind in die Bildungsstandards beider Sekundarstufe eingegangen (KMK 2003; KMK 2012). Diese offizielle Datenorientierung ist Ergebnis eines langen Wandlungsprozesses von einem allein auf der Wahrscheinlichkeitsanalyse beruhendem, stark kombinatorisch geprägtem Stochastikcurriculum hin zu eben dieser Datenorientierung (Eichler 2005). Grundlage dieses Wandels sind sicherlich die Anstöße aus den angelsächsischen Ländern (kumuliert etwa in Moore 1997; NCTM 2000; Garfield und Ben-Zvi 2004). Ebenso aber haben auch die intensiven nationalen Bemühungen, das Stochastikcurriculum stärker auf der Datenanalyse aufzubauen, ihren Einfluss entfaltet (z.B. Biehler 1982; 1995; 1997; Borovcnik 1987).

Eine zentrale Eigenschaft eines datenorientierten Stochastikcurriculums ist es, dass wesentliche Ideen der Stochastik, die Variabilität statistischer Daten wie auch die sich in großen Datenmengen zeigenden Muster (Wild und Pfannkuch 1999), rein deskriptiv bereits zu Beginn des Curriculums thematisiert und später stetig weiter verfolgt werden (Eichler und Vogel 2009). Die statistischen Daten repräsentieren dabei in der *Rückschau* Phänomene zufälliger Vorgänge, während die in ihnen enthaltenen Muster als *Prognose*modell zukünftiger Daten, das durch Wahrscheinlichkeiten ausgedrückt wird, dienen können (Vogel und Eichler 2011). Obwohl hier eine plausible und hinreichend scharfe Trennung von dem Blick in die Vergangenheit (empirische Welt der Daten) und dem Blick in die Zukunft (theoretische Modellwelt der Wahrscheinlichkeiten) besteht, scheint der Schritt von den konkreten zu den abstrakten, in Wahrscheinlichkeitsverteilungen versteckten Daten einerseits für Lernende groß und andererseits im Stochastikunterricht nicht sehr stark verhaftet zu sein (Eichler 2008). In diesem Beitrag soll daher der Nutzen simulierter Daten diskutiert werden, Mittler zwischen empirischen Daten und abstrakten Wahrscheinlichkeiten zu sein (vgl. auch Biehler 1991; Biehler und Maxara 2007).

Vorbereitend soll dazu der Übergang von der deskriptiven Datenanalyse zu Prognosemodellen der Wahrscheinlichkeitsrechnung verdeutlicht werden. Anschließend soll anhand verschiedener einfacher und potentiell bekannter Beispiele die Mächtigkeit von Simulationen im Stochastikunterricht diskutiert werden.

2 Vorbereitung für das Nutzen von Simulationen

2.1 Das Wechselspiel von Empirie und Theorie

Das Aufstellen eines wahrscheinlichkeitstheoretischen Modells kann einerseits auf Symmetrieüberlegungen, andererseits auf der Untersuchung empirischer Daten basieren. Ein Standardbeispiel für diese Modellbildung, die gleichsam den klassischen und frequentistischen Ansatz von Wahrscheinlichkeit umfasst, besteht in der Untersuchung der beiden in Abbildung 19.1 dargestellten Würfel, des normalen Spielwürfels und des Riemer-Quaders (Riemer 1991).

Abb. 19.1 Riemer-Quader und normaler Spielwürfel

Für den normalen Spielwürfel nimmt man nach dem Prinzip des unzureichen-den Grundes ab einem gewissen Alter an, die Wahrscheinlichkeiten seien gleichmäßig verteilt (Riemer 1991). Das Sammeln empirischer Daten durch das Werfen des Würfels hat damit die Funktion, das gewählte Modell (der Gleich-verteilung) zu untersuchen und im Extremfall auch zu modifizieren, etwa bei einem gezinkten Würfel. Als Ergebnis des Werfens eines normalen Spielwürfels ergibt sich, dass sich bei Kumulation der Ergebnisse die relativen Häufigkeiten gemäß des empirischen Gesetztes der großen Zahlen stabilisieren und zumindest annähernd bei dem im Modell zugrunde liegenden Wert von 1/6 einpendeln (Abb. 19.2 links). Diese Erkenntnis zum empirischen Gesetz der großen Zahlen kann für das Aufstellen eines Modells für nichtsymmetrische Zufallsgeneratoren wie den Riemer-Würfel wichtig sein (Eichler und Vogel 2011). Der Transfer besteht darin, auch hier den relativen Häufigkeiten nach längeren Versuchsserien die Eigen-schaft zuzubilligen, die dem Zufallsgenerator innewohnenden Wahrscheinlich-keiten zumindest annähernd schätzen zu können. So kann sich aus längeren Ver-suchsserien eine adäquate Schätzung der Wahrscheinlichkeiten des Riemer-Quaders, etwa der Augenzahlen 3 und 4, ergeben, wobei sich die klassische Betrachtung – die 3 hat aus Symmetriegründen die gleiche Wahrscheinlichkeit wie die 4 – und die frequentistische Betrachtung geeignet mischen und die *Schätzung* einer Wahrscheinlichkeit ausgehend von einem empirischen Ergebnis deutlich wird (Abb. 19.2, rechts). Zurückblickend hat damit das Werfen des normalen Würfels zwei Funktionen: Das Modell der Gleichwahrscheinlichkeit bezogen auf verschieden lange Serien des Würfelwurfs zu untersuchen und ebenso das Ver-trauen in die Schätzgüte langer Versuchsserien eines unbekannten Zufallsgenera-tors zu unterstützen.

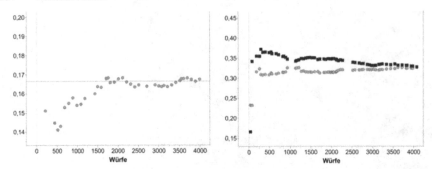

Abb. 19.2 Reale Wurfserien des normalen Würfels (links) und des Riemer-Quaders (rechts). Vermeintliche Lücken entstehen durch hohe Wurfanzahlen einzelner Gruppen.

Entscheidend bei der Betrachtung der beiden Würfel ist das Wechselspiel von Theorie und Empirie (Abb. 19.3). Die Daten repräsentieren die realen Auswirkungen eines Modells (normaler Spielwürfel) und das Modell selbst ermöglicht eine Prognose zukünftiger Häufigkeiten insbesondere in längeren Versuchsserien. Darüber hinaus ergibt sich erst aus der Empirie eine Grundlage, zunächst ein Modell zu schätzen (Riemer-Quader), dieses zu untersuchen oder gegebenenfalls auch zu modifizieren. Auf der Basis eines so erzeugten Modells ist wiederum die Prognose zukünftiger Daten möglich.

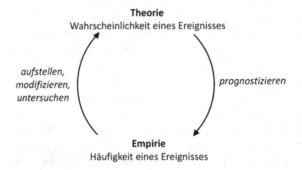

Abb. 19.3 Wechselspiel von Empirie und Theorie

Das Vertrauen in die Schätzung eines Modells durch längere Versuchsserien kann durch die Betrachtung empirischer Ergebnisse des normalen Spielwürfels, bei dem das Modell der Gleichverteilung plausibel ist, geschaffen werden, aber auch zusätzlich dadurch vorbereitet werden, dass genügend Erfahrungen mit eigenen Datensammlungen gemacht und die Aspekte der vorherrschenden Variabilität statistischer Daten in kleinen Stichproben sowie die sich stärker herausbildenden Muster in großen Stichproben betrachtet wurden (Wild und Pfannkuch 1999). Die für eine adäquate Schätzung sowie das Entdecken des empirischen Gesetzes der großen Zahlen sinnvollen Anzahlen von Versuchswiederholungen können bei eigenen Datensammlungen in Befragungen, Beobachtungen oder Experimenten auch in der Schule erzeugt werden (Eichler und Vogel 2009). Etwa liegen der Schätzung der Wahrscheinlichkeiten für die Augenzahlen 3 und 4 des Riemer-Quaders Ergebnisse von 66 Gruppen zugrunde, die den Würfel zwischen 10 und 255 und insgesamt 4072 Mal gewürfelt haben.

Dennoch sind der vertieften Untersuchung eines Modells, bestehend aus Wahrscheinlichkeiten, oder auch des empirischen Gesetzes der großen Zahlen praktische Grenzen gesetzt. Beispielsweise ist man bezogen auf reale Daten in aller Regel auf eine Versuchsserie beschränkt, anhand derer Phänomene im Zusammenhang mit dem empirischen Gesetz der großen Zahlen entdeckt werden können. Ebenso ist es kaum möglich, ein auf Schätzungen beruhendes Modell durchzuspielen, etwa das zum Riemer-Quader erzeugte Modell. So ist es höchstens möglich, den ursprünglichen Zufallsgenerator selbst (nicht aber das Modell) weitere Daten produzieren zu lassen. Aus der Perspektive der Lehrenden mag das auch nicht notwendig sein. Hier ist beispielsweise klar, dass die aufgestellten Modelle Schätzungen auf lange Sicht darstellen, die Streuung (etwa die Standardabweichung bezogen auf den Erwartungswert) relativ gesehen mit zunehmender Stichprobengröße abnimmt, oder auch das Überschreiten jedes beliebigen Abstand von tatsächlicher Wahrscheinlichkeit und der empirischen Häufigkeit für beliebig lange Versuchsserien die Wahrscheinlichkeit 0 hat. Aus der Perspektive der Lernenden sind das aber nicht erlebte, wenig anschauliche Aussagen, die empirisch kaum untersucht werden können. Diese können aber dann anschaulich werden, wenn durch Simulationen mit der Wirkung von Modellen experimentiert wird und Muster und Streuung in wenigen und vielen Wiederholungen einer Simulation analysiert werden. Die Simulation ermöglicht dabei Schülerinnen und Schülern einen virtuellen Blick in die Zukunft, der die stochastische Begriffsbildung fördern kann und der daher den notwendigen Aufwand für das Erarbeiten von Simulationen rechtfertigt.

2.2 Von händischen Simulationen zu Rechnersimulationen

Tatsächlich ist die Simulation zunächst unabhängig vom Rechner und sollte dies auch sein. Vielmehr geht es darum, den Vorgang der Simulation (der Rechner würfelt, wirft eine Münze) zunächst einmal selber auszuführen, um einschätzen zu können, zu was für einem Vorgang der Rechner anschließend virtuelle Ergebnisse erzeugt. Zusammen mit dem Vergleich händischer und Rechner-erzeugter Ergebnisse von Simulationen kann so Vertrauen in die letzteren ermöglicht werden (Vogel und Eichler 2011). Dieses Vertrauen ist notwendig, um das Potential der Simulation mit dem Rechner in vielfältigen Zusammenhängen und insbesondere hohen Wiederholungszahlen nutzen zu können.

Ein Beispiel für den Übergang von realen Daten zu Daten einer händischen Simulation ist die Untersuchung von M&M-Tüten. Real wurden dabei 27 Tüten ausgezählt. Die zusammenfassende Analyse der Farbverteilungen in den Tüten ist in Abbildung 19.4 links, einzelne Farbverteilungen sind rechts numerisch dargestellt.

	blau	braun	gelb	grün	orar
1	3	1	3	4	3
2	2	1	3	3	3
3	4	3	3	5	3
4	3	5	2	5	1
5	3	2	6	3	1
6	3	2	4	3	2
7	3	4	2	4	3

Abb. 19.4 Zusammenfassende Darstellung der Farbverteilung in M&M-Tüten (links) sowie Ergebnisse zu einzelnen Tüten (rechts), Boxplots mit eingezeichneten Mittelwerten

Aus der Analyse der realen Daten kann eine Modellannahme für M&M-Tüten abgeleitet werden, nachdem in einer Modell-Tüte 18 Kugeln enthalten sind, wobei die Farben im Mittel gleichverteilt sind. Simulieren lässt sich das Modell der Farbverteilung für die M&M-Tüten durch einen Zufallsversuch, das auf einem strukturgleichen Modell beruht. Da die M&M-Tüten sechs verschiedene Farben enthält, kann mit dem 18fachen Wurf eines normalen Spielwürfels die Farbverteilung einer Modell-Tüte *simuliert* werden (ebenso könnten auch Tüten mit mehr oder weniger als 18 Kugeln mit einem Würfel simuliert werden).

Das Vertrauen in die händische Simulation kann dabei auf zwei Wegen erzeugt werden. So lässt sich einmal der Gesamtversuch mit dem Würfel simulieren und dessen Ähnlichkeit mit dem empirischen Ergebnis zu den realen Daten erfahren (Abb. 19.5). Zudem erhält man auf der Ebene von Einzeltüten simulierte Ergebnisse, die ebenfalls den realen Tüten gleichen, selbst wenn diese scheinbar eine dem Modell widersprechende Ungleichverteilung zeigen.

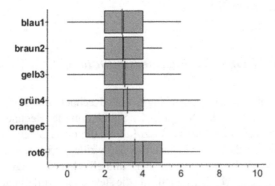

Abb. 19.5 Händische Simulation von M&M-Farben in 27 Tüten mit jeweils 18 Kugeln durch den 18fachen Wurf des Würfels, Boxplots mit eingezeichneten Mittelwerten

Ist Schülerinnen und Schülern der Übertrag von einem Modell, das aus realen Daten erzeugt wurde (M&M-Tüten), zu einem Simulationsmodell (Würfel) einsichtig, ist es möglich und sinnvoll Realmodelle auch mit dem Rechner zu simulieren. Sinnvoll insbesondere daher, da im Gegensatz zum händischen Simulieren nahezu beliebig geartete Modelle simuliert werden können und zudem die

Wiederholungszahl der Simulation zufälliger Vorgänge erheblich erhöht werden kann. Erst diese Möglichkeit, sehr viele Simulationswiederholungen ausführen zu können, schöpft das Potential aus, Modelle auch im Sinne einer stochastischen Begriffsbildung nutzen zu können. Diese hier propagierte Funktion der Simulation, nämlich als Laufenlassen eines bestehenden, im Grunde genommen bekannten Modells, die Biehler und Maxara (2007, S.45) als „Simulation zur Repräsentation" bezeichnen und die der stochastischen Begriffsbildung dienen kann, ist nicht die einzig mögliche. In der Realität vermutlich deutlich wichtiger als in der Schule ist die Simulation eines unbekannten Modells, das virtuelle, sonst gar nicht oder nur schwer erzeugbare Erkenntnisse bewirken kann. In diesem Beitrag geht es allerdings überwiegend um die Veranschaulichung von etwas prinzipiell Bekanntem.

3 Simulationen mit dem Rechner zur stochastischen Begriffsbildung

Überträgt man die Simulation von der Hand auf den Rechner, so kommt stets die Frage der geeigneten Software (subjektiv ist dies in der Regel diejenige, die man bereits kennt) oder auch einer geeigneten Programmiersprache auf. Für diesen Beitrag ist aber nicht das Mittel für die Simulation entscheidend, sondern allein die Möglichkeiten der Simulation für die stochastische Begriffsbildung zu nutzen. Hier geht es also nicht um die Frage, mit welcher Software bestimmte Simulationen geeigneter oder leichter ausgeführt werden können (in diesem Beitrag werden Excel, Fathom und der NSpire CAS verwendet). Vielmehr geht es darum, zu bestimmten Themengebieten deutlich zu machen, dass es wertvoll sein kann zu wissen, was alles in der Zukunft passieren könnte. Vorheriges Vertrauen schaffen kann dabei eine dem Rechner übertragene Simulation von Bekanntem, etwa des normalen Spielwürfels oder des (bereits modellierten) Riemer-Quaders. Simulierte Daten, die in gleicher Weise kumuliert werden wie die realen Daten, zeigen hier ähnliche Verläufe (Abb. 19.8).

3.1 Der stochastisch geprägte Unterschied zwischen wenig und viel

Eine wichtige Erkenntnis in der Stochastik ist die, dass erst viele (simulierte) Daten ein Muster zeigen, während bei wenigen Daten allein der Variabilität ins Auge fällt (Wild und Pfannkuch 1999). Diese Einsicht im Zusammenhang mit dem empirischen Gesetz, das in der Sekundarstufe II potentiell auch mathematisch belegt werden kann, ist nicht nur für ein adäquates stochastisches Denken entscheidend, sondern kann erheblich durch Simulation befördert werden (Moore 1997; Eichler und Vogel 2009; Prömmel 2012). Drei kurze Beispiele sollen dies illustrieren.

Chaos im Kleinen – Muster im Großen

Werden etwa M&M-Daten mit dem Modell der Gleichverteilung der sechs Farben mit dem Rechner simuliert und dabei jeweils eine simulierte Tüte den kumulierten Farbverteilungen vieler Tüten gegenübergestellt (Abb. 19.6), so kann man Zweier-lei erkennen: Die Einzeltüten erzeugen fast beliebige Farb-Verteilungen, bei der mal die eine, mal die andere Farbe bevorzugt ist. Probiert man nur lange genug, so wird jede in einer Klasse real beobachtete Farb-Verteilung einer Tüte auftauchen. Kumuliert bildet sich dagegen die Gleichverteilung heraus und zwar je besser, je mehr Wiederholungen durchgeführt werden. Aus dem Chaos im Kleinen wird dabei ein stabiles Muster im Großen.

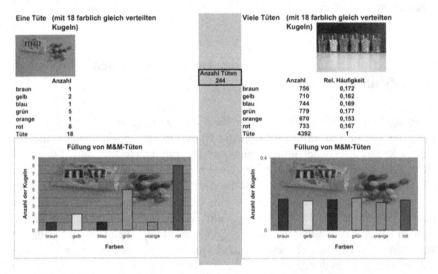

Abb. 19.6 Gegenüberstellung einer Einzelsimulation und der Kumulation von simulierten M&M-Tüten (Eichler und Vogel 2009). Schrittweise wird links die Farbverteilung der aktuell simulierten Tüte, rechts die Kumulation aller bisher simulierten Tüten dargestellt.

Prognosefähigkeit kurzer Serien

Die Startphase des Spiels „Siedler von Catan" ist oft entscheidend für den weiteren Spielverlauf. In dieser Phase ist es wichtig, dass die Summe eines dop-pelten Würfelwurfs diejenigen Zahlen ergeben, die auf einem Feld mit „Rohstof-fen" liegen, um mit den dafür gewonnenen Rohstoffen weitere gewinnbringende Aktionen ausführen zu können. Mit Glück und Geschick hat ein Spieler unter an-derem den Zugriff auf ein Rohstoff-Feld mit der Nummer 8 erhalten, eine Gegen-spielerin ein Feld mit der Nummer 3. Dass ein Doppelwurf die Augensumme 8 ergibt, ist mit 5/36 mehr als doppelt so wahrscheinlich wie für die Augensumme 3 mit 2/36. Dennoch erscheinen in der Startphase zur Enttäuschung des ersten Spiel-ers häufiger eine 3 als eine 8. Wie ist das möglich? Was in der Spielsituation mit

Pech, ungnädigen Würfeln und Ähnlichem begründet werden könnte, ist die noch große Musterlosigkeit in kurzen Versuchsserien, die man per Simulation offenlegen kann. Diese zeigt, was in der Zukunft in einer, aber auch vielen Startphasen des Spiels passieren kann. Nimmt man etwa an, dass die Startphase des Spiels 20 Doppelwürfe umfasst, so lässt sich simulieren, wie viele Startphasen vorkommen, die die 3 anstatt der 8 bevorzugen.

In Abbildung 19.7 ist die Anzahl der Spiele bei 1000 Simulationswiederholungen zu sehen, die in einer Startphase von 20 Doppelwürfen eine bestimmte Differenz $H_{20}(X = 8) - H_{20}(X = 3)$, also die Differenz der Häufigkeiten der 8 und der 3, erzeugt hat. Tatsächlich war in 130 der 1000 simulierten Startphasen die Differenz kleiner als 0, die 3 der 8 also überlegen. D.h. in immerhin 13% der Startphasen wäre also die oben geschilderte Enttäuschung eines Spielers berechtigt gewesen.

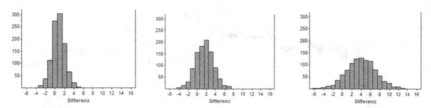

Abb. 19.7 Anzahl der Differenzen H_20 (X=8)-H_20 (X=3) bei 10 (links), 20 (Mitte) und 50 (rechts) Doppelwürfen des Würfels in einer Simulation mit 1000 Wiederholungen

Dieses Phänomen ist abhängig von der Annahme der 20 Doppelwürfe einer Startphase. Ausgehend von 10 Doppelwürfen, ergibt eine Simulation gar in 16% der Fälle die Überlegenheit der 3 gegenüber der 8. Ausgehend von 50 Doppelwürfen ergibt dagegen eine Simulation, dass die 3 der 8 nur noch in etwa 6% der Fälle überlegen ist. Bei noch mehr Doppelwürfen hat die 3 praktisch keine Chance mehr, häufiger gewürfelt zu werden als die 8 (schon bei 200 Doppelwürfen zeigt eine Simulation mit 1000 Wiederholungen nur noch 2 Differenzen kleiner als 0 an). Gänzlich losgelöst von einer numerischen Behandlung kann hier also eine einfache Simulation zeigen, dass sich, alltagssprachlich ausgedrückt, eine größere Wahrscheinlichkeit erst auf lange Sicht gegen eine kleinere durchsetzt. Ergänzend wäre es bei diesem Beispiel sinnvoll, durch vielfache Wiederholung der oben dargestellten Simulation eine Schätzung für die Wahrscheinlichkeit vorzunehmen, die die Überlegenheit der 3 gegenüber der 8 bei einer vorher festgelegten Anzahl von Doppelwürfen in einer Starphase des Spiels beschreibt.

Detaillierte Betrachtung langer Versuchsserien

In beiden Beispielen zuvor ist eher der Aspekt angesprochen worden, wie groß die Variabilität der auf einem festen Modell beruhenden statistischen Daten sein kann. In gleichem Maße kann die Simulation aber auch vertiefend verdeutlichen, warum man dem empirischen Gesetz der großen Zahlen, also der Stabilisierung relativer Häufigkeiten vertrauen kann, die in Abb. 19.2 in Versuchsreichen mit dem Riemer-Quader angedeutet ist.

In Abb. 19.8 wird etwa in der Simulation des Würfels der Gegensatz einer einzelnen Serie mit n Würfen und der dort erzielten Häufigkeit der 6 mit den kumulierten relativen Häufigkeiten vergleichen. Hier ist wiederum der Gegensatz von der Variabilität statistischer und hier simulierter Daten im Kleinen und dem sich abzeichnenden Muster im Großen deutlich. Weiter ist aber auch zu erkennen, dass die relativen Häufigkeiten in einem recht klar umrissenen Intervall zu streuen scheinen und sich dieses Intervall verkleinert, wenn man die Anzahl der Würfe in einer Serie erhöht (etwa von 50 in der Abbildung 19.8 links auf 100 in der Abbildung 19.8 rechts).

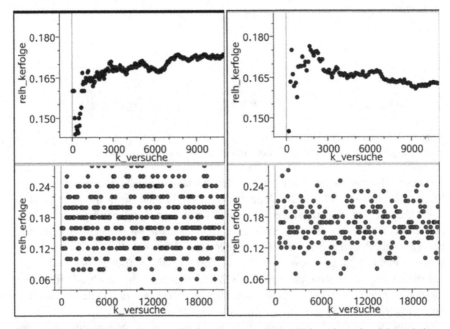

Abb. 19.8 Stabilisierung relativer Häufigkeiten (oben) und Schwankungsbereiche relativer Häufigkeiten in Wurfserien der Länge 50 (links) und 100 (rechts). Die „k_versuche" bezeichnen die insgesamt ausgeführten Versuche, bei den relativen Häufigkeiten wird zwischen den Häufigkeiten zu einer Wurfserie (unten) und den kumulierten Häufigkeiten (oben) unterschieden.

Sind solche Phänomene anhand simulierter Daten erfahren – mit selbst erhobenen, realen Daten wird dies schon schwierig – kann etwa das zuletzt genannte Phänomen als $1/\sqrt{n}$-Gesetz oder auch das Streuintervall formalisiert werden. Während beispielsweise das Intervall, in dem bei 50 Würfen – durch Auszählung der Daten – rund 95% der simulierten Daten liegen, $I_{50} = [0,06; 0,27]$ ist, ist dies für 100 Würfe $I_{100} = [0,09; 0,24]$. Das Verhältnis der Intervall-Längen 0,21 und 0,15 ist 1,4. Untersucht man weitere Verhältnisse, etwa für Serien der Länge 10,

100, 10000, so wird sich allmählich herausschälen, dass sich bei Erhöhung der Wurfserien um den Faktor *m* die Intervall-Längen um $\sqrt{m}$ verkürzen.

3.2 Simulation inferenzstatistischer Situationen

Das Prinzip, durch Simulationen einen vielfach wiederholten virtuellen Einblick in die Zukunft zu erhalten, lässt sich auch in der beurteilenden Statistik nutzen. Real ist sowohl zum Testen als auch zum Schätzen nur *ein*, aus empirisch gewonnenen Daten erzeugter Messwert vorhanden, der zu einer Entscheidung hinsichtlich eines zu überprüfenden Modells (Test) oder zu einer Intervallschätzung führt. Simulationen können hier zwar nicht die Beschränktheit auf einen empirischen Datensatz zum Testen oder Schätzen aufheben, sie können aber durch ihre fast beliebige Wiederholbarkeit dennoch die Grundideen des Testens und Schätzens vermitteln. Das soll anhand zweier Beispiele illustriert werden.

Testen mit Simulationen

Nimmt man den in der Sekundarstufe II üblichen Hypothesentest auf der Basis des Binomialmodells, so lässt sich etwa die auf der Nullhypothese basierende Binomialverteilung simulieren. Hier kann es von Vorteil sein, per Simulation zu schauen, was für Ergebnisse tatsächlich bei einer Anzahl von Versuchen erzielt werden können, als abstrakt die theoretischen Häufigkeiten zu betrachten. Nicht die Genauigkeit, sondern die konkreteren Simulationsausgänge und wiederum deren Variabilität können hier potentiell Einsicht verschaffen (Meyfarth 2006).

Ebenso könnten solche Tests per Simulation bearbeitet werden, die mathematisch Schülerinnen und Schülern eher verschlossen bleiben, wie etwa ein Unabhängigkeitstest. Etwa könnte man durch Randomisierung zweier empirisch erhobener Merkmale (im Beispiel in Abb. 19.7 Rauchen und Geschlecht von Studierenden in Freiburg) virtuell die Unabhängigkeit herstellen und auf der Basis dieses Modells Simulationen durchführen, deren Ergebnis mit Bezug auf das tatsächlich beobachtete empirische Ergebnis als Entscheidungsgrundlage für die Ablehnung der Unabhängigkeitshypothese dienen kann (Tab. 19.1 und Abb. 19.9; Eichler und Vogel 2011).

Tab. 19.1 Empirisches Ergebnis

	männlich (M)	weiblich (W)	Summe
Raucher (R)	29	29	58
Nichtraucher (NR)	74	157	131
Summe	103	186	289

Messgröße: $A = h(R|M) - h(R|W) \approx 0,12$.

Bei Randomisierung beträgt Anteil der simulierten Messgrößen, die größer 0,125 oder kleiner - 0,125 sind bei 1000 Simulationen 1,4% (vgl. Abb. 19.9).

Auf der Basis der Simulation kann also die Unabhängigkeitshypothese abgelehnt werden.

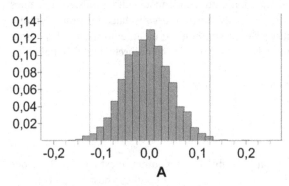

Abb. 19.9 Simulierte Messgröße A

Interpretation von Konfidenzintervallen

Bezogen auf Konfidenzintervalle lässt sich neben deren Konstruktion (Eichler und Vogel 2011) insbesondere auch deren für Lernende schwierige frequentistische Interpretation per Simulation verdeutlichen: Man hat die Konstruktion eines Konfidenzintervalls mitsamt dem Konfidenzniveau, etwa 95% festgelegt, das Konfidenzintervall zum Parameter p einer Binomialverteilung wird an der relativen Häufigkeit der Erfolge verankert. Die frequentistische Interpretation ist hier die, dass gerade 95% der so gebildeten Konfidenzintervalle den wahren Parameter enthalten, die anderen nicht.

Diese Interpretation lässt sich etwa durch eine Serie von Intervallkonstruktionen simulieren, wobei diese besonders eindrücklich ist, wenn man die Konstruktion animiert bzw. die Intervalle schrittweise konstruiert. In Abbildung 19.10 ist allein die vollständig abgelaufene Simulation zu sehen. In der Animation, in der die Konfidenzintervalle mit kurzem zeitlichen Abstand hintereinander erzeugt werden, ist dagegen schrittweise nachvollziehbar, wann eines der zufällig erzeugten Konfidenzintervall den wahren Parameter nicht überdeckt.

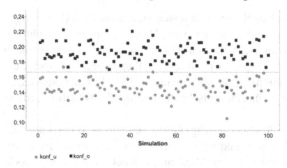

Abb. 19.10 Animierte Simulation von Konfidenzintervallen, bei denen bei rund 50 Wiederholungen bereits vier Intervalle den wahren Parameter nicht enthalten

In der Simulation sind beispielsweise drei Konfidenzintervalle „zu hoch" erzeugt worden, d.h. der Punkt als untere Grenze des Konfidenzintervalls liegt oberhalb der Geraden. Ebenfalls sind zwei Konfidenzintervalle „zu niedrig" erzeugt worden. Hier liegen die oberen Grenzen der Konfidenzintervalle unterhalb des Wertes 1/6. Nicht nur die frequentistische Interpretation von Konfidenzintervallen kann anhand dieser Simulation veranschaulicht werden, sondern auch der Blick auf die Anwendung von Konfidenzintervallen in der Realität geschärft werden, bei denen nicht 100 Versuche, sondern nur einer verwendet wird.

4 Rückblick

Kernanliegen des Beitrags ist es gewesen, den Nutzen von Simulationen für die Einsicht in die Verbindung von Theorie und Empirie bzw. der Modellwelt der Wahrscheinlichkeitsanalyse und der empirischen Welt der Datenanalyse zu verdeutlichen.

So sind dem in Abbildung 19.3 eingeführten Übergang von der Theorie zur Empirie in der Praxis Grenzen gesetzt:

- Ein wahrscheinlichkeitstheoretisches Modell kann per se in der Praxis nicht ausgeführt werden bzw. das Modell keine Daten erzeugen. Beispielsweise führt man mit dem erneuten Öffnen von M&M-Tüten nicht das Modell der Gleichverteilung aus, sondern man sammelt lediglich Daten, die auf dem unbekannten Zufallsmechanismus der Befüllung von M&M-Tüten basieren.
- Ein auf realen Daten basierendes Modell kann damit entweder rein wahrscheinlichkeitstheoretisch untersucht oder simuliert werden. Eine Simulation ist mit klassischen Zufallsgeneratoren wie Würfel und Münze möglich, da man hier davon ausgeht, dass die mit diesen Zufallsgeneratoren erzeugten Daten dem Modell der Gleichverteilung genügen.
- Beim händischen Simulieren ist die Anzahl der Versuche begrenzt (wie auch bei der Sammlung realer Daten). Beispielsweise kann man mit dem Würfel die

27 M&M-Tüten mit je 18 Kugeln noch händisch simulieren. Wiederholungen der gesamten Simulation sind dagegen von Hand kaum mehr möglich. Beispielsweise ließe sich im Beispiel zum Testen zwar eine, vielleicht auch zwei oder drei Randomisierungen der Merkmale Rauchverhalten und Geschlecht von Hand durchführen, sicher aber nicht die 1000 Simulationswiederholungen, die als Grundlage für eine Entscheidung in der Testsituation gedient haben.

- Damit ist erst die mit dem Rechner durchgeführte Simulation in der Lage, die Auswirkung von Modellen auf zukünftige Daten quasi-empirisch zu untersuchen. Insofern ist die Simulation eine Mittlerin im Wechselspiel von Theorie und Empirie, deren unmittelbare Verbindung für Schülerinnen und Schüler nur brüchig vorhanden ist (Abb. 19.11).

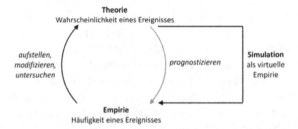

Abb. 19.11 Simulation als Mittlerin zwischen Theorie und Empirie

Das Wertschätzen von Simulationen ist dabei fachlich nicht notwendig, da man einmal erzeugte Modelle theoretisch beherrschen kann. Für das Lernen von Schülerinnen und Schülern verfolgt man aber mit der Simulation den Gedanken eines datenorientierten Stochastikcurriculums konsequent weiter. So sollen Schülerinnen und Schüler in der Datenanalyse lernen, reale Phänomene im *Rückblick* zu beschreiben und die dafür eingesetzten Methoden wertzuschätzen. Konsequenterweise können Schülerinnen und Schüler durch Simulation lernen, zukünftige, durch Modelle festgesetzte reale Phänomene anhand virtueller Daten zu beschreiben, um damit Möglichkeiten und Grenzen wahrscheinlichkeitstheoretischer Methoden einschätzen zu können, wie hier insbesondere anhand des Gegensatzes von kurzen und langen Serien zufälliger Vorgänge, aber auch der Beurteilung eines Modells (Test) bzw. der Untersuchung von Intervallschätzungen deutlich gemacht wurde. Dieses Handeln im Futur II, d.h. die Betrachtung, was alles in der Zukunft auf der Basis eines Modells passiert gewesen sein könnte, kann wertvolle Einsichten zu stochastischen Begriffen verschaffen, die bei einer rein theoretischen Behandlung abstrakt und vielleicht verborgen bleiben.

Literaturverzeichnis

Biehler, R. (1982). *Explorative Datenanalyse – Eine Untersuchung aus der Perspektive einer deskriptiv-empirischen Wissenschaftstheorie.* IDM Materialien und Studien, Band. 24. Bielefeld: IDM.

Biehler, R. (1991). Computers in probability education. In R. Kapadia, & M. Borovcnik (Hrsg.), *Chance encounters: Probability in education* (S. 169 – 212), Dordrecht: Kluver.

Biehler, R. (1995). Explorative Datenanalyse. *Computer und Unterricht, 17,* 56-66.

Biehler, R. (1997). Auf Entdeckungsreise in Daten. *Mathematiklehren, 97,* 4-5.

Biehler, R., & Maxara, C. (2007). Integration von stochastischer Simulation in den Stochastikunterricht mit Hilfe von Werkzeugsoftware. *Der Mathematikunterricht 53*(3), 45-61.

Borovcnik, M. (1987). Zur Rolle der beschreibende Statistik. *mathematica didactica 10*(2), 101-120.

Eichler, A. (2005). *Individuelle Stochastikcurricula von Lehrerinnen und Lehrern.* Hildesheim: Franzbecker.

Eichler, A. (2008). Teachers' classroom practice in statistics courses and students' learning. In C. Batanero, G. Burrill,C. Reading, & A. Rossman (Hrsg.), *Proceedings of the Joint ICMI /IASE Study teaching Statistics in School Mathematics. Challenges for Teaching and Teacher Education.* Monterrey, Mexico: ICMI and IASE.

Eichler, A., & Vogel, M. (2009). *Leitidee Daten und Zufall.* Wiesbaden: Vieweg + Teubner.

Eichler, A., & Vogel, M. (2011). *Leitfaden Stochastik.* Wiesbaden: Vieweg + Teubner.

Garfield, J., & Ben-Zvi, D. (2004). *Developing students' statistical reasoning.* New York: Springer.

Henze, N. (2010). *Stochastik für Einsteiger* (8. Aufl.). Wiesbaden: Vieweg+Teubner.

KMK, Sekretariat der Ständigen Konferenz der Kultusminister der Länder in der Bundesrepublik Deutschland (2003). *Bildungsstandards im Fach Mathematik für den Mittleren Schulabschluss.* München: Luchterhand.

KMK, Sekretariat der Ständigen Konferenz der Kultusminister der Länder in der Bundesrepublik Deutschland (2012). *Bildungsstandards im Fach Mathematik für den Mittleren Schulabschluss.* München: Luchterhand.

Meyfarth, T. (2006). *Ein computergestütztes Kurskonzept für den Stochastik-Leistungskurs mit kontinuierlicher Verwendung der Software Fathom – Didaktisch kommentierte Unterrichtsmatierialien. Kasseler Online-Schriften zur Didaktik der Stochastik, Band 2.* Abgerufen von http://nbn-resolving.org/urn:nbn:de:hebis:34-2006092214683.

Moore, D. S. (1997). New pedagogy and new content: The case of statistics. *International Staistical Review,* 65(2), 123-137.

The National Coucil of Teachers of Mathematics [NCTM] (2000). Principles and standards for school mathematics. Reston, VA: NCTM. (Übersetzung von C. Bescherer und J. Engel (2000). Prinzipien und Standards für Schulmathematik: Datenanalyse und Wahrscheinlichkeit). In M. Borovcnik, J. Engel, & D. Wickmann (Hrsg.), *Anregungen zum Stochastikunterricht* (S. 11-42). Hildesheim: Franzbecker.

Riemer, W. (1991). *Stochastische Probleme aus elementarer Sicht.* Mannheim: BI-Wissenschaftsverlag.

Vogel, A., & Eichler, A. (2011). Das kann doch kein Zufall sein! - Wahrscheinlichkeitsmuster in Daten finden. *Praxis der Mathematik, 34*(3), 2-8.

Wild, C., & Pfannkuch, M. (1999). Statistical Thinking in Empirical Enquiry. *International Statistical Review* 67(3), 223-248.

Chapter 20
Modelling and Experiments – An Interactive Approach towards Probability and Statistics

Manfred Borovcnik

Alpen-Adria Universität Klagenfurt

Abstract Modelling enriches the picture of mathematics and has a strong formative potential for probabilistic and statistical concepts. Experiments and case studies are presented and discussed here to corroborate this view. Activities and inter-activity help to integrate learners into the processes and are vital for a modelling approach. Modelling is learnt as a handicraft by doing. Essential features of it are missed if it were passively 'consumed'. One case study is extensively discussed to transmit a modelling-entrenched view on this field of mathematics, which is extremely prolific for modelling. Usually notions arise more or less out of an abstraction from real world. Here, the concepts are often the result of an imputation of an idea to reality. This artificial character is best learned by returning to the roots, i.e., by modelling. We describe modelling in probability and statistics, the basic competencies required, and our experiences with teachers.

1 A perspective of modelling

The following justifications to teach applications and modelling are listed by Blum (2012, pp. 29): i. *"Pragmatic"*: In order to understand and master real world situations, applications and modelling have to be explicitly treated. ii. *"Formative"*: Competencies can be advanced also by engaging in modelling activities. iii. *"Cultural"*: Relations to real world are indispensable for an adequate picture of mathematics. iv. *"Psychological"*: Real world examples may contribute to raise students' interest in mathematics and may be used to motivate or structure mathematical content.

1.1 Modelling in probability and statistics

Probability models have become increasingly important. Modern physicists build their theories completely on randomness (e.g., Styer 2000). Risk is a basic element of life. Statistical inference has become the standard method to generalise conclusions from limited data. Empirical research builds on a sound understanding of statistical conclusions going beyond the simple representation of data. Modelling, therefore, is a worthwhile goal in teaching probability and statistics.

We will illustrate the steps of modelling in empirical research exemplarily by case studies using authentic data 'produced' by the learners (Section 2). In these case studies, the hypotheses investigated emerge out of the context. This link to context will increase the motivation of learners and at the same time enhance the steps to undergo and interpret the results and difficulties encountered.

For probability modelling we have to refer the reader to Borovcnik (2006), Borovcnik (2009), and Borovcnik and Kapadia (2011) who focus on innovative examples and key properties of the concepts; we can only briefly summarize the main ideas here. The models are used to choose one action (from several possible ones) that "is better" than the others with regard to a criterion of success. Vital to this modelling approach is to search for crucial parameters (of the used model), which strongly influence the result. The key properties of the concepts used (e.g., a specific distribution) describe the 'internal structure' of a situation by the model's inherent assumptions. Such a key property allows for a direct check whether the model in question is adequate for a situation, i.e., whether it supplies the relevant 'structure' for the situation to be modelled.

In a modelling approach, more concepts than usual have to be developed and used. As the context also requires attention, it may be wise to handle the mathematical details more informally supported via simulation and animated visualizations. In this way, the focus may be directed towards the applications and the underlying ideas rather than on technicalities.

Modelling is a process with the general goal of improving a model and to make it fit more adequately to the target situation. The stages of this process include

- To get a sensible model to represent the situation at hand adequately;
- To derive a solution within the model;
- To transfer this solution to the original situation; and
- To evaluate the final solution including a critical appraisal of questions left open.

Several cycles may be needed to derive a satisfactory model. For judging the fit of a model, patterns for translating between the two 'worlds' are necessary: i. External structure of the model: interpret probability as relative frequency and judge the fit. This is done by statistical tests. ii. Internal structure of the model or the probability distribution, which is represented by its inherent assumptions. Here, the approach of key ideas associated to each of the probability distributions might be helpful (see Borovcnik and Kapadia 2011).

Alternative to models, the conception of a scenario may be useful: A scenario is a 'model' that is used heuristically to investigate the situation on an 'as-if' basis; a scenario might lack a perfect fit and yet yield relevant information for tackling the problem. For significant examples of scenarios see Borovcnik (2006), for an elaboration of the concept see Borovcnik (2012).

1.2 Competencies for modelling

Applications are a potential driving force for developing teaching units. However, *genuine* applications are generally judged as inappropriate for teaching:
i. Knowledge about the context is essential. ii. More than one model is needed to compare and describe the problem situation satisfactorily. iii. The modes of work involve organization of co-operation and combine results not all known in detail. These divergent demands mark the challenge of applications (in business, industries, etc.); to organize planned steps of teaching – even if it is mimicked and not genuine application (with the risk of ending up with no satisfying result) – seems unsuitable both for teachers and learners. Technology may be used to simplify the approach towards mathematics by visualization and simulation but requires a familiarity with the software used so that the difficulties may only be transferred from one to another level.

To foster ideas, project oriented teaching has been advocated (e.g., Krainer 1993), developing questions, working together to find partial answers using mathematical procedures, and learning how the concepts may help to make progress on the answers. In authentic applications, the initial phase of systems analysis is often too vague and sometimes leads to abandoning the project. Case studies may be used instead as a didactical tool to reduce the required efforts. Vital for the practical work is that questions are developed in the early phases of a case study, which may be answered by further steps of modelling the situation. A carefully chosen context helps for motivation and for the ongoing work: to interpret intermediate results helps to decide about further steps. Formulating an explicit aim is vital for guiding and shaping the joint endeavour.

1.3 Modelling and experiments with teachers

The author has used modelling, experiments, and case studies since long with groups of teachers in in-service courses. Hereby, an interactive approach towards probability and statistics is characterized by a cyclic mode of operation: If hypotheses arise naturally from the context and arouse interest, this increases not only the motivation but also makes it easier to understand and interpret single steps of modelling and intermediate results. Such a connection to context (and commonsense) also helps to revise first approaches if this is prompted by crucial questions emerging from the ongoing steps of analysis.

In the following, the "memory test experiment" will be analysed in more detail (Section 2) with the data from a workshop that was organized by John O'Donoghue at the University of Limerick with 46 teachers (27 female, 19 male). In this workshop several other experiments were discussed including "breaking the spaghettis", "motivation in competitions", and "placebo effect and regression to the mean" (see Borovcnik and Kapadia 2012). We will briefly summarize some reactions of the workshop, which suggest that these teachers appreciated the experimental approach despite its challenges and the fact that it is so different from what they usually do in teaching mathematics.

The pedagogy of such an approach differs in a number of ways from the pedagogy of pure mathematics, as taught to older pupils. In mathematics, the 'rule' is that a task has a correct answer. In statistics, there are many interpretations and usually there remains some degree of freedom in the modelling and in the interpretation of answers; in the end it is not a question of whether an answer is correct or not. While many can accept this ambiguity for statistical questions, it is even more difficult to perceive and acknowledge that it applies to modelling probabilistic situations as well (see Borovcnik and Kapadia 2011). Such openness might cause problems and difficulties for teachers as it changes their traditional role as authority figure in the classroom.

Despite these challenges, the teachers were receptive to the level of interactivity in work and responded well to the ideas discussed. This nourishes the hope that possible obstacles to modelling approaches in the classroom can be overcome. Teachers in the session found that group work was an especially effective approach. The setting of the experiments naturally lent to discussing ideas and exploring possibilities. Technology was crucial for collating results and for displaying diagrams in a variety of ways. This aided discussion and analysis.

That teachers valued the session is evident from their questions and sustained interest and concentration (as may be seen from the videotape of the workshop). They were excited by the new ways to generate data and use technology to produce results, which lead to valuable discussion and further steps of modelling. They expressed their intention to use some of the ideas in class.

The teachers appreciated the materials as authentic and interesting and contrasted this with the common (and rather dull) surveys of personal characteristics of pupils, such as times to travel to school or shoe size. It is unknown how pupils would react to these materials and how the experiments have to be adapted for them. However, in discussion, teachers were optimistic that the basic ideas of the experimental approach could be transferred to their classroom work.

2 Case studies in modelling

In statistics there are at least two lines of approach: one is to compare the results of the present sample against hypotheses about a parameter using statistical tests or confidence intervals. The other is to produce data – in the form of a quasi random sample – to investigate how a target variable depends on influential factors with the aim of describing such relations and gaining insight into underlying processes.

We will use case studies where the "statistical modellers" are involved in the process of model building as the hypotheses refer to them, data are personally anchored, and the results have to be interpreted in personal terms. This approach increases the motivation to learn and naturally leads to an interactive process where new conjectures are inspired by intermediate analyses and further questions are generated by the final conclusions. This illustrates authentically how statistics is used in the process of acquiring evidence-based knowledge in research.

2.1 A memory test

In the memory test, participants memorize items that they must reproduce later. Obviously, the variable under scrutiny is the success – how many items can be correctly retrieved? According to the two basic statistical approaches there are two ways for investigation:

In the first, we compare the achievement to other groups or findings from psychological experiments. In fact, based on corresponding experiments, Miller (1956) formulated a law according to which people can retrieve more or less 7 (plus or minus 2) completely unrelated items. This psychological law was used in the memory test activity as a way to motivate the group to do their best by arousing their sense of competition. Important features of this task are that the hypotheses are readily accessed and easy to interpret in terms of the context and there is a natural curiosity to perform the theoretical analysis to get an answer to the basic question: are we as a group better than such a law would predict?

The other statistical approach is to investigate potential factors that influence the success in the memory test. This leads into statistics as systems analysis: What are decisive factors that influence the success of retrieval? These factors may be related to the items, to the persons, or to the test situation. Can we measure them reliably and how can we generalize the results of our investigations? Basic to this approach is also the question for potential confounders of the influence factors. Confounders signify further influential variables, which correlate to factors under scrutiny that can bias the results found enormously if they are not controlled for by the design of the study.

If confounders are detected it may have an impact on the inferential part from above. For example, if the items are not unrelated as presumed, this may make it much easier to memorize and retrieve them correctly with the consequence that the group does much better than the law would predict. The interpretation that the group is better would not hold in this case despite the significant outcome of the test. The search for potential confounders is fundamental for applied statistics and has to be met within the systems analysis phase *prior* to the 'production' of data. Once such variables are neglected and data is not recorded, it will normally not be possible to track it back and re-establish the missing data if any suspect for confounders will occur later in the analysis.

2.2 Comparing the results to hypotheses or to other groups

We start with an exposition of the inferential part as this was the initial motive to introduce the context to the group. The aim of the activity is twofold: i. Analyse whether our group is better than a benchmark hypothesis. ii. Develop methods to perform such a comparison. A main characteristic of our approach is to focus on the link between the formal concept and the contextual situation. The methods introduced should be made accessible by its embodiment in the context and mutually, the methods should enhance the context.

The context was introduced by the following question: How good are people in general and how good are *we* at memorizing things? To answer it we could compare results of our group to other groups. For scientific progress, a usual way to evaluate observations is to compare them to data from a *general group*, i.e., to acknowledged law-like relations derived from comparable settings. Psychologists have done similar experiments in the 1950's and formulated their findings in a law of the magic 7 (Miller 1956): People can retrieve 7 out of completely unrelated items. It is important to emphasize that the items have to lack any relation among them as this is acknowledged as a substantial factor of influence. Of course, it may be difficult to judge whether chosen items fulfil such a requirement. A further interesting result of research is that the number of correctly retrieved items is independent of the number of offered items – it has no substantial influence if it were 15 or 100.

We decided to use 15 words. Each word was displayed for one second, with a short pause between. The instructions were clear, no notes were allowed, and the target was to investigate whether we were better than Miller's law would predict. A comparable test may be downloaded from PositScience (n.d.). In our experiment we followed a suggestion from Richardson and Reischman (2011).

The analysis of our 'research question' was put forward by various steps progressing from informal ways to evaluate it towards more formal methods: judge our performance by a stem-and-leaf diagram; evaluate parameters like mean or median and compare to the law, and introduce the sign test for corroboration of the previous results.

Are we better than the law of magic 7?

If we highlight the margin plus or minus 2 of the law of 7 within the stem-and-leaf diagram of the data on the number of correctly retrieved words, we see a marked shift in our data as compared to the benchmark (Figure 20.1): while only 1 person is below, 19 are above, with 56% fitting to the rule. A clear indication that our group is better than the law! The quality of the argument may be improved by using more advanced methods: The central tendency of our achievement (mean equals 8.93, median is 9.00) coincides with the upper limit of the law, which is 7 plus 2. If we consider the standard deviation of 2.37, we can state that the magic 7 is 'nearly' one standard unit below the observed mean, which marks a considerable shift as compared to the law.

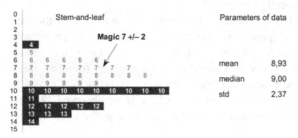

Fig. 20.1 Display of data

Yet, the argument has to be put on a firmer basis as we ignore the size of our group up to now: we know that random fluctuation is quite high for a small sample size n and gets smaller only by a rate of $1/\sqrt{n}$. That means that 100 data leads to a random fluctuation of $\pm 10\,\%$ if a probability has to be estimated. While the size of random fluctuation for our variable is a bit more complicated, the reference suggests that more analysis is needed to corroborate that our group is better than the rule.

Are we better than hypothetical data based on a *model* of the law of magic 7?

A more refined approach is based on the central question "What can be expected under the law of the magic 7?" With this question we enter the realm of hypotheses. The hypothesis hidden in the *magic 7* is one about a capacity to retrieve correctly words from memory, which is now a random variable. We could model its variation by a normal distribution with an expected value of 7 and apply the standard deviation of 2 to end up with a t test. Instead, we use a more informal approach.

If we interpret the 7 of the law as median of our (random) capacity variable, we can attribute to a person to be higher, lower, or equal to 7. To make the further analysis easier, we omit the 7s at this stage as they do not contribute to clarify our question whether we are *better* or *worse* than the law. The question is now about do we have more persons above or below. More precisely, do we have more persons above 7 than we can expect under the stated conditions? What can we expect, or how can we develop a model that we can judge what we can expect even in more extreme cases, which can and do occur occasionally? Furthermore, can we derive a threshold within our model, which marks the separation between the normal range of fluctuation and extreme cases?

The basic assumption for the comparison between our group and the law is now: Can we perceive our group as a random sample from a distribution of our theoretical capacity variable, which has a median of 7? If so, the model can be embodied in repeated coin tossing with an ordinary coin (with $p = \frac{1}{2}$ for heads).

The attribution of *above* (1) and *below* (0) the magic 7 fulfils the criteria of a Bernoulli chain with probability of ½, i.e., we have repeated trials with success probability ½, which are independently performed.

To revisit our question "Are we better than the magic 7?" we compare the data of our group *as if* it had been 'produced' under the conditions of the formulated model. With equal probabilities ½ the person is above as well as below the median of 7. The probabilities follow a binomial distribution with $n = 39$ (omitting the 7 persons of all 46 that retrieved exactly 7 words correctly) and $p = ½$.

The expected value according to this model is 19.5 while we observed 32. This is far off but we have to develop the logic in more detail to judge "how far" this is.

Providing the basis of comparison and the limits of acceptable variation

On the basis of our binomial model we can make it more precise what can be expected. We already know that the expected value is only a benchmark of what can happen that signifies the centre; it gives no clue about variation. We can derive a central interval of what can happen if the hypothesis applies. That involves a probability distribution (the binomial distribution) and an adequate interpretation of probabilities. The graph of the binomial distribution in Figure 20.2 left shows that the data of our group is far out of the centre of the distribution, which sheds substantial doubt on the underlying assumptions.

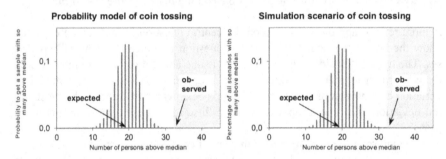

Fig. 20.2 Left. Potential results. Right. Simulated results – both subject to coin tossing hypothesis

It is much more vivid to show probability in action by simulating the conditions of the model used repeatedly to get one sample of 39 data and determine the number of persons above (and below) the median 7. And then repeat the scenario very often. Figure 20.2 right shows the results of a simulation scenario with 3,000 repetitions: we can see that 32 (or more) "successes" are far out; with 3,000 samples (like that of our group) not a single one has shown such a high number of 32. We calculate a probability of 0.00004 from the binomial model. We expect to see one group with such an excellent score in investigating 30,000 groups.

We clearly get the impression that the conditions of our model are doubtful. While still it is possible that such a result may stem from the presumed hypothesis, we decide that our group retrieves significantly more than 7 words correctly. The method we use here to test for a specific value of the median is known as sign test; each data gets a plus or minus sign with equal probabilities indicating whether it is above or below the assumed median which reflects the basic property of the concept of the median.

Logic of a significance test

While rarely any outcome of a sample can be excluded on the basis of an underlying model of a probability distribution, in practice a decision has to be made 'for' or 'against' the used probability model. The usual way to prepare such a decision is to identify an interval in the centre of the distribution, which marks the 'normal' range of fluctuation from one to another sample. Beyond the margins of this interval very rare outcomes are combined. The idea behind is that a target value (the centre) is missed by (random or non-random) fluctuations. A value beyond the thresholds misses the target by far and is thus awkward. Moreover, it marks a very rare outcome as the risk seems negligible that such an outcome would be 'produced' under regular conditions.

The further out of the centre of the distribution of potential results under the (coin tossing) hypothesis, the stronger the evidence against it. Thus, in case of an observation beyond the limits of the central interval, the data is judged as empirical evidence that the conditions (i.e., the underlying hypothesis) are violated.

In our memory test, to find (at least) 32 among 39 persons above the median corresponds to 32 heads in tossing a fair coin 39 times, which is a very rare event. The hypothesis that leads to our model of coin tossing was "The teachers are in line with the magical 7 law". Accordingly, we will reject this hypothesis and instead state that our teachers are *significantly* better than this law. We prefer this alternative rather than stick to the possibility that such a rare event actually happened. That is the logic of a significance test.

The conclusion about our group being better than the law of the magic 7 may still be questioned in different ways. Why it may be that our group is not a random sample from the population (with a median of 7)?

Further considerations would touch the conditions of the experiment or the experiment–persons relations. What if the words were known in advance to (some of) the group? What if the persons did make notes during the process of presenting the words? Then the observed achievement would represent something different from their capacity to retrieve words. The conditions of experiments have to be clearly stated and monitored. A further potential confounder for the recorded achievement would be that the persons did mutually look at their neighbour's list and thus combine their retrieved words.

There may be two persons highly better than the others while the rest of the group complies with the law. This is not the case here but would – if applicable – lead to an attribution of excellence (as a consequence of the rejection of the law of the magic 7) to the whole group while it were due only to these two. As such confounders are excluded here, we sum up our conclusion: the group of teachers is judged as significantly better than the law of magic 7.

Finding data to establish a significant deviation from the hypothesis is often misperceived as an empirical proof that the underlying hypothesis is false and thus the logical complement of it (here: we are better than the law of the magic 7) is *proven*. Such a significant result, however, is only a challenge to think about the contextual question and search for reasons for it to hold or explain the result (or explain it by reference to confounders).

Still is an open question whether the 'pattern' found may be generalized to all adult persons today as compared to people in the 1950's when Miller established the law. Here, a contextual explanation may be given for the finding: i. the teachers were especially interested as they attended an in-service course during their summer holidays and ii. they were selected to act as multipliers to spread new ideas at their schools. This justifies stating that this group as a whole can claim to outperform the law as this type of 'snowball teachers' (in Ireland) seem to be trained to memorize stuff. At the same time this description of the group makes clear that it is not representative of today's population and thus the findings may characterize the group's peculiarities rather than they can be generalized to state that today (in Ireland) people are better than the law of magic 7.

Significance tests may be seen as filter to increase the proportion of worthwhile hypotheses to investigate from the context in order to learn more about relevant interrelations in the subject.

Epilogue

If modelling is not an exercise per se it will be done within a framework with a data 'production' phase included. In reality not only the assumptions of random samples are violated – at earlier steps – there are many traps waiting for their chance. Missing data is one such trap. Always there will be missing data without a possibility to amend them. One approach is to omit a whole statistical object if any of its data is missing; a newer approach is to estimate the values and replace the missing values. Both ways have their merits but also severe disadvantages.

We have analysed the data of the correctly retrieved number of words without those worksheets that were blank on this part of the work. The conclusion was that the group of teachers under scrutiny is significantly better than the law of the magic 7. To remember, the mean number was 8.93 and we had 39 conclusive data (above or below the median) with 32 being above median. The probability to have such an extreme observation (32 or more) in one specific sample was calculated to 0.00004 – to be expected once within 30,000 comparable studies.

We have to note that 7 of the worksheets were not filled in for the memory test (6 female, 1 male). What to do with these persons? In our previous approach we omitted them. However, is there any correlation between "have not filled in" and success? Were participants who were unwilling to undergo the concentration task of the test worse than average? Were some of the good too nervous to compete and show their colleagues their present achievement? Did some of those who worked on the test not deliver their results as they were poor? We have no idea about the reasons behind. Can we preserve our findings despite this vagueness? What about studying the effect of the worst-case? If we assume 6 of them below and one hitting the median, we would have an artificially amended result of 45 decisive data with 32 above median; as before we calculate a probability to see such an extreme result in our sample (32 or more above median) of 0.0033. Still very rare as we would expect to see such a result on average in roughly 300 comparable studies.

As a suggestion for the practice, avoid missing data by careful monitoring the data production phase. This is in fact a good advice but in reality missing data is an everyday problem of a practical statistician and will remain a crucial issue despite any precautions taken. The findings are often highly sensitive to how such data is treated. Sometimes the whole project can be dismissed as it will get complicated to justify the findings. Here we help us out of the dilemma by a nearly worst-case analysis, which preserves the finding while it looks a bit less impressive by the end. Our teachers are significantly better than the law of the magic 7.

2.3 Factors influencing success and behaviour in the memory test

The target variable is success in correctly retrieving items from memory. To investigate potential influence factors may help for orientation how to learn effectively and retrieve from memory on demand if a task requires it. It may also help to filter out law-like relations between such influential factors on achievement as we have seen above.

Systems analysis of the involved factors

It is wise to identify hierarchically the units involved and reflect on their impact and interrelations. The players are the persons tested, the items, the testing situation and interrelations between them. We will just name a few more details about the linked processes, which are very vague in practice and have to be investigated under restrictions of time and money as the resources are always limited.

The persons. Age (children, grown-ups), gender, educational background including training in memory techniques, grade of familiarity with using memory.

Items. Single items have a context (maybe funny, recall bloody associations etc.), may be short or long, may look strange (words from Greek); even if items are intended to be unrelated some may be linked together thus conflicting with the assumption of being unrelated, which might increase the retrieval capacity. One interrelation between items offered is *time*: they have to be placed in a sequence – time of placement might influence success and the order of retrieval.

Test situation. The experimental situation might directly take an influence on the participants: they may take it serious, or are not interested and give arbitrary answers. The time of the day when the experiment is performed might influence success (after lunch e.g.); the medium of presentation might influence success (auditive or visual display).

Interrelations. There might be cross-relations between persons and time patterns or context. Women may establish meaningful interpretations linking words in a way that is markedly distinguished from an approach that men would use. Simply context of the words might have a different impact on men and women. As far as we have data on characteristics that might influence other variables we can study the direction and size of influence. If such data is missing, an influence could later be presumed but no more studied as the data may not be filled in at this stage. Such factors are then called confounders. If a result found is not cross-checked against potential confounders it may be wrongly generalized and only later be 'corrected' in follow-up studies.

Below we report some results of an analysis of the influence of context of words, time of placement of words, and gender on success and retrieval behaviour (for details see Borovcnik and Kapadia 2012).

Does the context of words influence success of retrieval? Ranking the words to success shows that context counts though its pattern is far from clear except that there are distinct top and bottom-placed words.

The time when words are presented influences the success of retrieval. The percentage of correct retrieval shows a clear pattern of dependence on the serial number with middle times being related to lowest achievement rates, which suggests introducing a variable "depth" (minimal distance to beginning or end). Such findings are not only in congruence with common-sense but may also be explained by psychological relations. The concentration will fall back towards the middle of the experiment; the final words have better chances to be retrieved from memory.

Does the time of placement influence the time of retrieval? Do people memorize the words in the sequence of presentation and – in due consequence – do they retrieve them in this order from their memory? While time of retrieval shows a moderate, depth has a strong connection to serial number ($R = 0.42$ and 0.63).

Cross-relations between gender and other influential factors. There is a remarkable difference of roughly 1 word more for women (yet not significant) while a worst-case analysis (correcting for missing values) ends up with a somewhat negligible difference of 0.36. For context and sequel of time, the results indicate that man and women follow different strategies in memorizing; men orientate themselves more sequentially, women more contextually.

What we can learn from the patterns detected by the analysis

The findings have the status of law-like relations. See also Ehrenberg (1981) for the aim to filter out such relations from data and how to investigate under which conditions they can be generalized. It is important to state that we performed multiple tests; the analysis was driven interactively by intermediate results of our re-search. Many findings in empirical research have no better claim for generaliza-tion – replication would be a magic key to split artefacts from evidence but is rare-ly applied. Confirmation by replication studies would be inline with the logic of science by Popper (1935). Thus, it would be better to speak of *hypotheses* that may be formulated out of such a project. They await further scrutiny; some of them may be corroborated by ongoing research. This is the basic task and direc-tion of empirical research that accumulates evidence-based insight.

The final discussion of results of statisticians with the experts from the context should comprise: an evaluation of results; an integration of results into a body of knowledge; a discussion of possible conflicts of results to long-term experience; and a synopsis of questions, which are aroused by the new knowledge.

3 Discussion and conclusions

Here, the field of psychological experiments is introduced as a motivating context for learning probability and statistics to embed case studies discussed before into a broader frame. We end up with the role of technology in modelling and with its formative power to enhance and establish a meaning to the concepts used.

A case for modelling: psychological laws versus random effects. Psychological experiments are a valuable source for teaching. From these, descriptions of human behaviour are derived that claim the status of general laws. It is motivating to use the tension and interest raised by such laws and find data within the group of learners to investigate the validity of them.

Psychologists used also music (single notes), binary digits, and letters with the general finding that 7 plus or minus 2 remain in the short-term memory. Is it really true that we can memorize only 7 out of unrelated units? Telephone numbers usu-ally are built of 7 digits, the occidental music divides the scales in seven intervals, and even the Likert scale uses up to 7 points to measure the grade of attitudes. Is there an archaic law of our potential to discriminate and memorize items behind? Other laws are based on only 4 information units (Bachelder 2001), which may be reflected by the split of telephone numbers in groups of four and three. Learning is – to some extent – characterized by organizing connections (fictitious or hidden) in order to overcome our restrictions. With the popular memory cards game, play-ers have an advantage if they are able to create a (thrilling) story between the pic-tures. For working groups, Hall (1986) has established a "law" of an optimal size of 8 to 12 participants and justifies this with decreasing engagement in larger groups.

Such discussions and the complementary pair of opposites of scientific laws and mere randomness motivate and may clarify at the same time the way how findings are corroborated statistically till we can speak of empirical evidence.

The role of technology for modelling. A modelling approach to statistics and probability shows more openly the purpose of those concepts and their contribution to answer research questions and questions of everyday relevance. However, the complexity of used mathematics increases as several models have to be compared and handled. Mathematics can no more be used schematically but should contribute to improve decisions in the real situation according to some – not arbitrarily chosen – criterion. For such a purpose the concepts have to be permanently re-interpreted in terms of the context, the final answers (plural!) have to be evaluated to the extent to which they meet their intended purpose. Difficult tasks as we have to know more mathematics *and* the context.

To support the approach, technology may be used extensively. Relative to the level and experience of the learners, options like Excel, Fathom (Clements 2007), RExcel (a hybrid between Excel and R, Heiberger and Neuwirth 2009), or the language R may be favoured. The increasing functionality of the geometrically oriented GeoGebra may change preferences (Hohenwarter and Hohenwarter 2012). Excel has some advantages with respect to future workplaces of the students.

Technology allows not only an easy access to tedious calculations but also to graphical displays, which played a central role in our analysis of the memory test and may form the driving force of applied work. Technology allows also for the simulation of presumed hypotheses to get artificial data based on them. While the approach generally is indispensable especially if the range of used methods is increased, there is one serious reminder: especially if presented graphically, simulated results attain the character of *facts*; a careful discussion is needed to show that they amount to *one* among several scenarios and their validity is restricted to the specific case that these assumptions hold. Models are progressively used in an "as if" manner resembling the idea of scenarios referred to in Borovcnik (2011).

The formative potential of modelling. Overall, modelling may be used as a hinge between the learner, the real situation, and the mathematics used. It sheds light on the *process* of building models, it enhances the real situation, and it fills mathematical concepts with life. Scientific concepts are always designed to some purpose; applying them accordingly might orientate about this purpose. Though Kapadia and Borovcnik (1991) analysed chance encounters from various perspectives – the perspective of mathematics and the learner included, it did not focus yet on modelling. Modelling and simulation, their mutual interaction and their potential to enhance learning probability and statistics are a central topic in Girard (2008) and Chaput and Girard (2008). Blum (2012) refers to the *formative power* of modelling: By integrating an attitude of modelling into teaching, the value and significance of the concepts is not only conveyed but established. Specific modelling makes sense (or not), modelling as an approach to learning shows the stakeholders in action. Modelling enhances the ingredients and it contributes to motivate learners to undergo the task of learning.

References

Bachelder, B. L. (2001). The magical number 4 = 7: Span theory on capacity limitations. *Behavioral and Brain Sciences, 24,* 116–117.

Blum, W. (2012, July). *Quality teaching of mathematical modelling – what do we know, what can we do?* Plenary lecture at ICME 12, Seoul.

Borovcnik, M. (2006). Probabilistic and statistical thinking. In M. Bosch (Ed.), *European Research in Mathematics Education* IV (pp 484-506). Barcelona: IQS Fundemi. Online: http://ermeweb.free.fr/CERME4/.

Borovcnik, M. (2009). Aufgaben in der Stochastik – Chancen jenseits von Motivation. *Didaktik-Reihe der Österreichischen Mathematischen Gesellschaft, 42,* 1-23. http://www.oemg.ac.at/DK/Didaktikhefte/2009%20Band%2042/VortragBorovcnik.pdf.

Borovcnik, M. (2011). Strengthening the role of probability within statistics curricula. In C. Batanero, G. Burrill, & C. Reading (Eds.), *Teaching statistics in school mathematics. Challenges for teaching and teacher education: A joint ICMI/IASE Study* (pp. 71-83). New York: Springer.

Borovcnik, M., & Kapadia, R. (2011). Modelling in probability and statistics–key ideas and innovative examples. In J. Maaß, & J. O'Donoghue (Eds.), *Real-world problems for secondary school students–Case studies* (pp. 1-44). Rotterdam: Sense.

Borovcnik, M., & Kapadia, R. (2012). Applications of probability: The Limerick experiments. Topic Study Group 17 'Mathematical Applications and Modelling in the Teaching and Learning of Mathematics'. ICME 12, Seoul. Online: http://www.icme12.org/sub/tsg/tsg_last_view.asp?tsg_param=17.

Chaput, B., Girard, J. C., & Henry, M. (2008). Modeling and simulations in statistics education. In C. Batanero, G. Burrill, C. Reading, & A. Rossman (Eds.), *Joint ICMI/IASE Study: Teaching Statistics in School Mathematics. Challenges for Teaching and Teacher Education.* Monterrey: ICMI and IASE. Online: http://iase-web.org/Conference_Proceedings.php.

Clements, C. (2007). *Exploring Statistics with Fathom.* Emeryville, CA: Key Curriculum Press.

Ehrenberg, A.S.C. (1981). *Data reduction.* New York: Wiley.

Girard, J. C. (2008). The interplay of probability and statistics in teaching and in training the teachers in France. In C. Batanero, G. Burrill, C. Reading, & A. Rossman (Eds.), *Joint ICMI/IASE Study: Teaching Statistics in School Mathematics. Challenges for Teaching and Teacher Education.* Monterrey: ICMI. Online: http://iase-web.org/Conference_Proceedings.php .

Hall, E. T. (1981). *Beyond Culture.* New York: Anchor Books Edition.

Heiberger, R. M., & Neuwirth, E. (2009). *R through Excel: A spreadsheet interface for statistics, data analysis, and graphics.* New York: Springer.

Hohenwarter, J., & Hohenwarter, M. (2012). *An introduction to GeoGebra.* Online: http://www.geogebra.org/book/intro-en.pdf.

Kapadia, R., & Borovcnik, M. (1991). *Chance encounters: probability in education.* Dordrecht: Kluwer.

Krainer, K. (1993). Powerful tasks: A contribution to a high level of acting and reflecting in mathematics instruction. *Educational Studies in Mathematics, 24*(1), 65-93.

Miller, G. (1956). The magical number seven, plus or minus two: Some limits on our capacity for processing information. *The Psychological Review, 63*(2), 81-97. Online: http://www.musanim.com/miller1956.

Popper, K. (2005). *Logik der Forschung* (Hrsg. H. Keuth, M. Siebeck), Tübingen: Mohr Siebeck. (Original work published 1935 by Springer in Vienna).

PositScience (n.d.). Brain Training Software. http://www.positscience.com/.

Richardson, M., & Reischman, D. (2011). The magical number 7. *Teaching Statistics, 33*(1), 17-19.

Styer, D. F. (2000). *The strange world of quantum mechanics.* Cambridge: Cambridge University Press.

Chapter 21
Using the software FATHOM for learning and teaching statistics in Germany – A review on the research activities of Rolf Biehler's working group over the past ten years

Tobias Hofmann

Theodor-Heuss-Schule Baunatal

Carmen Maxara

Friedrichsgymnasium Kassel

Thorsten Meyfarth

Jacob-Grimm-Schule Kassel

Andreas Prömmel

Gymnasium Ernestinum Gotha

Abstract In this paper we want to reflect the research activities of the working group of Rolf Biehler on the use of stochastic simulations with the educational tool software FATHOM. Over the past ten years the simulation potential of this tool software for teaching and learning statistics was analyzed. Didactical concepts for the use of simulations with FATHOM were developed, implemented and evaluated. Several studies with students at schools and universities were performed to attain an optimization of the didactical approach. In order to tying the results together in a unified whole this article shows the most important studies, outcomes, and implications of the research work with FATHOM: a prior analysis of FATHOM's simulation capabilities, a study on using simulations with FATHOM at university, a study on a computer-aided stochastic course at school, a study on the learning environment eFATHOM and a study on using simulation plan schemata and worked examples. Finally, we outline recent developments and activities.

1 Introduction

Technology has had a great impact on content, pedagogy and course format in teaching and learning statistics at school in the past 20 years. A contemporary course on stochastics, which adequately takes data and chance into account, is not feasible without the use of technology. However, the appropriate and adequate use of technology in the classroom is critical. Therefore, Biehler (1993; 1997) developed requirements for software tools. FATHOM is an educational tool software, which conforms to such requirements (Finzer 2001, http://www.keypress.com). For that reason FATHOM has been adapted for German schools and universities, and was made available via Springer (Biehler et al. 2006). As a dynamic educational tool software, FATHOM enables teachers and students to use, modify, and develop for themselves embedded microworlds. In such microworlds, the student has access to multiply-linked representations, as well as the possibility to model and run simulations, and use interactive explorative features, like sliders.

For the implementation of using FATHOM in schools, the research group of Rolf Biehler has developed the GESIM concept in the past ten years. The GESIM research concept (see Figure 21.1) consists of three levels: the level of tool software research, the level of design classroom experiments and the level of empirical research. The term GESIM means "*general entry* in learning probability and statistics with *sim*ulations". The GESIM material, developed in our research group, contains different learning units with a teacher manual for a probability and statistics course at the level of upper secondary school (age 17-18) and university. Studying stochastic concepts is linked to the acquisition of competencies of simulation and the acquisition of skills in handling FATHOM.

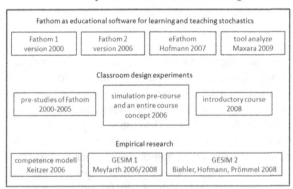

Fig. 21.1 GESIM concept as Design-Based Research (cf. Prömmel 2013)

The potential of computer-based stochastic simulations for learning and teaching statistics was recognized early (e.g. Biehler 1991; Schupp et al. 1992). Accordingly, simulation is a fundamental tool that allows students an experimental approach to adequate stochastic conceptions. Students become familiar with the frequency interpretation of probability. The intuitive understanding of random variability in empirical data can be developed more easily. As a consequence of the discussion about the national educational standards (KMK 2004) and the efforts of the working group "Stochastics in school" (AK Stochastik 2003; 2010) more attention is given to simulation in the new curricula of the federal states in Germany. Biehler and Maxara (2007) characterized three different aspects of the use of simulation in stochastic courses: Simulation as a representation tool of random experiments, simulation as an interaction tool between theoretical calculations and empirical methods and simulation as a "sui generis" tool. These three aspects were taken into account for the design of tasks and working sheets in our classroom design experiments (Meyfarth 2008; Prömmel 2013). Beside the important aspects of simulation, computer-based simulation can never replace the thinking about stochastic problems. Therefore students' activities in modeling random phenomena and the interpretation of simulation results become more important in classroom experiments. On the other hand teachers and students need didactical and instructional support for using computer-based simulations, in order to arrange teaching and learning statistics more comprehensively, and in a more differentiating, more interesting and more motivating way.

The four authors of this paper were members of Rolf Biehler's working group. Most of the studies of our working group can be classified as Design-Based Research, as it is described by Kelly (2003) or represented in a similar way by Wittmann (1995) in German didactics of mathematics. The development, testing and analyzing of innovative designs of learning are typical of Design Based Research studies. The challenge in our researches was the simultaneous development of three strands of students' competencies: the competence in probability and statistics, the competence of doing simulations (theoretical competencies), and the competence of using FATHOM for simulating random experiments (practical competencies). The research outcomes of the studies have been used to improve the GESIM concept with respect to theory and empiricism, as we present in the following sections.

2 First basic research about stochastic simulations and microworlds with FATHOM

The first basic research regarding a meaningful implementation of simulations with FATHOM in German statistics education were carried out in the context of an introductory course on probability and statistics for future mathematics teachers by Rolf Biehler at the University of Kassel. In the course, FATHOM was used continuously. This course started with exploratory data analysis and descriptive statistics, not only to teach basic techniques and concepts, but to implement statistical thinking by real data and authentic tasks (Biehler and Kombrink 2004). In the second part of the course, simulation as a method was systematically introduced at once in addressing the concept of probability. Corresponding to the German tradition, more emphasis was given to modeling and simulation in elementary probability than in inferential statistics and sampling distributions. This didactical innovation was developed, evaluated and optimized on the basis of intensive research.

In the following the analysis of FATHOM's simulation capabilities as well as the first didactical approaches for the implementation of simulation in an introductory course are presented. Afterwards some results of the first case study with students are pointed out.

A priori analysis of FATHOM's simulation capabilities and didactical approaches

Students might use FATHOM as a cognitive tool for modeling probabilistic situations. That would mean that students would have to develop a kind of mental model of the tool with regard to simulation problems. As a preparation for a didactical implementation, Maxara (2006) analyzed in detail the different ways to simulate random experiments with FATHOM and the working group of Biehler developed several concepts and notions for the students. In the course we distinguish three types of simulations: the *simultaneous simulation,* in which each single experiment corresponds to different columns, the *sequential simulation,* in which each single experiment corresponds to different rows and the *simulation by sampling* (Maxara 2006). Each type of simulation has different advantages and disadvantages due to the specifics of how the software simulates a random experiment (see Maxara (2009) for more details and an elaborated analysis). As a part of the prior analysis, Maxara developed a guidebook: "An introduction into simulation with FATHOM", for the students with a comprehensive overview of the possibilities of simulation in FATHOM and many simulation-microworlds. In addition to the three types of simulation, two further didactical components were essential: a three step process for a stochastic simulation and the use of a 'simulation plan'. The three step process consists of (1) setting up a stochastic model, (2) writing a *simulation plan* and (3) the realization in FATHOM. In the first step, the aim is that students describe the random situation for example by a concrete urn-model or in a more abstract way. They specify the probability distribution, the set of possible results, and random variables and events of interest (Maxara and Biehler 2006). In the second step – writing a *simulation plan* – the students transfer the analyzed random situation into a simulation plan for FATHOM. As a result of some empirical studies the simulation plan (built on ideas by Gnanadesikan et al. 1987) was redesigned several times and led to a 'four-step design as a guideline for stochastic modeling' (Table 21.1, Maxara and Biehler 2007). The students could do this transformation (stochastic model → simulation plan) step by step, because the two columns *probabilistic concepts* and *FATHOM objects & operations* correspond to each other in each step (Tab. 21.1, the right column is consistent with the simulation plan). The difficulties that students had with realizing simulations in FATHOM along these steps are identified in Maxara and Biehler (2006; 2007). To these four basic steps for a simulation, you can add two further steps: the modeling of the real situation at the beginning and the interpretation and validation of the results after the simulation. In this case, you get a six-step plan.

By using the simulation plan we focus on following pedagogical intentions: students could structure their simulations, reflect about their simulations, and document their simulations. Based on the study of Meyfarth (2008), we have had the hypothesis that the simulation plan could be a helpful metacognitive tool for students (Maxara and Biehler 2007; Biehler and Maxara 2007).

Table 21.1 Four-step design as a guideline for stochastic modeling (Maxara and Biehler 2007)

Step	Probabilistic concepts	FATHOM objects & operations
1	Construct the model, the random experiment	Choose type of simulation; define a (randomly generated) collection, simulate the random experiment
2	Identify events and random variables of interest (Events and random variables as bridging concepts)	Express events and random variables as "measures" of the collection
3	Repeat the model experiment and collect data on events and random variables	Collect measures and generate a new collection with values of the measures
4	Analyze data: relative frequency (events); empirical distribution (random variables)	Use FATHOM as a data analysis software

Maxara (2009) analyzed in her dissertation the possibilities of simulation with FATHOM more systematically, in more detail, and more exhaustively, than for the students' guidebook, so the analysis could be used for further didactical utilizations. With the analysis of the tool 'FATHOM', Maxara further developed a more general instrument that could also be used in a similar systematic way to analyze the simulation potential of other tool software such as EXCEL, Tinkerplots or TI-Nspire.

A case study with students – some findings

In the second part of her dissertation Maxara analyzed the processes of interaction and communication of pairs of students (of the introductory course on probability and statistics for future mathematics teachers) simulating the following random experiment with FATHOM:

> Mister Becker has to wear a black suit during his working hours, but he can choose the tie himself. 7 different ties hang in his wardrobe. Every morning he randomly takes one tie out of the wardrobe and puts it back in the evening.

1. What is the probability that Mister Becker will wear 5 different ties in his five-day working week?

2. What is the probability that Mister Becker will wear at least two identical ties in his five-day working week?
3. How many different ties does Mister Becker wear on average in his five-day working week?

The previous research that had been available at that time suggested doing a qualitatively-oriented explorative case study. At that time it was the first study of this kind regarding students constructing simulations with the software FATHOM. The students were video recorded while solving the problem. This material was transcribed and analyzed. In a first step of interpretation the solution processes were described and reconstructed relatively closely to the material to give an outline and interpretative background. In a second analysis stage "simulation process diagrams" were developed, with which the structures and operations of the solution processes on the basis of the components of the simulation plan were visualized. This representation of the solution process permitted instructive comparisons between the pairs of students. In a third step, the transcripts were coded and evaluated strictly according to the methods of qualitative content analysis (Mayring 2003). Based on works of the working group of Biehler, Maxara developed definable dimensions (high – middle – low) of the FATHOM specific simulation competencies (*general FATHOM competence, competence of formulas, competence of simulation* and *strategic competencies*) and the stochastic competencies (*mathematical modelling/using of mathematical terms, knowing of the meaning of simulated results*) of the students (Maxara 2009). A result of this study is therefore a nuanced view of the students' competencies as well as a conceptualization of different competencies and their interdependency, as they arise during the work on simulating random experiments with the educational software FATHOM. Due to the detailed analysis Maxara could determine that of major impact on the implementation of the simulation were the shortcomings in the competencies with formulas and simulations. Students' difficulties were mostly associated with employing a haphazard approach and not knowing formulas. At the stochastic level students showed shortcomings in the interpretation of the simulated result: no precise view on relative frequencies and probabilities and an insufficient interpretation of mean and expected value.

This first and extensive study pointed out many relevant and important details of students' simulation activities and competencies. Basically the students accepted simulation as a method of problem solving, but they needed more guidance to link the simulation activities with the stochastic content. Furthermore, the simulation process diagrams showed that the simulation plan might be used to steer students through their simulation process. Further studies dealing with the use of the simulation plan schema are Biehler and Prömmel (2010) and Prömmel (2013). Maxara's study suggests the following subject matters for further research:

- a further development of the simulation plan

  - to support students linking simulation activities and stochastic terms,
  - to support students to document their simulations,

- to support students to structure their simulations,
- a conception, an implementation and an evaluation of a stochastic course with simulations at school,
- an effective instruction to introduce FATHOM.

3 A teaching concept of the stochastic course in secondary schools – An empirical study about the simulation pre-course

Based on the innovative concepts and material for learning statistics developed by the working group of Rolf Biehler (Biehler 2003; Biehler and Kombrink 2004; Maxara 2006), the group wanted to introduce the method of stochastic simulation systematically in secondary schools. For the German didactics of mathematics it was the first time to develop a course concept with the continuous use of stochastic simulation in school.

Meyfarth designed a complete teaching concept for the advanced stochastic course in secondary schools, which uses the support of computer simulations and computer microworlds with the software FATHOM over the period of half a year. With regards to content the concept of the course follows the curriculum of the German federal state of Hessen for the advanced stochastic course (Kultusministerium Hessen 2003). The material and the teaching units of well-established German schoolbooks (e.g. Barth and Haller 1983; Strick 1998; Griesel et al. 2003) provide a further basis for developing the advanced stochastic course. Furthermore the tasks and the teaching arrangements of English-language schoolbooks (Wild and Seber 2000; Rossman et al. 2001; Erickson 2002) have been considered. In addition to the use of computer simulations and FATHOM-Microworlds, the teaching concept picks up present findings of the German didactics of mathematics, particularly with regards to choosing application-oriented problems and tasks to arouse the students' interest.

The teaching concept contains three key aspects:

At the beginning there is a "*simulation pre-course*": In this pre-course the students are learning on three levels. The students acquire "*tool competence*" in using the software FATHOM, they acquire *simulation competencies* in modeling the stochastic problems, and they acquire basic *competencies in probability and statistics*.

The second key aspect is the *Binomial distribution*. The Binomial distribution is the fundamental probability distribution in German stochastic courses at secondary school and it is used for explaining the concept of "probability distribution" exemplarily. Therefore the teaching of the Binomial distribution with support of computer simulations and dynamic FATHOM-Microworlds is reinforced with priority.

The third key aspect is the *Testing of Hypotheses*. In school this is a very demanding subject matter in the field of inferential statistics, and so it is also supported by the use of computer simulations and dynamic FATHOM-Microworlds.

The whole teaching concept is available as a free download on the „Kasseler Online-Schriften zur Didaktik der Stochastik" (Meyfarth 2006). The developed teaching material about the three key aspects is described in detail: the text contains the chosen stochastic problems with solutions (theoretical and by computer-simulation), the worksheets for the students and the FATHOM-Microworlds. Furthermore the material is supplemented by didactical annotations for teachers, who want to use the concept in their own class. In particular the subject matter of the "Binomial-Distribution" and of "Hypothesis Testing" is described substantially.

The teaching concept has been successfully tested several times at the secondary schools Albert-Schweitzer-Schule and Jacob-Grimm-Schule in Kassel and has been discussed and improved successively. In connection with testing the teaching concept at school, we implemented local training of teachers, where the developed teaching material was presented and an introduction to FATHOM was given. In addition to this training of the involved teachers, we discussed and evaluated the teaching material, aiming at an improvement of the concept. Furthermore this training of teachers was continued within the project „mathematik anders machen" (Törner 2007), which was sponsored by the Telekom foundation. These training activities in computer simulations and dynamic FATHOM-Microworlds in statistics education have been offered and held nationwide in Germany on several occasions.

A first empirical study about the simulation pre-course

An essential idea of the computer-assisted teaching concept is to start with a simulation pre-course. In this pre-course the students are instructed to solve interesting stochastic problems by using computer simulation, like several dice problems, the Coupon-Collector Problem, the Birthday Paradox and other phenomena. The pre-course starts with the following Multiple-Choice-Test Problem:

> Look at the following tests: Test A includes 10 questions, which can be answered with yes or no. Test B includes 20 questions, which can be answered with yes or no. Both tests are passed, if at least 60% of the questions were answered correctly. Which test could be passed more easily, if one only guesses? a) Test A, b) Test B, c) the probability for both tests is equal, or d) I do not know.

The problems are picked up theoretical in the sequel to the course. The acquisition of the simulation and the FATHOM competencies is linked with the frequentist concept of probability, with the Law of Large Numbers, as well as with the classical definition of probability in countable sample spaces. An important support in creating a computer simulation with FATHOM is offered by the simulation plan

(Table 21.1), which was developed by Maxara (2009) to give a frame for creating a computer simulation with FATHOM.

The simulation pre-course is the essential part of an empirical study by Meyfarth (2008). This Design Research Study concentrates on a course given by Meyfarth himself and on a second course given in a parallel class at the same time and with the same teaching material.

Following the concept of Design Research the study on the one hand aimed to capture und analyze the different levels of students' learning: simulation and FATHOM competencies, stochastic competencies, the students' attitude and motivation. On the other hand, the study ought to evaluate the effectiveness of the concept and the teaching material with regard to the development of the aspired competencies as well as to the students' single-handed learning, always with the objective of improving the teaching concept. Different empirical investigations were evaluated and possible improvements optimized. One focus has been on video sequences taken with screen capture software, recorded during the students' group work at the computer.

The study's results have shown many positive aspects of the simulation pre-course; the intentions of the pre-course could be achieved to a large extent. A marked class exercise and two competence tests have demonstrated the successful acquirement of simulation and FATHOM competencies as well as basic stochastic competencies. A questionnaire has substantiated a very positive improvement in attitude and motivation of most students. The transcripts of the screen-capture videos have shown the development of the pupils' competencies and how the pupils were dealing with the problems in detail. All in all the results of the study provided evidence that the design of the simulation pre-course could be adopted successfully in daily lessons, even considering the general conditions at school regarding the final examinations in Hessen.

The transcripts of the screen-capture videos however have shown deeper problems. For example problems with complex formulas like *uniqueValues()* or the *if()*-statement and with the idea of the *measure* on the level of FATHOM competencies. On the stochastic level the students showed problems in making a distinction between relative frequency and probability. On the modeling level the videos showed an unsatisfying handling of interpretive and verbal phrasing instructions by the students.

Furthermore the study teased out some important points for further development of the teaching concept:

- The concept needs a stronger link between the simulation and FATHOM competencies on the one hand and the basic stochastic subject matter already in the beginning of the simulation pre-course.
- The concept needs a stronger link between the simulation and FATHOM competencies on the one hand and the modeling competence on the other hand.
- The single-handed development of the simulation and FATHOM competencies has to be improved for example by means of a multimedia-based learning environment or a more intensive use of an enhanced simulation scheme.

- The effectiveness of the simulation pre-course in improving the students' performance in the rest of the course has to be studied.

4 Developing eFathom

The findings of the Maxara and Meyfarth study (see the last two sections) show impressively that FATHOM can support very well the learning process and a deeper understanding of stochastics at school and at university. However, a further need for additional support in learning how to use FATHOM became clear at various points. This is what researchers call "instrumental genesis": The "tool" has to be acquired as an "instrument" by every student to work on specific problems (Trouche 2004).

To address learners' initial difficulties in dealing with FATHOM, which had been identified (Meyfarth 2008; Maxara and Biehler 2006). Hofmann (2012) developed the multimedia learning environment eFATHOM as part of his dissertation. It offers the FATHOM-novice an example-oriented, interactive introduction to this software. Following the principle of instrumental genesis, the aim was that learners become able to use FATHOM as a tool for context-specific, problem-oriented applications. In addition to FATHOM-functions in the sense-creating context, stochastic skills are taught.

As an online-learning device, eFATHOM is available for nearly every FATHOM-user (e.g., via http://efathom.math.uni-paderborn.de). For those who want to work with FATHOM in their Stochastic-lessons, the acquisition of necessary FATHOM-skills can take place beyond the classroom as part of the homework. By means of this, self-directed learning is supported at one's own pace. Main topics of the learning environment are the introduction to *data analysis* and the introduction to the simulation with FATHOM. The technical and didactic treatment of the selected content is based on a broad range of operational experience using the software at school and at university. Furthermore, it includes results from empirical studies on the acquisition of skills and simulation and is based on them (Meyfarth 2008; Maxara 2009).

Simulations with Fathom

 Rolling two dice

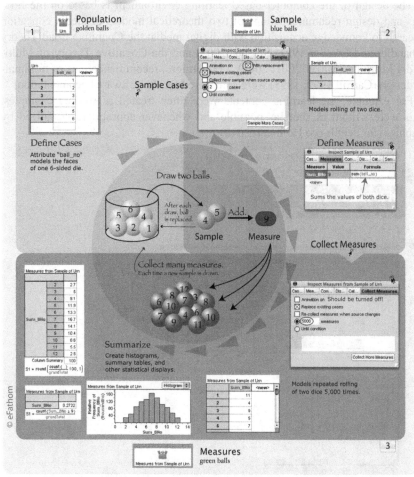

Key: **Right click** on a collection to open the Context menu and choose a command.
Double click on a collection to open the inspector.

Fig. 21.2 Graphic simulation plan; handout for the acquisition of the sampling-based simulation

The teaching of simulation competence focuses on two types of simulation: simultaneous simulation and sampling-based-simulation. At the beginning, the Law of Large Numbers is introduced, which offers the frequentist concept of probability as a key to understanding why stochastic simulations work. Applying rules of thumb, the accuracy of simulations is discussed. Step by step students are introduced to the concepts of sampling-based-simulation. The simulation plan by

Maxara (2009) serves as a didactical tool and was further developed as a specific simulation scheme (see next section). Furthermore, the graphic simulation plan (see Figure 21.2) was especially created for eFATHOM. It is designed as a worked-example and is supposed to support students learning simulations while modeling with FATHOM. In eFATHOM it is explained how students can read the plan.

The design of the computer-based learning environment is based on the findings and design recommendations of two theoretical models of media education research to knowledge acquisition with digital media: the Cognitive Load Theory (Chandler and Sweller 1991; Sweller 2005) and the Cognitive Theory of Multimedia Learning (Mayer 2001; 2005). Furthermore, recommendations of minimalist design (Carroll 1990; 1998) were considered for the software documentation.

For example, the learning contents were often treated as worked examples, texts were written as short as possible and whenever appropriate visualized by graphics (for details see Hofmann 2012).

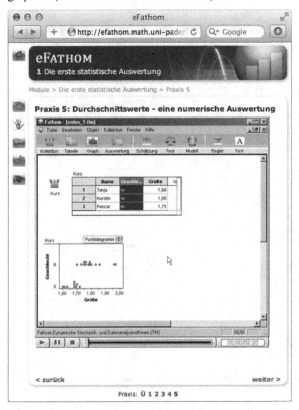

Fig. 21.3 Tutorial-video from eFATHOM

Additionally, design principles for eFATHOM have been developed. A substantial feature is teaching key concepts via tutorial videos (see Figure 21.3). In these, a tutor teaches in an action-oriented way how to acquire FATHOM competence using screen-casts and examples. The learner is encouraged to review the actions in

FATHOM. To make this possible and to reduce the cognitive load, the learning environment eFATHOM was designed so that it can be placed next to FATHOM without overlap on the screen. Other design features of the learning environment are little use of text, intuitive operation and an attractive and functional layout.

Empirical studies of the learning environment eFathom

The evaluation of the learning environment eFATHOM consisted of qualitative and quantitative empirical studies. The exploratory focus was the detection of difficulties in working with eFATHOM, the use of this learning environment by students, the acceptance of eFATHOM by different user groups, and the identification of conveyed FATHOM competences.

A total of 570 students participated in five different studies at school and at college and another 2000 people took part in a Web-study (for details see Hofmann 2012).

Difficulties dealing with the learning environment were identified at early stages of development and considered in the further development. Regarding the user habits it became clear that students follow a linear path through the several modules and implement the demonstrated actions in the tutorial videos into FATHOM at the same time.

The learning environment is equally well-received by school students as well as by university students and is appreciated as a simple and effective introduction to the software FATHOM. This confirms the results from a study of 350 university students and 39 school students. Having worked with eFATHOM, the students replied using a Likert scale from 1 = "not acceptance" to 5 "full acceptance" to value various items about eFATHOM (see Table 21.2).

Table 21.2 Some results from the acceptance study (Likert scale from 1 to 5 (not to full acceptance))

Item: eFATHOM ...	Mean score university students	Mean score School students
... is a good introduction into FATHOM.	4,44	4,25
... is efficient introduction into FATHOM.	4,33	4,08
... has a reasonable structure.	4,29	3,95
Split screen arrangement is helpful.	4,32	3,89
Tutorial videos are understandable.	4,69	4,43
Working with eFATHOM is fun.	3,59	3,46

It is difficult to measure to which extent the students acquire FATHOM competencies and statistical competencies using eFATHOM because the learning conditions of students cannot be controlled in that treatment. Therefore a small number of students (n = 7) without controlled group were observed in a qualitative study while working with eFATHOM. Their activities were recorded with a screen-capture software. Almost all students were able to solve the problems in the basic task and in the application task successfully. Moreover, they were able to deal with simulation problems more independently after using eFATHOM. The learning environment eFATHOM was evaluated in an introductory course of stochastics at schools (Biehler and Prömmel 2010). The research outcomes give some evidence that the heterogeneity of FATHOM competencies were substantially reduced and that the learning time could be used for developing statistical competencies instead of software competencies (cf. Biehler and Hofmann 2011).

The success of eFATHOM has led to a new German edition of FATHOM for schools that is provided with eFATHOM together with activities and instructional material for grade 7 to 12 (Biehler et al. 2011).

5 Development of student's modelling and simulation competencies by simulation plan schemas and worked examples

Based on the evaluation results of the simulation pre-course (Meyfarth 2006; 2008), the introductory course was redesigned and new course elements were developed, implemented and comprehensively analyzed (Biehler et al. 2008; Prömmel 2013):

- Students learn FATHOM in homework with eFATHOM, because it seemed that the learning of FATHOM in classroom could be improved.
- Students use prepared simulation plan schema and worked examples because the simulation plan and the self-generated documentation might be improved.
- The worksheets were changed by asking students for their expectations and interpretations because students worked predominantly on technical aspects.

The redesigned introductory course was aimed at the furtherance of students' knowledge in three domains: data analysis, random experiments and probability, and sampling distributions. The course design emphasizes simulations, the Law of Large Numbers and the role of sample size. For that reason, it was hoped that students' knowledge about distributions would develop through own experiences from the beginning. Simulations are a practical informal start in that direction. This section will focus on the use of the simulation plan schema and worked examples.

The use of the simulation plan schema and worked examples

It is well-known from other studies, that students have difficulties in dealing with simulations at different levels. De Jong and van Joolingen (1998) have identified four areas for explorative learning with computer simulations which can be transferred in modified form to simulations with FATHOM: formulating of hypotheses, designing of experiments, interpreting of data, and self-control of learning. From the results of the Meyfarth study (2008), the Biehler working group has designed a special simulation plan schema to assist simulation novices (Figure 21.4). The idea of designing such a schema was born in reaction to the observation that some students had difficulties abstracting the schema and its steps and to remember them well enough in the simulation process. The structure and the content are a realization of the general 6-step plan for stochastic modeling and simulation of Biehler and Maxara (2007). The table is split into two columns. The left grey column shows the individual steps for the simulation method *simulation by sampling from a box* like a guideline. The right column provides a predefined structure. When filling out the simulation plan schema, students have to consider specific properties of FATHOM as well as theoretical aspects. Especially step [3] - „description" and „possible values of the measure" - requires theoretical considerations. Such matters help to prevent a technical handling and documentation of FATHOM terms.

For that reason, such a schema can fulfill different functions, such as supporting computer-supported collaborative learning, and promoting a targeted acquisition of knowledge. One can distinguish three functions: planning, guidance, documentation (cf. Biehler and Prömmel 2010).

1. Planning represents an essential part in carrying out experiments. The simulation of stochastic phenomena by modeling random experiments, the review and evaluation under certain aspects are experimental situations. Therefore the simulation plan schema can be used as a planning tool, because multiple or single steps may be planned in advance.
2. The simulation plan schema has a guidance function through its design and through its structuring according to the steps in the simulation process. All essential steps are given and have to carry out one step after the other. Therefore it also serves as guidance. Such a schema can support the learning of complex simulation methods.
3. The notation of the simulation results, after a simulation with FATHOM, is called documentation function. Usual are two steps: the analysis of the simulation results and the reflective interpretation.

simulation by sampling
from a box

random experiment: *sampling from a box filled with two cases*

questions: *With which probability does someone have at least 8 correct guesses out of 12?*

[1] define box collection	content of box: *correct, incorrect* name of attribute: *guess* Fathom formula (if applicable):
[2] sampling	☒ with replacement ☐ without replacement sample size: *12*
[3] define measures	description: *number of correct guess* possible values of the measure: *0, 1, 2, ..., 12* name of measure: *number.success* Fathom formula: *count(guess = "correct")*
[4] collect measures	number of repetitions: *5000*
[5] analysis of results: distribution, rel. frequency, mean,...	*count(number.success ≥ 8)/grandTotal·100 = 19.32 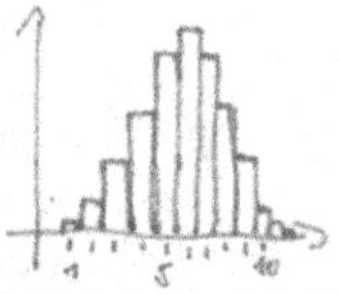 %*

interpretation: *The probability for at least 8 correct guesses out of 12 is about 20 percent. This means that you can get a price in 1 of 5 cases when you guess only.*

Fathom file:_____

Fig. 21.4 Filled-out simulation plan schema usable as a worked example

Completely filled out simulation plan schemas can be elaborate solution patterns for more simulations. Examples of such solutions preserve procedural knowledge and these afford a further acquisition of knowledge through the learning of expert solutions and self-generated solutions (Atkinson et al. 2000). A completely filled out simulation schema satisfies these requirements and it is called a process-oriented worked example (van Gog et al. 2004; 2008). In Figure 21.4, such a filled out form is shown for the following problem that was posed in one lesson of the introductory course:

> A test person, who receives 12 pieces of music randomly, has to identify whether it has CD-quality or MP3-quality. The person wins a prize, if she/he gets at least 8 correct. With which probability does someone get a prize by just guessing the sound quality (CD or MP3)?

The research design and some selected results

The intro-course was implemented in two courses at the Jacob-Grimm-School in Kassel and in one course at the Albert-Schweitzer-School in Kassel. For the evaluation of the introductory course, data were collected with several instruments, e.g. observer protocols of all lessons, recordings (screen capture videos plus audio) from all small group working phases.

How the simulation plan schema affected the work of the students with the simulation problems was one of the research questions. Specific research methods were developed to analyze this subject matter, both qualitative and quantitative. These methods focus on the analysis of the products, the processes and the communication of the student pairs. In a special comparison of treatment student pairs had to work on tasks in one lesson: one group as consecutive group, separates that work in an offline planning phase and an online computer phase, and another group as integrative group, works in only one online computer phase.

In the consecutive way, students have to fill out the simulation plan schema before the computer work as much as possible and then to switch to the computer. The simulation plan schema thus serves as a planning tool. In the integrative way students work immediately on their computer. These pairs of students have to fill out the simulation plan schema as an aside. The following research questions were of special interest:

1. How successfully did the student pairs solve the simulation tasks?
2. Did the consecutive pairs of students have fewer problems in the implementation with FATHOM than the integrative group?
3. Did the consecutive pairs of students need less help from outside than the integrative group?

In the following tables the major findings of 18 pairs of students (9 consecutive, 9 integrative) in this special treatment are summarized (cf. Prömmel 2013).

Table 21.3 Analysis of products – Simulation plan schema and FATHOM-Files

	Products (% of max. score)	
	Simulation plan schema	FATHOM- Files
consecutive	76 %	93 %
integrative	75 %	91 %

Table 21.4 Simulation plan schema along the simulation steps (% of max. score)

Simulation plan schema (steps)	M	S1	S2	S3	S4	I
consecutive	33 %	100 %	86 %	100 %	53 %	44 %
integrative	22 %	99 %	88 %	100 %	51 %	46 %

Table 21.5 Analysis of processes – problematic phases and help from outside

	Processes (absolute frequency)	
	Problematic phases	Help from outside
consecutive	28	29
integrative	59	61

Generally you can assess that the simulation plan schema has supported the students in the acquirement of procedural knowledge for the simulation by sampling. It offers a clear orientation for students of necessary steps in development, implementation and evaluation of complex simulations. The results of the empirical study show that the majority of student pairs without offline planning phase plan their work with the schema. This means that simulation novices filling out the simulation plan schema as far as possible before they implement the simulation in FATHOM. However, the change of planning and documentation is not anchored enough in some students' minds. Sometimes erroneous entries are not changed or missing entries are not supplemented on the simulation plan schema. Incorrect or incomplete documentations are not suitable as worked example for more simulations.

Nevertheless most of the student pairs successfully accomplished their solution process without help from the outside. In this sense, the simulation plan schema has proved as an instructional support measure. In particular the structured form of the simulation plan schema has helped the students to document their simulations efficiently and clearly. However, such a schema also has limits. With this design it can be used only for one class of simulations (simulation by sampling from a box). But generally it serves simulation novices also to foster an autonomous construction of simulations.

6 Summary

The Design-Based implementation of the educational tool software FATHOM for teaching and learning statistics at German schools and universities was one of the most important aims of our research group. Therefore we developed material, evaluated the material in classroom experiments and improved it on the basis of the research outcomes. The results of our studies showed that it is feasible to implement FATHOM in stochastic courses at schools and university, in order to demonstrate stochastic phenomena as well as to realize more students' activities in learning stochastics with simulation. In respect to other studies (Urhahne and Harms 2006) students need instructional support for doing simulation activities with FATHOM. Therefore our research focused on the development of the design of instructional support. The major outcomes are:

- The analysis of the software underlined the didactical potential of FATHOM for teaching and learning statistic with simulations.
- The simulation plan supports students to structure their simulation activities.
- Well-designed course concepts with simulation tasks and process-oriented worksheets show the feasibility of FATHOM in schools.
- The multimedia learning environment eFATHOM is effective for learning FATHOM.
- The simulation plan schema supports students to document their simulations.

Apart from these positive results, a lot of didactic material and teaching concepts for schools were developed (Maxara 2006; Meyfarth 2006; 2008; Hofmann 2012; Prömmel 2013). Some of the material is summarized in Biehler et al. (2011) and prepared with worksheets and FATHOM microworlds especially for teaching at schools. This book is also aimed at teachers, who still do not have much experience with using computers in classrooms. In addition the learning environment eFATHOM provides an elementary introduction into FATHOM. We hope that this book will give interesting and exciting innovations for stochastics education at school and university, which positively affects the learning processes of students. Complex mathematical concepts can be visualized and students are able to experiment with methods and data. In addition to the two strands "data analysis" and "probability distributions and inferential statistics", there is also a large strand of "simulation" in this book, in which the three types of simulations and the simulation plan schemata are introduced. Each chapter is divided into two parts: FATHOM basics are discussed in the first section, and in the further sections teaching ideas, worksheets and FATHOM microworlds are presented. The didactical material, eFATHOM and the simulation plan schema may support teachers and students in using new technology for enhanced teaching and learning probability and statistics in an effective way.

References

Arbeitskreis Stochastik der GDM (2003). Empfehlung zu Zielen und zur Gestaltung des Stochastikunterrichts. *Stochastik in der Schule, 23*(3), 21-26.

Arbeitskreis Stochastik der GDM (2010). *Leitidee Daten und Zufall für die Sekundarstufe II – Kompetenzmodelle für die Bildungsstandards aus Sicht der Stochastik und ihrer Didaktik.* Retrieved from http://stochastik-in-der-schule.de/Dokumente/Leitidee_Daten_und_Zufall_SekII.pdf.

Atkinson, R. K., Derry, S. J., Renkl, A., & Wortham, D. W. (2000). Learning from examples: instructional principles from the worked examples research. *Review of Educational Research, 70,* 181-214.

Barth, F., & Haller, R. (1983). *Stochastik Leistungskurs.* München: Ehrenwirth.

Biehler, R. (1991). Computers in Probability Education. In R. Kapadia, & M. Borovcnic (Eds.), *Chance Encounters: Probability in Education* (pp. 169-211). Dordrecht: Kluwer.

Biehler, R. (1993). Software tools and mathematics education: the case of statistics. In C. Keitel, & K. Ruthven (Eds.), *Learning From Computers: Mathematics Education and Technology* (pp. 68-100). Berlin: Springer.

Biehler, R. (1997). Software for learning and for doing statistics. *International Statistical Review, 65*(2), 167-189.

Biehler, R. (2003). Simulation als systematischer Strang im Stochastikcurriculum. H.-W. Henn (Eds.), *Beiträge zum Mathematikunterricht 2003* (pp. 109-112). Hildesheim: Franzbecker.

Biehler, R., & Hofmann, T. (2011, August). *Designing and evaluating an e-learning environment for supporting students' problem-oriented use of statistical tool software.* Paper presented at the 58th ISI Session, Dublin, Ireland.

Biehler, R., & Kombrink, K. (2004). Elementare Stochastik interaktiv - Einführende Stochastikausbildung mit Unterstützung didaktisch orientierter Werkzeugsoftware. In R. Biehler, J. Engel, & J. Meyer (Eds.), *Neue Medien und innermathematische Vernetzungen in der Stochastik.* (Anregungen zum Stochastikunterricht, Band 2, pp.151-168). Hildesheim: Franzbecker.

Biehler, R., Hofmann, T., Maxara, C., & Prömmel, A. (2006). *Fathom 2. Eine Einführung.* Berlin, Heidelberg: Springer.

Biehler, R. & Maxara, C. (2007). Integration von stochastischer Simulation in den Stochastikunterricht mit Hilfe von Werkzeugsoftware. *Der Mathematikunterricht, 53*(3), 45–62.

Biehler, R., Hofmann, T., Maxara, C., & Prömmel, A. (2011). *Daten und Zufall mit Fathom – Unterrichtsideen für die S1 und S2 mit Software-Einführung.* Braunschweig: Schroedel.

Biehler, R., Hofmann, T., & Prömmel, A. (2008). *Das GESIM-Konzept – Ein Ganzheitlicher Einstieg in die Stochastik in der gymnasialen Oberstufe mit computergestützter Simulation: Unterrichtsmaterialien und didaktische Anleitungen* (unpublished manuscript). Kassel: Universität Kassel.

Biehler, R., & Prömmel, A. (2010). Developing students' computer-supported simulation and modelling competencies by means of carefully designed working environments. In C. Reading (Ed.), *Proceedings of ICOTS 8, Ljubljana, Slovenia* [CD-ROM]. Retrieved from http://www.stat.auckland.ac.nz/~iase/publications/icots8/ICOTS8_8D3_BIEHLER.pdf.

Carroll, J. M. (1990). *The Nurnberg funnel: designing minimalist instruction for practical computer skill.* Cambridge, MA: MIT Press.

Carroll, J. M. (Eds.). (1998). *Minimalism Beyond the Nurnberg Funnel.* Massachusetts: MIT Press.

Chandler, P., & Sweller, J. (1991). Cognitive load theory and the format of instruction. *Cognition and Instruction, 8,* 293-332.

de Jong, T., & van Joolingen, W. R. (1998). Scientific Discovery Learning with Computer Simulations of Conceptual Domains. *Review of Educational Research, 68*(2), 179-201.

Erickson, T. (2002). *Fifty FATHOMs.* Oakland: eeps media.

Finzer, W. (2001). *Fathom Dynamic Statistics* (v.1.0) [Current version is 2.1]. Key Curriculum Press.

Gnanadesikan, M., Schaeffer, R. L., & Swift, J. (1987). *The Art and Techniques of Simulation*. Palo Alto: Dale Seymore.

Griesel, H., Postel, H., Strick, H.-K., & Suhr, F. (Eds.) (2003). *Leistungskurs Stochastik*. Elemente der Mathematik. Hannover: Schroedel.

Hofmann, T. (2012). *eFathom: Entwicklung und Evaluation einer multimedialen Lernumgebung für einen selbstständigen Einstieg in die Werkzeugsoftware Fathom*. Wiesbaden: Springer Spektrum.

Kelly, A. E. (2003). Theme Issue: The Role of Design in Educational Research. *Educational Researcher, 32*(1), 3-4.

Kultusministerium Hessen. (2003). *Lehrplan für die Bildungsgänge Hauptschule, Realschule, Gymnasium*. Wiesbaden: HeLP.

KMK (Ed.) (2004). *Bildungsstandards für das Fach Mathematik für den mittleren Bildungsabschluss – Beschluss der Kultusministerkonferenz vom 4.12.2003*. München: Wolters Kluwer.

Maxara, C. (2006). *Einführung in die stochastische Simulation mit FATHOM. KaDiSto – Kasseler Online-Schriften zur Didaktik der Stochastik, Band 1*. Universität Kassel. Retrieved from http://nbn-resolving.de/urn:nbn:de:hebis:34-2006082514477.

Maxara, C. (2009). *Stochastische Simulation von Zufallsexperimenten mit FATHOM. Eine theoretische Werkzeuganalyse und explorative Fallstudie*. Hildesheim: Franzbecker.

Maxara, C., & Biehler, R. (2006). Students' Probabilistic Simulation and Modelling Competence after a Computer-Intensive Elementary Course in Statistics and Probability. In A. Rossman, & B. Chance (Eds.), *Proceedings of the Seventh International Conference on Teaching Statistics*. [CD-ROM]. Voorburg, The Netherlands: International Statistical Institute. Retrieved from http://www.stat.auckland.ac.nz/~iase/publications/17/7C1_MAXA.pdf.

Maxara, C., & Biehler, R. (2007). Constructing stochastic simulations with a computer tool – students' competencies and difficulties. In D. Pitta-Pantazi, & G. Philippou (Hrsg.), *Proceedings of the Fifth Congress of the European Society for Research in Mathematics Education* (pp. 762-771). Nicosia, Zypern: Department of Education, University of Cyprus. http://www.erme.unito.it/CERME5b/WG5.pdf#page=79. Abgerufen am 20.05.2013.

Mayer, R. E. (2001). *Multimedia Learning*. Cambridge: Cambridge University Press.

Mayer, R. E. (2005). Cognitive Theory of Multimedia learning. In R. E. Mayer (Eds.), *The Cambridge Handbook of Multimedia Learning* (pp. 31-48). Cambridge, MA: Cambridge University Press.

Mayring, P. (2003). *Qualitative Inhaltsanalyse. Grundlagen und Techniken* (8. Auflage). Weinheim : Beltz UTB.

Meyfarth, T. (2006). *Ein computergestütztes Kurskonzept für den Stochastik-Leistungskurs mit kontinuierlicher Verwendung der Software FATHOM - Didaktisch kommentierte Unterrichtsmaterialien. KaDiSto - Kasseler Online-Schriften zur Didaktik der Stochastik, Band 2*. Kassel: Universität Kassel. http://nbn-resolving.de/urn:nbn:de:hebis:34-2006092214683.

Meyfarth, T. (2008). *Die Konzeption, Durchführung und Analyse eines simulationsintensiven Einstiegs in das Kurshalbjahr Stochastik der gymnasialen Oberstufe: Eine explorative Entwicklungsstudie*. Hildesheim: Franzbecker.

Prömmel, A. (2013). *Das GESIM-Konzept: Rekonstruktion von Schülerwissen beim Einstieg in die Stochastik mit Simulationen*. Wiesbaden: Springer Spektrum.

Rossman, A., Chance, B., & Barr von Oehsen, J. (2001). *Workshop Statistics: Discovery with Data*. New York: Key College Publishing.

Schupp, H., Berg, G., & Dabrock, H. (1992). *ProSto: Programme für den Stochastikunterricht*, Bonn: Dümmler.

Strick, H. K. (1998). *Einführung in die Beurteilende Statistik*. Hannover, Schroedel.

Sweller, J. (2005). Implications of cognitive load theory for multimedia learning. In R. E. Mayer (Ed.), *The Cambridge Handbook of Multimedia Learning* (pp. 19-30). Cambridge: Cambridge University Press.

Törner, G. (2007). Der wichtige Blick über den Tellerrand: Professor Günter Törner über das Projekt 'Mathematik Anders Machen'. *Schule im Blickpunkt, 2008*(4), 7-10.

Trouche, L. (2004). Managing the Complexity of Human/Machine Interactions in Computerized Learning Environments: Guiding Students' Command Process Through Instrumental Orchestrations. *International Journal of Computers for Mathematical Learning, 9*(3), 281-307.

Urhahne, D., & Harms, U. (2006). Instruktionale Unterstützung beim Lernen mit Computersimulationen. *Unterrichtswissenschaft, 34*(4), 358-377.

Van Gog, T., Paas, F., & Van Marienboer, J. (2004). Process-Oriented Worked Examples: Improving Transfer Performance Through Enhanced Understanding. *Instructional Science, 32*, 83-98.

Van Gog, T., Paas, F., & Van Marienboer, J. (2008). Effects of studying sequences of process-oriented and product-oriented worked examples on troubleshooting transfer efficiency. *Learning and Instruction, 18*, 211-222.

Wild, C. und G. A. F. Seber (2000). *Chance Encounters: A First Course in Data Analysis and Inference*. New York: J. Wiley.

Wittmann, E. Ch. (1995). Mathematics Education as a "Design Sience". *Educational Studies in Mathematics, 29*(4), 355-374.

Software

FATHOM 2.0 (2006). Springer

Kapitel 22
Konfektionsgrößen näher betrachtet – Ein Vorschlag zur Lehrerausbildung in Stochastik

Katja Krüger

Universität Paderborn

Abstract Diese Arbeit soll einen Anstoß geben, wie das Planen und Durchführen einer statistischen Erhebung in die Lehramtsausbildung Mathematik integriert werden kann. Das Thema Konfektionsgrößen liefert dabei einen Sachkontext, der gleichermaßen für Studierende wie für Schüler interessant sein dürfte und der zum Nachfragen herausfordert. Der Ausgangspunkt ist sicherlich persönlich privat, wenn man sich bei der Anprobe wundert: Warum sitzen Pullis oft so schlecht? Mit Hilfe von selbst erhobenen und ausgewerteten Daten können Studierende die Erfahrung machen, dass dieses Problem nicht singulär ist, sondern statistisch betrachtet viele Menschen Schwierigkeiten mit der Passform von Konfektionskleidung haben dürften. Wie sich eine solche Erhebung im Rahmen einer Lehrveranstaltung zur Stochastikdidaktik einbetten lässt und was Studierende dabei lernen können, davon handelt dieser Beitrag.

Einleitung

Im Rahmen der Leitidee „Daten und Zufall" fordern die KMK-Bildungsstandards im Fach Mathematik für den mittleren Schulabschluss, dass Schülerinnen und Schüler lernen sollen, eine statistische Erhebung selbstständig zu planen und durchzuführen. Dabei lassen sich Computer in sinnvoller Weise zur Darstellung und Auswertung der selbst erhobenen Daten nutzen.

Mit Blick auf die spätere Unterrichtstätigkeit sollten angehende Mathematiklehrerinnen und -lehrer bereits während ihres Studiums Erfahrungen mit einer statistischen Erhebung sammeln. Dass Planung, Durchführung, Auswertung und Interpretation in aller Regel erheblichen Gedanken- und Zeitaufwand erfordern, sollten sie wenigstens einmal durchlebt haben. Hierfür eignen sich Übungen zu einer Vorlesung in Stochastik(didaktik). Studierenden muss dabei verdeutlicht werden, dass eine statistische Erhebung mit einer echten Fragestellung beginnt und man sich schon einiges überlegen muss, bis man überhaupt passende Daten erheben kann. In dieser Arbeit soll ein Vorschlag vorgestellt werden, wie in aufeinander abgestimmten Präsenzübungen, Hausaufgaben und deren Nachbesprechungen Studierende alle Phasen einer statistischen Erhebung erleben und gemeinsam reflektieren können. Wichtig ist dabei, dass ein (perspektiv auch Schülern) zugänglicher Sachkontext gewählt wird, in dem die selbst erhobenen Daten spannende Informationen liefern. Reale Daten aus dem Umfeld der Schüler eignen sich gut, um das Suchen nach Zusammenhängen und Interpretieren von Daten lernen zu können (vgl. Stellungnahme AK Stochastik 2002, S. 2).

Wie kann nun Lehramtsstudierenden die Durchführung einer statistischen Erhebung von einer reizvollen Ausgangsfrage bis zu anregenden Anschlussfragen aufgrund gewachsener Sachkenntnisse nahe gebracht werden? Wir haben es mit einem Ausschnitt des volkswirtschaftlich durchaus bemerkenswerten Themenkreises "Konfektionsgrößen" versucht. Dabei lässt sich das Messen von Körpermaßen in einen weniger problematischen Sachkontext einbetten als etwa bei Körperhöhe- und Körpergewicht-Messungen.

Konfektionsgrößen

Konfektions- oder Kleidergrößen sind die Größen moderner, industriell in Serie gefertigter Bekleidung (Konfektion). „Die entscheidende Voraussetzung für die Konfektionierung von Kleidung ist die Standardisierung von Körpergrößen" (Schmidt und Janalik 2011, S. 72). Konfektionskleidung greift dabei nicht mehr auf individuelle Kleidergrößen zurück, sondern basiert auf wenigen Standardgrößen. Diese werden statistisch gewonnen und richten sich im Wesentlichen nach der Körperhöhe und dem Brustumfang.

Als grundlegende Körpermaße der Bekleidungsindustrie gelten:
– Brustumfang bzw. Oberweite
– Taillen- bzw. Bundumfang
– Hüft- bzw. Gesäßumfang
– Körperhöhe
– Taillenhöhe
– Beininnenlänge
– Armlänge
– Schulterbreite
– Oberarmweite
– Halsweite

Abb. 22.1 Körpermaße der Bekleidungsindustrie (Bildquelle: SizeGermany 2009)

Die europäischen Konfektionsgrößen für Männer werden in der Regel nach der folgenden Formel ermittelt: *Brustumfang : 2*.[1]

---

[1] Laut Döring 2009 lässt sich dieser Bezug von Konfektionsgrößen und dem halben Brustumfang auf das Verfahren des Zuschnitts von Kleidungsstücken beim Übergang von geschneiderter zu Konfektions-Kleidung im 19. Jahrhundert verorten. Als Basis für den Zuschnitt eines Kleides oder Mantels genügt ein Schnittmuster für eine Körperhälfte. Die halbierte Oberweite bildet dabei als halbierte Bekleidungsbreite das Grundmaß des Schnittmusters. (vgl. Döring 2009, S. 158ff.)

Bei Frauen wird diese Formel leicht abgewandelt, um die Konfektionsgröße zu ermitteln: *Brustumfang : 2 – 6*.

Zum Beispiel ergibt ein Brustumfang von 88 cm bei einem Mann die Konfektionsgröße 44 und bei einer Frau die Konfektionsgröße 38. Die „normalen" Kleidergrößen 44 bis 56 der Herren-Konfektion beziehen sich dabei auf 1,68 bis 1,84m große Männer. Die Damen-Normgrößen sind dagegen nur auf eine durchschnittliche Körpergröße von 168 cm angelegt und gehen davon aus, dass eine Frau zwischen 164 und 170 cm groß ist. Für kleinere bzw. größere Männer und Frauen gibt es außerdem ein eigenes Kurz- sowie ein Langgrößensystem (vgl. Eberle 2001, S. 176).

Tab. 22.1 Auszug aus der Burda-Maßtabelle (Nähjournal BurdaStyle 2012, S. 2)

Normale Herrengrößen								
Größe		44	46	48	50	52	54	56
Körpergröße	cm	168	171	174	177	180	182	184
Oberweite	cm	88	92	96	100	104	108	112
Bundweite	cm	78	82	86	90	94	98	104
Gesäßweite	cm	90	94	98	102	106	110	115
Rückenlänge	cm	42	43	43,5	44,5	45	45,5	46
Ärmellänge	cm	61	62	63	64	65	66	67
Halsweite	cm	37	38	39	40	41	42	43

Tab. 22.2 Auszug aus der Burda-Maßtabelle (Nähjournal BurdaStyle 2012, S. 2)

Normale Damengrößen bei Körpergröße 168 cm											
Größe	cm	34	36	38	40	42	44	46	48	50	52
Oberweite	cm	80	84	88	92	96	100	104	110	116	122
Taillenweite	cm	62	66	70	74	78	82	86	92	98	104
Hüftweite	cm	86	90	94	98	102	106	110	116	122	128
Rückenlänge	cm	41	41	42	42	43	43	44	44	45	45
Armlänge	cm	59	59	60	60	61	61	61	61	62	62
Schulterbreite	cm	12	12	12	13	13	13	13	14	14	14

Die beiden Maßtabellen stammen aus dem Nähjournal der Zeitschrift Burda-Style (11/2012) und zeigen eine Auswahl an Normgrößen. Von großer Bedeutung sind dabei die sogenannten „Sprungwerte", die Differenz zwischen zwei Körpermaßen benachbarter Konfektionsgrößen z.B. jeweils 4 cm bei den Oberweiten. Betrachtet man die Burda-Maßtabellen genauer, so fällt auf, dass einige Körpermaße linear mit der Konfektionsgröße wachsen, d.h. konstante Sprungwerte aufweisen wie z.B. die Rückenlänge, Ärmellänge und Halsweite bei den Herren oder die Taillen- und Hüftweite bei den Damen (bis Kleidergröße 46). Bei einigen Körpermaßen findet man dagegen nur stückweise konstantes Wachstum wie z.B. bei den Armlängen und Schulterbreiten der Frauen.[2]

Die Bekleidungsindustrie möchte natürlich mit ihren Konfektionsgrößen den Marktbedarf gut abdecken. Diese sollen daher die Körperformen möglichst vieler Menschen passgenau abbilden. Größentabellen liefern die Grundlage für die Konstruktion von Schnittmustern zur Herstellung von Konfektionsbekleidung. Nur wenn Kleidung gut sitzt, wird sie auch gekauft. Darüber hinaus ist auch der Bekleidungshandel an Marktanteilen interessiert. Wie viele Kleidungsstücke einer Größe sollen hergestellt bzw. eingekauft und gelagert werden? Es besteht daher ein großer Bedarf der Bekleidungswirtschaft an aktuellen Daten. Seit den 1960er Jahren führen die Hohenstein Institute in Deutschland Reihenmessungen durch, um die Körperformen von Männern, Frauen und Kindern statistisch zu erfassen und mit Hilfe dieser Massendaten Größentabellen für Konfektionsbekleidung zu erstellen. Bei der letzten großen Reihenmessung „SizeGERMANY" 2007/8 wurden mit einem 3D-Körperscan die Körpermaße von mehr als 13.000 Personen verschiedener Altersstufen erfasst. Bei der Beschreibung der Datenerhebung wird leider nicht explizit erwähnt, ob ein repräsentatives Stichprobenverfahren[3] zugrunde gelegt wurde (vgl. Abschlusspräsentation SizeGermany 2009).

Die Originaldaten sind nicht frei und kostenlos zugänglich, aber man findet immerhin einige interessante Auswertungen dieser Reihenuntersuchung (Abschlusspräsentation SizeGermany 2009). So konnten (wieder einmal) zeitbedingte Veränderungen der Maße des weiblichen und männlichen Körpers festgestellt werden. Die Durchschnittsmaße der für die Kleidergrößen relevanten „Primärmaße" Körperhöhe, Brust-, Hüft- und Taillenumfang haben sich gegenüber der letzten Reihenmessung deutlich vergrößert (vgl. Abb. 22.2). Dabei zeigt sich, dass nicht nur die Körperhöhe, sondern auch die Umfangsmaße deutlich angewachsen sind. Bei den Frauen sticht die Zunahme des durchschnittlichen Taillenumfangs um rund 5% innerhalb der letzten 15 Jahre hervor, während bei den Männern der Brustumfang in den letzten 30 Jahren um gut 7% zugelegt hat. Die Aussagekraft dieser Durchschnittswerte ist jedoch begrenzt, da gerade die Umfangsmaße altersbedingt variieren. Je älter ein Mann oder eine Frau ist, desto größer ist beispielsweise der Brustumfang (ebenda). Aufgrund der Alterung der deutschen Bevölkerung ist die Zunahme der durchschnittlichen Umfangsmaße somit nicht

---

[2] Möchte man diese Größenangaben genauer wissen, so sollte man besser die Damenoberbekleidungs-Gößentabelle (kurz DOB-Größentabelle) einer möglichst aktuellen Reihenuntersuchung der Hohenstein Institute verwenden. Auszüge daraus findet man z.B. in der Einleitung zu Stiegler 2009. Leider sind die aus Reihenuntersuchungen gewonnenen Größentabellen im Allgemeinen nicht frei zugänglich, so dass man sich für Unterrichtszwecke besser mit der Burda-Maßtabelle behilft.

[3] In der DDR wurde in den 1980er Jahren eine wissenschaftliche Reihenuntersuchung von der Humboldt-Universität durchgeführt. Die Biologin und Medizinerin Holle Greil vermaß im Rahmen ihrer Habilitation über 7000 Menschen verschiedenen Alters und Geschlechts in einer Querschnittstudie in Brandenburg und lieferte damit eine repräsentative Stichprobe, mit der sie den „Körperbau von Erwachsenen" in der DDR statistisch erfasste.

verwunderlich. In den Anfängen der Konfektion entwickelte sich dementsprechend ein Größensystem, das sich zunächst an Altersstufen orientierte (vgl. die Stern-Größen in der weiblichen Konfektion zu Beginn des 20. Jahrhunderts nach Döring 2009, S. 178f.).

Abb. 22.2 Veränderungen in den Durchschnittsmaßen im Vergleich zu den Reihenmessungen 1994 (Frauen) und 1980 (Männer). Bildquelle: SizeGERMANY 2009

Wollte man die Konfektionsgrößen der in Abb. 22.2 abgebildeten Durchschnittsfrau und des Durchschnittsmannes nach SizeGERMANY mit Hilfe der primären Körpermaße der Burda-Maßtabellen ermitteln, so tritt hier das Problem der mangelnden Passform auf. Der Durchschnittsmann hätte laut Körperhöhe und Gesäßweite die Konfektionsgröße 50, laut Oberweite und Bundumfang die Größe 52. Die Durchschnittsfrau müsste sich zwischen den Größen 42 und 44 entscheiden. Was ist also die passende Kleidergröße? Damit bietet das Thema Konfektionsgrößen einen durchaus anregenden Sachkontext aus dem Umfeld von Schülern. Jede/r hat beim Einkauf vielfältige Erfahrungen mit Konfektionsgrößen und der Passform von Bekleidung gemacht. Für den Stochastikunterricht muss man jetzt nur noch eine attraktive Fragestellung finden, die sich als Ausgangspunkt einer statistischen Erhebung zu diesem Thema eignet.

Die Problemstellung

Am Beginn einer statistischen Erhebung sollte ein für Schüler gleichermaßen interessantes und relevantes Problem stehen. Es muss ja schließlich einen Zweck haben, wenn man mit einigem Aufwand Daten im Mathematikunterricht erhebt und auswertet. Bereits in dieser Anfangsphase lässt sich in natürlicher Weise Modellbildung betreiben, denn häufig ist eine interessante Problemstellung nicht direkt einer Bearbeitung mit mathematischen Mitteln zugänglich d.h. es müssen zunächst geeignete Fragen gefunden werden, die sich mit Hilfe von Daten beantworten lassen. Um diesen Modellierungsschritt Lehramtsstudierenden bewusst zu machen, dient eine erste „Präsenzübung", in der die Aufgabe in Abb. 22.3 gemeinsam bearbeitet und Lösungsvorschläge diskutiert werden. Als Informationsmaterial eignen sich die Ausführungen über Konfektionsgrößen aus dem vorhergehenden Abschnitt. Zur Vorbereitung auf die Präsenzübung hat es sich bewährt, dass Studierende sich vorab eigenständig mit dem Aufgabenteil a) befassen und ihre Antworten mit Hilfe von Stichpunkten schriftlich fixieren.

Präsenzübung 1

Stellen Sie sich vor, Sie möchten im Rahmen einer Unterrichtsreihe zur beschreibenden Statistik mit Schülerinnen und Schülern der JGST 9/10 oder 11 eine statische Erhebung zum Thema Konfektionsgrößen durchführen.

a) Verschaffen Sie sich mit der BURDA-Maßanleitung und den beiliegenden Informationen über Konfektionsgrößen einen ersten Einblick in das Thema.
    Welche Problemstellungen aus diesem Kontext könnten für Schülerinnen und Schüler gleichermaßen interessant und relevant sein? Formulieren Sie geeignete Fragen.
b) Wählen Sie eine Ihrer Fragen aus dem Aufgabenteil a) und planen Sie gemeinsam eine statistische Erhebung, mit der diese Frage beantwortet werden soll. Berücksichtigen Sie dazu die in der Vorlesung gegebenen Hinweise zur Planung statistischer Erhebungen.

Abb. 22.3 Präsenzübung zum Finden einer Problemstellung und erste Planungsüberlegungen

Die nachfolgende Auswahl von Fragestellungen aus einer solchen Präsenzübung zeigt, wie ideenreich diese Aufgabe von Studierenden (Lehramt Mathematik an Gymnasien, Gesamtschulen sowie an Haupt- und Realschulen) bearbeitet wurde:

• Wie viele Schülerinnen und Schüler tragen laut der Formel: „Brustumfang halbe = Konfektionsgröße (-6 cm bei Frauen)" die passende Kleidung?
• Unterschiedliche Marken – unterschiedliche Größen? Wie kommt es, dass bei gleicher Kleidergröße, aber anderen Marken die Hose mal passt und mal nicht?
• Warum passen Pullis oft so schlecht? Entweder sind sie in den Ärmeln zu kurz oder in den Schultern zu lang.

Hier ist Gruppenarbeit im Rahmen der Präsenzübung besonders geeignet, da sich in der gemeinsamen Diskussion eher als in Einzelarbeit eine interessante Fragestellung herausschälen lässt und packend formuliert werden kann. Nach der Präsentation der Gruppenergebnisse lassen sich die Vor- und Nachteile der einzelnen Vorschläge diskutieren. Von den Lehramtsstudierenden wurde beispielsweise lebhaft erörtert, ob „sensible" Körpermaße im Unterricht direkt am Körper gemessen werden dürften, wie z.B. die Beininnenlänge, die bei Jeansgrößen eine wichtige Rolle spielt, oder der Brustumfang bei Schülerinnen. In der in diesem Beitrag beschriebenen Präsenzübung entschieden sich die Studierenden daher die in der

Aufzählung zuletzt genannte Frage, die häufig schlechte Passform von Pullovern, zu untersuchen, da das Messen von Armlängen und Schulterbreiten als weniger problematisch angesehen wurde.

SO NEHMEN SIE RICHTIG MASS

Für einen gut sitzenden Schnitt ist exaktes Maßnehmen Voraussetzung. Unsere Tipps: Messen Sie direkt auf der Unterwäsche. Lassen Sie sich dabei helfen! Bitte fixieren Sie mit einem festen Band Ihre Taille. Stehen Sie gerade, aber entspannt.

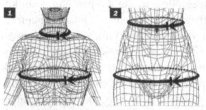

3 Vordere Taillenlänge Messen Sie vom seitlichen Halsansatz über die Brustspitze bis zur Unterkante des Taillenbandes. **Brusttiefe** Wird wie die vordere Taillenlänge gemessen, aber nur bis zur Brustspitze.
4 Rückenlänge Messen Sie von dem etwas vorstehenden Halswirbelknochen entlang der Rückenmitte bis zur Unterkante des Taillenbandes. **Rückenbreite** Waagrecht von Armansatz bis Armansatz messen. Wichtig: eine aufrechte Haltung!

1 Oberweite Das Maßband vorn auf die stärkste Stelle der Brust legen, unter den Armen hindurchführen, am Rücken leicht ansteigend. **Halsweite** Am Halsansatz direkt über dem Schlüsselbein, messen.
2 Taillenweite Gemessen wird an der schmalsten Stelle des Rumpfes. Hier wird auch das Taillenband fixiert. **Hüftweite** Maßband waagrecht über die stärkste Stelle des Gesäßes legen.

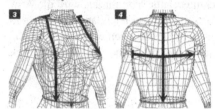

5 Schulterbreite Vom Halsansatz bis zum Schulterpunkt messen. **Oberarmweite** An der stärksten Stelle des Oberarms messen.
6 Armlänge vom Schulterpunkt bis zur Handwurzel, bei leicht angewinkeltem Ellbogen.

Wichtig: Vergleichen Sie Ihre gemessenen Körpermaße mit denen der Burda-Maßtabellen. Wählen Sie die Größe, die Ihren Maßen am nächsten kommt. Als Faustregel gilt: Blusen, Kleider, Jacken, Mäntel nach der Oberweite, Röcke und Hosen nach der Hüftweite.

Abb. 22.4 Maßanleitung nach dem Nähjournal BurdaStyle (2012, S. 6)

Planung der Erhebung

Um das Problem der schlecht sitzenden Pullis einer statistischen Untersuchung zugänglich zu machen, verwendeten wir die in den Burda-Tabellen aufgeführten Körpermaße Armlänge und Schulterbreite. Wie hängen Armlänge und die Schulterbreite eigentlich zusammen? Wie gut werden diese beiden Körpermaße durch Konfektionsgrößen erfasst? Mit der zunehmenden Präzisierung der Problemstellung in Form von Fragen, die sich mittels Daten untersuchen lassen, sind wir mitten in der Planung einer statischen Erhebung angelangt. Um die erhobenen Daten einordnen zu können, wurden die (überwiegend getragene) Konfektionsgröße, die Körpergröße sowie das Geschlecht abgefragt. Da die Studierenden vertrauter mit den internationalen Konfektionsgrößen waren, haben wir dieses Merkmal erhoben.

Tab. 22.3 Umrechnung des europäischen in das internationale Größensystem (Quelle: http://www.ihr-schneideratelier.de/sites/konfektionsgroessen.php, Zugriff: 1.3.2013)

Internationale Größen	XS	S	M	L	XL
Damen		32/34	36/38	40/42	44/46	48/50
Herren		40/42	44/46	48/50	52/54	56/58

Während das Geschlecht, die Konfektionsgröße und die Körperhöhe – ein kategoriales, ein rangskaliertes und ein metrisches Merkmal – einfach abgefragt werden können, müssen Armlänge und Schulterbreite direkt am Körper gemessen werden. Wie aber misst man diese beiden Körpermaße ausreichend genau? Richtiges Maßnehmen will gelernt sein. Nicht von ungefähr ist das Schneiderhandwerk ein Ausbildungsberuf mit einer langen Tradition. Hier bietet ebenfalls die Burda-Maßanleitung eine erste Orientierung (vgl. Abb. 22.4). Wichtig beim Maßnehmen ist, dass man den Anfangs- und Endpunkt korrekt ansetzt und sich Maßband und Körper in der richtigen Haltung befinden. Was ist in der Burda-Anleitung mit „Schulterpunkt" gemeint? Die Schulterhöhe (*Acromion*) beispielsweise bildet beim Menschen den höchsten Punkt des Schulterblatts. Von ihr aus wird die Armlänge bis zur Handwurzel gemessen. Der Arm sollte dabei leicht angewinkelt sein. Die Schulterbreite wird vom Halsansatz zur Schulterhöhe gemessen. Dabei sollte die Schulter nicht hochgezogen werden und das Maßband am Körper direkt anliegen

Datenerhebung

So einfach wie bei der Köperhöhe ist das Messen trotz der genauen Burda-Anleitung allerdings nicht. Während sich die Schulterhöhe und der Beginn der Handwurzelknochen noch relativ gut ertasten lassen, ist das mit dem Halsansatz nicht ganz so einfach, insbesondere bei stark ausgeprägter Schulter- und Halsmuskulatur. Wo fängt der Hals eigentlich genau an? Eine Hilfe bietet der Vergleich mit der Schulterbreite des getragenen Kleidungsstückes. Dabei ist darauf zu achten, ob bei dem Kleidungsstück, das die Studentin oder der Student trägt, die Schulterbreite richtig sitzt (vgl. Stiegler 2009, S. 16). Bei den gemeinsamen Messungen wird deutlich, wie wichtig die vorherige Festlegung statistischer Merkmale ist, wenn die eigenen Daten später vergleichbar und aussagekräftig sein sollen. An einem weiteren Übungstermin werden wieder in Form einer Präsenzübung die Daten erhoben. Dabei sollen die Studierenden die Durchführung der Messungen genau beobachten und auf mögliche Fehlerquellen achten.

Präsenzübung 2

Warum passen Pullis immer so schlecht? Oft sind sie in den Ärmeln zu kurz oder in den Schultern zu lang. Wie hängen eigentlich die Armlänge und Schulterbreite zusammen? Zur Untersuchung dieser Fragen messen bzw. fragen Sie die folgenden Merkmale ab:
- Armlänge (vom Schulterpunkt über den Ellenbogen bis zum Handgelenk)
- Schulterbreite (vom Halsansatz zum Schulterpunkt)
- Geschlecht
- Konfektionsgröße (S, XS, M, L, XL), die überwiegend bei der Oberbekleidung getragen wird.
- Körpergröße

Abb. 22.5 Präsenzübung zur Durchführung der Messung

Bei der Datenerhebung steht schnell die Frage im Raum: Wie genau lassen sich eigentlich die Schulterbreite und der Armlänge messen? Wir versuchen es zunächst auf halbe cm genau. Das sollte bei Längen zwischen 10 und 60 cm möglich sein. Schnell wird klar, dass das jedoch nicht ganz so einfach ist. Werden die Schultern etwas angehoben oder der Arm weniger stark angewinkelt, ändert sich die gemessene Länge um ein paar mm. Wie machen das eigentlich Maßschneider? Die Nachfrage einer Studentin bei einer Schneiderin ergab, dass sie Körpermaße mit dem Maßband nur auf cm genau messe. Wichtig sei es, den Bewegungsspielraum mit zu berücksichtigen.

Für die Datenerhebung bietet sich ein Messprotokoll in Tabellenform an. So können die Daten anschließend leicht in ein Computerprogramm zur Auswertung eingegeben werden.[4] Insgesamt wurden z. B. in einer Lehrveranstaltung die Daten von 30 Studentinnen und 20 Studenten erhoben (vgl. Tab. 22.4).

Tab. 22.4 Ausschnitt aus dem Messprotokoll

Körpergröße	Armlänge	Schulterbreite	Geschlecht	Kleidergröße
164	60	15	w	XS
172	62	15	w	L
173	59	14	m	L

Auswertung und Interpretation

Die gemeinsam erhobenen Daten lassen sich den Studierenden in elektronischer Form als Excel-Tabelle oder Fathom-Datei zur Verfügung stellen. Entsprechend ihrer Vorkenntnisse suchen sie sich ein Programm aus, mit dem sie eigenständig

---

[4] Der nachfolgend ausgewertete Datensatz ist unter „Ergänzende Materialien" zum Schulbuch Neue Wege Stochastik, Kapitel 3, Dateinummer 3211, online als Excel- und Fathomdatei verfügbar (vgl. Lergenmüller et al. 2012).

den vorliegenden Datensatz auswerten. Diese Aufgabe bietet sich als schriftliche Hausaufgabe an (vgl. Abb. 22.6).

Hausaufgabe 1: Analyse der Körpermaßdaten
Die in der Präsenzübung der vergangenen Woche erhobenen Körpermaßdaten haben wir in einer Excel-Tabelle sowie in einer Fathom-Datei zusammengefasst. Werten Sie die Daten im Hinblick auf unsere gemeinsam entwickelte Fragestellung aus. Verwenden Sie dazu geeignete statistische Darstellungen und Kennzahlen. Welche Schlussfolgerungen und weiterführenden Fragen ergeben sich aus Ihrer Datenanalyse? Erstellen Sie einen Bericht über Ihre Untersuchungsergebnisse.

Abb. 22.6 Hausaufgabe zur Auswertung und Interpretation der Daten

Im Folgenden werden drei verschiedene Auswertungen vorgestellt, die auf elementaren Methoden der Beschreibenden Statistik beruhen (Darstellung von Verteilungen im Boxplot, Ermittlung von relativen Häufigkeiten, Streudiagramme und Korrelationen). In der Nachbesprechung der Hausaufgabe sollten möglichst verschiedene Bearbeitungen präsentiert werden. So können Studierende erleben, dass es in der Phase der Auswertung und Interpretation nicht nur den einen richtigen Weg gibt.

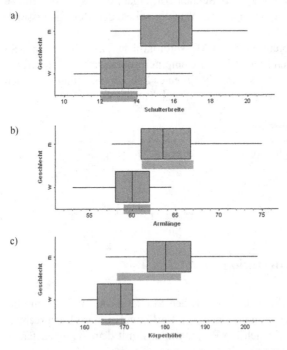

Abb 22.7 a – c Verteilung Schulterbreiten, Armlängen und Körperhöhen nach Geschlecht. Die Bereiche der in Tab. 22.1 und 22.2 angegebenen Normgrößen nach Burda sind unterhalb der Boxplots markiert (Angaben zur Schulterbreite bei Männern fehlen).

- Ein erster Blick auf die Daten: Verteilungen der Körpermaße

Zunächst einmal kann man sich mit Hilfe von Boxplots einen Überblick über die erhobenen Körpermaße nach Geschlechtern getrennt verschaffen (vgl. Abb. 22.7). Unterhalb der Boxplots wurden, soweit in der Burda-Maßtabelle angegeben, die Spannbreiten der normalen Körpermaße markiert.

Während sich die Studentinnen von den Studenten offensichtlich in ihrer Körperhöhe unterscheiden, die Rechtecke der beiden Boxplots überlappen sich nicht, ist das bei der Armlänge und der Schulterbreite nicht ganz so deutlich der Fall. Vergleicht man die Lage des oberen und unteren Quartils mit den „normalen" Armlängenbereichen aus der Burda-Maßtabelle, 59 bis 62 cm bei den Frauen sowie 61 bis 67 cm bei den Männern (vgl. Abb. 22.7 b), so stimmt die Lage der jeweiligen Box in etwa damit überein. D.h. etwa die Hälfte der Armlängen der Studierenden liegt im Bereich der Normgrößen. Diese Anteile lassen sich mit Programmunterstützung genauer berechnen (z.B. für die Studentinnen in Tab. 22.5).

Tab. 22.5 Anteile der Körpermaßdaten im Bereich der normalen Damen-Konfektionsgrößen

Studentinnen		
Normgrößen-Bereich	Anzahl	Anteil
59 cm ≤ Armlänge ≤ 62 cm	18	60%
12cm ≤ Schulterbreite ≤ 14 cm	17	57%
164 cm ≤ Körperhöhe ≤ 170 cm	9	30%

Die Auswertung mittels relativer Häufigkeiten zeigt, dass jeweils mindestens 40% der Studentinnen Körpermaße aufweisen, die nicht über die normalen Konfektionsgrößen abgedeckt werden. Für die Studenten erhält man ein vergleichbares Ergebnis. Hier lässt sich vermuten, dass es bei der Passform von Pullovern zu Problemen kommt.

- Wie hängen eigentlich die Armlänge und Schulterbreite zusammen?

Um diese Frage auszuwerten, bietet es sich an Streudiagramme zu erstellen und die Korrelationen der jeweiligen Datenwolken zu ermitteln. Dabei zeigt bereits die Form der beiden getrennt nach Geschlecht dargestellten Punktwolken, dass die Armlänge und Schulterbreite jeweils nur einen schwachen positiven linearen Zusammenhang aufweisen (vgl. Abb. 22.8 a und b). Berechnet man die Korrelationskoeffizienten nach Geschlecht getrennt, so erhält man bei den Männern 0,3 und bei den Frauen 0,27. Bei dieser Auswertung tauchte in der Übungsnachbesprechung die Frage auf, ob man nicht besser den Korrelationskoeffizienten der Gesamtgruppe zur Auswertung verwenden solle, der mit 0,5 deutlich höher liegt. Was ist hier eigentlich der „richtige" Korrelationskoeffizient? Diese Überlegung führt auf das Problem einer Scheinkorrelation, die durch die inhomogene Zusammensetzung der Gesamtgruppe verursacht wird. Beide Untergruppen weisen für sich genommen nur eine niedrige Korrelation auf. Erst bei der Betrachtung des gesamten Datensatzes wird durch den Lageunterschied der beiden Punktwolken scheinbar eine höhere Korrelation erzielt (vgl. Abb. 22.9). Daher muss in jedem Fall das Merkmal Geschlecht in der Auswertung berücksichtigt werden.

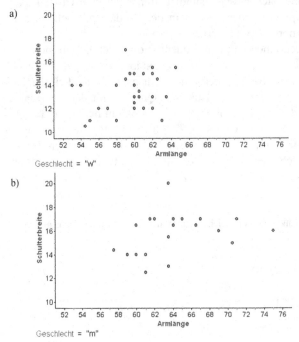

Abb. 22.8 a und b Streudiagramm von Armlänge und Schulterbreite nach Geschlecht

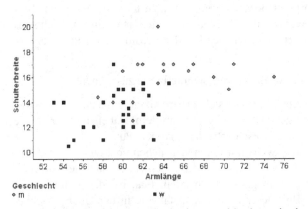

Abb. 22.9 Körpermaßdaten von Studentinnen und Studenten in einem Streudiagramm

Die schwach ausgeprägten Korrelationen der beiden Teilgruppen sind insofern bemerkenswert, weil andere Körpermaße von Menschen viel stärker miteinander korrelieren so z.B. die Armlänge und die Körpergröße. Hier lässt sich auch bei unseren Daten im Streudiagramm viel deutlicher ein positiver linearer Zusammenhang nachweisen. Die Korrelationskoeffizienten liegen mit knapp 0,8 bei den Männern und knapp 0,5 bei den Frauen vergleichsweise höher.

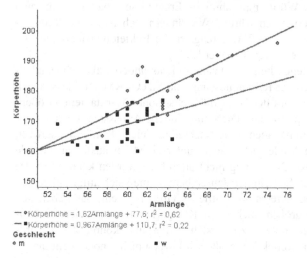

— ◇Körperhöhe = 1,62Armlänge + 77,6; r^2 = 0,62
— ■Körperhöhe = 0,967Armlänge + 110,7, r^2 = 0,22
Geschlecht
◇ m   ■ w

Abb. 22.10 Linearer Zusammenhang zwischen Armlänge und Körperhöhe

Wie lassen sich die Parameter der Regressionsgeraden interpretieren? In Abb. 22.10 erkennt man deutlich: Je länger der Arm, desto größer die betreffende Person. Bei den Studenten wächst die Körperhöhe um durchschnittlich 1,6 cm, wenn die Armlänge um einen cm zunimmt. Interessant ist ein Vergleich mit den Normgrößen der Burda-Maßtabelle. Dort nimmt die Körperhöhe von Männern um jeweils 3 cm pro einen cm Zuwachs der Armlänge zu. Offenbar passen die Normgrößen nach der Burda-Tabelle mit Blick auf den Zusammenhang von Körperhöhe und Armlänge nicht gut zu den Körpermaßen der Studenten in unserer Stichprobe. Für die Studentinnen lässt sich keine vergleichbare Interpretation angeben, da sich hier die Normgrößen für die Konfektion auf eine durchschnittliche Körperlänge von 168 cm beziehen (vgl. Tab. 22.2).

Fazit und Ausblick: Warum passen Pullis häufig so schlecht?

Was sagen uns diese statistischen Darstellungen und Kennzahlen über die Passform von Pullovern? Die geringen Korrelationen zeigen an, dass die Armlänge und Schulterbreite bei den in unserer Stichprobe vermessenen Studierenden nur schwach linear zusammenhängen. Die Datenwolken in den Streudiagrammen streuen stark. Kein Wunder, dass bei Pullovern, Blusen oder Hemden Schulterbreite und Armlänge oft nicht recht zusammen passen wollen. Nur ein gewisser Anteil der von uns vermessenen Studierenden liegt mit seinen Körpermaßen im Bereich der normalen Konfektionsgrößen. Dieses Problem, das sich hier im Kleinen zeigt, macht auf eine grundlegende Schwierigkeit der Bekleidungsindustrie im

Großen aufmerksam: Wie lässt sich bei der Herstellung von Konfektionsware eine möglichst gute Passform für möglichst viele Kunden erzielen? Die Ergebnisse regen zum Weiterfragen an: Würde man ähnliche Ergebnisse erhalten, wenn man mehr Personen genauer vermessen würde? Wie ändern sich Körpermaße oder - proportionen im Lauf eines Lebens? Wie reagiert die Bekleidungsindustrie, um solchen zeitlichen Änderungen gerecht zu werden?

Konfektionsgrößen festlegen heißt im Prinzip nichts anderes als Größentabellen angeben. Aus den Daten der Reihenmessung von SizeGERMANY lassen sich für ausgewählte Zielgruppen auf der Basis möglichst hoher Marktanteile Größentabellen computergestützt optimieren.[5] Diese individuelle Anpassung von Größentabellen durch Unternehmen hat auch seine Nachteile. Gleiche Konfektionsgrößen können bei verschiedenen Modemarken ganz unterschiedlich ausfallen. Gerade bei Online-Bestellungen, wo Bekleidung nicht anprobiert werden kann, führt das zu unerwünschten Retouren. Seit einigen Jahren bemüht man sich aufgrund dessen in der EU um ein differenzierteres, einheitliches Größensystem, das neben den Kurz-, Normal- und Langgrößen auch noch Kategorien für verschiedene Umfangstypen (schmal, normal, untersetzt) vorsieht. Allerdings konnte die Neuordnung der Etikettierung von Konfektionsgrößen in Europa bisher noch nicht umgesetzt werden.

So hat schließlich die hier vorgestellte statistische Erhebung zum Thema Konfektionsgrößen den Studierenden nicht nur reflektierte Erfahrungen mit den verschiedenen Phasen einer Statistischen Erhebung, sondern außerdem einen Einblick in ein reales und relevantes Problem der Bekleidungsindustrie ermöglicht.

Literaturverzeichnis

Arbeitskreis Stochastik der GDM (2002). *Empfehlungen zu Zielen und zur Gestaltung des Stochastikunterrichts*. Abgerufen 01. April 2012, von http://www.mathematik.uni-dortmund.de/ak-stoch/stellung.html.

KMK, Sekretariat der Ständigen Kultusministerkonferenz der Länder in der BRD (2003). *Bildungsstandards im Fach Mathematik für den Mittleren Schulabschluss*. München: Luchterhand.

Döring, D. (2009). *Zeugende Zahlen. Mittelmaß und Durchschnittstypen in Proportion, Statistik und Konfektion*. Berlin: Kulturverlag Kadmos.

Eberle, H. (2001). *Fachwissen Bekleidung* (6. Aufl.). Haan-Gruiten: Verlag Europa-Lehrmittel, Nourney, Vollmer.

Hohenstein-Institute & Human Solutions (2009). *SizeGERMANY – Die deutsche Reihenmessung. Abschlusspräsentation anlässlich der Pressekonferenz vom 21.4.2009 in Köln*. Abgerufen 04. Januar 2013, von http://www.sizegermany.de/pdf/SG_Abschlusspraesentation_2009.pdf.

Lergenmüller, A., Schmidt, G., & Krüger, K. (2012). *Stochastik-Oberstufenband Neue Wege. Arbeitsbuch für Gymnasien*. Braunschweig: Schroedel.

---

[5] vgl. das Tool isize der Firma HumanSolutions unter http://www.human-solutions.com/fashion/front_content.php?idcat=151, Zugriff am 1.2.2012.

Nähjournal der Zeitschrift BurdaStyle aus Heft 11 (2012) München: Verlag Aenne Burda GmbH & Co. KG.

Schmidt, D., & Janalik, H. (2011). *Kleidung, Körper, Gesundheit.* Baltmannsweiler: Schneider-Verl. Hohengehren.

Stiegler, M., Deutsche Bekleidungs-Akademie München (2009). *Schnittkonstruktionen für Kleider und Blusen: System M. Müller & Sohn* (24. Auflage). München: Rundschau-Verlag.

Kapitel 23
Konzeptualisierung unterschiedlicher Kompetenzen und ihrer Wechselwirkungen, wie sie bei der Bearbeitung von stochastischen Simulationsaufgaben mit dem Computer auftreten

Carmen Maxara

Friedrichsgymnasium Kassel

Abstract Es ist notwendig mehrere Bestandteile von Werkzeugkompetenzen zu unterscheiden, um differenzierte Aussagen zu Anforderungen und Diagnose von Schülerkompetenzen bei komplexen, computerbasierten Aufgabenstellungen machen zu können. Dies gilt auch im speziellen für die Kompetenzen, die benötigt werden, um eine Simulationsaufgabe mit der Werkzeugsoftware FATHOM zu erstellen und zu lösen. Aufbauend auf Vorarbeiten der AG Biehler ist es gelungen, Dimensionen FATHOM-spezifischer Simulationskompetenzen voneinander abzugrenzen, so dass eine reliable Kodierung im Sinne der qualitativen Inhaltsanalyse möglich wird und zu interessanten Ergebnissen führt. Neben den FATHOM-spezifischen Simulationskompetenzen werden auch stochastische Kompetenzen in einem ersten Ansatz in zwei inhaltlichen Kategorien unterschieden und mit den Werkzeugkompetenzen in Verbindung gesetzt. Die hier vorgestellten Ansätze bieten eine Grundlage für einen methodischen Zugriff zur Diagnose differenzierter Schülerkompetenzen, wie sie bei Lösungsprozessen bei komplexen, computerbasierten Aufgabenstellungen auftreten.

Einleitung

Die Konzeptualisierung der unterschiedlichen Kompetenzen ist ein Ergebnis, das im Rahmen meiner Dissertation (Maxara 2009) entstanden ist. In diesem Artikel sollen nicht nur die wesentlichen Ergebnisse dieses Parts vorgestellt, sondern zunächst auch kurz die Einbettung in den Forschungszusammenhang der Arbeitsgruppe Biehler erläutert werden (vgl. dazu Hofmann et al. 2013, in diesem Buch).

Die Notwendigkeit Schülerkompetenzen erfassen und beschreiben zu können, wie sie bei komplexen, computerbasierten Simulationsaufgaben gefordert sind, entstand aus dem Forschungszusammenhang der Arbeitsgruppe Biehler heraus, der sich mit dem Einsatz von Stochastiksoftware in der universitären und schulischen Stochastikausbildung beschäftigt. Er kann als Voraussetzung für eine didaktische Weiterentwicklung des Softwareeinsatzes in der Stochastikausbildung gesehen werden. Die AG Biehler entwickelte Konzepte und Materialien für einen didaktisch sinnvollen Einsatz von computergestützter Simulation in der Stochastikausbildung und den Stochastikunterricht (Biehler 2003; 2006). Die Vorlesung „Elementare Stochastik" für GHR-Studenten wurde erstmals im WS 01/02 neu konzipiert, wobei zunächst das Forschungsinteresse auf der (Explorativen) Datenanalyse lag (Biehler und Kombrink 2004), und später auch die Simulation erstmals als systematischer Strang in das Curriculum integriert wurde (Biehler Vorlesungsskript 2006/07; Maxara 2009, S. 265ff.). Einerseits sollten die Studierenden lernen, selbständig stochastische Probleme zu modellieren und zu simulieren. Anderseits können mit Simulationen Zufallsphänomene dynamisch veranschaulicht werden, auf die sich dann die Theoriebildung bezieht. Die Äquivalenz von Zufallsgeräten zur Simulation zu erkennen ist ein wichtiger Schritt in der Entwicklung der stochastischen Begriffsbildung (vgl. Biehler und Maxara 2007, S. 45). Die Simulation systematisch zunächst in die universitäre Stochastikausbildung, dann in den Stochastikunterricht der Schule zu integrieren, sollte forschungsgestützt geschehen. Daher wurden Materialien und didaktische Konzepte entwickelt, evaluiert und optimiert. Eine Grundlage dazu war die Konzeptualisierung und Analyse von Kompetenzen, die Lernende benötigen um Simulationen von Zufallsexperimenten mit einer Werkzeugsoftware durchzuführen.

Für ein solches Vorhaben (Schüler sollen aktiv und eigenständig Daten analysieren, simulieren und modellieren), ist es notwendig eine angemessene, unterstützende Werkzeugsoftware zu nutzen. Biehler hat Anforderungsprofile für Werkzeugsoftware entwickelt, die das Lernen sowie das Anwenden von Statistik und Stochastik unterstützt (Biehler 1991; 1997; Biehler et al. 2013). Die in den USA entwickelte Software FATHOM (Key Curriculum Press) entspricht am besten diesen Anforderungsprofilen für eine Werkzeugsoftware (vgl. Biehler 1997) und wurde daher auch von der Arbeitsgruppe Biehler ins Deutsche adaptiert (Biehler et al. 2006). FATHOM wurde und wird aus diesen Gründen kontinuierlich in diversen Lehrveranstaltungen und Schulen eingesetzt (z.B. Meyfarth 2006; 2008; Prömmel und Biehler 2009).

Im Gegensatz zu anderen (internationalen) Studien, die vor allem auf das Verständnis von Stichprobenverteilungen oder Inferenzstatistik abheben (z.B.: Sedlmeier 1999; Saldanha und Thompson 2003; Chance und Rossman 2006; delMas et al. 1999), sollte der Schwerpunkt bei der Konzeption hier auf der Modellierung und Simulation von Zufallsexperimenten liegen, wie es der deutschen Tradition entspricht. Da es bisher keine Konzepte und Lehrvorschläge zur Integration stochastischer Simulation mit dem Fokus auf der elementaren Wahrscheinlichkeitsrechnung und Modellierung gab, war es notwendig, zum einen die Software FATHOM auf ihr Simulationspotential hin zu untersuchen und zum anderen didaktische Konzepte zur Implementation zu entwickeln. So wurden als Vorbereitung auf die didaktische Integration von Simulationen in die Stochastikausbildung ausführlich die Möglichkeiten untersucht, wie man mit FATHOM Zufallsexperimente simulieren kann, und für die Studierenden aufbereitet (Maxara 2006; Maxara

2009). Des Weiteren wurde ein Aufgabensystem mit Handreichungen und ein dreigegliedertes didaktisches Konzept mit Simulationsplan in der AG Biehler entwickelt und implementiert (Maxara 2006; Hofmann et al. 2013; Maxara und Biehler 2006; 2007).

Bis dato wurde das Potential von Simulationen in der Didaktik gesehen (Schupp et al. 1992; Biehler 1991) und didaktische Anregungen zum Einsatz von Simulationen veröffentlicht (z.B. Trauerstein 1990; Lipson 1997; Engel 2002). Allerdings gab es kaum empirische Untersuchungen, insbesondere zur Simulation von Zufallsexperimenten mit FATHOM (Mills 2002; Maxara 2009). Des Weiteren sollte die didaktische Implementation in die Lehrveranstaltung evaluiert werden. Dazu war es nun notwendig zu untersuchen, welche stochastischen, simulations- und softwarebezogenen Kompetenzen die Studierenden für eine Simulation mit FATHOM brauchen und welche sie nach dem Besuch der Veranstaltung erworben haben. So entstand aus einer der ersten Studien zum Umgang mit der Software FATHOM zur Simulation von Zufallsexperimenten bei Studierenden eine Konzeptualisierung und Analyse unterschiedlicher Kompetenzen, die hier vorgestellt werden soll.

Die Studie

Nach einer der ersten Veranstaltungen, die auf das entwickelte didaktische Konzept aufbaute und die Simulation von Zufallsexperimenten parallel zum Konzept der Wahrscheinlichkeit einführte, wurde im WS 03/04 eine Fallstudie mit 8 Studierenden durchgeführt.[1] Die Studierenden besuchten alle die Veranstaltung „Elementare Stochastik" in der FATHOM durchgehend zum Daten Analysieren, Simulieren, Modellieren sowie zum Erproben von Methoden und Konzepten eingesetzt wurde. Die Studierenden sollten einerseits Wahrscheinlichkeitsmodelle mathematisch beschreiben können, sowie über Ereignisse und Zufallsgrößen explizite Annahmen treffen können. Andererseits mussten sie auch in FATHOM Wahrscheinlichkeitsmodelle bilden und die FATHOM-Sprache zur Definition von Ereignissen und Zufallsgrößen verwenden. Ebenso nutzten die Studierenden FATHOM als Datenanalysewerkzeug und zur Dokumentation ihrer Arbeit und ihrer Ergebnisse (Maxara und Biehler 2007, S. 762f.). Die beiden Ansätze *mathematisch formal* und *simulativ* wurden anhand typischer Problemstellungen (z.B. Würfeln mit 2 Würfeln; Problem des Chevalier de Méré) thematisiert und in Übungen vertieft. Soweit wie möglich, wurden die beiden Ansätze immer parallel verfolgt (vgl. Maxara 2009, S. 264ff.). Nach dem Ende der Vorlesung wurde 8 Studierenden in Partnerarbeit eine Simulationsaufgabe (Modell: Urnenziehung mit Zurücklegen, Aufgabentext und detaillierte Analyse in Maxara 2009, S. 251ff.) mit FATHOM gegeben, die sie modellieren, lösen und dokumentieren sollten. Der Lö-

---

[1] Das zugrundeliegende didaktische Konzept wurde erst später veröffentlicht (vgl. Maxara 2006; Biehler & Maxara 2007).

sungsprozess sowie die Kommunikation der Studierenden wurden videographiert, die Aktivitäten am Computer mit einer Screencapturesoftware aufgezeichnet. Anschließend wurde mit den Studierenden einzeln das Video des Lösungsprozesses noch einmal angeschaut, wobei sich die Studierenden in einem Methodenmix aus stimulated recall und halbstrukturiertem Interview zu dem Lösungsprozess äußern konnten. Das gesamte Videomaterial wurde transkribiert.

Um mit FATHOM zu modellieren, simulieren und zu experimentieren, brauchen die Studierenden bestimmte Kompetenzen. Die Fragen, die wir durch die Studie u.a. beantworten wollten, die aber nicht alle in diesem Artikel aufgegriffen werden, waren daher zum einen: Welche Kompetenzen brauchen die Studierenden bezüglich FATHOM, um eine solche Aufgabe zu lösen? Lassen sich diese Kompetenzen trennscharf formulieren und differenzieren? Und in welchem Maße haben die Studierenden diese Kompetenzen am Ende der Veranstaltung erworben? Außerdem ist es möglich, mit Hilfe von Simulationen Aufgaben zu lösen, die mathematisch für die Studierenden noch nicht lösbar sind. Daher wollten wir zum anderen wissen: In wieweit nutzen die Studierenden die Chance sich während des Simulationsprozesses auf die stochastischen Konzepte zu konzentrieren? In wieweit verknüpfen die Studierenden die stochastischen Konzepte mit den FATHOM-Objekten und -Operationen?

Zur Auswertung dieser Fallstudie wurden unterschiedliche Methoden für die verschiedenen Fragestellungen, die hier nicht im Detail aufgeführt werden, angewendet (für eine detaillierte Ausführung siehe Maxara 2009). Für die Konzeptualisierung und Analyse der verschiedenen Kompetenzen war es aber hilfreich, zunächst die Lösungsprozesse der Paare zu rekonstruieren und interpretieren. Dies lieferte einen wesentlichen Anschauungs- und Interpretationshintergrund für die nächsten Stufen methodischer Auswertung und sollte dem Leser den Lösungsprozess transparent und nachvollziehbar darlegen. Wesentlicher Teil der Auswertung war dann die Entwicklung und Anwendung von zwei Kategoriensystemen: Kategoriensystem 1 bezieht sich auf die FATHOM-bezogenen Kompetenzen, Kategoriensystem 2 bezieht sich auf die stochastischen Kompetenzen.

Konzeptualisierung von Kompetenzen

Ziel war es solche Kompetenzen von Studierenden zu erfassen und einzuschätzen, die speziell zur Simulation mit FATHOM benötigt werden und die einen Bezug zum stochastischen Inhalt möglich machen. Als Auswertungsmethode wurde die skalierende Strukturierung, die innerhalb der qualitativen Inhaltsanalyse zu verorten ist, gewählt (Mayring 2003, S. 82ff.; Maxara 2009, S. 284ff.).

Für die Analyse ist, neben der Festlegung des Ausgangsmaterials, die Konstruktion der Kategoriensysteme der erste und wesentliche Schritt. Nach Mayring (2005, S. 11) werden bei der hier angewandten *Strukturierung* die Kategorien theoriegeleitet entwickelt und anschließend an das Material herangetragen. Eine Grundlage für die softwarebezogenen Kompetenzen stellte dabei das in der Arbeit von Keitzer (2006) innerhalb der AG Biehler entwickelte Kategoriensystem zu den FATHOM-spezifischen Simulationskompetenzen dar. Die dort definierten vier „Kompetenzstufen" (*Allgemeine* FATHOM-*Kompetenz, Formelkompetenz, Simulationskompetenz, strategische und generalisierende Kompetenzen mit* FATHOM) wurden als Kompetenzbereiche adaptiert und am Material geprüft, überarbeitet und ausgeschärft. Jeder der Kompetenzbereiche wurde genauer definiert und in die drei Ausprägungen *hoch - mittel - niedrig* eingeteilt. Zur Erfassung der stochastischen Kompetenzen wurden die Kategorien *Mathematisierung/Verwendung und Verständnis stochastischer Fachbegriffe* und *Verständnis über die Aussagekraft von Simulationsergebnissen* - mit induktiven Anteilen - neu entwickelt. Sie umfassen ebenfalls die drei Ausprägungen *hoch - mittel - niedrig.* Durch die Entwicklung und Definition des zweiten Kategoriensystems *stochastische Kompetenzen* sollen gerade auch Beziehungen zwischen den stochastischen Kompetenzen und den softwarebezogenen Kompetenzen untersucht werden können.

Im Folgenden werden nun die beiden Kategoriensysteme genauer vorgestellt. Die Ausschärfung der Ausprägungsgrade ist in Maxara (2009, S. 291ff.) nachzulesen.

Die FATHOM-spezifischen Simulationskompetenzen

Allgemeine FATHOM-*Kompetenz*: Dieser Bereich umfasst vor allem die technischen Kompetenzen im Umgang mit den grundlegenden Objekten und Werkzeugen in FATHOM - mit Ausnahme des Formeleditors. Mit dazu zählt: die Kenntnis und das Nutzen können des grundsätzlichen Bildschirmaufbaus mit der Menü- und Symbolleiste, dem Arbeitsbereich und den Möglichkeiten, Objekte in FATHOM zu erstellen, sowie diese miteinander zu verlinken, die Kenntnis und das Arbeiten können mit Kollektionen, Merkmalen, Messgrößen, Datentabellen, Graphen und Auswertungstabellen, inklusive Info-Fenstern, die Fähigkeit sinnvolle Bezeichnungen zu vergeben und den Bildschirm übersichtlich zu gestalten.

Formelkompetenz: Die Formelkompetenz umfasst alle Kenntnisse und Kompetenzen bezüglich des Formeleditors und des Einsatzes von Funktionen und Formeln zur Definition von Merkmalen und Messgrößen oder zur Auswertung. Die Bedienung des Formeleditors, die Semantik und Syntax von Funktionen sowie ein Basiswissen zu bestimmten Funktionen (*random, randomPick, randomInteger, caseIndex, if, switch, count, uniqueValues, concat, mean, proportion, median* - diese wurden immer wieder in der Vorlesung und den Übungen verwendet)[2] ist Bestandteil der Formelkompetenz.

Simulationskompetenz: Die Simulationskompetenz kann in drei Bereiche unterteilt werden: das Wissen und die geeignete Wahl einer der drei verschiedenen Si-

---

[2] Die Studie wurde mit der englischsprachigen FATHOM -Version durchgeführt, da zu diesem Zeitpunkt noch keine deutschsprachige Version vorlag.

mulationsarten, die (gedankliche) Planung der Simulation sowie die Modellierung und Interpretation auf den lokalen Ebenen. Die Planung der Simulation umfasst die Fähigkeit, den Aufbau der Simulation in FATHOM schon im Voraus in Schritten durchdacht zu haben und zu wissen, welche Simulationsschritte welchen Objekten oder Tätigkeiten in FATHOM entsprechen. Die Modellierung der stochastischen Situation in FATHOM-Objekte und die anschließende Interpretation von Objekten oder Tätigkeiten, kann man beispielsweise daran festmachen, ob die Studenten wissen, was die Inhalte einer Kollektion oder Messgrößenkollektion inhaltlich auf ihre Aufgabe bezogen darstellen oder ob die Studenten wissen, dass sie jetzt eine Messgröße definieren müssen und was diese genau berechnen soll.

Strategische Kompetenzen: Unter den strategischen Kompetenzen werden Kontroll- und Debugging-Strategien verstanden. Das Überprüfen von Zwischen- und Endergebnissen sowie das Auffinden, Diagnostizieren und Beheben von Fehlern sind dabei die wesentlichen Aspekte. Das Überprüfen von Ergebnissen ist so zu verstehen, dass die Studenten ihre Simulationsergebnisse direkt mit ihren Erwartungen abgleichen sollten und somit überprüfen, ob die gewählte Formel auch das gewünschte Ergebnis liefert. Dieser Kompetenzbereich umfasst in gewissem Sinne Meta-Strategien, die auch bei anderer Software erforderlich sind.

Die stochastischen Kompetenzen

Verwendung/Verfügbarkeit von Fachbegriffen: Die Verwendung, Verfügbarkeit und das Verständnis stochastischer Fachbegriffe wie Ergebnis, Ergebnismenge, Ereignis, Zufallsgröße, stochastisch unabhängig, Gleichverteilung werden in dieser Kategorie erfasst. Eine hohe Ausprägung wurde vergeben, wenn diese Begriffe richtig verwendet, im Kontext abgerufen und mühelos in den Sprachgebrauch integriert wurden. Die jeweilige mathematische Bedeutung der Begriffe ist klar. Bei der mittleren Ausprägung ist dagegen die Bedeutung der Begriffe nicht eindeutig klar. Die Begriffe werden weniger oft und unsicher verwendet. In der niedrigen Ausprägung werden die Begriffe gar nicht oder falsch verwendet.

Verständnis über die Aussagekraft von Simulationsergebnissen: Diese Kompetenz soll das Verständnis und die Anwendung des empirischen Gesetzes der großen Zahl erfassen. Die Bedeutung der simulierten relativen Häufigkeit als Schätzer für die gesuchte Wahrscheinlichkeit, die Abhängigkeit der Genauigkeit des Schätzers von der Anzahl der durchgeführten Wiederholungen sowie die Schätzung des Erwartungswertes durch das arithmetische Mittel sind Kernpunkte dieser Kategorie.

Übertragbarkeit und Nutzen

Die oben aufgeführten vier Kompetenzbereiche zur FATHOM-spezifischen Simulationskompetenz lassen sich prinzipiell auch auf andere Software und Anwendungsbereiche übertragen. Die Allgemeinen FATHOM-Kompetenzen, die Formelkompetenz sowie die Strategischen Kompetenzen lassen sich ohne Weiteres auf eine andere Werkzeugsoftware wie z.B. EXCEL übertragen. Dann würden beispielsweise die Allgemeinen EXCEL-Kompetenzen die technischen Kompetenzen im Umgang mit den grundlegenden Objekten und Werkzeugen in EXCEL umfassen usw. Die Simulationskompetenz wäre bei Simulationsaufgaben auch übertragbar, wobei zunächst das Simulationspotential der jeweiligen Software analysiert oder für Lernende didaktisch aufbereitet werden müsste. Evtl. müsste dann nicht zwischen verschiedenen Simulationsmöglichkeiten gewählt werden, aber die Planung der Simulation sowie die Modellierung und Interpretation auf den lokalen Ebenen würde bestehen bleiben. Auch für andere Aufgabentypen wären die Kompetenzbereiche nutzbar, wenn die Simulationskompetenz durch einen entsprechenden Kompetenzbereich ersetzt werden würde.

Die stochastischen Kompetenzen sind von ihrer Anlage her nicht so ausgereift wie die FATHOM-spezifischen Simulationskompetenzen und daher erweiter- und überarbeitbar, z.B. auf Grundlage der Vorschläge des Arbeitskreises Stochastik der GDM (2010). Sie dienten vor allem dazu, einen Bezug zwischen den FATHOM-spezifischen Simulationskompetenzen und den stochastischen Inhalten herzustellen. Bei anderen Aufgabenformaten oder Untersuchungszielen liegen die Schwerpunkte auf anderen mathematischen Inhalten und anderen Kompetenzen, so dass diese neu entwickelt oder auf vorhandene zurückgegriffen werden müssten.

Methodisches Vorgehen

Das konkrete methodische Vorgehen wird ausführlich in Maxara (2009, S. 301 ff.) erläutert. Unter anderem wurden Häufigkeitsauszählungen zu den vergebenen Kodierungen erstellt. Diese quantitativen Auswertungen können eine Vergleichsgrundlage unter den Paaren bilden und ermöglichen eine prägnante Kompetenzeinordnung. Weiterhin können eventuelle Strukturen bei den Kompetenzbereichen erkannt werden. Aus den Häufigkeitsauswertungen wurden Kompetenzprofile erstellt, die   jeweils die FATHOM-spezifischen Simulationskompetenzen und die stochastischen Kompetenzen in je einer Graphik zusammenfassen (siehe Abb. 23.1).

Die Anzahl der Kodierungen einer Kategorie bilden dabei jeweils die 100%-Marke, die dann auf die Ausprägungen hoch, mittel und niedrig aufgeteilt wurden. Jeder Kategorie liegt also eine unterschiedliche Anzahl an Kodierungen zugrunde. Dabei wurde darauf geachtet, dass die jeweilige Kompetenzausprägung nicht von der mehrfachen Kodierung kleinerer Aspekte zu sehr beeinflusst wird. Es ist aber zu berücksichtigen, dass Schwierigkeiten, ebenso wie ihr Fehlen, Einfluss auf die Häufigkeiten und Vergabe von Kodierungen haben. Wenn also ein Paar Schwierigkeiten hatte eine bestimmte Formel zu finden, häufen sich bei diesem Kompetenzausprägungen zur niedrigen Formelkompetenz. Umgekehrt, wenn ein Paar keine Schwierigkeiten bei der Formelfindung hatte, finden sich dort eher wenige Kodierungen zur hohen Formelkompetenz. Auch die verschiedenen Dokumente haben bei der Kodierung jeweils unterschiedliche Anzahlen von Kodes erhalten, so dass sie mit unterschiedlicher Gewichtung in die Profildiagramm eingehen. (Die Anteile der verschiedenen Dokumente sind in etwa: Transkripte der Lösungsprozesse: 62%, Transkripte der anschließenden Interviews: 28%, Dokumentationen und FATHOM-Dokumente: 10%.) Auf die gleiche Weise wurde zu den stochastischen Kompetenzen ein Profildiagramm erstellt (vgl. Abb. 23.2), für das die gleichen Konstruktionskriterien gelten wie für die FATHOM-spezifischen Simulationskompetenzen.

Zusätzlich zu der Erstellung der Profildiagramme wurden die kodierten Stellen der verschiedenen Kategorien zu jedem Paar einzeln detailliert und qualitativ interpretiert. Dadurch entstand eine qualitative und quantitative, prägnante Beschreibung der FATHOM-spezifischen Simulationskompetenzen sowie der stochastischen Kompetenzen, die inhaltlich transparent ist, es ermöglicht, die Kompetenzen der einzelnen Paare miteinander zu vergleichen sowie mögliche Strukturen aufzudecken und Hypothesen über die Zusammenhänge der Kompetenzen zu generieren.

Möchte man die Auswertungsmethode noch etwas verbessern, so scheint es sinnvoll, die *Produkte* der Arbeit der Studierenden (FATHOM-Datei mit der Simulation, Dokumentation der Simulation) sowie die Transkripte der anschließenden Interviews getrennt von den Transkripten der Lösungs*prozesse* zu kodieren, um diese dann mit den Ergebnissen der Lösungs*prozesse* in Beziehung zu setzen. Eventuell ergeben sich durch eine differenzierte Auswertung der unterschiedlichen Dokumentarten neue interessante Resultate. Des Weiteren könnte man überlegen, ob man bestimmte Passagen des Simulationsprozesses stärker gewichten möchte als andere.

Eine beispielhafte Analyse und Interpretation der Kompetenzen von Paul und Paula

Um die möglichen Ergebnisse dieser Auswertungsmethode zu verdeutlichen, werden beispielhaft die Auswertungen und Ergebnisse zu dem Paar Paul/Paula kurz dargestellt. Da es der Umfang dieses Artikels nicht zulässt die Aufgabenstellung sowie die interpretative Beschreibung des Lösungsprozesses wiederzugeben, wird versucht die Ergebnisse nicht aufgabenbezogen, sondern möglichst kontextunabhängig zu beschreiben.

Die FATHOM-spezifischen Simulationskompetenzen von Paul und Paula sind übersichtlich im Profildiagramm in Abb. 23.1 zu sehen. Es folgt nun eine kurze qualitative Interpretation der Kategorien auf inhaltlicher Ebene (für eine ausführlichere Beschreibung und Interpretation, siehe Maxara 2009, S. 394ff.).

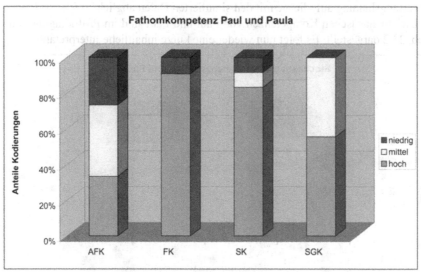

Abb. 23.1 Profildiagramm zu den FATHOM-spezifischen Simulationskompetenzen von Paul und Paula (AFK: Allgemeine FATHOM-Kompetenz, FK: Formelkompetenz, SK: Simulationskompetenz, SGK: Strategische Kompetenzen)

Allgemeine FATHOM-*Kompetenz*: Paul und Paula waren die Zugriffsweisen auf die FATHOM-Objekte nicht immer gleich präsent. Sie kannten aber die Funktionsweise der Objekte. (Dies wurde auch in anderen Studien beobachtet, vgl. Meyfarth 2008). So mussten sie beispielsweise kurz suchen, wo sich die Stichprobenkollektion erzeugen lässt. Ähnliche Probleme traten beim Sammeln von Messgrößen auf. Sie bezeichneten Objekte und Messgrößen naheliegend, aber für eine Interpretation nicht sonderlich hilfreich. Die Bildschirmgestaltung war dagegen sehr übersichtlich.

Formelkompetenz: Obwohl sie Schwierigkeiten hatten, FATHOM-Objekte zu finden und zu erzeugen, wusste vor allem Paula die Formeln zur Definition der Messgrößen (*uniqueValues( ) = 5, uniqueValues( ) ≤ 4, uniqueValues(Krawatte)*) und Auswertung sofort.

Simulationskompetenz: Vor allem Paula erklärte Paul immer wieder die Simulationsschritte, die sie gerade in FATHOM durchführen mussten. Sie gab aus einer Meta-Sicht Einblick in das Vorgehen und zeigte eine sehr reflektierte Sicht auf den Simulationsprozess. So erklärte sie Paul beispielsweise: „Also, wir haben jetzt die Simulation mit der Urnenziehung gewählt. Wir haben sieben Objekte in die Kollektion getan und jetzt würde ich eine Stichprobe vom Umfang 5 ziehen, weil es sich auf die 5 Tage einer Woche bezieht." Im weiteren Verlauf interpretierten die beiden immer wieder die FATHOM-Objekte und -Inhalte auf den Aufgabenkontext bezogen.

Strategische Kompetenzen: Paul und Paula überprüften oft ihre erhaltenen Ergebnisse und konnten so eventuelle Fehler berichtigen. Vor allem Paula baute eine Erwartungshaltung auf, die sie mit den simulierten Daten abglich.

Die stochastischen Kompetenzen von Paul und Paul sind im Profildiagramm in Abb. 23.2 dargestellt. Es folgt nun wieder eine kurze inhaltliche Interpretation.

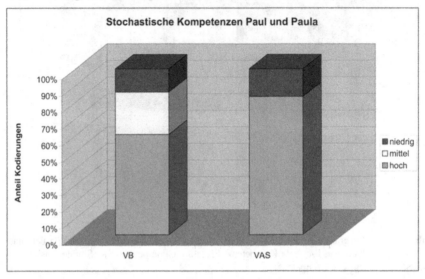

Abb. 23.2 Profildiagramm zu den stochastischen Kompetenzen von Paul und Paula (VB: Verständnis/Verfügbarkeit stochastischer Begriffe, VAS: Aussagekraft von Simulationsergebnissen)

Mathematisierung/Verwendung von Begriffen: Paul und Paula verwendeten stochastische Fachbegriffe und beschränkten sich bei der Aufgabenbearbeitung nicht auf die technische oder Softwareebene. Sie argumentierten auch auf stochastischer und inhaltlicher Ebene und zeigten dabei ein fundiertes Verständnis der stochastischen Inhalte. Paul zeigte leichte Unsicherheiten bei der Verwendung der Begriffe Ereignis und Zufallsgröße.

Verständnis über die Aussagekraft von Simulationsergebnissen: Die beiden waren die Einzigen der vier Paare, die bei der Dokumentation ihrer Ergebnisse explizit auf die Schätzung der gesuchten Wahrscheinlichkeit hingewiesen haben: „Man kann die relativen Häufigkeiten der Measures als Annäherung für die Wahrscheinlichkeiten ansehen. Die Wahrscheinlichkeit (...) beträgt dann etwa 16,6%." Schwierigkeiten hatten sie (wie alle anderen auch) bei der Interpretation des mean, weil der berechnete Wert eine Dezimalzahl war, bei der Aufgabenstellung die Messgröße aber nur ganze Werte als mögliche Ergebnisse annehmen konnte, so dass sie für die Interpretation, den mean auf die nächste ganze Zahl aufrundeten.

Durch die Konstruktion differenzierter Kategoriensysteme zu den FATHOM-spezifischen Simulationskompetenzen sowie den stochastischen Kompetenzen lassen sich auf Grundlage stringenten methodischen Vorgehens reliable und inhaltlich interpretierbare Kompetenzbeschreibungen zu Studierenden bei der Bearbeitung computerbasierter, komplexer Aufgaben erstellen. Die Zusammenfassung der Kompetenzen in Profildiagrammen ermöglicht eine prägnante Einschätzung und bietet die Möglichkeit die Kompetenzen unter den Paaren miteinander zu vergleichen sowie eventuelle Strukturen aufzudecken, um daraus Fördermaßnahmen ableiten zu können.

Strukturen und Zusammenhänge

Vergleicht man nun die Profildiagramme der FATHOM-spezifischen Simulationskompetenzen der vier Paare untereinander, lassen sich drei Strukturen unterscheiden, wobei Struktur 1 zweimal auftrat (vgl. Abb. 23.3).

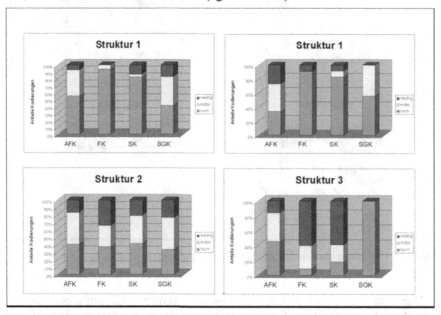

Abb. 23.3 Profildiagramme zu den FATHOM-spezifischen Simulationskompetenzen, nach erkennbaren Strukturen sortiert

Struktur 1: Allgemeine FATHOM-Kompetenz und strategische Kompetenzen lassen sich als hoch-mittel einstufen, Formel- und Simulationskompetenz als hoch.
Struktur 2: Alle Kompetenzen haben etwa gleiche Anteile der Ausprägungen hoch-mittel-niedrig. *Struktur 3*: Bei der Formel- und Simulationskompetenz dominiert die niedrige Ausprägung, die anderen beiden Kompetenzen schwanken zwischen hoch-mittel. Struktur 3 kann fast als „Komplementärstruktur" zu Struktur 1 bezeichnet werden. Da man aus der Analyse von vier Paaren keine schon verallgemeinerbaren Strukturen ableiten kann, wäre es für weitere Studien interessant, ob sich diese Strukturen auch bei anderen Paaren wiederfinden und verfesti-

gen lassen oder durch weitere Strukturen ergänzt werden können. Die stärksten Auswirkungen auf eine erfolgreiche Simulation mit FATHOM haben die beiden Kompetenzbereiche Formel- und Simulationskompetenz. Mängel in diesen Bereichen waren vor allem fehlendes Formelwissen sowie die Vermischung von Simulationsmöglichkeiten und planloses Vorgehen.

Betrachtet man die stochastischen Kompetenzen lassen sich ebenfalls wieder drei Strukturen erkennen, die sich aber von ihrer Art her ähneln, weil die Ausprägungen hoch-mittel-niedrig in beiden Kompetenzen jeweils ähnlich ausgeprägt sind (vgl. Abb. 23.4).[3]

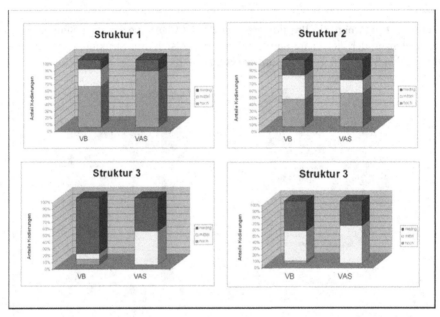

Abb. 23.4 Profildiagramme zu den stochastischen Kompetenzen, nach erkennbaren Strukturen sortiert

Man könnte also auch von nur einer Struktur sprechen, die mit unterschiedlichen Ausprägungen auftritt. Schwierigkeiten auf der stochastischen Ebene treten vor allem bei der Interpretation der simulierten Daten auf: relative Häufigkeiten - Wahrscheinlichkeiten, Interpretation des arithmetischen Mittels. Insgesamt bleiben die Ergebnisse der stochastischen Kompetenzauswertungen hinter den Erwartungen zurück.

---

[3] Struktur 3 besitzt (fast) keine hohen Kompetenzausprägungen. Es überwiegen die niedrigen und mittleren Kompetenzausprägungen.

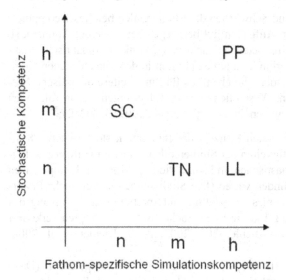

Abb. 23.5 Einordnung der vier Paare mit zusammengefassten stochastischen und FATHOM-spezifischen Simulationskompetenzen

Um nun mögliche Zusammenhänge zwischen den zwei Kompetenzklassen aufzudecken, kann man die jeweiligen Kompetenzausprägungen pauschal zusammenfassen, so dass die Struktur 1 bei der FATHOM-spezifischen Simulationskompetenz als hoch eingestuft wird, Struktur 2 als mittel und Struktur 3 als niedrig. Ebenso verfährt man bei den stochastischen Kompetenzen und vergibt für Struktur 1 hoch, für Struktur 2 mittel und für Struktur 3 niedrig. Trägt man die Zusammenfassungen in ein Streudiagramm (Abb. 23.5) auf, kann man erkennen, dass verschiedene Kombinationen der Kompetenzen auftreten, aber kein direkter Zusammenhang auszumachen ist. Paare mit hohen FATHOM-spezifischen Simulationskompetenz können hohe wie niedrige stochastische Kompetenzen besitzen. Für weittragendere Aussagen müsste man die Fallzahl erhöhen.

Resümee

Als Ergebnis kann man festhalten, dass die Studierenden die Simulation mit FATHOM als Problemlösemethode akzeptiert haben, da sie ihre simulierten Ergebnisse als Lösungen für die Aufgabenstellung annehmen, und FATHOM auch geeignet einsetzen konnten. Allerdings war die Integration stochastischer Konzepte in die Simulationstätigkeiten aus unserer Sicht unzureichend. Die Kompetenzanalyse zu den werkzeugbezogenen und mathematischen Kompetenzen ist zwar in ihrer Entwicklung aufwendig und komplex gewesen, lässt sich nun aber in der Weiterentwicklung des Kurskonzeptes leichter anwenden oder auf andere Werkzeugsoftware und Aufgabentypen übertragen. Es hat sich gezeigt, dass durch diese

Kompetenzanalyse Stärken und Schwächen der Studierenden bei dieser computer-basierten, komplexen Art von Aufgaben gut herausgearbeitet werden konnten, so dass man gezielt an diesen Kompetenzen für weitere, konkrete didaktische Maßnahmen ansetzen kann. Weiterhin konnten Strukturen in den verschiedenen Kompetenzklassen ausgemacht werden, die ebenfalls für eine weitere didaktische Nutzung von Vorteil sein können. Wesentliche Schlussfolgerungen, die in folgenden Studien aufgegriffen wurden, waren daher (vgl. Maxara 2009, S. 431ff.):

- die Forderung nach einem neuen Konzept der Integration stochastischer Konzepte in die Simulationstätigkeiten der Studierenden, da kaum Hinweise auf eine Verknüpfung von stochastischen und simulationsspezifischen Kompetenzen bei den Studierenden zu finden waren (Die Studierenden sollten mehr Erwartungen aufbauen und Ergebnisse sorgfältiger interpretieren. Die Nutzung des Werkzeugs FATHOM sollte stärker mit den stochastischen Konzepten verknüpft werden. Z.B. Prömmel und Biehler 2006/2007; 2009; Biehler et al. 2008; Prömmel 2010.),
- eine Verbesserung der FATHOM-spezifischen Simulationskompetenzen (Dazu scheint es sinnvoll den Simulationsplan zu überarbeiten und die Struktur der Simulation dadurch noch transparenter zu machen. Ebenfalls könnte den Studierenden ein Handout mit wichtigen Formeln zur Verfügung gestellt werden. Z.B. Hofmann 2010),
- einer stärkere Fokussierung auf die Genauigkeit der Simulationsergebnisse und das empirische Gesetz der großen Zahl.

Literaturverzeichnis

Arbeitskreis Stochastik der GDM. (2010). *Leitidee Daten und Zufall für die Sekundarstufe II – Kompetenzmodelle für die Bildungsstandards aus Sicht der Stochastik und ihrer Didaktik.* http://stochastik-in-der-schule.de/Dokumente/Leitidee_Daten_und_Zufall_SekII.pdf. Abgerufen am 20.05.2013.

Biehler, R. (1991). Computers in Probability Education. In R. Kapadia, & M. Borovcnic (Hrsg.), *Chance Encounters: Probability in Education* (S. 169-211). Dordrecht: Kluwer.

Biehler, R. (1997). Software for learning and for doing statistics. *International Statistical Review, 65*(2), 167-189.

Biehler, R. (2003). Simulation als systematischer Strang im Stochastikcurriculum. In H.-W. Henn (Hrsg.), *Beiträge zum Mathematikunterricht 2003* (S. 109-112). Hildesheim: Franzbecker.

Biehler, R. (2006). Working Styles and Obstacles: Computer-Supported Collaborative Learning in Statistics. In A. Rossmann, & B. Chance (Hrsg.), *Working Cooperatively in Statistics Education. Proceedings of ICoTS 7. Salvador da Bahia, Brazil* [CD-ROM]. Voorburg, The Netherlands: International Statistical Institute.

Biehler, R. (2006/07). *Vorlesungsskript "Elementare Stochastik", WS 06/07* (unveröffentlichtes Skript). Universität Kassel.

Biehler, R., & Kombrink, K. (2004). Elementare Stochastik interaktiv - Einführende Stochastikausbildung mit Unterstützung didaktisch orientierter Werkzeugsoftware. In R. Biehler, J. Engel, & J. Meyer (Hrsg.), *Neue Medien und innermathematische Vernetzungen in der Stochastik.* (Anregungen zum Stochastikunterricht, Band 2, S. 151-168). Hildesheim: Franzbecker.

Biehler, R., Ben-Zvi, D., Bakker, A., & K. Makar (2013). Technology for Enhancing Statistical Reasoning at the School Level. In K. Clements, A. J. Bishop, C. Keitel, J. Kilpatrick, & F. Leung (Hrsg.), *Third International Handbook of Mathematics Education* (S. 643-689). New York: Springer.

Biehler, R., Hofmann, T., & Prömmel, A. (2008). *Das GESIM-Konzept – Ein Ganzheitlicher Einstieg in die Stochastik in der gymnasialen Oberstufe mit computergestützter Simulation: Unterrichtsmaterialien und didaktische Anleitungen.* (unveröffentlichte Schrift). Kassel: Universität Kassel.

Chance, B., & Rossman, A. (2006). Using Simulation to Teach and Learn Statistics. In A. Rossmann, & B. Chance, (Hrsg.), *Working Cooperatively in Statistics Education. Proceedings of ICoTS 7. Salvador da Bahia, Brazil* [CD-ROM]. Voorburg, The Netherlands: International Statistical Institute. http://www.stat.auckland.ac.nz/~iase/publications/17/7E1_CHAN.pdf.   Abgerufen   am 20.05.2013.

DelMas, R. C., Garfield, J., & Chance, B. (1999). *Exploring the Role of Computer Simulations in Developing Understanding of Sampling Distributions.* Paper presented at the AERA Annual Meeting. Montreal.

Engel, J. (2002). Activity-based Statistics, Computer Simulation and Formal Mathematics. In B. Phillips (Hrsg.), *Proceedings of ICoTS 6. Cape Town, South Africa [CD-ROM].* Voorburg, The   Netherlands: International   Statistical   Institute. http://icots6.haifa.ac.il/PAPERS/5A3_ENGE.PDF. Abgerufen am 22.05.2013.

Hofmann, T., Maxara, C., Meyfarth, T., & Prömmel, A. (2013). Using the software Fathom for learning and teaching statistics in Germany – A review on the research activities of Rolf Biehler's working group over the past ten years. In T. Wassong, D. Frischemeier, P. R. Fischer, R. Hochmuth, & P. Bender (Hrsg.), *Mit Werkzeugen Mathematik und Stochastik lernen – Using Tools for Learning Mathematics and Statistics* (pp. 283–304). Wiesbaden: Springer Spektrum.

Keitzer, C. (2006). *Selbständig-kooperative Bearbeitung von stochastischen Simulationsaufgaben am Computer - Qualitative Analysen zu Schülerkompetenzen und Arbeitsweisen* (unveröffentlichte Examensarbeit). Kassel: Universität Kassel.

Lipson, K. (1997). What do students gain from simulation exercises? An evaluation of activities designed to develop an understanding of the sampling distribution of a proportion. In J. Garfield, & G. Burrill (Eds), *Research on the Role of Technology in Teaching and Learning Statistics. Proceedings of the 1996 IASE Round Table Conference University of Granada, Spain, 23-27 July, 1996* (S. 137–150). Voorburg, Niederlande: International Statistical Institute. http://www.stat.auckland.ac.nz/~iase/publications/8/12.Lipson.pdf.   Abgerufen   am 20.05.2013.

Maxara, C. (2006). *Einführung in die stochastische Simulation mit Fathom. KaDiSto - Kasseler Online-Schriften zur Didaktik der Stochastik, Band 1.* Kassel: Universität Kassel. http://nbn-resolving.de/urn:nbn:de:hebis:34-2006082514477.

Maxara, C. (2009). *Stochastische Simulation von Zufallsexperimenten mit FATHOM. Eine theoretische Werkzeuganalyse und explorative Fallstudie.* Hildesheim: Franzbecker.

Maxara, C., & Biehler, R. (2006). Students' Probabilistic Simulation and Modelling Competence after a Computer-Intensive Elementary Course in Statistics and Probability. In A. Rossman, & B. Chance (Hrsg.), *Working Cooperatively in Statistics Education. Proceedings of IcoTS 7. Salvador da Bahia, Brazil* [CD-ROM]. Voorburg, The Netherlands: International Statistical Institute. http://www.stat.auckland.ac.nz/~iase/publications/17/7C1_MAXA.pdf. Abgerufen am 20.05.2013.

Maxara, C., & Biehler, R. (2007). Constructing stochastic simulations with a computer tool – students' competencies and difficulties. In D. Pitta-Pantazi, & G. Philippou (Hrsg.), *Proceedings of the Fifth Congress of the European Society for Research in Mathematics Education* (S. 762-771). Nicosia, Zypern: Department of Education, University of Cyprus. http://www.erme.unito.it/CERME5b/WG5.pdf#page=79. Abgerufen am 20.05.2013.

336   Carmen Maxara

Mayring, P. (2003). Qualitative Inhaltsanalyse. In U. Flick, E. v. Kardorff & I. Steinke (Hrsg), *Qualitative Forschung. Ein Handbuch* (S. 468-474). Hamburg: rowohlts.

Mayring, P. (2003). *Qualitative Inhaltsanalyse. Grundlagen und Techniken*. Weinheim: Beltz Verlag.

Mayring, P. (2005). Neuere Entwicklungen in der qualitativen Forschung und der Qualitativen Inhaltsanalyse. In P. Mayring, & M. Gläser-Zikuda (Hrsg.), *Die Praxis der Qualitativen Inhaltsanalyse* (S. 7-19). Weinheim: Beltz.

Meyfarth, T. (2006). *Ein computergestütztes Kurskonzept für den Stochastik-Leistungskurs mit kontinuierlicher Verwendung der Software* FATHOM *- Didaktisch kommentierte Unterrichtsmaterialien. KaDiSto - Kasseler Online-Schriften zur Didaktik der Stochastik, Band 2*. Kassel: Universität Kassel. http://nbn-resolving.de/urn:nbn:de:hebis:34-2006092214683.

Meyfarth, T. (2008). *Die Konzeption, Durchführung und Analyse eines simulationsintensiven Einstiegs in das Kurshalbjahr Stochastik der gymnasialen Oberstufe: Eine explorative Entwicklungsstudie*. Hildesheim: Franzbecker.

Mills, J. D. (2002). Using Computer Simulation Methods to Teach Statistics: A Review of the Literature. *Journal of Statistics Education 10*(1). http://www.amstat.org/publications/jse/v10n1/mills.html.

Prömmel, A., &. Biehler, R. (2006/07). Einführung in die Stochastik in der Sekundarstufe I mit Hilfe von Simulationen unter Einsatz der Werkzeugsoftware FATHOM. In A. Eichler, & J. Meyer (Hrsg.), *Schulbuchkonzepte; Daten und Zufall als Leitidee für die Sekundarstufe I* (Anregungen zum Stochastikunterricht, Band 4, S. 137-158). Hildesheim: Franzbecker.

Prömmel, A., & Biehler, R. (2009). Instruktionale Unterstützung selbständigen Lernens in der gymnasialen Oberstufe beim Einstieg in die Stochastik. In M. Neubrand (Hrsg.), *Beiträge zum Mathematikunterricht 2009* (S. 799-802), Münster: WTM-Verlag.

Saldanha, L., &. Thompson, P. W. (2003). Conceptions of sample and their relationship to statistical inference. *Educational Studies in Mathematics 51*(3), 257-270.

Schupp, H., Berg, G., & Dabrock, H. (1992). *ProSto: Programme für den Stochastikunterricht*, Bonn: Dümmler.

Sedlmeier, P. (1999*). Improving Statistical Reasoning. Theoretical Models and Practical Implications*. Mahwah, New Jersey: Lawrence Erlbaum Associates.

Trauerstein, H. (1990). Zur Simulation mit Zufallsziffern im Mathematikunterricht der Sekundarstufe I. *Stochastik in der Schule 10*(2), 2-30.

Kapitel 24
Explorative Datenanalyse und stochastische Simulationen mit TinkerPlots – erste Einsätze in Kassel & Paderborn

Daniel Frischemeier, Susanne Podworny

Universität Paderborn

Abstract Die in den USA entwickelte Datenanalyse- und Simulationssoftware TinkerPlots, vorgesehen für den Schuleinsatz in den Klassen 4-8, liegt nach der Migration durch die Arbeitsgruppe Biehler nun auch in deutscher Version vor. In diesem Artikel wird die Arbeit mit der Software in der Arbeitsgruppe Biehler beschrieben. Zunächst werden die verschiedenen Einsätze der Software in Lehrveranstaltungen an den Universitäten Kassel und später in Paderborn vorgestellt. Anschließend gibt es Einblicke in empirische Studien, die im Rahmen von universitären Lehrveranstaltungen stattgefunden haben mit zwei verschiedenen Schwerpunkten: Zum einen zur „explorativen Datenanalyse mit TinkerPlots" und zum anderen zu „stochastischen Simulationen mit TinkerPlots".

Einleitung

Im Zuge der Einführung der Bildungsstandards (Blum et al. 2006) in Deutschland mit der Betonung der Leitideen „Daten, Häufigkeit und Wahrscheinlichkeit" (Primarstufe) und „Daten und Zufall" (Sekundarstufe) wurde der Stochastik im Schulunterricht eine größere Bedeutung zugemessen. Als zukünftige mündige Bürgerinnen und Bürger benötigen die Schülerinnen und Schüler die grundlegende Fähigkeit, mit Daten sicher umzugehen und Unsicherheit einzuschätzen. Betrachtet man alleine den Zweig „Daten", so gibt es dort viele Kompetenzen, die man für den Alltag mitbringen sollte: Zum Beispiel ist das Lesen, Beschreiben und Interpretieren von statistischen Darstellungen eine fundamentale Kompetenz, ebenso wie das Erstellen eigener statistischer Graphiken. Im Schulunterricht und in den Lehrplänen wird hier besonders der Einsatz realer und für die Schülerinnen und Schüler motivierender Datensätze gefordert, die sehr oft aufgrund ihrer Vielfalt und ihres Umfangs nur mit passender Software analysiert werden kann. Es ergeben sich die Fragen: Welche Software ist geeignet für einen Einsatz in Primarbereich - bzw. in der „unteren" Sekundarstufe I? Welche Voraussetzungen sollte diese Software mit sich bringen? Bakker (2002) unterscheidet zwischen „route-type" und „landscape-type" Software. Während bei einer Software vom Typ „route-type" nur begrenzte Anwendungen zur Verfügung stehen, bietet eine Software vom Typ „landscape" eine ganze Landschaft und Vielfalt an Features sowie die Möglichkeit, diese durch „freie Erkundung" dieser Landschaft kennenzulernen. TinkerPlots 2.0 (Konold und Miller 2011), eine in den USA entwickelte, dynamische Datenanalyse- und Simulationssoftware, die für den Stochastikunterricht in den Klassen 3-8 vorgesehen ist (Biehler 2007a; 2007b), ist eine solche Software vom Typ „landscape". Wesentliche Basisanforderungen an eine statistische Werkzeugsoftware für den Unterricht sind in Biehler (1997) nachzulesen. TinkerPlots erfüllt einen Großteil dieser Anforderungen, ist zusätzlich gekennzeichnet durch Farbenfrohheit, leicht bedienbare Elemente, Differenzierungsfähigkeit[1] und benötigt wenig Formeln und kaum Programmierkenntnisse. TinkerPlots scheint nicht nur für den Einsatz in der Schule, sondern auch für die Lehrerausbildung geeignet zu sein. Das Potential der Software umfasst zwei Aspekte der Stochastik: Explorative Datenanalyse und Simulation von Zufallsexperimenten. Zentrale Konzepte zur Datenanalyse bei TinkerPlots sind das Arbeiten mit Datenkarten und das selbstständige Erzeugen von Graphiken mithilfe von drei Grundoperationen: „Stapeln", „Trennen" und „Ordnen". Die Simulation geschieht mit einer graphischen Zufallsmaschine, mit der auch komplexe Zufallsexperimente durchgeführt werden können. Bei der Entwicklung der Maschine ist darauf geachtet worden, dass möglichst viele reale Situationen ohne Verwendung von Formeln abbildbar sind. Eine erste ausführliche Vorstellung der Software in der deutschen mathematikdidaktischen Kommunität findet sich bei Biehler (2007a; 2007b). Einen guten Überblick über das Potenzial der Software für einen Einsatz in der Primarstufe bietet außerdem Wagner (2006). Studien zum Einsatz von TinkerPlots im Mathematikunterricht sowie Unterrichtsvorschläge und evaluierte Lernumgebungen liegen in Konold (2007), Watson und Donne (2009) u.a. vor. Als ein Ergebnis sei hervorgehoben, dass Werkzeuge wie TinkerPlots es erlauben, den Prozess der Datenanalyse bzw. den Prozess der stochastischen Simulation mit dem Erlernen der Software zu synchronisieren.

TinkerPlots in Lehrveranstaltungen an den Universitäten Kassel und Paderborn

Kinderuni Kassel „Spannende Statistik: Mit Zahlen spielen und die Welt erklären": Der Einsatz der Software TinkerPlots unter Rolf Biehler reicht bis ins Jahr 2008 zurück. In Kassel fanden in jenem Jahr im sechsten Semester Veranstaltungen der Kinderuni[2] statt. Zum ersten Mal beteiligte sich der Fachbereich Mathematik/Informatik der Universität Kassel an der Veranstaltungsreihe, und am 21.

---

[1] Alle Funktionen und Werkzeuge lassen sich – je nach Einsatz der Software passend zum Anforderungsniveau in der jeweiligen Jahrgangsstufe - ein- und ausschalten, sodass der Funktionsumfang je nach Bedarf differenziert werden kann.

[2] Siehe http://www.die-kinder-uni.de/html/kassarchiv.html

November stand mit dem Vortrag *Spannende Statistik: Mit Zahlen spielen und die Welt erklären* Rolf Biehler als Dozent vor den jungen Schülerinnen und Schülern. In seinem Vortrag über lebendige Statistik fand auch das Programm TinkerPlots Anwendung als Übertragung der realen Vorgänge im Hörsaal auf den Computer. Im Hörsaal ordneten sich die anwesenden Kinder zum Beispiel (im Sinne von „Lebendiger Statistik") nach ihrer Körpergröße, ermitteln durch Abzählen den Median und erkunden so spielerisch die Welt der Daten. TinkerPlots, damals in der englischen Beta Version 2.01, diente hier als Beispiel für ein Statistikprogramm, mit dem Daten anschaulich aufbereitet und visualisiert werden können. Für diesen Vortrag wurde auch der heute noch genutzte Datensatz Kinder-Uni2008.tp aus den Fragebögen der zum Vortrag angemeldeten Kinder verwendet. Dieser enthält Angaben von 285 „Kindern", die zu 28 Merkmalen (z.B. zu ihren Freizeitgewohnheiten, zum Schulweg, etc.) befragt wurden.

Die Lehrveranstaltung „Mathematische Anwendungen": Im Wintersemester 2008/09 wurde TinkerPlots an der Universität Kassel erstmals systematisch in der Veranstaltung „Mathematische Anwendungen" eingesetzt. Für diese Vorlesung mit Übung wurde von Susanne Podworny ein vierstündiger Einführungskurs in TinkerPlots entwickelt. Dieser fand in mehreren Durchgängen statt und bereitete die Studierenden auf den eigenständigen Umgang mit dem Programm während des Semesters vor. Im Themenkomplex Datenanalyse und Wahrscheinlichkeitsrechnung der Vorlesung wurde TinkerPlots als Softwareunterstützung eingesetzt.

Die Lehrveranstaltung „Leitidee Daten und Zufall von Klasse 3 bis 8": Nach dem Wechsel Rolf Biehlers an die Universität Paderborn wurde TinkerPlots dort erstmals im Wintersemester 2010/11 eingesetzt. Im Rahmen der Dissertationsprojekte von Daniel Frischemeier und Susanne Podworny fand ein gemeinsames fachdidaktisches Seminar „Leitidee Daten und Zufall von Klasse 3 bis 8" unter der Leitung von Rolf Biehler statt. In sechs Projektsitzungen bearbeiteten die Teilnehmenden Fragestellungen zur explorativen Datenanalyse und zur stochastischen Simulation mit TinkerPlots. Zusätzlich gab es Literaturvorträge über den Einsatz von TinkerPlots in der Schule. Ausführlichere Informationen zu den Sitzungen zur explorativen Datenanalyse finden sich bei Frischemeier und Biehler (2011a). Hieraus wurden die zwei folgenden Seminare entwickelt.

Die Lehrveranstaltung „Statistisch denken und forschen lernen": Das Seminar „Statistisch denken und forschen lernen" wurde im Wintersemester 2011/12 als fachwissenschaftliches Vertiefungsseminar für Studierende des Lehramts Mathematik an Grund-, Haupt-, Real- und Gesamtschulen an der Universität Paderborn angeboten. Das Grundkonzept dabei war, dass die Teilnehmer eigenständig statistisch arbeiten, in dem Sinne, dass sie selbst statistische Fragestellungen formulierten, ein geeignetes Erhebungsinstrument (in Form eines Fragebogens) entwickelten, sowie die Auswertung ihrer Fragestellungen anhand der erhobenen Daten vornahmen und somit einen kompletten Datenanalyse-Zyklus (nach Wild und Pfannkuch, 1999) durchliefen. Die Ergebnisse und Interpretationen wurde in Reports festgehalten. Die Exploration und Auswertung der Daten anhand vertiefender Fragestellungen erfolgte mit TinkerPlots. Genauere Informationen zu den Inhalten des Seminars finden sich in Frischemeier und Biehler (2012). Einen Überblick zu dem Seminar zeigt die Tabelle 24.1.

Tab. 24.1 Inhalte des Seminars „Statistisch denken und forschen lernen"

Sitzung	Inhalte
1	Organisatorisches, Eingangstest, Eingangsbefragung
2-3	Planung einer Datenerhebung, Statistische Fragestellungen generieren, Einführung in die Konstruktion von Fragebögen, Erhebung von Daten in der Veranstaltung „Kultur der Mathematik"
4-6	Einführung in die Software TinkerPlots, Erste Graphiken mit TinkerPlots erstellen, Informelles Schließen I
7	Auswertung von kategorialen Variablen: Mehrfeldertafeln in TinkerPlots
8-10	Verteilungen eines numerischen Merkmals; Vergleich zweier Verteilungen eines numerischen Merkmals (Gruppenvergleich) mit TinkerPlots
11-12	Untersuchung des Zusammenhangs zweier numerischer Variablen mit TinkerPlots
13-14	Informelles Schließen II: Randomisierungstest und P-Wert, Ausgangstest, Ausgangsbefragung

Die Lehrveranstaltung „Angewandte Stochastik": Im Seminar (Sommersemester 2012 an der Universität Paderborn) „Angewandte Stochastik – Mit Simulationen komplexe Probleme verstehen und lösen" standen Simulationen im Vordergrund. Mit vielen Anwendungsaufgaben wurden Themen der Wahrscheinlichkeitsrechnung bis hin zur beurteilenden Statistik erarbeitet. Ebenso wurde die Genauigkeit von Simulationen mit TinkerPlots thematisiert. Wie im vorigen Seminar stellten Randomisierungstests den Abschluss der Veranstaltung dar. Die Tabelle 24.2 liefert einen Überblick über das Seminar.

Tab. 24.2 Inhalte des Seminars "Angewandte Stochastik"

Sitzung	Inhalte
1	Organisatorisches, Eingangstest, Eingangsbefragung
2	Einführung in die Datenanalyse mit TinkerPlots
3-5	Einführung in die Simulation mit TinkerPlots; Alltagsprobleme lösen mit Simulationen
6-7	Schließen aus Stichproben; Genauigkeit von Simulationen; $1/\sqrt{n}$ - Gesetz
8-9	Stochastische Un-/Abhängigkeit: Schließen vom Modell auf Daten und umgekehrt mit TinkerPlots
10-13	Beurteilende Statistik mit P-Wert Tests und Randomisierungstests mit TinkerPlots
14	Ausgangstest, Ausgangsbefragung

Erste Einblicke in empirische Studien

Nach diesem Überblick über den Einsatz der Software an den Universitäten Paderborn und Kassel möchten wir nun exemplarisch einige Ergebnisse unserer empirischen Studien auf Basis der drei Seminare an der Universität Paderborn beschreiben. Im Zuge des oben beschriebenen Seminars „Leitidee Daten und Zufall in Klasse 3-8" wurde eine qualitative Fallstudie zur Datenanalysekompetenz der Teilnehmenden durchgeführt. Generell gibt es in Deutschland, aber auch international, nur wenige Studien, die den Einsatz und das Potential (beispielsweise auf die Entwicklung einer Datenanalysekompetenz) von TinkerPlots in der Lehrerausbildung thematisieren. Hervorzuheben sind hier Hammerman und Rubin (2004) und Monteiro et al. (2010). Wir wollen bei unseren ersten explorativen Studien, die in den oben beschriebenen Lehrveranstaltungen an der Universität Paderborn durchgeführt worden sind, folgenden Fragestellungen nachgehen:

1. Bei der Datenanalyse: Wie kann TinkerPlots den Vergleich zweier Verteilungen mit quantitativen Merkmalen unterstützen? Wie vergleichen die Teilnehmenden zwei Verteilungen mit TinkerPlots? Welche Werkzeuge benutzen sie dabei?
2. Bei der Simulation: Wie schnell sind Studierende befähigt, eigenständig Simulationen durchzuführen? Welche Funktionen nutzen sie? Können Sie nach einer 90 minütigen Einführung eine Aufgabe selbständig lösen?

1. Forschungseinblicke zur Datenanalyse mit TinkerPlots: Diesen Fragen (siehe 1.) sind wir in einer Fallstudie im WS 2010/11 nachgegangen. Nachdem die Teilnehmenden in der ersten Sitzung in die Datenanalyse mit TinkerPlots eingeführt worden sind und in der zweiten Sitzung Verteilungsvergleiche mit TinkerPlots basierend auf statistischen Fragestellungen zum Muffins-Datensatz (Biehler et al. 2003) durchgeführt haben, sollte in der dritten und abschließenden Sitzung ein statistisches Projekt bearbeitet werden. Dieses enthielt mehrere Fragen (zum Muffins-Datensatz), die durch Exploration der Daten mit TinkerPlots beantwortet werden sollten. Fundamentaler Aspekt dieses statistischen Projekts waren Gruppenvergleiche und somit Fragen wie: Inwiefern investieren Jungen mehr Zeit in Hausaufgaben als Mädchen? Der Arbeitsauftrag für die Teilnehmenden lautete, einen kurzen statistischen Report unter Einbezug ihrer in TinkerPlots erstellten Graphiken zu verfassen. Dabei arbeiteten die Teilnehmenden in Zweierteams. Zum Zweck der späteren Rekonstruktion ihrer Gedanken und Vorgehensweisen sollten sie möglichst viel und offen in Form des „lauten Denkens" (Weidle und Wagner 1982) miteinander kommunizieren. Die Bildschirmaktivitäten sowie die Kommunikation der Probanden wurden mit der Software Camtasia aufgenommen, die Mitschriften und TinkerPlots-Graphiken in Form von Word-Dokumenten eingesammelt. Im Anschluss fand eine Analyse der Daten mit qualitativen Methoden auf zwei Ebenen statt: auf einer lokalen interpretativen und einer globalen deskriptiven Ebene.

Zunächst zur lokalen Ebene: Hier wollen wir lediglich einige der entstandenen Graphiken analysieren. Die Abbildung 24.1 zeigt die Graphik von Anna und Ben zu der Fragestellung, inwieweit sich Schülerinnen und Schüler in ihrem Leseverhalten unterscheiden (Variable Zeit_Lesen zur Frage: Wie viele Stunden gehst du pro Woche der Aktivität Lesen (ohne Schulbücher) nach?). Anna und Ben haben einen Boxplot sowie den Median (rotes, umgedrehtes T) und das arithmetische Mittel (blaues Dreieck) eingezeichnet. Diese Graphik heben wir als eine „best-practice"-Graphik für einen Gruppenvergleich in TinkerPlots hervor. Viele relevante Aspekte können aus dieser Graphik interpretiert werden.

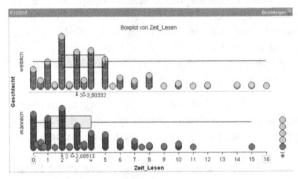

Abb. 24.1 TinkerPlots-Graphik von Anna & Ben

Bei einer weiteren Graphik (von Tim & Tom, Abbildung 24.2), geht es ebenfalls um die Unterschiede im geschlechtsspezifischen Computernutzungsverhalten (Variable: Zeit_Comp).

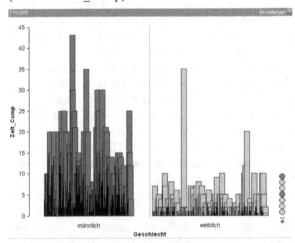

Abb. 24.2 TinkerPlots Graphik von Tim & Tom

Die Graphik 2 ist ein Beispiel für eine weniger gut geeignete Darstellung. Die verwendeten Value Bars (dt. Wertebalken) lassen in diesem hier vorliegenden, ungeordneten Zustand allenfalls vage Aussagen über Lage und Variabilitäten zu.

Bei der Betrachtung der Transskripte fällt noch auf, dass unabhängig von der Graphik ausschließlich mit den arithmetischen Mitteln beider Verteilungen argumentiert wird.

Nun zur globalen Analyse: Diese soll einen Überblick über die bei den Reports verwendeten TinkerPlots-Features und statistischen Kennzahlen sowie schnelle Vergleiche zwischen den teilnehmenden Gruppen ermöglichen. Zu diesem Zweck haben wir die Matrix von Fitzallen und Watson (2010) modifiziert. Der Tabelle 24.3 kann entnommen werden, welche Features die Teilnehmer tatsächlich nutzen.

Tab. 24.3 Überblick über die verwendeten Tinkerplots-Werkzeuge und statistischen Kennzahlen

Gruppe	Punkt-diagramm	Histo-gramm	Mittelwert	Boxplot	Dividers	Wertebalken
1	Ja	Ja	aM	Nein	Nein	Nein
2	Ja	Nein	aM	Nein	Nein	Ja
3	Ja	Nein	Med, aM	Ja	Nein	Nein
4	Ja	Nein	Med, aM	Ja	Nein	Nein
5	Ja	Nein	aM	Nein	Nein	Ja
6	Ja	Nein	Med	Nein	Nein	Nein
7	Ja	Nein	Mod, aM	Nein	Nein	Nein
8	Ja	Nein	aM	Nein	Nein	Nein
9	Nein	Ja	Keiner	Ja	Nein	Ja

Eine Erläuterung der einzelnen Kategorien findet sich in Frischemeier und Biehler (2011b). In Tabelle 24.1 wird ersichtlich, dass überwiegend Mittelwerte, vor allem das arithmetische Mittel, zum Vergleich benutzt werden. Boxplots werden, obwohl in der Seminarsitzung zuvor noch als Vergleichsinstrument hervorgehoben, nur in einem Drittel der Fälle genutzt. Dividers, ein Feature in TinkerPlots, welches ermöglicht weitere Fragen an eine Verteilung zu stellen, werden gar nicht verwendet. Immerhin wird in drei von neun Fällen mit „value bars" gearbeitet, die allerdings durchgängig ungeordnet sind und somit zu wenig aussagekräftigen Graphiken führten, ähnlich wie in Abbildung 24.2. Insgesamt bestätigen sich zwei Eindrücke: zum einen der „Rückfall auf Mittelwerte" beim Vergleich von Verteilungen (Biehler 2007c) und zum anderen die Konstruktion von Graphiken, die allerdings nur unzureichend beschrieben oder kommentiert wurden (ähnliche Beobachtungen bei Heaton und Mickelson (2002) und bei Francis (2005)).

2. Forschungseinblicke zur Simulation mit TinkerPlots: Ausgehend von der These, dass lediglich eine kurze Einführung in das Simulieren mit der Zufallsmaschine von TinkerPlots nötig ist, begannen Studierenden bereits ab der zweiten Sitzung zur Simulation mit dem selbstständigen Bearbeiten von Aufgaben. Einen Einblick in die ersten Simulations-„fähigkeiten" lieferte die Bearbeitung der Hausaufgabe „Glücksrad auf dem Weihnachtsmarkt", die nach der ersten Einführungssitzung gestellt wurde. Gearbeitet wurde mit der englischen Betaversion von TinkerPlots.

Aufgabe: Das Glücksrad auf dem Weihnachtsmarkt

Die Klasse 8b möchte auf dem Weihnachtsmarkt ein Glücksrad aufbauen. Die Schüler wollen pro Spiel im Mittel 50 Cent für die Klassenkasse einnehmen. Pro Einsatz darf der Besucher dreimal drehen. Die Aufteilung des Glücksrads lautet wie folgt:

In 1 % der Fälle gewinnt man 50 Euro.

In 45 % der Fälle gewinnt man 1 Euro.

In 54 % der Fälle gewinnt man nichts.

a) Wie viel muss ein Spiel kosten, damit die Klasse pro Spiel im Mittel 50 Cent einnimmt?

b) Wie wahrscheinlich ist es, dass ein zufälliger Besucher nichts gewinnt?

Abb. 24.3 Aufgabe: Glücksrad auf dem Weihnachtsmarkt

Die Abbildung 24.4 zeigt eine Musterlösung der Aufgabe. Folgende Schritte müssen durchgeführt werden, wobei Variationen möglich sind:

- Modellieren der Zufallssituation mit der Zufallsmaschine von TinkerPlots: In der Beispiellösung wird das Bauteil Kreisel (andere Modellierungen sind ebenfalls möglich) entsprechend der gegebenen Werte eingesetzt, und die Anzahl der Ziehungen symbolisiert die drei Drehungen.
- Wiederholtes Durchführen des Zufallsexperiments, modelliert durch die Anzahl der Durchgänge (Stichwort Gesetz der großen Zahlen). Im Beispiel 1000 Mal.
- Die interessierende Größe ist die Auszählung nach drei Drehungen, in der Tabelle heißt das zugehörige Merkmal „Summe" und wird automatisch über eine interne Formel berechnet.
- Darstellen und Auswerten der interessierenden Größe in einem Graphen: Um Teilaufgabe a) beantworten zu können, wird der Mittelwert der Auszählungen benötigt, die in der Spalte „Summe" zu finden sind.
- Für Teilaufgabe b) ist das interessierende Ereignis, dass in drei Drehungen insgesamt nichts erspielt wurde. Dies lässt sich entweder durch ein weiteres Auswertungsmerkmal in der Tabelle (s. o. „KeinGewinn") oder über die bereits erstellte Graphik zur „Summe" und Einblenden der dem Ereignis {0} zugehörigen relativen Häufigkeit erzielen.

Aus diesen Schritten zur Lösung der Aufgabe wurden neun Kategorien entwickelt. Die Abbildung 24.4 zeigt die entsprechenden relativen Lösungshäufigkeiten bei n = 18 Teilnehmern, die diese Aufgabe bearbeitet haben.

Die Kategorie „Sampler richtig erstellt": Ein richtig erstellter Sampler (deutsch: Zufallsmaschine) ist trivialerweise nötig, um korrekt weiterarbeiten zu können. In allen Bearbeitungen erstellten Studierende die Zufallsmaschine richtig.

Die Kategorie „Repeats ≥ 1000": Es ist nicht vorgegeben, wie häufig die Simulation durchgeführt werden soll. Im Seminar selbst wurden zwei Beispiele mit 1000 Wiederholungen (Repeats) vorgeführt. Allerdings wird das Thema „Genauigkeit von Simulationen" erst später thematisiert. Von den Studierenden benutzen 15 eine Wiederholungszahl von 1000 oder mehr (bis zu 100.000), drei arbeiten mit 100 Durchgängen. Diese Kategorie führt nicht dazu, dass die Aufgabe als nicht gelöst eingestuft wird.

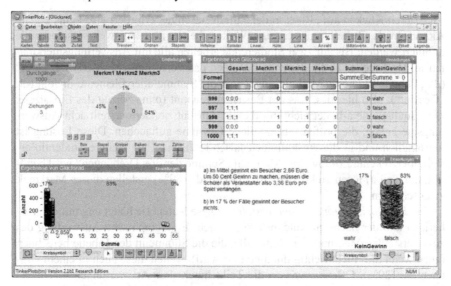

Abb. 24.4 Lösung der Glücksrad-Aufgabe in TinkerPlots

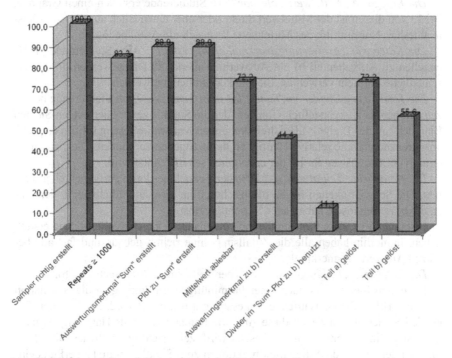

Abb. 24.5 Erfolgsanteile nach Kategorien bei der Glücksrad-Aufgabe

Die Kategorie „Auswertungsmerkmal 'Sum' erstellt": Eine Tabelle mit den Simulationsergebnissen findet sich bei allen Studierenden. Dies geschieht automatisch durch Klicken auf „Run" in der Zufallsmaschine, erfordert also keine eigene Aktion. In der Tabelle müssen zum Weiterarbeiten neue Auswertungsmerkmale definiert werden. Bis auf zwei Studierende arbeiten alle mit Auswertungsmerkmalen in der Tabelle. 16 Studierende benutzen das vorformulierte Merkmal „Sum" (dt. „Summe"). Dieses Merkmal berechnet die Summe aller Elemente in Join (dt. „Gesamt"), also hier das, was ein Besucher gewinnt (ohne Abzug des Einsatzes). Eine Studierende arbeitet nicht mit diesem Attribut, sondern erstellt acht Auswertungsmerkmale, die prüfen, ob mögliche Gewinne auftauchen. Diese Lösung ist umständlich und in diesem Fall auch falsch realisiert. Eine weitere Studentin erstellt hierzu keine Lösung. Das Merkmal „Summe der drei Drehungen" wird benötigt, um Teil a) lösen zu können. 16 Studierende (89%) erstellen dieses Merkmal, zwei scheitern.

Die Kategorie „Plot zu ‚Sum' erstellt": Diese Kategorie hängt eng mit der vorherigen und der nächsten Kategorie zusammen. Eine graphische Auswertung des zuvor erstellten Merkmals ist nötig. Alle, die die Summe in der Tabelle berechnen (auch wenn dies nicht richtig durchgeführt wird), stellen diese auch in einem Graphen dar (89%). Offenbar wissen die Teilnehmer, dass Auswertungen in Tinker-Plots über Graphen geschehen.

Die Kategorie „Mittelwert ablesbar": 16 Studierende erstellen einen Graphen, bei 13 von diesen ist der Mittelwert direkt oder über einen Blick auf die Statusleiste ablesbar. Nur mithilfe des Mittelwerts lässt sich die Aufgabe lösen.

Die Kategorie „Auswertungsmerkmal zu b) erstellt": Acht Studierende (44%) erstellen zu Aufgabenteil b) ein Auswertungsmerkmal. Dabei werden drei unterschiedliche Formeln verwendet, die alle korrekt benutzt werden.

Die Kategorie „Divider im 'Sum'-Plot zu b) benutzt": Zwei Studierende benutzen den bereits zu a) erstellten Graphen, der die Summe zeigt. Hier nutzen Sie den Einteiler, um das Ereignis „Bei drei Drehungen wurden insgesamt 0 Euro erspielt" abzugrenzen und die relative Häufigkeit dafür angeben zu können. Dies ist ein alternativer Lösungsweg für b), der erstaunlich selten genutzt wird.

Die Kategorie „Teil a) gelöst": 72% der Teilnehmer lösen Aufgabenteil a) korrekt. Die Schwierigkeiten liegen offensichtlich im zweiten Aufgabenteil, denn alle, die b) richtig lösen, haben auch a) richtig gelöst.

Erstaunlich ist, dass fünf Studierende überhaupt keine sinnvolle Lösung erstellt haben. Immerhin haben alle die Zufallsmaschine richtig gebaut, und fast alle benutzen Auswertungsmerkmale.

Die Kategorie „Teil b) gelöst": 55% der Teilnehmer lösen auch Teil b).

Erfreulich bleibt, dass nach einer 90-minütigen Einführung in die Simulation mit TinkerPlots die Hausaufgabe insgesamt gut bearbeitet wird. Auch in den folgenden Semestern, in denen diese Aufgabe immer wieder als Hausaufgabe nach der ersten Simulationssitzung eingesetzt wird, zeigt sich ein ähnliches Bild. Dies führt zu der These, dass die Grundfunktionen zum Simulieren in TinkerPlots einfach und innerhalb von kurzer Zeit zu erlernen sind. Dies wird ausführlich im Dissertationsprojekt von Susanne Podworny diskutiert werden.

Über Einteiler, Zufallsmaschinen und Kreisel: Eine deutsche TinkerPlots Version

Anfang des Jahres 2010 gibt es erste Überlegungen in der AG Biehler, eine deutsche Version von TinkerPlots zu entwickeln. Der Einsatz der Software, besonders in den unteren Klassen, dürfte durch die englische Benutzerführung wesentlich behindert werden. Aber auch von Studierenden wird immer wieder die Sprache als großes Hindernis für die Verwendung von TinkerPlots genannt. Noch im selben Jahr wird das Projekt „TinkerPlots-Übersetzung" von der Gruppe Biehler, Frischemeier, Podworny gestartet. Mehrere Layoutwechsel und Funktionsänderungen von Seiten der Entwickler und zwei Jahre Übersetzungsarbeit später liegt nun eine erste komplett deutsche Beta-Version von TinkerPlots vor, die im Wintersemester 2012/13 in universitären Lehrveranstaltungen in Paderborn eingesetzt wird.

Weitere (Forschungs-) Vorhaben

Nachdem TinkerPlots nun in deutscher Sprache vorliegt, soll das Programm regelmäßig an der Universität Paderborn in der Lehrerausbildung eingesetzt werden. Angedacht ist der Einsatz der Software in der Pflichtveranstaltung „Elemente der Stochastik" und in Workshops zum Beispiel von Lehrerfortbildungen. Es gilt nun, in den zugehörigen Dissertationsvorhaben das Lernen stochastischer Inhalte mit TinkerPlots genauer zu untersuchen. Im Sinne der internationalen Diskussion um das *Informal Inferential Reasoning (IIR)* soll untersucht werden, inwieweit TinkerPlots informelles Schließen fördert. Darüber hinaus sollen auf der Basis der gewonnenen Erkenntnisse neue Lernumgebungen konstruiert werden. Entwicklungsforschung für den Einsatz von TinkerPlots in der Primar- und in der Sekundarstufe ist für das Frühjahr 2013 geplant. Im Laufe des Jahres 2013 soll dann die deutsche TinkerPlots-Version veröffentlicht werden. Entsprechendes Begleit- und Lehrmaterial ist in Arbeit, so dass TinkerPlots sowohl in der Schule als auch an Universitäten vielfältig genutzt werden kann.

Literatur

Atkinson, R. K., Derry, S. J., Renkl, A., & Wortham, D. W. (2000). Learning from examples: instructional principles from the worked examples research. *Review of Educational Research, 70*, 181-214.

Bakker, A. (2002). Route-type and landscape-type software for learning statistical data analysis. In *Proceedings of ICOTS 6*. Abgerufen von
http://www.stat.auckland.ac.nz/~iase/publications/1/7f1_bakk.pdf.

Biehler, R. (1997). Software for learning and for doing statistics. *International Statistical Review 65*(2), 167-189.

Biehler,R. (2007a). TINKERPLOTS: Eine Software zur Förderung der Datenkompetenz in Primar- und früher Sekundarstufe. *Stochastik in der Schule 27*(3), 34-42.

Biehler, R. (2007b). Arbeitsumgebungen zur Entwicklung von Datenkompetenz ab Klasse 1 – Das Potential der Software Tinkerplots. In *Beiträge zum Mathematikunterricht 2007* (S. 480-483). Hildesheim: Franzbecker. Abgerufen von http://www.mathematik.tu-dortmund.de/ieem/cms/media/BzMU/BzMU2007/Biehler.pdf.

Biehler, R. (2007c). Students' strategies of comparing distributions in an exploratory data analysis context. In *Proceedings of 56th Session of the International Statistical Institute* [CD-ROM]. http://www.stat.auckland.ac.nz/~iase/publications/isi56/IPM37_Biehler.pdf.

Biehler, R., Frischemeier, D., & Podworny, S. (2012). *tinkerplots.de*. Abgerufen am 10. November 2012 von http://lama.uni-paderborn.de/personen/rolf-biehler/projekte/tinkerplots.html.

Biehler, R., Kombrink, K., & Schweynoch, S. (2003). MUFFINS – Statistik mit komplexen Datensätzen – Freizeitgestaltung und Mediennutzung von Jugendlichen. *Stochastik in der Schule, 23*(1), 11-25.

Blum, W., Drüke-Noe, C., Hartung, R., & Köller, O. (2006). *Bildungsstandards Mathematik: konkret*. Berlin: Cornelsen Verlag.

Fitzallen, N. & Watson, J. (2010). Developing statistical reasoning facilitated by TinkerPlots. In C. Reading (Hrsg.), *Proceedings of ICoTS 8*. Ljubljana, July 2010 [CD-ROM]. Voorburg: IASE.

Francis, G. (2005). An Approach to Report writing in statistics courses. In B. Chance, & B. Philipps (Hrsg.), *Proceedings of the IASE/ISI Satellite Conference on Statistics Education*, 4-5 April 2005, Sydney, New South Wales, Australia.

Frischemeier, D., & Biehler, R. (2011a). Spielerisches Erlernen von Datenanalyse mit der Software TinkerPlots - Ergebnisse einer Pilotstudie. In R. Haug, & L. Holzapfel (Hrsg.), *Beiträge zum Mathematikunterricht 2011* (S. 275-278). Münster: WTM.

Frischemeier, D., & Biehler, R. (2011b, July). Learning Data Analysis with TinkerPlots - A University Course for Student Teachers. Paper presented at the Seventh International Research Forum on Statistical Reasoning, Thinking, and Literacy (SRTL-7). Texel Island, The Netherlands.

Frischemeier, D., & Biehler, R. (2012). Statistisch denken und forschen lernen mit der Software TinkerPlots. In M. Ludwig, & M. Kleine (Hrsg.), *Beiträge zum Mathematikunterricht 2012* (S. 257- 260). Münster: WTM.

Garfield, J. B., & Ben-Zvi, D. (2008). *Developing students' statistical reasoning: Connecting research and teaching practice*. New York: Springer.

Hammerman, J. K., & Rubin, A. (2004). Strategies for Managing Statistical Compexity with New Software Tools. *Statistics Education Research Journal 3*(2), 17-41.

Heaton, R. M., & Mickelson, W. T. (2002). The learning and teaching of statistical investigation in teaching and teacher education. *Journal of Mathematics Teacher Education, 5*(1), 35-59.

Konold, C., & Miller, C. (2011). *TinkerPlots 2.0*. Emeryville, CA: Key Curriculum Press.

Konold, C. (2007). Designing a Data Analysis Tool for Learners. In M. Lovett, & P. Shah (Hrsg.), *Thinking with data: The 33rd Annual Carnegie Symposium on Cognition*. (S. 267-292). Hillside, NJ: Lawrence Erlbaum Associates.

Monteiro, C., Asseker, A., Carvalho, L., & Campos, T. (2010). Student teachers developing their knowledge about data handling using TinkerPlots. In C. Reading (Hrsg.), *Proceedings of ICoTS 8*. Ljubljana, July 2010 [CD-ROM]. Voorburg: IASE.

Rossman, A. (2008). Reasoning about Informal Statistical Inference: A Statistician's View. *Statistics Education Research Journal 7*(2), 5-19.

Wagner, A. (2006). Entwicklung und Förderung von Datenkompetenz in den Klassen 1-6. *KaDiSto-Kasseler Online-Schriften zur Didaktik der Stochastik, Band 4*. Kassel: Universität Kassel. Abgerufen von nbn:de:hebis:34-2006092214690.

Watson, J. & Donne, J. (2009). TinkerPlots as a Research Tool to Explore Student Understanding. *Technology Innovations in Statistics Education, 3*(1), 1-35. Abgerufen von http://www.escholarship.org/uc/item/8dp5t34t.

Weidle, R., & Wagner, A. C. (1982). Die Methode des Lauten Denkens. In G. L. Huber, & H. Mandl (Hrsg.), *Verbale Daten. Eine Einführung in die Grundlagen und Methoden der Erhebung und Auswertung* (S. 81-103). Weinheim: Beltz.

Wild, C. J. & Pfannkuch, M. (1999). Statistical Thinking in Empirical Enquiry. *International Statistical Review 67*(3)*, 223-265.

Chapter 25
Wondering, Wandering or Unwavering?
Learners' Statistical Investigations with Fathom

Katie Makar

The University of Queensland

Jere Confrey

North Carolina State University

Abstract Dynamic statistical software is often promoted as a way to develop students' investigative capabilities, particularly in contexts of exploratory data analysis (EDA). However, there is some question about whether learners use these software packages as intended. This study explored the styles of 18 prospective mathematics and science teachers as they conducted a semi-structured statistical investigation in Fathom. Interviews and computer screen capture were collected as data and analyzed using an approach based on Land and Hannafin's (1996) intention-action-feedback cycles. The paper reports on patterns among observations, evaluations and conclusions drawn for three investigative approaches: Wondering, Wandering and Unwavering. Results suggest measurable differences in the patterns of investigation and discuss the benefits and drawback of each approach.

Biehler's Contributions to Exploratory Data Analysis

"Exploratory data analysis is an attitude, a flexibility, and a reliance on display, NOT a bundle of techniques, and should be so taught" (Tukey 1980, p. 23). Since John Tukey first introduced Exploratory Data Analysis (EDA) in the 1970s, there has been a dramatic shift in how statistics education has been conceptualized. Previously, teaching statistics focused solely on formal concepts underpinning statistical inference. While the practice of school statistics continues to be on formalities (Sorto 2004), EDA has slowly moved statistics education towards more active investigations of data.

One of the leading advocates of EDA has been Professor Rolf Biehler. For more than three decades, Biehler has argued that this shift requires more than just a different set of tasks, it requires a new way of thinking about statistics (Biehler 1993; 1994a). His work has focused on moving from the ideals of EDA towards building empirical-theoretical links that support its implementation. To make substantive changes in how learners experience mathematics and statistics, Biehler (1994b) argued that new ways of theorizing pedagogy is needed, including research into the critical supports for teacher education to develop prospective teachers as innovators of change. His work has argued theoretically for reorienting the field to bridge the gap between human activity and theoretical concepts (Bakker et al. 2005; Biehler 1994b) and putting theoretical ideas into practice (Biehler and Prömmel 2010; Maxara and Biehler 2007).

EDA activities on their own do not guarantee that students make the links between formal concepts and applications. Biehler (1994a) argues:

> Students are never confronted with a more unstructured situation where they have to do initial exploratory data analysis, and then decide about whether to construct a probability model or to critically question discoveries ... This may lead to the problem that EDA experiences are cognitively stored in a separate compartment, and strategies and concepts of data analysis are never reflected and adjusted according to the concepts and experiences in probability and statistical inference. This compartmentalization is hardly a desirable outcome of education. (p. 14)

Authentic data-based tasks "push students to participate in the definition of the problem space (e.g., picking the attributes to be measured), the selection of representational tools, and the process of coordinating empirical evidence and theory" (Corredor 2008, p. 34). With rapid development in software for learning statistics, the application of technologies has greatly enhanced the power of EDA to address complex problems (Biehler et al. 2013; Finzer et al. 2007). The approaches that teachers and students take to engage in technology-supported investigations based on authentic data are important to understand to move the field forward towards the kinds of changes advocated by Biehler and his colleagues (e.g., Biehler 1994b; Biehler and Leiss, 2010). This chapter explores the ways that learners use dynamic statistical software to conduct a semi-structured investigation based on an authentic, ill-structured task. The research question is, *"How do prospective secondary mathematics and science teachers use dynamic statistical software to conduct statistical investigations?"*

Investigations Using Dynamic Technology

Dynamic technology environments support visualization, inductive reasoning (connecting informal and formal mathematics), and engage students as creators of mathematical knowledge (Olive and Makar, 2010). In a dynamic environment, visual representations can be manipulated by students and through their actions and reflections on actions, they can receive immediate feedback (Zbiek et al. 2007).

Dynamic statistical software builds on an ethos of exploratory data analysis (EDA) to seek connections, explain variation and look at the data (Biehler 1994a; Biehler and Prömmel 2010). As these tools were evolving, Biehler (1997) developed criteria to provide designers and evaluators with critical areas that would be needed to build a culture of EDA. He argued that software should be designed so that students can: (1) practice graphing and analytic approaches that engender an exploratory working style; (2) create their own models in order to study and test them; (3) engage in statistical research that allows them to compare multiple statistical methods; and (4) use, modify or create "microworlds" (e.g., visualisations, simulations) to conduct interactive experiments. These criteria are still valid today as one can use them to assess newer software tools such as Fathom and Tinkerplots to understand their benefits for learners (Biehler et al. 2013).

One of the hallmarks of dynamic visualization software is its ability to engage learners in data investigations with easy-to-create drag-and-drop graphs, support "what-if" questions, enable visualization of statistical tendency through linked data representations, and use simulations to assess relationships in data. However, Hollebrands (2007) suggested that dynamic software, with its ease of creating and manipulating objects, may encourage reactive rather than proactive strategies. Do learners use the software as intended in practice? This study investigated the ways that learners used Fathom to investigate data.

Method

The data presented in this paper are from a larger study designed to examine pre-service teachers' development of statistical inquiry through exploration of standardized assessment data (Confrey et al. 2004; Makar 2004). The participants were 18 tertiary students enrolled in a secondary mathematics and science preservice teacher education course in the United States. The students were majoring in mathematics or science, but many had no coursework in statistics. Fathom was used in the course as a tool to highlight how data provided unique insights into equity issues in assessment. Statistical concepts were taught as needed in the investigations (for examples, see Confrey et al. 2004 or Makar 2004).

Data Collection

The prospective teachers were interviewed as they undertook an investigation in Fathom. Interviews were videotaped with their work on the computer digitally recorded. Before they began their investigations, they were told they would be investigating data of 273 randomly selected Hispanic students in grade 10 (age 16) from urban and rural schools across Texas (data source: Texas Education Agency). The data contained 14 variables including demographic data and raw and scaled test scores of students in grades 8 and 10 on the state mathematics and reading tests. The prospective teachers were asked to first state a conjecture about the relative performance of Hispanic students in urban and rural schools. They then used Fathom and the data set provided to investigate their conjecture until they felt they had enough evidence to state a conclusion. They could ask for technical assistance, if needed, but the interviewer did not guide their investigation. The prospective teachers had to create any tables, graphs, or summaries of the data. Most began with a representation showing the grade 10 Mathematics scaled test score (MTLI) of urban (upper) and rural (lower) students in Texas (Fig. 25.1), and marked the means (vertical lines) on the figure.

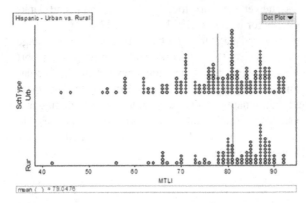

Fig. 25.1 Initial graph created by most participants

Data Analysis

Data analysis went through two phases: (1) content analysis and open-coding; and (2) focused analysis adapted from Land and Hannafin's (1996) intention-action-feedback cycles. The purpose of the first phase was to understand approaches that participants took in their investigations. Transcripts were annotated for content then subjected to open coding (Flick 2009) to generate initial aspects to consider, such as evidence chosen to support findings, focus of investigations, and links between evidence and conjecture. From this phase, the three investigative behavior types were identified: Wondering, Wandering, and Unwavering.

Using Land and Hannafin's (1996) work on Open-Ended Learning Environments (OELE), categories were identified to document patterns in the investigations for these three behavior types; those containing an observation, evaluation, or conclusion were collated for each participant (Table 25.1). Because of the small sample size, it did not make sense to test for significance in the differences found between each group, so only basic statistics (means, sd, percentages) are reported.

Table 25.1 Description of codes used in finer analysis of interview transcripts

Code	Description	Example
Observations	factual statements not related to the conjecture	*The sample size of the urbans is larger.* (April)
Evaluations	evaluative statements about results related to a conjecture or interim conjecture	*So there were more students not passing in the urban schools, by, uh, percentage, right?* (Carmen)
Conclusions	conclusions drawn, based on the results, to support or refute a conjecture	*It doesn't matter where the Hispanic students come from, an urban area or a rural area, they perform at about the same level.* (Gabriela)

Results

Overview

Table 25.2 General decision path of each investigation approach

Investigation approach	n	Theory	Evidence linked to theory and conclusion	Conclusion	Possible investigation driver
Wondering	6	Multiple	Clear	Multiple	"I wonder" questions
Wandering	8	Single	Not clear	Multiple	Search for interesting results
Unwavering	4	Single	Clear	Single or Multiple	Solution

The prospective teachers were able to state a measurable conjecture, seek appropriate evidence, and state a logical conclusion. This suggests that the teachers had sufficient facility with the software to navigate the conjecture-evidence-conclusion process in a semi-structured task.

The content analysis suggested that the prospective mathematics and science teachers approached their investigations with one of three distinct sets of approaches, which we describe as *wondering, wandering,* and *unwavering.* These categories were determined qualitatively based on the decision path each prospective teacher took from conjecture → evidence → conclusion (Table 25.2). An overall summary of each investigative approach is given in Table 25.3 in relation to the number and types of statements made and the length of their investigation.

Table 25.3 Statement categories and time taken by each investigative approach

Mean (sd) number of statements	Wondering (n=6)	Wandering (n=8)	Unwavering (n=4)	Overall
Observations	14.5 (12.0)	10.5 (5.4)	4.8 (2.2)	10.6 (8.3)
Evaluations	11.3 (7.3)	5.1 (3.6)	3.0 (2.2)	6.7 (5.8)
Conclusions	5.3 (2.9)	4.3 (1.8)	2.5 (0.6)	4.2 (2.2)
Mean (sd) total statements	31.2 (19.5)	19.9 (9.4)	10.3 (4.8)	21.5 (14.7)
Mean (sd) time (minutes)	26.2 (12.1)	12.6 (4.5)	5.7 (2.6)	15.6 (10.9)

Below, we first discuss each of these approaches and patterns that emerged, using metaphors of the place of theory and evidence in their investigation. In each case, we further discuss measures that highlight practices distinguishing each approach.

Wondering Approach

A wondering approach was one in which prospective teachers were guided during their investigation by "I wonder" questions that emerged as they tested their initial conjecture. Results encouraged the creation of hunches (speculative explanatory theories) which started the cycle again (Fig. 25.2). The use of the technology in this approach was as a tool for inquiry, one that would support the inquirer in the process of testing, evaluating, and generating new emerging theories.

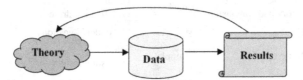

Fig. 25.2 Model of Wondering Approach

In seeking evidence to support, refine and extend their theories, those taking a wondering approach spent the longest time on average investigating their conjecture ($\bar{x}$ = 26.2 minutes, s = 12.1). This approach took more than twice as long as the other two approaches, however there was a greater balance towards evaluations rather than observations relative to the other two approaches. This agrees with a goal-oriented characterization of this approach.

Wandering Approach

A wandering approach was identified by a tendency to look through the data to see if anything "popped out" at them, rather than going into the data with particular evidence in mind. The wandering investigations included a conjecture, evidence, and a conclusion like those of the other approaches, but their time was often spent wandering through the variables looking for patterns to emerge.

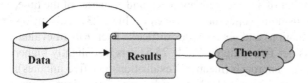

Fig. 25.3 Model of Wandering Approach

Unlike a wondering approach which was driven by internal theories, a wandering approach appeared to be driven by results as they emerged (Fig. 25.3). In this approach, the prospective teachers alternated between data and graphs until something useful appeared, leading them to develop a post-hoc explanation. The mean *percentage* of coded statements that were evaluations (Table 25.4) was lower (non-significant) for a wandering approach than that of prospective teachers taking a wondering approach.

Table 25.4 Percentage of coded statements that were observations and evaluations

Mean (sd) Percentage of coded statements	Wondering (n=6)	Wandering (n=8)	Unwavering (n=4)	Overall
Observations	46.5 (17.2)	52.4 (13.2)	46.8 (3.7)	47.2 (13.0)
Evaluations	34.5 (10.5)	24.9 (10.8)	24.4 (16.5)	28.0 (12.3)

Unwavering Approach

The prospective teachers demonstrating the unwavering approach used the software as a tool to directly locate particular evidence to test their conjecture and then were quick to draw a conclusion. This approach was identified by the decision pathway used: investigators looked for a particular piece of evidence to support or refute their original conjecture, and once they found it they were satisfied that they had answered the question put to them (Fig. 25.4). A shortcoming of this approach is that investigators appeared uninterested in understanding any underlying relationships that might help to explain the outcome.

Fig. 25.4 Model of Unwavering Approach

As might be expected, prospective teachers using the unwavering approach spent by far the least amount of time conducting their investigation, under six minutes on average — half that of a wandering approach and a quarter of the time, on average, spent by those using a wondering approach (Table 25.3). This makes comparing numbers of evaluations and observations difficult. Their *rate* of evaluations (mean 2.6 evaluations per five minutes, Table 25.5) was just slightly higher than those taking a wandering approach (mean 2.0 evaluations per five minutes); but their rate of observations (mean 4.4 observations per five minutes) was equivalent to those taking a wandering approach and much higher than those taking a wondering approach (mean 2.6 observations per five minutes). If one considers evaluative statements as more relevant to the investigation outcome than observations (which are not necessarily connected to their conclusion), then it appears that the short time between their conjecture and conclusion may have been less focused than might be assumed.

Table 25.5 The rate (number of statements per five minutes) of observations and evaluations made by each investigation approach

Mean (sd) number of statements per five minutes	Wondering (n=6)	Wandering (n=8)	Unwavering (n=4)	Overall
Observations	2.6 (1.2)	4.4 (2.4)	4.4 (2.0)	3.8 (2.1)
Evaluations	2.0 (0.8)	2.0 (1.2)	2.6 (1.8)	2.1 (1.2)

Discussion

The investigative approaches taken by the prospective teachers each had distinctive features. The wondering approach was characterized as theory-driven. Rather than driving findings, wandering strategies appeared less focused and devised their explanatory theories in *response* to findings. The unwavering, more direct approach aimed at efficiently testing and finding evidence for the initial conjecture. Each of these approaches had unique evidence, benefits and challenges.

The key feature of a wondering approach was the search for explanations which generated further theories which were tested, extended, revised and refined. This approach was likely one which the software designers were aiming for in creating the software, yet it was used by only about one-third of participants in the study. The benefits of this approach are likely in the ability to deepen understanding of the context under investigation and potentially develop rich content understandings and inquiry-based norms (Makar et al. 2011). Although it is not the intent that *all* technology-rich investigations should take a wondering approach, there is a need for learners to have *some* experience with this approach to build their ability to manage the unique reasoning it engenders and skills it develops to manage complex systems.

The wandering approach demonstrated evidence of being less goal-oriented. Some researchers have blamed the design of the software in creating drag-and-drop graphs with such ease as to encourage this approach (Hollebrands 2010). Others suggest, however, that a wandering approach has some benefits. Martin (2010), for example, argues that wandering releases creativity. Likely both perspectives have merit. Given that the wandering approach was the most common strategy taken among the participants, further research is needed to understand the benefits, drawbacks and characteristics of using dynamic technologies for wandering.

Finally, the unwavering approach distinguished itself by its focus and speed in responding to the question put to them. The patterns of behavior they exhibited create some question about whether their method might be considered efficient or rigid. That is, should this approach be encouraged or should those using such a direct approach be persuaded to take on more wondering behavior patterns? Yanchar et al. (2010) argued that mixing formal and more eclectic theories may allow for greater flexibility than formal theories alone, which can constrain creativity if applied rigidly. Further research into the unwavering approach would be valuable, particularly how those who use these approaches cope with problems that are ambiguous, ill-structured and must manage multiple contradictory issues.

This study identified three distinct approaches used in data-based investigations among competent users of Fathom. As a small exploratory study its aim was not to make broad generalizations about the patterns of behavior as to raise awareness that users take different pathways in software-supported semi-structured investigations, to highlight potential benefits and challenges related to each approach and to provoke questions for further research in this area. The designers of Fathom likely aimed to encourage learners to take a wondering approach (Finzer et al. 2007). In this study, however, that approach was found in only a minority of cases. The results of this study suggest a need to understand these investigative approaches in more depth and to make recommendations for teachers regarding the benefits and drawbacks of each approach. There is likely an assumption that wondering approaches are valued more highly and wandering approaches should be discouraged. Perhaps this assumption is made in haste, as some authors argue that wandering is a desirable and forgotten approach to learning (Martin 2010).

Post-script

This study would likely not have been conducted without the foundational work of Professor Rolf Biehler. Biehler has been an international campaigner of change in mathematics and statistics education for decades (e.g., Biehler at al. 1994). Biehler's work with Exploratory Data Analysis (EDA) in technological environments laid the groundwork for studies like this one and many others on every continent worldwide (e.g., Garfield and Ben-Zvi 2008; Healy 2006). Backed by his advocacy, extensive research program, and mentoring of countless researchers in statistics education, EDA has achieved international recognition as an approach which supports students to seek and gain meaning in data.

Biehler has long argued that students need to gain understanding of data by *doing* statistics and working closely with authentic data that has meaning for them. When I first met Biehler in 2001 at SRTL as a doctoral student, his insight into students' investigative approaches inspired me to continue pursuing this approach. For fellow EDA-advocates, his foundational work in developing criteria for software that would support this approach likely had a substantive impact (Biehler 1997). His recent evaluation of some of the most productive software packages for EDA using these same criteria (Biehler et al., 2013) highlight them as visionary and robust future-oriented standards to judge EDA software. His empirical work, including research with students and colleagues around the world, ensured that his advocacy went beyond lipservice (e.g., Bakker et al. 2005; Biehler and Prömmel 2010; Maxara and Biehler 2007).

Acknowledgments This research was originally funded by the National (U.S.) Science Foundation under their Collaborative for Excellence in Teacher Preparation program (DUE-9953187). This paper significantly extends outcomes previously presented at the ICMI-17 Study Conference in Hanoi, Vietnam (Makar and Confrey 2006); support for much of the extended writing was funded by The University of Queensland's Promoting Women Fellowship.

References

Bakker, A., Biehler, R., & Konold, C. (2005). Should young students learn about box plots? In G. Burrill, & M. Camden (Eds.), *Curricular Development in Statistics Education: International Association for Statistical Education (IASE) Roundtable, Lund, Sweden, 28 June-3 July 2004.* (pp. 163-173). Voorburg, The Netherlands: International Statistical Institute. http://www.stat.auckland.ac.nz/~iase/publications/rt04/4.2_Bakker_etal.pdf.

Biehler, R. (1993). Software tools and mathematics education: The case of statistics. In C. Keitel, & K. Ruthven (Eds.), *Learning from computers: Mathematics education and technology* (pp. 68-100). New York: Springer.

Biehler, R. (1994a). Probabilistic thinking, statistical reasoning, and the search for causes - Do we need a probabilistic revolution after we have taught data analysis? In J. Garfield (Eds.), *Research Papers from ICOTS 4, Marrakech 1994* (pp. 20-37). Minneapolis: University of Minnesota.

Biehler, R. (1994b). Teacher education and research on teaching. In R. Biehler, R. W. Scholz, R. Sträßer, & B. Winkelmann (Eds.), *Didactics of mathematics as a scientific discipline* (pp. 55-60). Dordrecht, The Netherlands: Kluwer Academic Publishers.

Biehler, R. (1997). Software for learning and for doing statistics. *International Statistical Review, 65*(2), 167-189.

Biehler, R., Ben-Zvi, D., Bakker, A., & Makar, K. (2013). Technological advances in developing statistical reasoning at the school level. In K. Clements, A. Bishop, C. Keitel, J. Kilpatrick, & J. Leung (Eds.), *Third international handbook of mathematics education* (pp. 643-689). New York: Springer.

Biehler, R., & Leiss, D. (2010). Empirical Research on Mathematical Modelling. *Journal für Mathematik-Didaktik, 31*(1), 5-8.

Biehler, R., & Prömmel, A. (2010).Developing students' computer-supported simulation and modelling competencies by means of carefully designed working environments. In C. Reading (Ed.), *Proceedings of ICoTS 8, Ljubljana, July 2010* [CD-ROM]. Voorburg: IASE. http://www.stat.auckland.ac.nz/~iase/publications/icots8/ICOTS8_8D3_BIEHLER.pdf.

Biehler, R., Scholz, R. W., Sträßer, R., & Winkelmann, B. (Eds.) (1994). *Didactics of mathematics as a scientific discipline*. Dordrecht, The Netherlands: Kluwer Academic Publishers.

Confrey, J., Makar, K., & Kazak, S. (2004). Undertaking data analysis of student outcomes as professional development for teachers. *ZDM International Reviews on Mathematics Education, 36*(1), 32-40.

Corredor, J. A. (2008). *Learning statistical inference through computer-supported simulation and data analysis* (Doctoral Dissertation). University of Pittsburgh.

Finzer, W., Erickson, T., Swenson, K., & Litwin, M. (2007). On getting more and better data into the classroom. *Technology Innovations in Statistics Education, 1*(1), Article 3.

Flick, U. (2009). *Introduction to qualitative research* (4th Ed.). Thousand Oaks: Sage Publications.

Garfield, J., & Ben-Zvi, D. (2008). *Developing students' statistical reasoning: Connecting research and teaching practice.* New York: Springer.

Healy, L. (2006, December). *A developing agenda for research into digital technologies and mathematics education: A view from Brazil.* Paper presented at the 17th Study for the International Congress on Mathematics Instruction. Hanoi, Vietnam.

Hollebrands, K. (2007). The role of a dynamic software program for geometry in the strategies high school mathematics students employ. *Journal for Research in Mathematics Education, 38*(2), 164-192.

Land, S. M., & Hannafin, M. J. (1996). A conceptual framework for the development of theories-in-action with open-ended learning environments. *Educational Technology Research and Development, 44*(3), 37-53.

Makar, K. (2004). *Developing statistical inquiry: Secondary preservice mathematics and science teachers' investigations of equity and fairness through analysis of accountability data* (Unpublished doctoral dissertation). College of Education, The University of Texas at Austin.

Makar, K., Bakker, A., & Ben-Zvi, D. (2011). The reasoning behind informal statistical inference. *Mathematical Thinking and Learning, 13*(1-2), 152-173.

Makar, K., & Confrey, J. (2006). *Dynamic statistical software: How are learners using it to conduct data-based investigations?* Paper presented at the Seventeenth Study of the International Commission on Mathematics Instruction. Hanoi University of Technology, Hanoi.

Martin, M. (2010). *Learning by wandering: An ancient Irish perspective for a digital world.* Bern, Switzerland: Peter Lang Publications.

Maxara, C., & Biehler, R. (2007). Constructing stochastic simulations with a computer tool – students' competencies and difficulties. In D. Pitta-Pantazi, & G. Philippou (Eds.), *Proceedings of the Fifth Congress of the European Society for Research in Mathematics Education* (pp. 762-771). Nicosia, Zypern: Department of Education, University of Cyprus. http://www.erme.unito.it/CERME5b/WG5.pdf#page=79.

Olive, J., & Makar, K. (2010). Mathematical knowledge and practices resulting from access to digital technologies. In C. Hoyles, & J.-B. Lagrange (Eds.), *Mathematics Education and Technology - Rethinking the Terrain* (pp. 133-177). New York: Springer.

Sorto, A. (2004). *Prospective middle school teachers' knowledge about data analysis and its application to teaching* (Doctoral dissertation). Department of Mathematics: Michigan State University.

Tukey, J. (1980). We need both exploratory and confirmatory. *The American Statistician, 34*(1), 23-25.

Yanchar, S. C., South, J. B., Williams, D. D., Allen, S., & Wilson, B. G. (2010). Struggling with theory? A qualitative investigation of conceptual tool use in instructional design. *Educational Technology Research and Development, 58*, 39–60.

Zbiek, R. M., Heid, M. K., Blume, G., & Dick, T. P. (2007). Research on technology in mathematics education: A perspective of constructs. In F. K. Lester Jr., (Ed.), *Second handbook of research on mathematics teaching and learning* (pp. 1169–1207). Charlotte, NC: NCTM and Information Age Publishing.

Chapter 26
Preparing teachers to teach conditional probability: a didactic situation based on the Monty Hall problem

Carmen Batanero, J. Miguel Contreras

University of Granada

Carmen Díaz

University of Huelva, Spain

Gustavo R. Cañadas

University of Granada

Abstract In this paper we reflect on the need for a better preparation of teachers to teach conditional probability, a topic in which a wide variety of misconceptions have been described. We also suggest a way to organize didactical situations that serves to improve the teachers' mathematical and pedagogical knowledge related to this topic. As an example, we analyze an activity based on the Monty Hall problem, for which there is a wide availability of Internet simulators and resources. We describe the mathematical objects underlying the correct solutions and common incorrect solution approaches. We conclude by analyzing the didactic suitability of this problem in the training of teachers.

1 Introduction

Conditional probability is included in the secondary education curricula (e.g., MEC 2007) because of its relevance in daily life and in the applications of statistics. This concept allows us to change our degree of belief about chance events as we acquire new information; however, extensive research literature suggests the existence of incorrect intuitions in its application (e.g., Falk 1986; Gras and Totohasina 1995; Díaz et al. 2010).

Moreover, several authors (e.g., Franklin and Mewborn 2006) suggest that few current programs train mathematics teachers adequately to teach statistics and probability and that misconceptions related to conditional probability may be shared by the prospective teachers who consequently may be unable to assess and correct these incorrect reasoning in their own students. As stated by Ball, Hill and Bass (2005, p.14) teachers' mathematical knowledge "is central to their capacity to use instructional materials wisely, to assess students' progress, and to make sound judgments about the presentation, emphasis, and sequencing of a topic" (Ball et al. 2005, p.14).

1.1 Components in Teachers' Knowledge

Many authors have analyzed the nature of knowledge needed by teachers to achieve effective teaching outcomes. For example, Ball and her colleagues (Ball et al. 2001; Hill et al. 2008) developed the notion of "mathematical knowledge for teaching" (MKT) in which they distinguished six main categories:

- *Common content knowledge (CCK),* or the mathematical knowledge shared by most educated adults, which includes basic skills and general knowledge of the subject.
- *Specialized content knowledge (SCK),* the particular way in which teachers master the subject matter that supports their activity in planning and handling classes and in assessing students' knowledge.
- *Knowledge in the mathematical horizon,* a view of the larger mathematical landscape, that teaching requires.
- *Knowledge of content and students (KCS),* including knowledge about common student conceptions and misconceptions, and their strategies to solve specific mathematics tasks.
- *Knowledge of content and teaching (KCT),* needed in the design of instruction, including how to choose examples and representations, and how to guide students towards accurate mathematical ideas.
- *Knowledge of the content and the curriculum:* how this content is included in the school curriculum.

Given the variety of knowledge needed by teachers and the scarce time available in their initial preparation, we need activities that serve to simultaneously increase several components in teachers' knowledge (Batanero et al. 2005). One possibility is using some paradoxical problems that would be posed to the prospective teachers, and explored with the help of simulation. Batanero et al. (2005) suggest that in helping teachers overcome some of their biases it may be helpful for teachers to first confront their wrong intuitions in probabilistic problems and then conduct experiments via simulation that provide counter evidence. After the teachers reach their solutions (which will likely include some wrong reasoning), they are then involved in a didactic analysis of the problem. This includes an analysis of the correct solutions as well as common incorrect approaches. This analysis may serve to help teachers acquire new knowledge of students' strategies and difficulties and of methods to help students overcome these difficulties. Below we reflect on some didactic features of simulation and then present an example of such a situation, based on the Monty Hall problem.

1.2 The role of simulation in teaching probability

In the perspective of "stochastics teaching", a sub-domain of mathematics comprising probability and statistics (Burrill and Biehler 2011), probability is not only the bases of statistics, but also is the mathematical branch that models nondeterministic relationships, random phenomena, and decisions under uncertainty. In this perspective, Heitele (1975) included simulation in his list of fundamental stochastic ideas because of its role, similar to that played by isomorphism in other mathematical areas. Through simulation we put two random experiments in correspondence in order to indirectly study one of them; in this way we overcome the difficulty of solving some probability problems by analytical or combinatorial methods. It is also possible to substitute formal proofs by more intuitive reasoning. Three additional positive points in simulation according to Biehler (1991) are: (1) the representations used may help students think through concrete models before being able to generalize to abstract probabilistic spaces; (2) most simulators include the data processing capability and facilitate the estimation of probability; and (3) the students must first construct a coherent understanding of the target situation as well a model of it before doing the computation. Biehler (1997) expands on this analysis and other roles of technology in the teaching of statistics, suggesting that simulation can serve to explore abstract probabilistic objects by creating virtual microworlds where students can experiment with the different variables involved. Having reflected on the didactic interest of simulation, we now analyze a didactic situation based on the Monty Hall problem and directed to increase the didactic knowledge of teachers.

2 A didactic situation based on the Monty Hall problem

As it is widely known, this problem arose from the TV game show "Let's Make a Deal". It was popular between 1963 and 1986 on American television and received its name from the game host, Monty Hall (Bohl et al. 1995). A controversy developed, which is still covered in both the popular media and academic circles regarding the decision faced by every contestant in a wide literature (e.g., Selvin 1975a; 1975b; Gillman 1992; Rosenhause 2009; Shaugnessy and Dick 1991; Eisenhauer 2000; Krauss and Wang 2003; Rosenhouse 2009; Borovcnik 2012). One possible formulation of the problem is reproduced below:

Suppose you are on a game show and are given the choice to select one of three doors. Behind one door there is a car and behind each of the other two doors there is a goat. Once you select a door, say No. 1 (which is closed), the host, who knows what is behind each door, opens another door (say No. 3), which contains a goat. You are now given the option of changing your selection to door No. 2 or sticking with door No. 1. What would you do?

In courses we imagine for teachers, the activity would start by asking the participants to solve the problem and decide whether the contestant should change the door or not, as well as to justify their responses with a probabilistic argument. The teacher's instructor would encourage them to figure out what kind of player is most likely to win the car, the one who changes the door initially selected or the one who sticks to the initial choice. If teachers fail to provide an analytical solution or give a wrong solution (which is common), they would be given the opportunity to simulate the game, perhaps using one of the applets available on the Internet, to get some experimental data that may help them find the correct solution. After some solutions have been provided, the class would conduct a didactic analysis, led by the instructor. Firstly, some possible correct solutions at different levels of formalization would be analyzed. Then the teachers would be asked to identify the mathematical objects used in each correct solution. Finally the incorrect reasoning behind the wrong solutions would be identified. Below we summarize these analyses.

2.1 Some solutions

When confronted with this problem, the majority of subjects (event those with statistical training) choose not to switch doors (Shaughnessy and Dick 1991). Apparently, the information about the open door with the goat can be used to eliminate this door and, since there were a priori equal probabilities for all the doors to contain the car, the remaining doors are viewed as equally likely. It is, however, possible to get to the correct analysis in an intuitive way (Solution 1). Before analyzing some solutions, we remind that the game host never picks a door randomly to show the contestant after they select the door. The host always shows a door with a goat behind it.

Since there are two doors with no prize and one door with a prize, the probability of choosing the door with the prize with no other information available is 1/3. If we do not change the initial solution, we just have 1/3 chance of winning and 2/3 of losing the prize. However, suppose after selecting a door and learning which of the two remaining doors does not contain the prize, we change our initial choice. Then the probability of winning is equal to the probability of having initially selected the door with no prize, which is 2/3. It is therefore advantageous to change the door.

In another solution (Solution 2) we use the compound experiment and a tree diagram (Figure 26.1) to help teachers visualize the situation. Let's first consider the experiment "door containing the prize" (each door has 1/3 probability). Next, a second experiment, consists of the door chosen by the contestant (again there is 1/3 probability of choosing each door). These two experiments are independent. The third experiment is the door opened by the host, which depends on the two previous experiments (Figure 26.1) because the host never opens the door with the car.

To compute the probability of winning when we do not change the door, we add the probabilities along all branches (choosing door 1 if the car is in behind door 1, choosing door 2, if the car is behind door 2 and to choosing door 3 if the car is behind that door). Each of these compound events has a probability of 1/9, so that the chances of winning are 1/3.

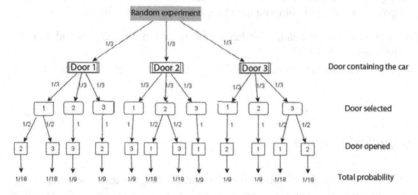

Fig. 26.1 Tree diagram showing the game structure

Suppose we change the door. If we choose a door with a goat, the host will open the door with the other goat. We change to the door with the car and win. For example, if the car is behind door 1, and we chose door 2, the host opens door 3 and we can only switch to door 1, which is the one with the car. This event has a probability of 1/9. The same reasoning would apply to the other doors.

If we choose the door with the car, the host shows us one of the two doors with the goats. For example, if the car is behind door 1 and we choose door 1, the host opens either door 2 or 3, each with probability 1/18, totaling 1/9. If we change the door, we lose with probability 1/9. Since there are three doors, the total probability of winning if we change is 2/3 and the probability of losing would be 1/3.

2.2 Formal solution

The above solutions may be approached more formally using, mathematical symbols and the properties of probability. For example we can use the following notation:

- C: The player selects the door containing the car.
- G: The player selects a door containing a goat.
- W: The player wins the car.

We compute P(W) for two different types of player, the one who changes the door and the one who maintains the initial selection.

Since $(C \cap G) = \emptyset$, $W = (W \cap C) \cup (W \cap G)$, and we can then apply the addition rule; since $\{C, G\}$ is a partition of Ω, the sample space of the experiment, therefore:

$$P(W) = P((W \cap C) \cup (W \cap G)) = P(W \cap C) + P(W \cap G)$$
$$= P(W|C)P(C) + P(W|G)P(G)$$

By applying the Laplace rule: $P(C) = 1/3$ and $P(G) = 2/3$ because there is a car and two goats. Finally, we compute the probability of winning for each player:

- If the player does not change his/her initial selection $P(W|G) = 0$ and $P(W|C) = 1$. Therefore, $P(W) = 1/3$.
- If the player changes the door, then $P(W|C) = 0$ and $P(W|G) = 1$. Therefore, $P(G) = 2/3$.

2.3 Empirical solution

By working with an applet or some other simulation tool, the data one collects can make the consequences of the two strategies (stick or switch) visible. If the teacher initially believed that it switching or sticking did not matter, the experimental results of a sufficiently large trial will provide contradictory evidence. This will, ideally, produce cognitive conflict that will motivate one of the intuitive analyses shown above.

Given that the results are random, we should repeat the game a considerable number of times (at least 100) so that the results are reasonably close to the expected percentage; this is easy since the computer allows a quick simulation of a large number of experiments. However, even though the applet or simulation tool provides supporting evidence in the form of data, to understand the results the teacher educator will still need to prompt the teachers to an analysis that is consistent with the results.

3 Mathematical objects involved in the problem

After having presented some possible intuitive, experimental and formal solutions for the Monty Hall problem, the teachers would be asked to analyze the mathematical components involved in each solution. We can discuss with the teachers the different categories of these components and how they are linked to various mathematical practices (Godino et al. 2007):

- *Problem-situation:* Applications, exercises, problems, actions that induce a mathematical activity. In the situation described in this paper, the problem is deciding the best strategy (in the sense of giving the highest probability of winning) in the Monty Hall game. *Language:* Since mathematical objects are immaterial, we use different representations of them; for example, in the solutions described we used tree diagrams, symbols, words. In the experimental solution we might also use iconic language or other graphs, depending on the simulator used.
- *Concepts:* When solving mathematical problems, we use mathematical concepts whose definitions need to be known and recalled. In the Monty Hall problems we use the ideas of randomness, sample space, event, simple, conditional, and, joint probability, independence.
- *Properties* of the concepts or relations between different concepts. In the solutions described for the Monty Hall problems, we had to remember that the sum of all the probabilities in the sample space adds one; in the experimental solution, students should grasp an intuitive idea of "convergence", by understanding the tendency of relative frequencies towards an underlying probability and that larger samples are more reliable than small samples.
- *Procedures:* operations, algorithms, rules. In some solutions we used Laplace's rule, the product and sum of probability rules, listing of events, construction of tree diagram or arithmetic operations.
- *Arguments:* Reasoning or proofs used to validate or explain the properties, procedures or solution to problems.

In Table 26.1 we summarize some of the mathematical objects involved in each of the solutions described in section 2. Depending on the solution provided, the solver may use more or less complex mathematical objects. Formal solutions use symbolic language are, which in general is complex to understand; an intuitive understanding of convergence and the difference between probability and relative frequency only appear in the experimental solution. The intuitive solution could still be simpler if we work only with natural frequencies instead of using probabilities and the more complex formal solution with the explicit use of the Bayes' theorem (Krauss and Wang 2003). Consequently, both the game and the type of solution reached determine the mathematical work in the classroom. This makes it possible to work at various levels of difficulty, depending on the type of student and their prior knowledge.

Table 26.1 Configurations of mathematical objects in five solutions to the Monty Hall problem

Type	Mathematical objects in the situation	Meaning in the situation	Int. Sol.1	Int. Sol.2	Emp. Sol.	Formal Sol
Pr.	Changing the door or not	Finding the best strategy	x	x	x	x
Language	Verbal	Explaining the situation	x	x	x	x
	Graphical	Tree diagram	x	x		
	Symbolic	Events and probabilities				x
	Numeric	Probabilities of each event	x	x		x
	Numeric	Relative frequencies			x	
	Iconic	Icons in the applet			x	
Concepts	Events; sample space	Door number; Wining/not	x	x	x	x
	Compound experiment	Composition of experiments	x	x	x	x
	Compound sample space	Cartesian product	x	x	x	x
	Relative frequency	N. of successes / n. of trials			x	
	Convergence	Frequency tends towards probability			x	
	Events union	Joining elements of two events A and B				
	Impossible event	Intersection of an event and its complementary				
	Classical probability	Proportion of favourable to possible cases	x	x		x
	Frequentist probability	Frequency limit			x	
	Conditional probability	Probability of one event conditioned on another event	x	x	x	x
Properties	Probability axioms	Formal rules				x
	Sum rule	Probability of winning the car	x	x		x
	Product rule	Joint probability; dependence	x	x		x
	Total probability theorem	Application to the situation				
	Relationship between simple and conditional probability	Restriction of sample space	x	x		x
	Frequency converges towards probability	Law of large numbers (empirical)			x	
	Intuitive calculus of probability	Applying intuitive rules	x	x		
Procedures	Formal calculus of probability	Apply formal rules				x
	Computing probability from frequencies	Estimating probability from frequency			x	
	Graphical representation	Building tree diagram	x	x		
Arg	Deductive reasoning	Proving the solution	x	x		x
	Empirical reasoning	Comparing strategies			x	

4 Students' possible difficulties

The complexity of the Monty Hall problem shown in the above analysis of math-ematical objects is also reflected in the extensive literature describing wrong solu-tions. For example, in Granberg and Brown's (1995) experimental study only 3% of participants correctly chose to change doors.

Analysis of incorrect reasoning should also be part of the teachers' knowledge of content (probability). Consequently, when working with the Monty Hall prob-lem in a course directed at teachers, the wrong solutions provided by the teachers themselves may serve to organize a discussion that help teachers increase their KCS. Below we describe some of usual incorrect reasoning (more examples can be found in Krauss and Wang 2003).

4.1 Assuming independence of experiments

An incorrect solution may be produced when the person does not perceive the de-pendence of successive actions (how the door opened by the host depend on the door selected by the contestant). Either the solver does not understand the struc-ture of the compound event or attributes an incorrect property (independence) to these events. To be specific, the solver believes that it makes no difference wheth-er you switch or stick reasoning (correctly) that the host showing them what is be-hind another door could not influence the probability (1/3) that the person selected the correct door in the first place. One way to explain this reasoning error is to say that the person fails to realize that it is possible to condition an event A by another event B that happens after A, and that this conditioning can change the initial probability of event A. Relevant new knowledge can, and should, affect the prob-abilities we assign to events. Otherwise, we would never subject ourselves to med-ical tests, which basically allow us to revise the probability that we have a certain condition.

This reasoning is explained by the "fallacy of the temporal axis" described by Falk (1986), where people mistakenly believe that a current piece of information (the door shown by the host) cannot affect the probability for an event that oc-curred before it (probability of the door containing the prize). This fallacy is partly caused by confusion between conditioning and causation.

From the point of view of probability, if an event A is the strict cause of an event B, whenever A happens, B will also happen, so that $P(B/A) = 1$. If an event A causes another event B, then B is dependent on A, but the opposite is not always true, since an event B can depend on another event A, and still neither of them is the strict cause of the other (Tversky and Kahneman 1982; Falk 1986; Díaz et al. 2010).

4.2 Incorrect enumeration of the sample space

Another source of potential errors in the Monty Hall problem is an incomplete enumeration of the sample space in one or more of the events involved. There is a failure in describing sample space which is specific in the conditioning situation, a failure which is, according to Gras and Totohasina (1995), frequent in solving conditional probability problems. We may intuitively believe that, after a door with no prize is opened, this door should no longer be taken into account in computing the probabilities. Accordingly, there are only two possible remaining doors (events), the door containing the prize and the one that does not. Therefore, the probability is 1/2 for each door and it does not matter whether you stick or switch.

The failure appears in not considering as relevant the door opened by the host (second step in the game); this door will depend on the contestant selection (first step in the game); therefore the sample space in the second stage is different, depending on the outcome of the first stage.

- If the player, in the first step, chooses the door containing the car (with a probability of 1/3) then the host in the second step can open either of the remaining two doors. Consequently, the sample space for the selection of those two doors by the host in the second step has two possibilities each with probability of being opened as 1/2.
- But if the player chooses a goat in the first step (with a probability of 2/3), in the second step the host only has the option of opening the remaining door containing a goat. In that case, the sample space in the second step has a single element.

4.3 Incorrect assignment of probabilities

Once the host opens a door, people are inclined to believe that this door is irrelevant, so the probabilities to win are equal ½ for the two remaining doors. This reasoning is incorrect as the three doors can split into two subsets (Granberg and Brown 1995): (1) Subset 1: the door selected by the contestant, and (2) Subset 2: the two doors not selected. People incorrectly apply the sum of the probabilities rule: although the opening of the door by the host does not affect the original choice (the probability that the prize is in subsets 1and 2 are still 1/3 and 2/3) it affects the other two non-selected doors. Once a door is opened and shown to have a goat, that door has a probability of 0 to have a car behind it. As the probability that the subset 2 contains the car is equal to 2/3 in the initial experiment and given that there is now only one door left in subset 2, the probability of that door having the prize becomes 2/3. That is, the probability of 1/3 associated with the opened door is entirely transferred to the door which was neither chosen nor opened by the host, because the host cannot open the door initially chosen. Added to the initial probability of 1/3, the probability that the unselected and unopened door has the car is now 2/3.

4.4 Incorrect interpretation of convergence

If the solver obtains, as a result of a series of simulations, a solution that is contrary to the correct solution of the problem (due to chance), these empirical results may reaffirm the solver's belief that switching or sticking makes no difference. He will be reassured in his "belief in the law of small numbers" (Tversky and Kahneman 1982), by which people believe that even small samples are representative of the underlying probabilities.

This possibility is greater when the number of experiments made with the applet is small, since the convergence of the relative frequencies to the underlying probability is achieved in the long term, but not in small series of trials. The teacher educator is responsible for encouraging the solver to increase the sample size and organize a discussion around the effect of sample size on the reliability of results.

Another difficulty is that not all students would agree that the best strategy is that would give the highest probability in the long run. Some may argue that you only play the game once, and that looking at what happens with multiple simulations is not relevant to a single decision. By this reasoning, they might still think that it is indifferent whether you switch or stick.

5 Final remarks

In this paper we have suggested a possible situation with a first stage where the teachers are asked to solve the Monty Hall problem. This problem, based on a paradox by Bertrand (1888), has a counterintuitive solution, and can be used either in teacher training or teaching of conditional probability to students. Its solution illustrates some basic fundamental ideas (Heitele 1975), including that of probability, sample space, addition and product rules, compound experiment, independence, sampling and convergence.

In courses directed to teachers, and once the problem is posed, the teachers' educator will provide some time to teachers until they reach a possible solution. Then, he will organize a debate of the correct and incorrect solutions, until the majority come to a consensus on the solution. Throughout the debate the teachers' educator would help the teachers analyze the causes of errors and would provide a summary of what was learned in the activity. In the case of resistance to the correct solution, the solutions can be compared with the empirical evidence produced by the applet or simulation tool so that teachers eventually realize their erroneous intuitions and revise them. Since the Monty Hall problem could also be posed to high school or university students, the teachers would also acquire new knowledge about instructional approaches to teach probability.

Similar activities may be based on other paradoxical problems (see, for example, the situation described in Batanero et al. 2004). These situations fulfill the criteria of didactic suitability proposed by Godino, Wilhelmi, and Bencomo (2005):

- Epistemic or mathematics suitability: This involves the amount and quality of mathematical content in the situation that we wish to teach. The Monty Hall problem is rich in this content, as shown in the analysis of the latent mathematical objects. Note that the epistemic suitability will depend on the type of solution you expect and not just on the problem. In general, formal solutions have greater suitability in a university and teacher training course; intuitive solutions may be sufficient at secondary school level.
- Cognitive suitability is the degree to which the situation is appropriate for students and whether it is likely to provoke learning.
- Interactional suitability: This is the degree to which the organization of learning allows for the identification and resolution of the participant's conflicts. This will depend on how the teachers' educator organizes the classroom work and the discussion of the solutions.
- Media suitability: This concerns the availability and adequacy of the resources needed for the teaching-learning process. Since not many resources are needed in the Monty Hall problem, it rates high on this dimension.
- Emotional suitability: This concerns the interest and motivation of students in the situation. In the Monty Hall problem, interest and motivation are high as indicated by their eagerness in playing the game, in arguing about the correct solution and in working to understand the solution.

Finally, a didactic analysis, similar to the one described in this paper, serves to increase the teachers' knowledge and awareness of probability, teaching probability, and the difficulties students are likely to have. The process could be improved if solutions given by actual students were available so that teachers could evaluate these solutions to detect the errors described.

References

Ball, D., Hill, H. C., & Bass, H. (2005). Knowing mathematics for teaching: who knows mathematics well enough to teach third grade, and how can we decide? *American Educator, 29*(1), 14-46.

Ball, D. L., Lubienski, S. T., & Mewborn, D. S. (2001). Research on teaching mathematics: The unsolved problem of teachers' mathematical knowledge. In V. Richardson (Ed.), *Handbook of research on teaching* (pp. 433-456). Washington, DC: American Educational Research Association.

Batanero C., Biehler R., Maxara C., Engel J., & Vogel M. (2005). Using simulation to bridge teachers' content and pedagogical knowledge in probability. In *15th ICMI Study Conference:*

The Professional education and development of teachers of mathematics. Retrieved from http://stwww.weizmann.ac.il/G-math/ICMI/strand1.html.

Batanero, C., Godino, J. D., & Roa, R. (2004). Training teachers to teach probability. *Journal of Statistics Education, 12*(1). Retrieved from http://www.amstat.org/publications/jse/.

Bertrand, J. (1888). *Calcul des probabilities.* Paris: Gauthier Villars.

Biehler, R. (1991). Computers in probability education. In R. Kapadia, & M. Brorvcnik (Eds.), *Chance encounters: probability in education* (pp. 169-211). Dordrecht: Kluwer Academic Publishers.

Biehler, R. (1997). Software for learning and for doing statistics. *International Statistical Review, 65*(2), 167-190.

Bohl, A. H., Liberatore, M. J., & Nydick, R. L. (1995). A tale of two goats and a car, or the importance of assumptions in problem solutions. *Journal of Recreational Mathematics, 27,* 1-9.

Borovcnik, M. (2012). Multiple perspectives on the concept of conditional probability. *Avances de Investigación en Educación Matemática, 2,* 5-27.

Burrill, G., & Biehler, R. (2011). Fundamental statistics ideas in the school curriculum and in training teachers. In C. Batanero, G. Burrill, & C. Reading (Eds.), *Teaching statistics in school mathematics-challenges for teaching and teacher education. A joint ICMI/IASE Study* (pp. 57-69). New York: Springer.

Díaz, C., Batanero, C., & Contreras, J, M. (2010). Teaching independence and conditional probability. *Boletín de Estadística e Investigación Operativa, 26*(2), 149-162.

Eisenhauer, J. (2000). The Monty Hall matrix. *Teaching Statistics, 22*(1), 17-20.

Falk, R. (1986). Conditional probabilities: insights and difficulties. In R. Davidson, & J. Swift (Eds.), *Proceedings of the Second International Conference on Teaching Statistics.* (pp. 292-297). Victoria, Canada: International Statistical Institute.

Franklin, C., & Mewborn, D. (2006). The statistical education of PreK-12 teachers: A shared responsibility. In G. Burrill (Ed.), *NCTM 2006 Yearbook: Thinking and reasoning with data and chance* (pp. 335-344). Reston, VA: NCTM.

Gillman, L. (1992). The car and the goats. *American Mathematical Monthly, 99,* 3-7.

Godino, J. D., Batanero, C., & Font, V. (2007). The onto-semiotic approach to research in mathematics education. *ZDM - The International Journal on Mathematics Education, 39*(1-2), 127-135.

Godino, J., Wilhelmi, M. R., & Bencomo, D. (2005). Suitability criteria of a mathematical instruction process. A teaching experience of the function notion. *Mediterranean Journal for Research in Mathematics Education, 4*(2), 1-26.

Granberg, D., & Brown, T. A. (1995). The Monty Hall dilemma. *Personality and Social Psychology Bulletin, 21,* 711–723.

Gras, R. & Totohasina, A. (1995). Chronologie et causalité, conceptions sources d'obstacles épistémologiques à la notion de probabilité conditionnelle. *Recherches en Didactique des Mathématiques, 15*(1), 49-95.

Heitele, D. (1975). An epistemological view on fundamental stochastic ideas. *Educational Studies in Mathematics, 6*(2), 187 - 205.

Hill, H., Ball, D. L., & Schilling, S. (2008). Unpacking pedagogical content knowledge: Conceptualizing and measuring teachers topic-specific knowledge of students. *Journal for Research in Mathematics Education, 39,* 372-400.

Krauss, S., & Wang, X. (2003). The psychology of the Monty Hall problem: Discovering psychological mechanisms for solving a tenacious brain teaser. *Journal of Experimental Psychology: General, 132*(1), 3-22.

MEC (2007). *Real Decreto 1513/2006, de 7 de diciembre, por el que se establecen las enseñanzas mínimas de la Educación Primaria.* España: Ministerio de Educación y Cultura.

Rosenhouse, J. (2009). *The Monty Hall problem: the remarkable story of math's most contentious brainteaser.* Oxford University Press.

Selvin, S. (1975a). A problem in probability. *American Statistician, 29*(1), 67.

Selvin, S. (1975b). On the Monty Hall problem. *American Statistician, 29*(3), 134.

Shaughnessy, J. M., & Dick, T. (1991). Monty's dilemma: Should you stick or switch? *Mathematics Teacher, 84,* 252-256.

Tversky, A. & Kahneman, D. (1982). Judgment under uncertainty: Heuristics and biases. In D. Kahneman, P. Slovic, & A. Tversky (Eds.*), Judgment under uncertainty: Heuristics and biases* (pp. 3-20). Cambridge: Cambridge University Press.

Chapter 27
Teaching Statistical Thinking in the Data Deluge

Robert Gould

Dept. of Statistics, UCLA

Mine Çetinkaya-Rundel

Dept. of Statistical Science, Duke University

Abstract Engaging with data is at the heart of statistical thinking, but readily-available data are now complex and large, and so statistical thinking is perhaps also more complex than in the recent past. We discuss an undergraduate event, DataFest, as a means for supplementing classroom approaches to teaching statistical thinking, and discuss lessons learned from observations during two DataFests held at two universities in the United States.

1 Introduction

Statistical thinking is, simply put, "thinking in the way a professional statistician thinks" (Garfield and Ben-Zvi 2007). More detailed definitions vary across academic tradition (see Chance 2002) and across time, in reaction to developments in statistical theory and technology (Pfannkuch and Wild 2004). Despite this variability, engaging with data remains at the core of most of these definitions. Beginning in the 1980s, educators and statisticians strove to place data at the center of the statistics curriculum (Hogg 1991; Marquardt 1987; Snee 1993; Singer and Willet 1990; Bradley 1982). The American Statistical Association's Undergraduate Statistics Education Initiative proposed that undergraduate degrees might prepare "data specialists" and that the discipline be more broadly framed as "data science" (Bryce et al. 2001). This idea goes back at least as far as Deming in 1940: "Above all, a statistician must be a scientist... [statisticians] must look carefully at the data, and take into account the conditions under which each observation arises." (Deming 1940).

Precisely what is meant by data, and precisely how students should "engage" has not been widely discussed, possibly because in recent years the nature and character of data have changed dramatically. We will argue that data have changed dramatically, and this change requires instructors to provide new learning experiences for students. We offer DataFest as one such experience. DataFest was created to provide students with the opportunity to engage in rich and complex data, and to develop statistical thinking skills that are, perhaps, not always fully developed in the classroom.

1.1 What Do We Mean by Data?

The data available for classroom instruction has changed as technology has changed. Singer and Willet (1990) advocated the use of "real data" in the classroom. Real data, they said, required several pedagogical features: authenticity, background information, interest and relevance, substantive properties, richness sufficient for multiple analyses, and "rawness". As Singer and Willet noted, use of real data in the classroom became feasible only because of the invention and widespread use of desktop computers. Toy datasets consisting of integers and easily rounded values had their place in helping ease students past computational tedium and towards a richer understanding, but the computer can now be put to this same purpose using real and complex data.

Before the invention and implementation of the World Wide Web in approximately 1991 (Berniers-Lee 2004), technology placed limits on the level of "reality" that classroom data could bear. For example, data had to be shared via floppy disk or emailed. Data were often owned by the researchers, which made finding real data sets much more difficult than today. This limited the size and scope of the data used in classrooms.

The datasets that appear in most current textbooks may seem quaint when compared to the massive data sets that challenge many practicing statisticians. This observation does not detract from the pedagogical value of textbook data, but while textbook datasets might be more real than they were in the 1980s, they can fail to prepare students for the challenges of working with massive data. Massive datasets are more complex and may even be dynamic in nature, in that they are continuously growing or changing. In some cases these datasets can serve as their own population. Thus, in addition to the challenge of accessing these massive datasets, educators should consider aspects of statistical thinking that address complex data structure and new issues of data collection.

1.2 What do we mean by "engage"?

One aspect of engagement is the ability to touch the data. For most textbook data, this involves downloading the data from a CD onto the computer, and then uploading into a statistical package. We believe this is a neglected skill worthy of class time, but it is a skill easily taught and easily mastered and so we do not consider it further here. On the other hand, access to massive data is a considerable challenge. One approach is to provide tools that deliver data to students. This is the approach taken by Fathom (Finzer et al. 2007) and StatCrunch (West 2009), both of which provide tools that allow students to upload HTML tables directly from the internet. Fathom provides random samples from the U.S. Census database, and StatCrunch allows students to import data about their Facebook friends and from Twitter feeds and to share data with other StatCrunch users. Both packages allow users to easily upload HTML tables directly from the source. Several R packages are available that assist in accessing data streams, for example RGoogleTrends (Temple Lang 2009) and RFlickr (http://www.omegahat.org/Rflickr/, Temple Lang 2008). Pruim et al. (2012) provide instructors with some potential activities and exercises using data with complex structures from within R. Ridgway et al. (2012) provide an overview of data on the "semantic web" as well as a discussion of data access issues.

Another approach is to teach students how to scrape their own data from the web, which requires more advanced data handling skills and computational knowledge. In 1999, the Undergraduate Statistics Education Initiative (USEI) of the American Statistical Association recommended a set of computer skills undergraduates should be taught: "familiarity with word processing, spreadsheets, web use, email [...] familiarity with a [...] statistical software package, familiarity with various aspects of statistical computing, and some experience handling and managing data." (Bryce et al. 1999). Since that time, the computational background required of statisticians has broadened considerably. Nolan and Temple Lang (2010) state that the "ability to reason about and express the computations [...] is involved in and supports all facets of a statistician's work." They propose topics for an undergraduate course for statistics students that include familiarity with R, visualization and graphics, fundamental data structures, the UNIX shell, relational databases and SQL, and HTML/XML programming.

Another aspect of engagement with data is analysis. Fast computation has allowed exploratory data analysis (EDA) to be taught at an early age, since students no longer need to concern themselves with, say, the details of histogram generation and can instead focus on what the histogram tells them about the data. Research into young learners' conceptions of statistics has produced software that makes EDA accessible to young students, for example, Tinkerplots (Konold and Miller 2005; Watson and Donne 2009), although appropriate caution must be taken (Bakker et al. 2004). With respect to statistical inference, implementations of randomization and resampling techniques reduce the level of mathematical skill needed to access and understand concepts such as sampling distributions, confidence intervals and hypothesis tests (Simon 1997; Cobb 2007). Kaplan argues that an algorithmic notation may be more appropriate for helping students communicate statistical concepts than mathematical notation (Kaplan 2007).

On the other hand, one can argue that the complex structures of massive data seem to require that students develop *more* mathematical skills. Very large data sets often reflect associations across time and space and are extremely hierarchical, thus requiring an understanding of statistical modeling in order to guide analysis. Brown and Kass (2009) define statistical thinking as primarily concerned with statistical modeling and the ability to evaluate the performance of models and inferential methods. In this respect, they offer a higher-level description than described by Chance (2002), and their definition places greater emphasis on the need for probability as a description of variation.

This tension between EDA and Probabilistic Modeling was identified by Biehler (1994), who characterized the "EDA Culture" as deterministic, believing that patterns in data reflect causes. The EDA culture considers data to be the reality, whereas models are, to quote Biehler, "tentative constructions". In contrast, the probability culture is non-deterministic, believing that patterns can occur solely through chance mechanisms. The model, from the probability culture's perspective, reflects a "deeper, although not completely known reality". Successful statistical thinking, Biehler argues, involves an interplay of both cultures. Statistical thinking should not be seen as concerned only with inference, Biehler argues, but also should be concerned with the need to think in terms of causes.

In Section 2, we outline some of the more pressing educational challenges caused by the Data Deluge. Section 3 describes the DataFest, an annual data competition which provides students with the opportunity to work with rich data sets that are beyond what they encounter in the classroom. To conclude, Section 4 discusses the types of statistical thinking we observed at DataFest and challenges educators to develop curricula that will prepare modern data analysts.

2 Statistical Thinking in the Data Deluge

Statistics education has long struggled to serve (at least) two audiences. On the one hand we have "consumers", who must learn to critically evaluate statistical reports and will not, one assumes, do statistical analyses themselves (Rumsey 2002; Utts 2003). On the other, we have "producers", who will one day become statisticians or work in professions in which they will analyze data. This dichotomy made sense in an age in which data were rare and private and access was difficult.

At some moment in time—we don't know when—a critical point was reached and, suddenly, data were *everywhere*. Certainly the internet played a role, as did the proliferation of smart phones and increasingly smaller, faster, and prettier computing devices. Some have referred to this as the Data Deluge, a term that may have first appeared in *The Economist* (February 25, 2010).

A key feature of the Data Deluge is accessibility, which is not always achieved but is always *desired*. The U.S. Government, for example, created the website data.gov as an earnest attempt to make public data available, even though not all data sets are easily accessed by amateurs. The British newspaper The Guardian started, in 2006, a "Free Our Data" campaign (Arthur and Cross 2006) and defines "open data" as data available in "a spreadsheet or a csv or some other machine-readable format which allows analysts to do something with it." (Rogers 2012). Accessibility refers not only to access to data, but also access to data analysis tools. Gapminder (http://www.gapminder.org) provides casual viewers with data sets as well as the means to easily interact with and visualize these data, and share the results. ManyEyes (http://www-958.ibm.com/software/data/cognos/manyeyes) provides tools for visualizing data in a variety of formats, and gives some advice for which visualizations are best. R (R Core Team 2012), while burdened by a formidable learning curve, is free to anyone, and environments such as RStudio (RStudio 2012) and R Commander (Fox 2005) make it friendlier. Other statistical packages make use of a GUI to make accessible complex statistical modeling without the need for training in the software. Fathom goes even further than drop-down menus by providing an intuitive interface that is nearly tactile and invites first-time users to explore data.

Not only is data accessibility growing, but data themselves are everywhere in our daily lives. Gould (2010) provides several examples that show how students interact, voluntarily or not, with data in their social lives. Besides providing a potentially rich source of classroom material, these data also challenge the statistics educator to demonstrate that statistics has value when applied to these every-day data. Brown and Kass (2009) speak to the dangers of losing good students due to a too-rigid curriculum and the inability to present statistics as a "deep" subject. We speak to the complementary problem: losing students by creating a disconnect between the data we analyze in class and the data that are analyzed by their music players and smartphones. Nolan and Temple Lang (2009) urges us to present statistics as "vibrant and relevant in the modern world as well as for the future." If we cannot teach a coherent system of statistical thought that encompasses both academic and "every day" data, how can we continue to remain relevant?

How, then, to broaden the curriculum to prepare students to engage with complex and rich data? While a complete restructuring of the curriculum may be called for, we offer an intermediate step, an event we call DataFest, that we have found provides students with a rich experience in computing with data.

Ostensibly for undergraduates, DataFest nonetheless serves to better prepare faculty to teach this statistical thinking by highlighting strengths and weaknesses of students (and therefore of the current curriculum) in ways not revealed by classroom assessment. Additionally, DataFest aids our educational efforts by forging ties with experts outside the academy who often have access to valuable and challenging data, and by creating a cooperative community within the department.

3 Celebrating Data

DataFest is an annual two-day competition in which teams of undergraduate students work to reveal insights into a rich and complex data set. The data set is a surprise for the students, who therefore have no means to prepare for the task. Three prizes are awarded by a panel of judges based on each teams' 7-minute presentation, where teams are allowed only two slides of content. Prize categories are relatively generic, such as Best Insight, Best Visualization, and Best Use of

External Data. The latter prize is meant to encourage students to go beyond the data provided and seek connections to the wider world. In this section, we begin with a general overview of the event, and then in subsections 3.1 through 3.3 we describe necessary features for a successful event and the benefits of this event for our students and faculty.

DataFest was first held at University of California, Los Angeles (UCLA) in 2011 (http://datafest.stat.ucla.edu), and in 2012 joint DataFests were held at UCLA and at Duke University (Durham, North Carolina) (http://stat.duke.edu/datafest). At Duke University 23 students participated in DataFest. Participation at UCLA grew from 30 students in 2011 to 70 in 2012, including 5 students from Pomona College, a private college about 50 miles east of UCLA campus.

Data for DataFest2011 were provided by the Los Angeles Police Department (LAPD), and included reports for every arrest in Los Angeles from 2005-2010. (There were almost 10 million reports.) These reports are filed by the arresting officer, and include details about the suspect and the nature of the alleged crime. The LAPD had geo-tagged (using a proprietary system) the reports wherever possible to indicate the location of the arrest, and had done some data cleaning. The data were presented to the students by Lieutenant Thomas Zak, Officer-in-Charge of the LAPD Strategic Crime Analysis Section, who challenged the teams to suggest policy changes that could improve public safety.

DataFest2012 data came from Kiva.org, a non-profit company that brokers micro-loans internationally. Visitors to the website are able to invest small amounts of money in loans awarded to entrepreneurs in developing countries, and can reinvest the money when the loan is repaid. Records are kept on every transaction, and Kiva.org makes these available via an API. The data were presented separately to students at Duke and UCLA by Kiva engineer Noah Balmer via Skype. The challenge was very broad; as Balmer explained, Kiva wished to know what outsiders would find interesting and useful, and so invited the teams to discover any insight or association they felt meaningful.

DataFest challenges students to engage with data in many ways that we have discussed above. While the students are provided the datasets, access to the data is not trivial. The datasets are quite large (millions of rows and hundreds of variables) requiring creativity when loading and working with the data. In addition, students are provided multiple datasets that they might need to match to each other. For example, at DataFest2012 separate data files containing information on lenders, loans, partner organizations were made available. Depending on the particular research question the teams decided to pursue, they needed to extract relevant information from each file and create their own customized dataset. Furthermore, teams were encouraged to integrate outside data sources and so needed to scrape their own data from the web and match these data to the existing records. (Encouragement took the form of a special prize for the team that made the best use of "external" data.) The data gathering and analysis processes in this context are often iterative, which is quite different than the workflow for data analysis that students do as part of their classwork.

As an example of the students' work, Figure 27.1 shows a slide from the winner of the best use of outside data award at Duke. The team investigated lending within the old boundaries of European empires, inspired by an article on how Facebook connections mirror old empires in *The Economist* (2012). They created choropleth maps similar to those in *The Economist* article, highlighting the amount of funds originating in empire countries such as Great Britain, Spain, etc. These maps unraveled lending patterns similar to those highlighted in the article between the global Facebook friendships and the old European empires. The judges were especially impressed by the use of this timely data from an article that appeared in the media just a month before DataFest 2012.

Another example of the students' work is given in Figure 27.2. This slide is from *The Data Wranglers,* who won the best visualization award at UCLA. The team examined patterns in funded loans over time and found spikes around "special events" such as media coverage and promotions. In addition, time series plots of length of time it takes to fully fund a loan revealed seasonal patterns such as quick funding of loans around the holidays. These findings were used to make recommendations to Kiva to increase their funding success.

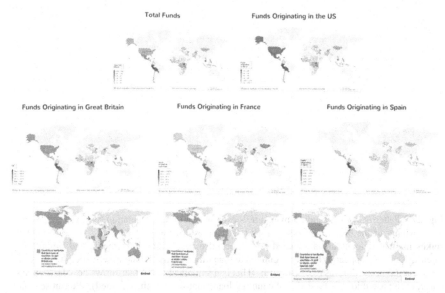

Fig. 27.1 Destination and amount of funds compared to the boundaries of the old European empires - excerpt from the presentation by the team The Statisto-nots, winner of the best use of outside data award at DataFest2012 at Duke University.

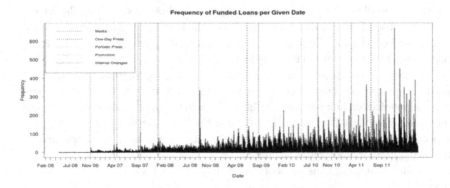

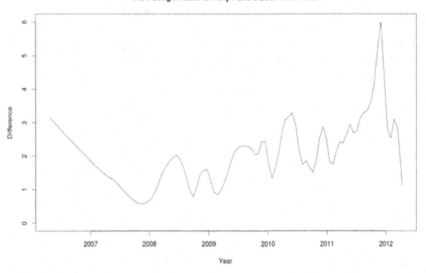

Fig. 27.2 Patterns in loan funding over time - excerpt from the presentation by The Data Wranglers, winner of the best visualization award at DataFest2012 at UCLA. **Top:** "We can see that spikes in funded loans are associated with 'special events', especially media and promotion. Their effects also appear to be more immediate than other forms. To increase funded loans, Kiva could capitalize on their usage." **Bottom:** "This is a smoothed curve that shows how fast, on average, each posted loan is funded (not counting loans that weren't fully funded). We can see that there are dips when special events occurred. There also seems to be a seasonal effect around the holidays, where loans are funded quicker."

These two challenges emphasize a feature that is common to data in this new age. Neither dataset was collected to serve a particular research agenda. As a result, one challenge was to ask questions of the data that, perhaps, the "owners" of the data did not know they could ask.

While tried-and-true visualization techniques were helpful and widely used, many teams found it necessary to extend these tools to aid in multivariate visualizations. Since very few of the students had formal experience with multivariate techniques beyond those provided in a multiple regression course, the visitors played an essential role reminding students of the primary purposes of data visualization (*show us what values and how many*) at times of struggle.

Multivariate analysis and visualizations rely strongly on computational skills. At Duke, UCLA, and Pomona College undergraduate statistics students are taught R, and fortunately, for both data sets, R was very useful, particularly for visualizations. However, students were allowed to use any software and resources they chose, and teams did make substantial progress using a variety of programming languages and software packages such as MATLAB, JMP, Stata, Python, and Many Eyes. Some teams learned, quite quickly, that a fruitful strategy was to utilize the strengths of different packages whenever possible. Another insight for some teams (particularly those with non-statistics students) was that Excel was quite useless in all but a few very basic situations.

3.1 Choosing Data

The choice of data is essential for a successful DataFest. We spent several months discussing possible data sets, and several weeks doing pre-cleaning and familiarizing ourselves with the data. In general, successful data for DataFest share the same qualities mentioned by Singer and Willet (1990), but in particular, have two key components:

1. *Familiarity.* The context of the data must be sufficiently familiar to students for them to generate their own hypotheses with a minimum amount of introduction to the context. Ideally, naive ideas should (sometimes) have a pay-off. The LAPD data certainly fit this criterion, but might not work as well for students outside of Los Angeles. The Kiva data was slightly less successful in this sense since few students were familiar with micro-loans. The real-life process that generated the data was complex, resulting in many-to-many relationships between lenders and borrowers, and hence understanding these relationships was tricky. At Duke, an expert on micro-loans was able to speak at the event, which helped guide some investigations, but UCLA was unable to obtain a substantive expert. The 2013 data set will be provided by the on-line dating service e-Harmony, and we expect students to have strong intuitive knowledge of the context.

2. *Urgency.* Students want an audience for their discoveries. Ideally, this requires close cooperation with an expert who actually wants to know what the students discover, as was the case with the LAPD. (Although we were also careful to temper the LAPD's expectations.) Another ingredient that contributes to this sense of urgency is to use timely data, which helps students understand that if they discover something, it could very well be an original finding.

3.2 Setting the Tone

A successful DataFest has several other somewhat intangible features.

1. *Inviting.* Data should be inviting in the sense that students feel that they can participate constructively regardless of their level or background. One approach to providing an inviting environment is to encourage teams to have a diverse composition in terms of the amount of statistical knowledge and computational skills. While the majority of students were statistics majors, engineering, social and natural sciences, and humanities students also participated in DataFest.
2. *Communal.* Success also depends on integrating DataFest within the local data analytic community. Faculty at host universities and other area colleges, "VIP" visitors from the local chapter of the American Statistical Association and the R Users Group, and faculty contacts visited during the event and interacted with the students, discussing approaches and answering questions. Graduate students took shifts throughout the event. Many volunteers and visitors (and even some faculty) were hesitant out of a concern that they might not be able to be helpful if the problem were too far out of their area of expertise. However this turned out to not be a problem. Most technical questions (*How can I animate a heat map in R?*) could be addressed by converting to a strategic question (*What do you expect to learn from this, and will it be worth the investment of time?*). Almost always, teams simply needed someone to listen to their ideas and question their strategy. In cases where the questions did require technical expertise, a resident expert could usually be found soon enough. If not, the students were quite comfortable knowing that it was their responsibility to solve any problems that arose.
3. *Fun.* Above all, DataFest should be fun. It is vitally important that the competition be friendly, particularly well into the second full day, when sleep deprivation causes wandering attention spans and short tempers. Food goes a long ways towards achieving this, and we made sure that students were supplied with a variety of snacks and with fun meals, and raffles of small prizes donated by local organizations.

3.3 Benefits

DataFest helped our faculty identify shortcomings in our curriculum. In particular, students were not as adept at statistical graphics as we assumed, and needed further training in spatial statistics. They also had difficulty implementing the R help system, in part because they did not have a good understanding of the structure of R.

Participating first and second year students were inspired to take more courses in statistics and computer science, and a few students at Duke and UCLA decided to pursue a statistics major as a result of their experience. Three senior students at

UCLA were offered jobs at a statistical consulting company after impressing one of the VIP visitors. Another three received internships at the Los Angeles Times data division, and later co-authored articles in the newspaper (Bensiger and Frank 2012). Many students later reported gaining a sense of purpose and accomplishment that helped them with later career decisions. At the conclusion of each DataFest all students are asked to fill out a questionnaire in which they are asked to list their three favorite and three least favorite things about the event. The quotes below are taken from these questionnaires and summarize well why we believe this is a valuable and rewarding experience:

> "DataFest2012 was an awesome experience. To me, the best part was working in a team of friends that I usually hung out with, but had not had a chance to work together intensively on a project. [...] The fact that we were given a huge amount of data really challenged us to come up with creative and practical approaches. Another important part was the presentation. Every team had to explain well to the judges their objectives and solutions." (Duke)
> "DataFest2011 was a glorious experience that I would recommend to anyone who wants to try a hand at actual analysis on real-world data. Unlike in a class project, you're given a very broad open-ended question and must get creative in how you use the data to answer it. You learn pragmatic and usable skills... I've actually been able to use parts of the code I wrote in jobs after graduating!" (UCLA)

4. Discussion

The tension between EDA and Probability cultures (Biehler 1994) seems heightened, at first glance, by the Data Deluge. The EDA culture is data-centered, and in the context of a data-engulfed world, seems in agreement with Nolan and Temple Lang (2010) that statistical thinking must include habits of mind for accessing, organizing, and summarizing rich data nestled within complex structures, as well as the ability to compute with data. Some in the popular culture have taken the extreme view that models, and even the scientific method, are obsolete when confronted with vast data that can serve as their own population (Anderson 2008). Brown and Kass (2010) speak for the probability culture, and argue for a greater emphasis on modeling with probability. Their definition of statistical thinking does not even mention data. In an age when vast numbers of hypothesis tests can be carried out in an instant on large numbers of data sets, probability modeling seems a necessary safeguard against Diaconis' "magical thinking" (Diaconis 1985) in which data analysts place too much meaning on chance-induced patterns. Konold and Kazak (2008) point out, quite reasonably, that probability cannot be cleanly separated from EDA, and that doing so perhaps exacerbates misconceptions in young learners. This does not mean, necessarily, that students need more mathematics to understand EDA, since tools such as Model Chance (Konold et al. 2007) and the visualization tools of Wild et al. (2011) can establish probabilistic thinking without the need to engage in mathematical probability.

Still, the problems described by these researchers are concerned mostly with reasoning about samples of data. In the Data Deluge, a common feature of data is that the data can serve as their own population. To our minds, this again calls into question the role that probability plays in an undergraduate statistics curriculum.

In order to help students resolve the conflict between the cultures, Biehler (1994) recommended that educators offer students the opportunity to work in an "unstructured" environment. Such environments, in which students must establish their own line of enquiry, determine whether to fit models and, if so, evaluate the fit, are a necessary part of a modern statistics education and extremely difficult to provide in the classroom. DataFest provides exactly that opportunity, and at the same time, provides an opportunity for educators to witness the types of statistical thinking our students employ and the gaps in their statistical thinking when confronted with modern data and modern analysis tools. From this vantage point, we can make several tentative observations about statistical thinking in this context.

- Data description is of utmost importance, but is difficult to carry out and difficult to learn in the context of these complex datasets. DataFest2011 participants wrestled with the ability to describe data that varied spatially and temporally, and DataFest2012 participants had the added complication of dealing with complex inter-relations between units of observation.
- When data are not produced to answer a small number of specific research questions, students must learn to be their own investigators. Successful teams reasoned about causes, suggesting hypotheses (*I bet there are more arrests in gang-related areas*) and modifying these hypotheses upon discovery (*There were fewer arrests than we thought, and so I wonder if we can find factors to explain this*).
- Data description requires careful thinking about the appropriate scale over which to smooth data (across time and space) and this thinking leads to questions about correlation structures. For example, students struggled with whether to describe crimes on the scale of days, or whether to aggregate across weeks or months. All such decisions imply some sort of a model, or at least have implications about how models will be structured. Hence, issues in statistical modeling are important to interpret and employ descriptions of data.
- Even large and rich data can benefit from being supplemented by other data sets. In DataFest2011, one team found data on gang-related violence, and employed maps of gang injunctions (roughly speaking, these maps describe gang territories) to help explain some of the variability in crimes in Los Angeles. In DataFest2012, many teams used data provided by the United Nations and other international development sites to understand the cultures and economies recorded in the Kiva data.
- Traditional inference (i.e., parameter estimation) is of less concern in these contexts, although issues involving model fitting and prediction are of greater concern. Students were quick to find confidence intervals for parameters in the L.A. crime data (because this is what classroom exercises had trained them to do), but the interpretation and usefulness of such intervals in a setting where

the entire population is known is not very clear. On the other hand, finding a model that "explains" crime and then convincing others of the accuracy of future predictions is of great interest.

As the name suggests, we initially conceived of DataFest as an activity belonging to the EDA culture. However, we observed students utilizing both cultures in their thinking. Students strove to describe and summarize data, but these summaries were motivated by discussions and considerations of causality, context, and the potential role of chance-induced variability. While statistical thinking in the Data Deluge should place great weight on the ability to reason and work with data, this ability should not be separated from the ability to reason with statistical models.

This is not much changed from Moore (1997), who, in reference to statistical thinking in the first course, advocated a "more balanced introduction to data analysis, data production, and inference". We claim this list needs only a little augmentation to apply to a general undergraduate statistics curriculum, suitable for students of the Data Deluge. In this balance, data analysis should get more weight, data production should be enhanced by learning to think about both causation (*What factors might explain variation?*) and data flow (*Where did these data originate, how were they pre-processed, and how do I get them into a useful format for my analysis?*), and statistical inference should be augmented with statistical modeling. In addition, statistical computation should be taught as it supports and enhances all of these activities.

The Data Deluge has provided educators with the possibility of truly placing data at the center of the curriculum. Our observations with DataFest show that statistics educators must think carefully about developing a curriculum that teaches students to employ computational thinking as a means for approaching both EDA and statistical modeling. While researchers have discussed the role of statistical thinking in the context of samples from a population, much work remains to be done when confronted with the challenge of teaching students to work with the complex and rich data that are now routinely accessible.

References

Anderson, C. (July 16th, 2008). *The End of Theory: The Data Deluge Makes the Scientific Method Obsolete*. Wired Magazine. Retrieved August, 30, 2012 from http://www.wired.com/science/discoveries/magazine/16-07/pb_theory.

Arthur, C., & Cross, M. (March 8th, 2006). Give us back our crown jewels. *The Guardian*. Retrieved August, 28, 2012 from http://www.guardian.co.uk/technology/free-our-data.

Bakker, A., Biehler, R., & Konold, C. (2005). Should young students learn about box plots? In G. Burrill, & M. Camden (Eds.), *Curricular Development in Statistics Education: International Association for Statistical Education (IASE) Roundtable, Lund, Sweden, 28 June-3 July 2004.* (pp. 163-173). Voorburg, The Netherlands: International Statistical Institute. http://www.stat.auckland.ac.nz/~iase/publications/rt04/4.2_Bakker_etal.pdf.

Bensiger, K., & Frank, E. (August 15th, 2012). Dealers' repeat sales of same used car surprisingly common. *Los Angeles Times.* Retrieved August, 26, 2012 from http://articles.latimes.com/2012/aug/15/business/la-fi-boomerang-cars-20120815.

Berniers-Lee, T. (2004). Pre-W3C Web and Internet Background, World Wide Web Consortium. Retrieved   August,   31,   2012   from   http://www.w3.org/2004/Talks/w3c10-HowItAllStarted/all.html.

Biehler, R. (1994). Probabilistic thinking, statistical reasoning, and the search for causes - Do we need a probabilistic revolution after we have taught data analysis? In J. Garfield (Eds.), *Research Papers from ICOTS 4, Marrakech 1994* (pp. 20-37). Minneapolis: University of Minnesota.

Bradley, R. (1982). The future of Statistics as a Discipline. *Journal of the American Statistical Association, 77*(377), 1-10.

Brown, E.N., & Kass, R.E.. (2009). What is Statistics? (with discussion). *American Statistician, 63*(2), 105-123.

Bryce, G.R., Gould, R., Notz, W., & Peck, R.L. (2001). Curriculum Guidelines for Bachelor of Science Degrees in Statistical Science. *The American Statistician, 55*(1), 7-13.

Chance, B. (2002). Components of Statistical Thinking and Implications for Instruction and Assessment. *Journal   of   Statistics   Education,   10*(3).   Retrieved   from http://www.amstat.org/publications/jse/v10n3/chance.html.

Cobb, G.W. (2007). The Introductory Statistics Course: A Ptolemaic Curriculum? *Technology Innovations in Statistics Education, 1*(1). Retrieved from http://escholarship.org/uc/item/6hb3k0nz.

Diaconis, P. (1985). Theories of Data Analysis: From Magical Thinking Through Classical Statistics. In D.C. Hoaglin, F. Mosteller, & J.W. Tukey (Eds.), *Understanding Robust and Exploratory Data Analysis*, (pp. 1-236). New York: Wiley.

Deming, E. (1940). Discussion of Professor Hotelling's paper. *The Annuals of Mathematics Statistics, 11*(4), 470-471.

The Economist (February 25th, 2010). Technology. *The data deluge: Businesses, governments and society are only starting to tap its vast potential.* Retrieved September, 11, 2012 from www.economist.com/node/15579717.

The Economist online (March 19th, 2012). *The sun never sets.* Retrieved September, 11, 2012 from www.economist.com/blogs/graphicdetail/2012/03/daily-chart-12.

Finzer, W., Erickson, T., Swenson, K., & Litwin, M. (2007). On Getting More and Better Data Into the Classroom. *Technology Innovations in Statistics Education, 1*(1). Retrieved from http://escholarship.org/uc/item/09w7699f.

Fox, J. (2005). The R Commander: A Basic Statistics Graphical User Interface to R. *Journal of Statistical Software, 14*(9), 1-42.

Garfield, J., & Ben-Zvi, D. (2007). How Students Learn Statistics Revisited: A Current Review of Research on Teaching and Learning Statistics. *International Statistical Review, 75*(3), 372-396.

Gould, R. (2010). Statistics and the Modern Student. *International Statistical Review, 78*(2), 297-315. Retrieved from doi:10.1111/j.1751-5823.2010.00117.x.

Hogg, R. (1991). Statistical Education: Improvements Are Badly Needed. *The American Statistician, 45*(4), 342-343.

Kaplan, D. (2007). Computing and Introductory Statistics. *Technology Innovations in Statistics Education 1*(1). Retrieved from http://escholarship.org/uc/item/3088k195.

Konold, C., Harradine, A., & Kazak, S. (2007). Understanding distributions by modeling them. *International Journal of Computers for Mathematical Learning, 12*(3), 217-230.

Konold, C., & Kazak, S. (2008). Reconnecting Data and Chance. *Technology Innovations in Statistics Education, 2*(1). Retrieved from http://escholarship.org/uc/item/38p7c94v.

Konold, C., & Miller, C.D. (2005). *TinkerPlots: Dynamic data exploration.* Emeryville, CA: Key Curriculum Press.

Marquardt, D. (1987). The Importance of Statisticians. *Journal of the American Statistical Association, 82*(397), 1-7.

Moore, D.S. (1997). New Pedagogy and New Content: The Case of Statistics. *International Statistical Review, 65*(2), 123-165. Retrieved from http://dx.doi.org/10.1111/j.1751-5823.1997.tb00390.x.

Nolan, D., & Temple Lang, D. (2010). Computing in the Statistics Curriculum. *The American Statistician, 64*(2), 97-107.

Nolan, D., & Temple Lang, D. (2009). Approaches to Broadening the Statistics Curricula. In M.C. Shelley, L.D. Yore, & B.B. Hand (Eds), *International Perspectives and Gold Standards* (pp. 357-381). Retrieved from http://dx.doi.org/10.1007/978-1-4020-8427-0_18.

Pfannkuch, M., & Wild, C. (2004). Towards an Understanding of Statistical Thinking. In D. Ben-Zvi, & J. Garfield (Eds), *The Challenge of Developing Statistical Literacy, Reasoning and Thinking* (pp. 17-46). Netherlands: Kluwer Academic Publishers.

Pruim, R., Horton, N., & Kaplan, D. (2012). *Teaching Statistics with R and R-Studio: An Instructors Guidebook* (unpublished, copyright Pruim, Horton & Kaplan).

R Core Team (2012). *R: A Language and Environment for Statistical Computing*. Vienna, Austria: R Foundation for Statistical Computing. Retrieved from http://www.R-project.org.

RStudio (2012). *RStudio: Integrated development environment for R* (Version 0.96.122). Boston, MA. Retrieved from www.rstudio.org.

Ridgway, J., Nicholson, J., & McCusker, S. (2012). The semantic web demands "new" statistics. In *Proceedings of the IASE 2012 Roundtable Conference, Cebu City, The Philippines, July 2-6 2012*. Retrieved from http://www.dur.ac.uk/resources/smart.centre/Publications/IASE2012RidgwayNicholsonMccusker_Semanticweb.pdf.

Rogers, S. (August 3rd, 2012). London 2012: is this the first open data Olympics?, Datablog, *The Guardian*. Retrieved August, 30, 2012. from www.guardian.co.uk/commentisfree/2012/aug/03/london-2012-olympics-open-data.

Rumsey, D. (2002). Statistical Literacy as a Goal for Introductory Statistics Courses. *Journal of Statistics Education 10*(3). Retrieved from www.amstat.org/publications/jse/v10n3/rumsey2.html.

Simon, J.L. (1997). *Resampling: The New Statistics* (2nd edition). Retrieved from http://www.resample.com/content/text/index.shtml.

Singer, J., & Willet, J. (1990). Improving the Teaching of Applied Statistics: Putting the Data Back into Data Analysis. *The American Statistician, 44*(3), 223-230.

Snee, R. (1993). What's Missing in Statistical Education? *The American Statistician, 47*(2), 149-154.

Temple Lang, D. (2008). *The Rflicker Package*. Retrieved from http://www.omegahat.org/Rflickr.

Temple Lang, D. (2009). *RGoogleTrends*, Retrieved from http://www.omegahat.org/RGoogleTrends.

Utts, J. (2003). What Educated Citizens Should Know About Statistics and Probability. *The American Statistician, 57*(2), 74-79.

Watson, J., & Donne, J. (2009). TinkerPlots as a Research Tool to Explore Student Understanding. *Technology Innovations in Statistics Education, 3*(1). Retrieved from http://escholarship.org/uc/item/8dp5t34t.

Wild, C.J., & Pfannkuch, M. (1999). Statistical Thinking in Empirical Enquiry. *International Statistical Review, 67*(3), 223-265.

West, W. (2009). Social Data Analysis with StatCrunch: Potential Benefits to Statistical Education. *Technology Innovations in Statistics Education, 3*(1). Retrieved from http://escholarship.org/uc/item/67j8j18s.

Wild, C. J., Pfannkuch, M., Regan, M., & Horton, N. J. (2011). Towards more accessible conceptions of statistical inference. *Journal of the Royal Statistical Society: Series A (Statistics in Society)*, 174(2), 247-295.

Chapter 28
Students' difficulties in practicing computer-supported statistical inference: Some hypothetical generalizations from a study

Maxine Pfannkuch, **Chris Wild**, **Matt Regan**

The University of Auckland, New Zealand

Abstract When introducing students to statistical inference using bootstrapping and randomization methods and new infrastructure such as dynamic visualizations, new conceptual development issues may be revealed. From a pilot study and a main study involving over 3000 students from the final year of high school and introductory university statistics, we use preliminary results to conjecture potential conceptual issues and obstacles. In imitation of an insightful paper of Biehler (1997), we identify seven problem areas and difficulties of students related to using bootstrapping and randomization inferential methods from our research. Although dynamic visualizations have the power to reveal chance variation and the depth of the conceptual structure underpinning the methods in ways that were not previously accessible, the identified areas indicate that attention to the necessity of precise verbal descriptions and the nature of the argumentation are important. In accord with Biehler we surmise that we may need to develop a habit of mind in students that is orientated towards a careful interpretation and understanding of graph visualizations.

Introduction

As one of the pioneers of statistics education research, Rolf Biehler has contributed extensively to the research knowledge base including intellectually challenging researchers to think more about the curriculum, technology, students' reasoning and conceptual progression (Bakker et al. 2005; Biehler 1995; Biehler 2011). Over the years we have often used his insights to think about our own research (e.g., Biehler 1989; 1994; 1995). Biehler (1997), in particular, has been critical in our thinking about the difficulties of explicating statistical concepts and thoughts. Using interviews of students while they were performing computer-based tasks, Rolf took the perspective of the expert and identified 25 problem areas related to elementary exploratory data analysis. As we have struggled to find ways to develop students' inferential reasoning, we have begun to appreciate and more fully understand the importance of his insights or what we call *rolfisms* (aphorisms) about learning statistics. Two rolfisms from Biehler (1997) have particularly resonated with us as we conducted research seeking to understand what student reasoning needed to be developed for statistical inference. The first rolfism is the need to attend to verbalizations in statistics learning:

> The description and interpretation of statistical graphs and other results [was] a difficult problem for interviewers and teachers. We must be more careful in developing language for this purpose … An adequate verbalization is difficult to achieve and the precise wording of it is often critical. There are profound problems to overcome in interpreting and verbally describing statistical graphs and tables that are related to the limited expressability of complex quantitative relations by means of common language (p. 176).

The second rolfism is:

> Interpretation of graphs and tables that are more than a mere reading off of coded information requires a rich conceptual repertoire (p. 189).

Until we started to explicate and develop for teaching the myriad of underpinning and interconnected concepts needed to interpret sample distribution displays, we did not appreciate the richness of the conceptual repertoire invoked when reasoning from plots.

In homage to the Biehler (1997) paper we will identify some problem areas and difficulties that we had in designing learning trajectories for bootstrapping and randomisation methods and that students in Year 13 (last year of secondary school in New Zealand) and introductory statistics at university had when introduced to them. We will imitate the basic structure of Biehler (1997) and draw attention to problem areas that are similar.

The context and methodology of the study

The research project, to which we refer, has not been completed and therefore the student data is based on a pilot study and some preliminary results from the main study. The project team has 33 members comprising statisticians, researchers, Year 13 teachers and university introductory course lecturers. Over 3000 students are involved. Data collected are pre and post-tests, student interviews on the pre and post-tests, interviews where students were given a computer-based task, videos of classroom implementation and our own reflections and observations on students' and our difficulties.

The idea of the project was to build on previous work that we had done on growing students' (aged 14 to 16) informal inferential reasoning through developing their ideas about sample, population, sampling variability, and how to make a call when comparing two groups (Arnold et al. 2011; Pfannkuch et al. 2010; Wild et al. 2011). We introduced Year 13 and introductory students to formal inferential reasoning via sample-to-population inference using a simple bootstrap method and causation inference for experiments using a randomisation method. Our choice of method is based on the principle that the method should mimic the data production process (cf. Hesterberg 2006). The introduction to these two quite distinct types of inference necessitated the development of hands-on activities, dynamic visualizations for learning and analysis, and the creation of new verbalizations. The dynamic visualizations come in the form of a graphics window in a specially created learning and analysis software package named *Visual Inference Tools* (VIT) (see: http://www.stat.auckland.ac.nz/~wild/VIT). We will now briefly describe the VIT package.

The VIT graphics window for the bootstrap method comprises three plots presented vertically: Sample, Re-sample and Bootstrap distribution (Fig. 28.1). The Sample plot displays the original sample, in this case the weights of a randomly selected sample of university students, as a dot plot and box plot. Note that the box plot is not displayed on its own as Biehler (1997, p. 178) reported that:

> The graphical conventions underlying the definition of the box plot are very different from conventions in other statistical displays. This can become an obstacle for students. Moreover, a conceptual interpretation of the box plot requires at least an intuitive conception of varying "density" of data. This is a concept that is not taught together with box plots.

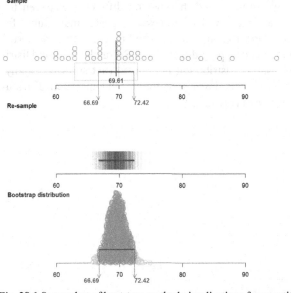

Fig. 28.1 Screenshot of bootstrap method visualizations for quantifying a confidence interval

A re-sample of the same size is taken with replacement from the original sample. The re-sampling is displayed dynamically. One of the observations is randomly selected and highlighted in the Sample plot and then a copy of this observation slowly drops into the Re-sample plot. This is repeated until the re-sample is of the required size, generating a temporary dot plot in the Re-sample plot. The re-sample mean is calculated and shown on the re-sample dot plot as a permanent vertical blue line. As each re-sample mean is displayed, a copy of it drops down onto the Bootstrap distribution plot. This process is repeated a large number of times building up a band of blue re-sample means in the Re-sample plot and a mounded dot plot of re-sample means in the Bootstrap distribution plot (Fig. 28.1). The speed and detail of this dynamic visualization can be controlled. Finally, a percentile bootstrap confidence interval is constructed using the central 95% of the bootstrap distribution and is displayed on the Bootstrap distribution plot. It then moves up and is finally displayed superimposed on all three plots as a means of connecting the representations and visually connecting the final inference made with the original data (Fig. 28.1).

The randomization method for causal inference in experiments on the VIT graphics window starts with comparing the results of two groups in the Data plot (Fig. 28.2) where, in this case, the participants have been randomly allocated to two treatment groups, target present (treatment) and target absent (control) for a standing jump, and the distance jumped in centimetres is recorded. To determine whether the observed difference includes an effect of "treatment" or could be produced by chance alone, students see what happens when chance is acting alone. Hence in the Re-randomised plot the original data drops down, temporarily losing any group membership identity until they are randomly re-assigned group labels (random re-allocation into two groups) and the ensuing difference between the group means is recorded with a red arrow. This process is repeated many times for students to see what can occur under chance alone. The process is then re-started with each difference in the means dropping down to the Re-randomisation Distribution plot until a re-randomisation distribution is built up (not shown). Finally, the original observed difference drops down onto the re-randomization distribution and the tail proportion is shown (Fig. 28.3).

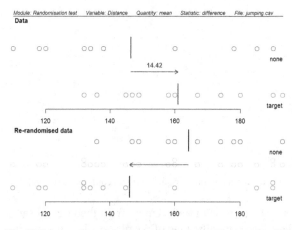

Fig. 28.2 Screenshot of randomisation test visualizations for differences in means

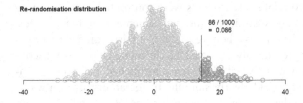

Fig. 28.3 Screenshot of re-randomisation distribution visualization for tail proportion

We are currently analysing the test responses and interviews of students from the perspective of eliciting issues that seem to be arising in their reasoning processes as a result of being exposed to the bootstrapping and randomization methods. Also we will reflect on conceptual issues that we found challenging when developing learning resources, software and new verbalizations. Hence, for this article, we will focus on intrinsic difficulties that seem to be arising with statistical inferential concepts and "where we can expect to encounter critical barriers to understanding" (Biehler 1997, p. 170). We will identify seven problem areas related to statistical inference.

Bootstrapping confidence intervals

In this section some inherent difficulties with understanding the nature of the variable being verbalized and bootstrap confidence interval argumentation will be explored. The analysis compares some students' responses to questions and compares this to how "statistical experts" would respond.

In the tests and computer-based task interview transcripts we noticed students tended not to verbalize terminology such as *sample means* and *population mean,* rather they used the word *it* or left the words unsaid. For example, a student stated: "for a large sample size you're going to get less variability." To demonstrate further she drew two bell-shaped curves with the one for the larger sample size narrower than the one for the smaller sample size and she did not label or talk about what the x-axis was measuring. This practice of not stating what one gets less variability in and not labelling the x-axis of the sampling distribution of the statistic was prevalent amongst the students. When a student constructed a bootstrap confidence interval using the software in a computer-based task interview he gave the following description:

> it is a fairly safe bet *it* is somewhere between here ... Rather than just saying *it* will be this many days. It's more certain that the value you will get from doing another *one* that *it* will fall in between the two rather than saying *it* will be that value.

All the students in the computer-based task interviews seem to understand that, usually, the population mean was within the confidence interval with comments such as "where we believe the mean value is of the entire population of guinea pigs." However, when the 10 pilot study students were confronted with answering true or false to the following confidence interval statement in the post-test (adapted from delMas et al. 2007): "We believe that it is a fairly safe bet that each cookie for this brand has approximately 18.6 to 21.3 chocolate chips", we discovered in the post-test interview that six out of the ten students had incorrect reasoning. Even though the students could clearly state the bootstrap distribution was a distribution of re-sample medians or means when looking at and talking about what they were seeing on screen, the idea of a distribution of a statistic seemed to slip from some of their minds when faced with a new situation.

We conjecture that the students' reluctance to verbalize sample mean and population mean and to label the x-axis for the bootstrap distribution is hindering their understanding and partially preventing them from transitioning from viewing a plot as a distribution of data to a distribution of a statistic. Another reason is that we did not pay sufficient attention to this key transition phase of a new concept of a familiar plot in our learning trajectory. We also noted that our resources have the measure in the title for the plots rather than as labels on the x-axes. We are still thinking about how to address this issue and how to get students to conjure up a mental image of the bootstrap process when dealing with a word-only context.

We can formulate the first problem area as:

1. The software tool supports students in reconstructing the meaning of the bootstrap distribution but students tend to forget the meaning when presented with a static display or in a context where software is not used (cf. Problem 8, Biehler 1997, p. 177).

A related problem area is the interpretation of confidence intervals. When a student was asked to interpret a bootstrap confidence interval in the post-test she wrote: "It is a fairly safe bet that the population of full time working New Zealanders with a bachelor's degree get paid between $693 and $937 weekly." In her interview she could state that the confidence interval contained an interval of plausible values for the *mean* weekly income. After some probing she could not see the conflict in her two statements; she did not seem to realise that omitting the word *mean* from her written statement was giving a different interpretation. A number of other students also did not appreciate the significance of omitting a single word.

2. There are profound problems to overcome in verbal description where precise wording is critical in an interpretation (cf. Problem 7, Biehler 1997, p. 176).

Our bootstrap software can be used to make the abstract concrete by providing "visual alternatives to classical procedures based on a cookbook of formulas" (Hesterberg 2006, p. 39). These visual alternatives have the potential to make the concepts and processes underpinning bootstrap inference transparent, more accessible, and connected to physical actions. But the seemingly easy accessibility to concepts may obscure the necessity of forming deeper conceptual understandings.

Despite teachers worrying about why the bootstrap method works in forming confidence intervals, students seemed to accept the method and did not query the underlying theory. Our difficulty was about how to convey the transformation of taking multiple samples from a population to form a sampling-variation band, which students had been exposed to in previous years, into the idea of taking one sample and forming an uncertainty band around the estimate through bootstrapping and hence deriving a confidence interval. That is, the expert wants to give the following inversion argument as a motivation:

> The sampling-variation band for a sample median captures almost all of the variation in sample medians observed when taking samples from a known population. It is rare for us to get a sample where the sample median is further from the true population median than revealed by the extent of this band. Rewording, it is very rare to get a sample where the population median is further from the sample median than shown by the extent of the sampling-variation band. Thus, if we had a good estimate of the width of the central 95% of the sampling-variation band, and placed it around a sample median, the interval so constructed would capture the true value for almost all samples taken (Pfannkuch et al. 2012, p. 906).

Our concern is about how and whether we should get students to appreciate the inversion argument. Some members of our research team do not want to introduce the inversion argument when students are initially introduced to the bootstrap idea, as they believe that it will result in cognitive overload and students will not appreciate the argument. They argue that the inversion argument should be introduced later in the course. Other members think it should be addressed in the initial introduction. We are still arguing.

3. Rather than analysing the end result, software tools can allow students to analyse directly the behaviour of a phenomenon and to conceive visually a statistical process that develops over time in ways that were not previously possible. An obstacle to overcome is that the phenomenon is perceived as a seemingly transparent and easy conceptualization, yet it can mask the depth of understanding required to understand the process at a deep conceptual level.

Randomisation test for experiments

The experiments we talk about are comparative experiments where there is an intervention and also random allocation to two groups is performed in an attempt to make the comparison fair. In this section some inherent difficulties with understanding experimental argumentation and the nature of causal inference will be explored.

In a pre and post-test question, students were asked to provide two main explanations for the observed difference in blood pressure reduction between people on a regular oil diet and a fish oil diet and were given the story and plots of the data. Our initial analysis compares *some* students' pre-test intuitive responses to this question and compares this to what an expert would have considered as the two main explanations before we move to some other considerations. Such an analysis will assist teachers to understand how students intuitively think and how to scaffold such thoughts towards expert thinking.

In response to the question asking for two main explanations for the observed difference, some students *read behind the data* (Shaughnessy 2007), seeking explanations for the reduction and gave a biological reason such as:

"Fish oil helps the regulation of cholesterol in the circulatory system"; "Fish oil cancels out the mercury content in the blood stream more effectively than regular oil"; "Fish oil is widely known as a good source of Omega 3 which lowers the cholesterol, which lowers blood pressure"; and "Fish is a good source of vitamins that help reduce blood pressure."

An expert would state that the data suggest that the fish oil treatment is effective in reducing blood pressure. Since the data do not measure cholesterol, mercury content or vitamin level, the expert knows such an inference is beyond the scope of the evidence presented. These students, intuitively, do not stop and think about whether the observed difference may be an indication that the fish oil is effective. Rather they skip to the next step and think *why* it is effective, that is, they are doing the thinking for the next study through speculation about potential biological mechanisms. Students need to put these ideas aside temporarily and think about the purpose of the experiment.

4. Students tend to use pre-internalised causal chains to make sense of the situation rather than using evidence from the data. A problem to overcome is for students to realize that statistical evidence is only one part that contributes to a larger problem area (cf. Problems 2 & 3, Biehler 1997, pp. 172-173).

Some students gave suggestions for the observed difference by thinking of other reasons that might explain the difference between the two groups such as the fish oil group were fitter, older, differed in life style or had extremely high blood pressure compared to the regular oil group. We were anticipating similar student explanations of "what if the treatment group happened to have ..." in the classroom and we realised we should use that opportunity to try to convince students that because of the design of the study we may label such explanations as *chance explanations*. An expert is aware that random allocation allows us to classify all explanations other than the difference-in-treatment explanation as chance explanations and that the randomisation test takes account of these chance imbalances.

5. Students need to perceive alternative explanations for an observed difference as chance explanations that are captured in the concept of chance is acting alone. When random allocation to two groups is misunderstood, inadequate concepts are formed.

In the pilot study several students, after the teaching intervention, indicated they could not comprehend what was meant by the words *chance is acting alone*. After much consideration of how to facilitate students' understanding of this concept, which is the crux of the randomisation test, we developed a dynamic visualization module that showed that randomly allocating data to groups will produce an apparent difference between the groups. In the class demonstration we used the module to randomly allocate the weights of some people to two groups with the difference between the group mean weights being recorded on the screen as randomised data and as a randomisation distribution similar to Figure 28.2. In this way students could see that the difference between the group mean weights could be up to 20 kg under chance alone. A student, who observed that the tail proportion was small in a class example giving evidence of the treatment being effective, queried the lecturer about whether the treatment is effective means that the observed difference in centres comprises chance and treatment effects. Such an insight is similar to the expert view of two possible explanations for the observed difference in the two means in blood pressure: (1) the variability can be entirely explained by chance factors alone (who happened to be randomly assigned to each group and measurement errors); and (2) the variability requires explanation by both chance factors and the treatment factor. (Pfannkuch et al. 2011).

6. An obstacle to interpreting an observed difference is to connect and integrate cognitively that the observed difference will always include chance effects and will also include treatment effects if they are present.

In the pilot study post-test students were asked what conclusion they should draw if the tail proportion was 0.3. Two responses were:

> Student A: "The result is higher than 10%, showing that it is likely that chance is acting alone" and Student B: "They cannot conclude anything, it means that there's no evidence against chance acting alone and that maybe some other factors could be acting along with chance."

The expert would give a response similar to Student B, whereas Student A exemplifies a common misconception that a large p-value is taken as evidence in favor of the chance-alone explanation (Nickerson 2004). In an interview another student had the idea that the tail proportion was a measure of how effective fish oil was, with a proportion in the region of 50% indicating that fish oil was definitely not useful, a proportion of around 27% indicating that fish oil may be useful, and that a tail of 10% indicating that fish oil may or may not be useful. Such a misconception has been consistently documented for hypothesis testing. Their responses, however, were not surprising since the tail proportion idea has not changed through our visualizations, only an appreciation of how the tail area is obtained has changed, that is, it is not a numerical value rather a part of an understandable distribution. Interpretation of a large tail proportion and the indirect nature of the logic of the argument seem to remain a problem with this method as it is with normal-based inference (Nickerson 2004). We think that the argumentation will continue to remain difficult as it appears to be an alien way of reasoning, particularly when overlaid with chance alone, the re-randomization distribution of differences in means, and tail proportion ideas.

7. Students need to overcome the obstacle that the argumentation is not about whether the treatment is effective or not, it is about whether a treatment effect is detectable under the obscuring effects of chance variation.

Conclusion

Introducing students to the bootstrapping and randomisation methods has presented many problem areas with respect to verbalization of structure and concepts in graph displays and word-only contexts. In accord with Biehler, (1997, p. 188) we agree that "the habit of careful and thorough reading and interpreting ... is difficult to develop" and it is also "difficult to write a report, that is, to produce written or oral descriptions and interpretations." We attempted to influence students' conceptual understanding and cognitive structure by exposing them to our especially designed dynamic visualizations. Surprisingly, we observed that our own perceived expert thinking grew on exposure to the visualizations and students' reasoning and we gained an even greater understanding of the "rich conceptual structure" (Biehler, 1997, p. 188) underpinning statistical inference. Thus, even experts experiencing thinking with new infrastructure can deepen their conceptual knowledge and through such experiences appreciate the powerful and rich ideas inherent in statistical thinking. Only further research will clarify and identify how to stipulate "adequate didactical provisions for overcoming these difficulties" (p. 189), deepen students' conceptualization of inference, and resolve our seven problem areas. We need to develop a habit of mind in students, which a *rolfism* describes as developing "a conceptual orientation for interpreting and using graphs and tables [visualizations]" (p. 189).

Acknowledgments   The work on this paper is supported in part by a grant from the Teaching and Learning Research Initiative (http://www.tlri.org.nz).

References

Arnold, P., Pfannkuch, M., Wild, C.J., Regan, M., & Budgett, S. (2011). Enhancing students' inferential reasoning: From hands on to "movies." *Journal of Statistics Education, 19*(2). Retrieved from http://www.amstat.org/publications/jse/v19n2/pfannkuch.pdf.

Bakker, A., Biehler, R., & Konold, C. (2005). Should young students learn about boxplots? In G. Burrill, & M. Camden (Eds.), *Curricular development in statistics education: International Association for Statistical Education (IASE) Roundtable, Lund, Sweden 28 June-3 July 2004*, (pp. 163-173). Voorburg, The Netherlands: International Statistical Institute. Retrieved from http://iase-web.org/documents/papers/rt2004/4.2_Bakker_etal.pdf.

Biehler, R. (1989). Educational perspectives on exploratory data analysis. In R. Morris (Ed.), *Studies in mathematics education: The teaching of statistics*, (Vol. 7, pp. 185-201). Paris: UNESCO.

Biehler, R. (1994). Cognitive technologies for statistics education. In L. Brunelli, & G. Cicchitelli (Eds.), *Proceedings of the first scientific meeting of the International Association for Statistical Education 1993* (pp. 173-190). Perugia: Universita di Perugia. Retrieved from http://iase-web.org/documents/papers/proc1993/173-190rec.pdf.

Biehler, R. (1995). Probabilistic thinking, statistical reasoning and the search for causes: Do we need a probabilistic revolution after we have taught data analysis? In J. Garfield (Ed.), *Research papers from the fourth International Conference On Teaching Statistics, Marrakech 1994* (pp. 20-37). Minneapolis, MN: University of Minnesota.

Biehler, R. (1997). Students' difficulties in practicing computer-supported data analysis: Some hypothetical generalizations from results of two exploratory studies. In J. Garfield, & G. Burrill (Eds.), *Research on the role of technology in teaching and learning statistics, International Association for Statistical Education Round Table Conference, Granada 1996* (pp. 169-190). Voorburg, The Netherlands: International Statistical Institute. Retrieved from http://iase-web.org/documents/papers/rt1996/14.Biehler.pdf.

Biehler, R. (2011). *Five questions on curricular issues concerning the stepwise development of reasoning from samples.* Discussant session presented at the seventh International Forum on Statistical Reasoning, Thinking and Literacy, 17-23 July 2011, Texel, The Netherlands.

delMas, R., Garfield, J., Ooms, A., & Chance, B. (2007). Assessing students' conceptual understanding after a first course in statistics. *Statistics Education Research Journal, 6* (2), 28-58. Retrieved from http://iase-web.org/documents/SERJ/SERJ6%282%29_delMas.pdf.

Hesterberg, T. (2006). Bootstrapping students' understanding of statistical concepts. In G. Burrill (Ed.), *Thinking and reasoning with data and chance. Sixty-eighth National Council of Teachers of Mathematics Yearbook* (pp. 391-416). Reston, VA: NCTM.

Nickerson, R. (2004). *Cognition and Chance.* Mahwah, NJ: Lawrence Erlbaum Associates.

Pfannkuch, M., Wild, C.J., & Parsonage, R. (2012). A conceptual path to confidence intervals. *ZDM – The International Journal on Mathematics Education 44*(7), 899-911. doi: 10.1007/s11858-012-0446-6.

Pfannkuch, M., Regan, M., Wild, C. J., & Horton, N. (2010). Telling data stories: essential dialogues for comparative reasoning. *Journal of Statistics Education, 18*(1). Retrieved from http://www.amstat.org/publications/jse/v18n1/pfannkuch.pdf.

Pfannkuch, M., Regan, M., Wild, C.J., Budgett, S., Forbes, S., Harraway, J., Parsonage, R. (2011). Inference and the introductory statistics course. *International Journal of Mathematical Education in Science and Technology, 42*(7), 903-913.

Shaughnessy, M. (2007). Research on statistics learning and reasoning. In F. Lester (Ed.), *Second handbook of research on the teaching and learning of mathematics,* (Vol. 2, pp. 957-1009). Charlotte, NC: Information Age Publishers.

Wild, C. J., Pfannkuch, M., Regan, M., & Horton, N. (2011). Towards more accessible conceptions of statistical inference. *Journal of the Royal Statistical Society: Series A (Statistics in Society), 174*(2), 247–295.

Chapter 29
Using TinkerPlots™ to develop tertiary students' statistical thinking in a modeling-based introductory statistics class

Robert delMas, Joan Garfield, Andrew Zieffler

University of Minnesota, USA, Department of Educational Psychology

Abstract This chapter describes the development of students' thinking as they experienced an innovative introductory statistics curriculum that replaced traditional content and methods with an approach based on simulation and resampling. The methods employed in the curriculum were based on a framework for inference that had students specify a chance model, draw repeated samples of simulated data, create a distribution of summary measures, and use the distribution to evaluate a claim. Students used TinkerPlots™ software to resample simulated data from chance processes and models, as well as to explore the distribution of summary measures. The software incorporates many features of a "Monte Carlo Workbench" (see Biehler, 1997a) that allows students to visualize the entire modeling process. Problem solving interviews were conducted with five students after five weeks of the curriculum. These interviews revealed that students were beginning to develop an understanding of important concepts underlying the process of statistical inference. The results suggest that students are able to create and use appropriate chance models and simulations to draw statistical inferences after only a few weeks of instruction in an introductory course. The interviews also suggest that TinkerPlots™ provides students with a memorable, visual medium to support the development of their thinking and reasoning.

1 Introduction

Technology has become an important component of learning statistics. Instructors can choose among a variety of software programs, Web applets, and graphing calculators (see Chance et al. 2007). The TinkerPlots™ software (Konold 2011) provides a unique approach to visualizing statistical concepts as well as modeling concepts. Tertiary students used TinkerPlots™ to model and simulate data throughout an innovative new curriculum that replaced traditional content and methods with a simulation and resampling approach (see Garfield et al. 2012). This chapter describes how TinkerPlots™ was used to develop students' thinking and presents results from the analysis of qualitative data from problem-solving interviews where students used TinkerPlots™ to make statistical inferences.

Modeling in the Introductory Statistics Curriculum

A fundamental aspect of statistical practice involves the use of models, for comparison with empirical data or to simulate data to make an estimate or test a hypothesis (Garfield and Ben-Zvi 2008). Part of developing statistical thinking is to develop ideas of statistical modeling (see Wild and Pfannkuch 1999) and realizing why Box's statement that all models are wrong, but some are useful (Box and Draper 1987) is so wise. Models are of particular importance when considering statistical inference. Inferences are made by comparing observed results to results produced by a model, typically producing a p-value.

Modeling approaches have been advocated in mathematics education to shift attention from finding a solution to a problem to creating a model that can be generalized to other problems (e.g., Doerr and English 2003). Konold, Harradine, and Kozak (2007) describe an instructional approach that helps students develop important statistical ideas by creating models to simulate observed data. Both types of modeling have become part of introductory college courses.

Technology Tools for Learning Ideas of Probability and Statistics

Rolf Biehler has long contributed to the research and theory on using technology tools to develop students' statistical thinking and reasoning. He has offered many insights and suggestions regarding the use of software tools, outlining several characteristics that statistical software would need to support future directions of statistics instruction (Biehler 1997a). One of those characteristics is to support a "Monte Carlo workbench" (Thisted 1986) to promote guided discovery approaches to understanding statistical methods and concepts, as well as modern randomization and bootstrapping methods. Biehler envisioned software that would allow instructors to create customized microworlds, primarily through a graphical user interface and with little need to use a programming language.

Biehler (1997a) suggested that a flexible microworld can be produced by software that offers a multiple window environment with multiple linked objects (e.g., plots, summaries, spreadsheets, annotations) that can be arranged by the user. Software supports exploration and experimentation when values and points in all objects are updated automatically upon changes made in one object, the results of a process become new data that can be analyzed or processed, and points selected in one plot are highlighted in all other plots and data tables.

Biehler (1997b) also identified characteristics of statistical software that have negative influences on students statistical thinking and reasoning. One of the major problems cited by Biehler is when students can produce plots and tables with little reflection. Biehler observed that when working on an elementary statistical problem, many students produced a large variety, if not all, of the artifacts that the software allowed. This is in contrast to a statistical expert who would first conceptualize the problem and then make reflective choices of what to produce. Similarly, Biehler (1997b) also observed that students sometimes wanted to create tables and plots that the software could not generate. Therefore, on one hand the software did not promote reflection, and on the other hand lacked the flexibility needed by statistical novices to build their experience and understanding.

Later research by Biehler and Prömmel (2010) examined the nature of student learning using innovative software. They found evidence that providing secondary students with a simulation scheme and prompting the students to first conduct a planning phase before building a simulation in Fathom® improved the quality of the students' reasoning process.

Innovative Technology Teaching at the Tertiary-Level

Today there are many technological tools that support and enhance student learning in the introductory tertiary-level statistics course. Two of the more innovative software tools are Fathom® (Key Curriculum Press 2012) and TinkerPlots™ (Konold 2011). Both offer flexible ways to explore and visualize data, provide the capability to model and simulate data, and have many of the features outlined by Biehler (1997a) that support student exploration and conceptual development.

Applications and uses of Fathom® and TinkerPlots™ to support students learning have been studied. For example, Watson and colleagues (Watson 2008; Watson and Wright 2008) have shown that the use of TinkerPlots™ was instrumental in helping students begin to build ideas of statistical inference. Other researchers have used TinkerPlots™ as a tool for developing students' reasoning about distributions (Konold et al. 2007; Konold and Lehrer 2008) and variability (Watson et al. 2008).

Despite the capabilities of these tools, research has not always shown a positive impact on student outcomes. Maxara and Biehler (2010), for example, found that many students' understanding of the Law of Large Numbers was fragile even after a simulation intensive course that used Fathom®. Research has also identified several challenges for students in learning to use Fathom®, especially when embedded in a modeling and simulation framework (Biehler 2003; Maxara and Biehler 2006; 2007; Biehler and Prömmel 2010).

The CATALST Curriculum

The CATALST curriculum is one example of a course that uses simulation and randomization-based methods to develop students' understanding of statistical inference. This student-centered course was designed around four foundational components: a focus on modeling, the use of simulation, an emphasis on statistical thinking, and attention toward the core logic of inference (Cobb 2007, p. 13).

When applied to randomized experiments and random samples, Cobb refers to the core logic of inference as the "three Rs": *randomize, repeat,* and *reject.* The CATALST project generalized this logic for a broader simulation-based approach to inference as follows:

- *Model.* Specify a model to generate data that reasonably approximates the variation in outcomes attributable to the random process.
- *Randomize & Repeat.* Use the model to generate simulated data for a single trial, specify a summary measure to be collected, then generate simulated data for *many trials,* each time collecting the summary measure.
- *Evaluate.* Examine the distribution of the resulting summary measures. Use this distribution to compare the behavior of the model to observed data, make predictions, etc.

TinkerPlots™ was selected for use in the CATALST curriculum because its highly visual and interactive approach to model building and simulated data generation allows students to initially spend their cognitive resources on content rather than the intricacies of learning software. This allowed the focus of the course to begin and remain rooted in the examination and evaluation of distributions of statistics in order to draw statistical inferences.

The current version of the CATALST curriculum consists of three units: (1) *Chance Models and Simulation,* (2) *Models for Comparing Groups,* and (3) *Estimating Models using Data.* The remainder of this section presents brief outlines of each instructional unit. Additional details about the three units of the curriculum are presented in Garfield, delMas, and Zieffler (2012).

Unit 1: Chance Models and Simulation

The activities in the *Chance Models and Simulation* unit have students build and explore probability models. Initial models are based on simple random devices (e.g., coins, dice) and are used for simple modeling of real-life phenomena (e.g., birthrates, 'blind guessing'). Throughout the unit, the overall purpose for modeling remains the same–students generate data from a model in order to judge whether a particular observed outcome is likely to have occurred by chance. Students learn how simulation results are used to examine conjectures or hypotheses, to examine the 'unusualness' of an observed result under a particular model, and to assess the strength or degree of evidence against the conjectured/hypothesized model.

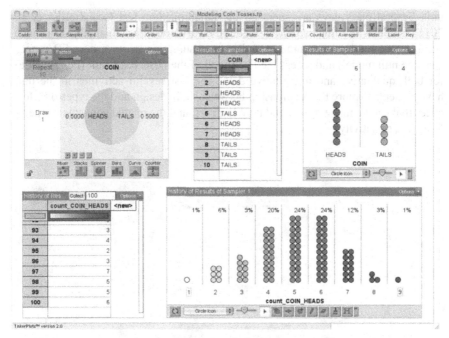

Fig. 29.1 A screenshot of the use of TinkerPlots™ in the Chance Models and Simulation unit. The example shows a simulation of the number of heads obtained from tossing a coin 10 times.

Figure 29.1 shows a screenshot of how the TinkerPlots™ software is used in the *Chance Models and Simulation* unit. The depicted simulation is from a simulation model that students set up early in the first unit to model tossing a fair coin ten times. Students use the simulation to explore the question, "How good are people at predicting random outcomes of common chance devices such as coins and dice?" Prior to building the simulation, students are asked to conjecture and respond to questions such as "How often–what percentage of the 100 sets of ten coins–would you expect to get a result of all ten heads?" and "Which result, two heads or eight heads, would you expect to see more often? Why?" The course builds students' understanding of how to construct simulations that allow them to build a probability model that corresponds to a statistical question, draw a random sample, plot summary information for a sample, collect summary statistics from multiple samples, and analyze a plot of the summary statistics to answer the statistical question.

Unit 2: Models for Comparing Groups

The second unit, *Models for Comparing Groups*, extends the ideas of modeling, simulation and hypothesis testing to comparing groups. Study design, random assignment and random sampling, and the role of variation play a central role in this unit. Several class days are spent on the randomization test (see Zieffler et al. 2011). Students use this method to model the variation in a statistic due to chance (in this case random assignment) under the assumption of no group differences (i.e., the null model) and assess the likelihood of the observed result (i.e., *p*-value) for both qualitative and quantitative data. The last several activities in the unit have students explore the relationship between study design and the types of inferences that can be made, type I and type II errors, and the related concepts of specificity and sensitivity.

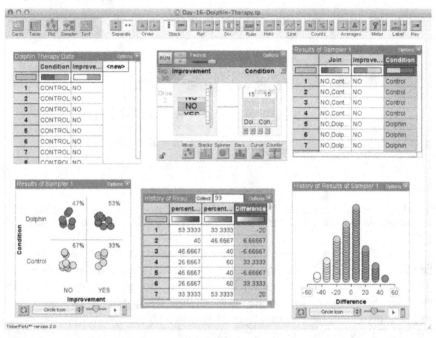

Fig. 29.2 A screenshot of the use of TinkerPlots™ in the Models for Comparing Groups unit. The example shows a simulation of the difference in improvement between treatment and control groups for 100 randomizations of the data.

Figure 29.2 shows an example of how TinkerPlots™ is used in the *Models for Comparing Groups* unit. The figure shows the observed data (upper-left) and model (upper-middle) that students simulate data from in order to answer whether swimming with dolphins is therapeutic for patients suffering from clinical depression. The model visually shows the null model associated with the randomization test (labels of improvement (subjects) are randomly assigned to either control or treatment). The results of one such randomization are shown in both tabular form (upper-right) and more visually (lower-left). The percentage of patients who improved are collected (first two columns of the table in the lower-middle) and the difference computed. This process is repeated many times. The differences for 100 such randomizations are plotted (lower-right) and students evaluate the results computed from the observed data using this distribution.

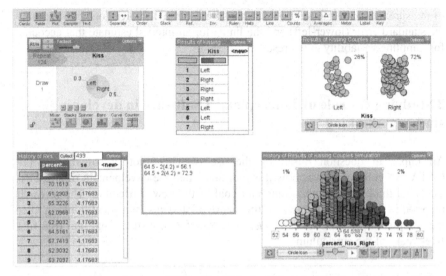

Fig. 29.3 A screenshot of the use of TinkerPlots™ in the Estimating Models using Data unit. The example shows a simulation of the percentage of couples that lean to the right when kissing. The initial model used to simulate data is based on observed summary data.

Unit 3: Estimating Models using Data

In the last unit, *Estimating Models using Data*, the focus is on estimating models using sample data. In the first unit activity, students explore how different sampling methods and sample sizes affect the precision of parameter estimates. Students are introduced to the nonparametric bootstrap to obtain an estimate of the standard error (for more detail of this methodology, see Efron (1981) or Efron and Tibshirani (1993)) and compute a margin of error using ±2SE to obtain an interval estimate that accounts for sampling variation. Students also learn about effect size via the two-group comparison. Under the assumption of group differences, students learn to bootstrap using a model that resamples from each group separately. The unit concludes with a transition to non-simulation based statistical methods and terms (e.g., *t*-test, confidence interval), so that students can see how the results from such methods are interpreted in a similar way to the results of the "no differences" tests and interval estimates they have learned to carry out in the CATALST curriculum.

Figure 29.3 shows an example of how TinkerPlots™ is used in the *Estimating Models using Data* unit. The figure shows the model (upper-left) that students use to simulate data in order to estimate the percentage of couples that lean their heads to the right when kissing. The results of a single run of the simulation are shown in both tabular form (upper-middle) and more visually (upper-right). The percentages of couples that lean to the right are collected (first column of the table in the lower-left) and the process is repeated many times. The results for 500 such trials of the simulation are plotted (lower-right). Students use TinkerPlots™ to estimate the standard error (lower-left) and the limits for an interval estimate that accounts for sampling variability in the results.

2 Studying the role of Tinkerplots™ Software in developing students' thinking

An important question raised by the use of Tinkerplots™ software in the CATALST curriculum was how this tool facilitated the development of students' statistical thinking, the most important goal of the new curriculum. The research reported in this chapter focuses on the nature of students' statistical thinking after the completion of the *Chance Models and Simulation* instructional unit (the first five-weeks of the curriculum).

Problem Solving Interview

Problem-solving interviews were used to explore how well students were able to draw statistical inferences (both informally and formally) after minimal instruction related to modeling and simulation. Based on the learning activities and assignments experienced in the first unit, students were expected to:

- Select an appropriate chance model for a given context;
- Understand what is meant by an unlikely outcome under a given model;
- Consider sampling variability when drawing statistical inferences;
- Anticipate how changing the underlying chance model would affect the distribution of sample statistics.

We also expected that students might be able to quantify the likelihood of an outcome, relate it to the concept of a *p*-value, and use the quantification to help draw a statistical inference.

A set of problem-solving tasks was designed to gather evidence about the extent to which the above expectations were met and the nature of students' thinking and reasoning at the end of Unit 1. The set of tasks, along with other details of the research methodology, are described in subsequent sections.

Interview Participants

Interviews were conducted during the fall of 2011. At the end of the first instructional unit, all students enrolled in the course were sent an email inviting them to participate in a one-hour problem-solving interview. Students were offered a gift certificate to a popular online store as an incentive for completing the interview. Seven students responded to the email invitation and were scheduled for an interview. However, only five of the students, one male and four females, showed up and completed an interview. Based on performance on homework assignments and class participation during the five weeks of instruction, three of the five students interviewed were judged to be in the top third of their respective course sections, and the remaining two were judged to be in the middle third of their respective course sections.

Interview Questions and Contexts

A set of interview questions were written to elicit student responses that would be conducive for exploring students' reasoning and understanding about key ideas related to samples and sampling in the context of informal inference. These questions were based on the informal inferential reasoning (IIR) framework and assessment tasks suggested by Zieffler, Garfield, delMas, and Reading (2008). The questions posed to the interviewees were organized into two primary contexts. The description of the context for each set of questions was read aloud to the student by the interviewer. Both of these contexts are described below.

Context 1: Spinners. The context for the first set of questions (spinners) was designed to be familiar to interviewees based on their course experiences. The interviewee was shown a spinner divided into four equal sections labeled 'a', 'b', 'c' and 'd'. The interviewee was told that when someone used the spinner 10 times, the letter 'b' resulted on five of the 10 spins. They were then asked to state how they would determine if this was an unusual outcome.

After responding to the first problem, the interviewee was presented with a second spinner that again had four divisions, but the letter 'b' represented 40% of the space while the other three letters each represented 20%. This was a somewhat novel situation, since the chance models students encountered in the instructional unit had equally likely outcomes. The interviewee was again asked to consider the same result (i.e., the letter 'b' occurring on five of the 10 spins) and determine how to decide if this was unusual.

Context 2: ESP. The second set of questions was based on a scenario that described a research setting used in studies of extra-sensory perception, or ESP, and other paranormal phenomena (see Utts 1991). The second context represented a much more complex and detailed design than the first context. This complexity was introduced purposefully to see if students could extract the relevant details from the context to build an appropriate chance model for testing whether or not a particular result was "unusual".

The first question in this context involved a scenario where there were 10 sets of four images. The student was told that a "sender" in an isolated room viewed a single randomly selected image from a set that was also randomly selected. The "receiver", in a different isolated room, then described the image that the sender was viewing. A "judge" rank ordered the four pictures in the selected set according to how well they matched the receiver's description. For a given set of images, if the judge's highest ranked picture was the same as the one shown to the sender, this was scored as a match. The interviewees were told that matches occurred on five out of 10 trials and were asked to describe how they would decide if this result was evidence that the receiver had ESP.

In the second question, the number of images was reduced to three. The same result was posed (5 matches out of 10 trials) and the student was asked how the evidence could be used to make a decision about whether the "receiver" had ESP.

Additional Tasks

In both contexts, interviewees were asked to actually carry out the simulation using TinkerPlots™ for each question. Prior to carrying out the simulation for the second question, interviewees were asked if and how they expected the distribution of simulation results to differ from the distribution of simulation results obtained for the first question. For additional detail on the interview process, including the complete protocol and set of questions, please contact the first author.

3 Results

This study employs a qualitative data analysis using the interview data to build inductively from specific statements made by each student to general themes and categories (Creswell 2009). Videos of each interview, which took place outside of class, were digitally recorded. In addition, the interviewees' actions on the computer were recorded using a screen-capturing tool. A transcript of the recordings was produced for each of the five interviews.

Analysis of the digitally recorded interviews was based on an adaptation of Frederick Erickson's recommendations for the analysis of recordings of social interactions (Erickson 2006). The interviews were viewed independently by two of the authors (Zieffler and delMas) to identify segments related to the students' statistical thinking. The two authors discussed the identified segments to determine an initial coding scheme for emergent themes. They also independently noted interesting examples of reasoning in the students' statements related to the expected learning outcomes. After completing independent analyses, these two authors met together to discuss each interview with respect to the themes that emerged, both within and across the interviews, and to reach consensus. A final review and coding of themes was then applied to each interview transcript.

Analysis of the transcripts and video of the students' statements and use of the TinkerPlots™ software during the problem-solving interviews identified four predominant themes related to the nature of students' statistical inferential thinking.

1. *Sampling Variation:* The extent to which a student considers sampling variation when drawing a statistical inference.
2. *Factors that Affect Sampling Variation:* The extent to which a student anticipated how changing the underlying model would affect the distribution of sample statistics and attribute the change to sampling variation.
3. *Quantification of "Unusualness":* Whether or not a student quantified the "unusualness" of an observed outcome and related that quantification to the chance model created by the student.
4. *Influence of TinkerPlots™ on Students' Thinking and Reasoning*

This section focuses on the fourth of these themes, the influence of TinkerPlots™ on students' thinking and reasoning. A more detailed analysis of all aspects of the student interviews, including detail related to the three other themes, can be found in delMas, Zieffler, and Garfield (in press).

Influence of TinkerPlots™ on Students' Thinking and Reasoning

This section presents themes that emerged from the transcript analysis that were not among the expected learning outcomes. Pseudonyms are used to identify each of the students. Bracketed italicized text is used within transcript excerpts to provide additional information for understanding the context.

As the interviews were reviewed across the five students, it became apparent that the students' reasoning often seemed to be facilitated by working with TinkerPlots™. For example, Jody appeared to change her understanding of the first spinner problem after she started to work with TinkerPlots™. After presenting the context and question for the first spinner problem, the interviewer asked Jody to describe how she would use TinkerPlots™ to answer the question.

> *Jody:* Um, Well first of all I would take out the plot and then I would add A, B, C, and D. And then I would change the, um, probability to 25% for each letter. And then, um, I think I would set the repeat to 10. So one trial would be 10 spins. And then I could see how many Bs, of each letter popped up.

Note that Jody did not state that she would collect the number of Bs from the trial, collect this statistic from numerous trials, and look at the distribution of the sample statistics. After creating the sampler and producing just one trial of 10 spins in TinkerPlots™, she indicated that she was done. The interviewer then asked how she would use the information to answer the question.

Jody: Okay. So I hit run and looks like I have 10 spins, and I have one, one B. So, there we go.
I: Okay. So that's it? You're done there?
Jody: Yeah.
I: Okay. How would you use that information to make a decision about, um, one person said five Bs is too many, one person said no that's not unusual. Now, you just set this up in TinkerPlots™ and done this one set of 10 trials. So, how can you use that?
Jody: Well, um, this probably is not enough. I probably have to run more trials in order to see how likely that is. Because you can't just base your, um, information off of just one trial.
I: Okay. So do you want to do that?
Jody: Sure.

While the interviewer's question most likely helped Jody realize she needed to run additional trials, it probably did so by having her consider what she had produced in TinkerPlots™, possibly prompting her to recall or visualize additional steps that were taken during class activities. Jody went on to collect the number of Bs out of 10 spins from 122 trials, and considered the distribution of the counts in her reasoning about the question. Jody indicated that five out of 122 represents an unlikely result, whereas 20 out of 100 represents a result that is very likely.

Additional examples of TinkerPlots™ facilitating students' thinking come from Audrey and James who changed their thinking about the chance model for the first ESP problem. Both Audrey and James initially conceived of a chance model with two linked chance devices, with Audrey not being certain of how to set up the second device. As each student worked with TinkerPlots™, they modified their original ideas for the chance model. Both students constructed a chance model with linked spinners where one spinner had a single outcome and the other was divided into four equal sections. James came to realize he could use a model with a single spinner, similar to the model he created for the first spinner problem.

In contrast, Audrey initially thought out loud about how she would model the ESP problem in TinkerPlots™. She found the situation to be complex and, while talking through different aspects of the context, decided to consider a less complex part of the problem by creating a model and generating a single trial. At that point in her thinking she stated:

Audrey: Um, at that point I'm not sure exactly how we would apply it to seeing the 10 matches. I guess we could look if there was a 50% chance for observed data that, um, you are going to pick the correct image. So we could see and compare the 50% chance to see if, um, using this one trial there was a 50% chance that we got the correct image.

Audrey had not visualized the model as representing a one in four chance of a correct match and she was not exactly sure where the approach would lead her. The interviewer then invited Audrey to try her ideas in TinkerPlots™, which she agreed to do. The next excerpt illustrates how Audrey's reasoning about the chance model changed as she worked to set up a sampler in TinkerPlots™.

Audrey: OK. So, [I'll] just to try to see if this might make it more clear for me, if how I can make a diagram, I'll try just showing for one trial. Um, OK. So, I'm going to link two mixers together so we can match what the Sender had sent, or had seen, at least to what the Receiver was showing. And so if we're saying that when he's judged, it's possible for the Receiver to have, for the top image to be picked would be one of four. So I'll put four elements in, like the Receiver's box. I'll just label that "Receiver". Um, and that would be A through C. And then if we're assuming the Sender is only seeing one image.

I: Is it A through C? Is that what you said?

Audrey: Oh, A through D for the four.

I: A through D. OK.

Audrey: And then assuming the Sender has only seen one image, there would be only chance for one item to be drawn here, because he's seeing the one image, A, and then the Receiver, um, has a 25% chance because his, whatever his words were, however he described it was getting matched to, um, these four pictures. So the judge has to, there's no option for the judge to say there's nothing here that looks like it doesn't match. So he has to pick one of the four to be the closest match, and then the second, third and fourth. So for that reasoning I would have four here. Um, then we draw two. And I guess we can. We could repeat this 10 times, actually. I think that is how it would work for the 10 trials. Because it doesn't exactly matter, um, that image A, because for this it doesn't necessarily mean that A is going to be the same image that is seen there. So if we repeat this 10 times, I think that would simulate the 10 images the Sender saw, and then the 10 options the Receiver did. So this way we can look more at seeing the number of matches and we can compare to the, um, 5 observed matches.

Audrey first created the model on the left in Figure 29.4, then modified it to create the second model on the right. Audrey referred to the second model when she said, "Because it doesn't exactly matter, um, that image A, because for this it doesn't necessarily mean that A is going to be the same image that is seen there."

Audrey went on to build the complete simulation, including the distribution of counts for the number of matches in 100 trials. She provided an explanation for why she was considering five or more matches, quantifying the probability. Audrey used the probability to make a decision by reasoning that the observed count of five out of 10 matches is "somewhat uncommon" under the chance model.

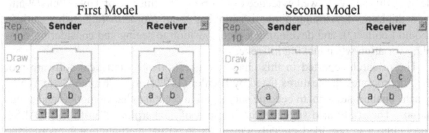

Fig. 29.4 The two models created by Audrey for the first ESP problem

4 Discussion and summary

The results indicate that the interviewed students had begun to develop an understanding of important concepts related to statistical inferential thinking during the first few weeks of the CATALST course. An interesting observation that emerged was the possible way that the software, TinkerPlots™, facilitated students' thinking and reasoning. When asked to describe how they would answer a question, students seemed to consider how to build the model in TinkerPlots™ and to reason and think through problems using the language of TinkerPlots™.

Jody did not initially demonstrate an understanding that a distribution of sample statistics was needed to address the question posed in the first spinner problem. When confronted with whether or not the results from a single trial in TinkerPlots™ could be used to answer the question, Jody realized that additional trials and a distribution of sample statistics was needed. While James articulated an appropriate model for the first ESP problem, he refined the model as he worked in TinkerPlots™, realizing similarities between the contexts of the first spinner and first ESP problems and reducing the linked spinner model to a single spinner model. Audrey came to an impasse in her thinking about the first ESP problem, but appeared to re-conceptualize the problem and produced an appropriate model while working in TinkerPlots™. TinkerPlots™ may have facilitated Audrey's thinking by helping her to visualize and keep track of the complexities of the ESP context. These observations suggest that the development of students' inferential reasoning after about five weeks in the course was supported by and somewhat dependent on the learning tool.

In contrast to other approaches where an emphasis is placed on understanding various types of distributions (population, sample, sampling distribution) and relationships among them, the CATALST course does not focus on the original sample or individual simulated samples. The CATALST approach of having students use TinkerPlots™ to set up the chance model, generate a simulated sample, identify a statistic to collect, generate and collect the statistic from many simulated samples, and examine the distribution of that statistic allowed the focus of the course to stay on the distribution of sample statistics from the start. The five students in this study showed evidence of incorporating this use of TinkerPlots™ into their thinking, developing norms of thinking about problems (e.g., what statistic should I collect?), and developing a predisposition to seeing and considering sampling variation when making inferences.

The research reported in this chapter suggests that a statistical software tool that incorporates the features listed by Biehler (1997a) for a "Monte Carlo Workbench" enables students to become successful at creating chance models and simulations. The study also suggests that the use of Tinkerplots™ in the CATALST curriculum also provided students a memorable, visual medium that supported the development of their thinking and reasoning about chance models and inference.

Acknowledgments We gratefully acknowledge the contributions of Beth Chance and Herle McGowan to this research, and the valuable contributions of our additional CATALST collaborators (George Cobb, Allan Rossman, John Holcomb and Rob Gould). We greatly appreciate funding from the National Science Foundation (DUE-0814433).

References

Biehler, R. (1997a). Software for learning and for doing statistics. *International Statistical Review, 65*(2), 167-189.

Biehler, R. (1997b). Students' difficulties in practicing computer-supported data analysis: Some hypothetical generalizations from results of two exploratory studies. In J. B. Garfield, & G. Burril (Eds.), *Research on the Role of Technology in Teaching and Learning Statistics: Proceedings of the 1996 International Association of Statistics Education (IASE) Round Table Conference* (pp. 169-190). University of Granada, Spain. Voorburg, The Netherlands: International Statistical Institute.

Biehler, R. (2003). Interrelated learning and working environments for supporting the use of computer tools in introductory courses. In International Statistical Institute (Ed.), *Proceedings of IASE Satellite conference on Teaching Statistics and the Internet* [CD-ROM]. Berlin: Max-Planck-Institute for Human Development. Retrieved February, 10, 2013 from http://iase-web.org/documents/papers/sat2003/Biehler.pdf.

Biehler, R., & Prömmel, A. (2010). Developing students' computer-supported simulation and modelling competencies by means of carefully designed working enviroments. In C. Reading (Ed.), *Proceedings of the eighth International Conference on Teaching Statistics (ICOTS-8, July 2010), Ljubljana, Slovenia*. Voorburg, The Netherlands: International Statistical Institute. Retrieved February, 10, 2013 from http://iase-web.org/documents/papers/icots8/ICOTS8_8D3_BIEHLER.pdf.

Box, G. E. P., & Draper, N. R. (1987). *Empirical model-building and response surfaces*. New York: Wiley.

Chance, B. L., Ben-Zvi, D., Garfield, J., & Medina, E. (2007). The role of technology in improving student learning of statistics. *Technology Innovations in Statistics Education Journal, 1*(1). Retrieved from http://repositories.cdlib.org/uclastat/cts/tise/vol1/iss1/art2/.

Cobb, G. W. (2007). The introductory statistics course: A ptolemaic curriculum? *Technology Innovations in Statistics Education, 1*(1). Retrieved September, 1, 2012 from http://escholarship.org/uc/item/6hb3k0nz#page-1.

Creswell, J. (2009). *Research design: Qualitative, quantitative, and mixed method approaches*. Thousand Oaks, CA: Sage Publications.

Doerr, H., & English, L. (2003). A modeling perspective on students' mathematical reasoning about data. *Journal for Research in Mathematics Education, 34*(2). 110-136.

delMas, R., Zieffler, A., & Garfield, J. (in press). Tertiary students' reasoning about samples and sampling variation in the context of a modeling and simulation approach to inference. *Educational Studies in Mathematics*.

Efron, B. (1981). Nonparametric standard errors and confidence intervals. *Canadian Journal of Statistics, 9*, 139–172.

Efron, B., & Tibshirani, R. J. (1993). *An introduction to the bootstrap*. New York: Chapman & Hall.

Erickson, F. (2006). Definition and analysis of data from videotape: Some research procedures and their rationales. In J. L. Green, G. Camilli & P. B. Elmore (Eds.), *Handbook of complementary methods in education research* (pp. 177-191). Mahwah, NJ: Erlbaum.

Garfield, J., & Ben-Zvi, D. (2008). *Developing students' statistical reasoning: Connecting research and teaching practice*. New York: Springer.

Garfield, J., delMas, R., & Zieffler, A. (2012). Developing statistical modelers and thinkers in an introductory, tertiary-level statistics course. *ZDM - The International Journal on Mathematics Education, 44*(7), 883-898. doi: 10.1007/s11858-012-0447-5

Key Curriculum Press (2012). *Fathom® dynamic data (v. 2L)* [computer software]. Emeryville, CA: Author.

Konold, C. (2011). *TinkerPlots™ Version 2* [computer software]. Emeryville, CA: Key Curriculum Press.

Konold, C., Harradine, A., & Kazak, S. (2007). Understanding distributions by modeling them. *International Journal of Computers for Mathematical Learning, 12.* 217-230.

Konold, C., & Lehrer, R. (2008). Technology and mathematics education: An essay in honor of Jim Kaput. In L. D. English (Ed.), *Handbook of international research in mathematics education* (2nd ed.) (pp. 49-71). New York: Routledge.

Maxara, C., & Biehler, R. (2006). Students' probabilistic simulation and modeling competence after a computer-intensive elementary course in statistics and probability. In A. Rossman, & B. Chance (Eds.), *Proceedings of the 7th International Conference on Teaching Statistics (ICOTS-7, July 2006), Bahia, Brazil.* Voorburg, The Netherlands: International Statistical Institute. Retrieved February, 10, 2013 from http://iase-web.org/documents/papers/icots7/7C1_MAXA.pdf.

Maxara, C., & Biehler, R. (2007). Constructing stochastic simulations with a computer tool— students' competencies and difficulties. In D. Pitta–Pantazi, & G. Philippou (Eds.), *Proceedings of the fifth Congress of the European Society for Research in Mathematics Education* (pp. 762-771). Larnaca, Cyprus. Retrieved September, 1, 2012 from http://www.erme.unito.it/.

Maxara, C., & Biehler, R. (2010). Students' understanding and reasoning about sample size and the law of large numbers after a computer-intensive introductory course on stochastics. In Reading, C. (Ed.), *Proceedings of ICoTS 8, Ljubljana, July 2010* [C D-ROM]. Voorburg: IASE. Retrieved February, 10, 2013 from http://iase-web.org/documents/papers/icots8/ICOTS8_3C2_MAXARA.pdf.

Thisted, R. A. (1986). Computing environment for data analysis. *Statistical Science, 1,* 259-275.

Utts, J. (1991). Replication and meta-analysis in parapsychology. *Statistical Science, 6*(4), 363-378.

Watson, J. (2008). Eye colour and reaction time: An opportunity for critical statistical reasoning. *Australian Mathematics Teacher, 64*(3), 30-40.

Watson, J. M., Fitzallen, N. E., Wilson, K. G., & Creed, J. F. (2008). The representational value of hats. *Mathematics Teaching in the Middle School, 14,* 4-10.

Watson, J., & Wright, S. (2008). Building informal inference with TinkerPlots in a measurement context. *Australian Mathematics Teacher, 64*(4), 31-40.

Wild, C. J., & Pfannkuch, M. (1999). Statistical thinking in empirical enquiry. *International Statistical Review, 67*(3), 223–265.

Zieffler, A., Garfield, J., delMas, R., & Reading, C. (2008). A framework to support research on informal inferential reasoning. *Statistics Education Research Journal, 7*(2). 40-58. Retrieved February, 10, 2013 from http://iase-web.org/documents/SERJ/SERJ7(2)_Zieffler.pdf.

Zieffler, A., Harring, J., & Long, J. (2011). *Comparing groups: Randomization and bootstrap methods using R.* New York: Wiley.

Chapter 30
TinkerPlots as an Interactive Tool for Learning about Resampling

Jane Watson

University of Tasmania

Abstract The question motivating this study was the feasibility of students in grade 10, aged about 15, assimilating the ideas of inference through resampling procedures. The tool to be used to achieve repeated random sampling was the software *TinkerPlots*, which includes a pseudo-random Sampler capable of reassigning data to treatments and keeping a history of such simulations to calculate frequencies of occurrence of particular statistics to compare with experimental values. Two grade 10 teachers volunteered their classes for 5 weeks of a unit on "statistics and probability" to take part in the study. Although lessons with detailed instructions were provided by the researcher, the project, because it took place in actual classrooms, became an action research project as the teachers adapted the material to their needs and the needs of their students. Although there were several research questions associated with the study, this chapter focusses on issues associated with the students' use of *TinkerPlots* and the goal of their appreciation of the nature of randomization and resampling.

1 Introduction

There has been debate for some time about the merits of replacing the formal inference procedures of the 20th century based on classical statistics, with random numerical procedures classified under headings such as resampling, bootstrapping, and randomization-based statistics. Cobb (2007) makes a strong case for such a move in suggesting that introductory statistics courses at the tertiary level should be taught based on the three R's: *randomize* data production, *repeat* by simulation to see what is typical, and *reject* any model that puts the data in its tail. The availability of fast computing capacity makes the randomizing and repeating an achievable goal without the theoretical baggage of z- or t-distributions. Examples of computer programs to carry out the procedures abound, including in *Excel* (Christie 2004), *R* (Arnholt 2007), *Minitab* (Taffe and Garnham 1996) and *Fathom* (Clements et al. 2007). The most recent software to make resampling possible is *TinkerPlots* (Konold and Miller 2011). Although not created for this purpose, the Sampler tool in the software allows random samples to be collected from populations that can then be structured to answer questions about the uniqueness of an experimental outcome in relation to random reallocations of the same data set.

The existence of tertiary courses successfully employing resampling procedures (e.g., Rossman and Chance 2008; Tintle et al. 2011) leads to the question of moving down in the curriculum to expose students at the secondary level to the concept of random resampling leading to the ability to make decisions in statistical contexts. Scheaffer and Tabor (2008) employed resampling in contexts based on boys and girls having curfews and on two methods of measuring the area of a triangle. Similarly Shaughnessy, Chance, and Kranendonk (2009) extended an application originally suggested by Kader in National Council of Teachers of Mathematics (2009) based on students memorizing meaningful and nonsense words. Shaughnessy et al. presented data and examples of classroom discussion surrounding the experience. At approximately the same time the Common Core State Standards for Mathematics (2010) appeared in the US. Among the recommendations for high school (grades 9 to 12) for Statistics and Probability under the heading "Make inferences and justify conclusions from sample surveys, experiments, and observational studies," was the following: "Use data from a randomized experiment to compare two treatments; use simulations to decide if differences between parameters are significant" (p. 82). Such examples and recommendations encouraged belief that a trial should take place in Australia where a new national curriculum was being developed (e.g., Australian Curriculum, Assessment and Reporting Authority [ACARA] 2012a).

The release of Version 2.0 of the constructivist statistical software *TinkerPlots*, developed for middle school students by Konold and Miller (2011), further encouraged the possibility of trialling material with school students. *TinkerPlots* is an example of landscape-type software in contrast to route-type software (Bakker 2002). Route-type graphing software has its operations hidden and the user must decide on an appropriate representation before asking the software to create it, whereas with landscape-type graphing software the user has a level of control over the operations and can use a process of trial and error to produce representations. The features of *TinkerPlots* satisfy many of the requirements set by Biehler (1997) for software to assist *learning* as well as *doing* statistics. More recent research on the use of technology for enhancing statistical reasoning (Biehler et al. 2013) confirms the importance of technology-enhanced learning environments and features *TinkerPlots* as one of two exemplary software packages.

TinkerPlots (Version 2.0) provides a Sampler, which allows for simulation of repeated random trials of chance devices with replacement after each trial and random sampling from designated populations without replacement, a procedure suitable for resampling. As well the software has a Ruler, a Measure tool that, for example, calculates differences in means or medians, and a History tool that creates a table to keep track of differences over many runs of the Sampler. This makes the process of resampling visual and intuitive. Watson (2013) has provided examples of the layout of the software procedures based in the context of measuring reaction time for school boys in different grade levels or for boys and girls in the same grade level. Further Watson and Chance (2012) have made a case for inclusion of resampling in the Australian senior mathematics curriculum, including examples using *TinkerPlots* with the data from Shaughnessy et al. (2009) on memorizing meaningful and nonsense works and from Rossman (2008) on an experiment where people suffering mild to moderate depression swam in the Caribbean either with bottlenose dolphins or alone, the same number of hours per day for 4 weeks.

The aim of the action research was to monitor the interaction of grade 10 students with the software and the material and assess their take-up of the ideas associated with simulation, randomization, sample size, resampling and inference. Similar classroom research has been carried out by Rolf Biehler and his colleagues (e.g., Biehler 2006; Biehler and Prömmel 2010; Maxara and Biehler 2006; 2010) with *Fathom* (Key Curriculum Press 2005), software of the same type and compatible with *TinkerPlots* but more advanced. The results are encouraging but Biehler and his colleagues were dealing with older students in a more intensive setting and not progressing as far as considering resampling for inference. The goal of this chapter is to tell a related story extended to resampling in a typical classroom.

2 Method

Two grade 10 mathematics classes took part in the study, constituting a convenience sample as the two teachers volunteered to replace their 5-week unit on Statistics and Probability with the material provided by the researcher. The classes were considered of average ability by their teachers, with a wide range of capacity. The classes met for 80 minutes on Mondays and Wednesdays and for 60 minutes on alternate Fridays. Although the school had recently purchased *TinkerPlots*, these students had not been exposed to it. Hence the materials provided for teachers included suggestions for introductory lessons using data sets from the resource, *Digging into Australian Data with TinkerPlots* (Watson et al. 2011).

To introduce the Sampler and ideas associated with increasing confidence in models with increasing sample size, several lessons related to probability models based on simulation were provided. To move from using probability sampling with replacement to population sampling without replacement, a lesson was provided based on sampling from the First Fleet of six ships with 780 convicts arriving in Australia in 1788. Finally two lessons were provided that focussed on resampling to test a hypothesis. The first was based on memorizing meaningful and nonsense words (Shaughnessy et al. 2009) and the second on the swimming with dolphins study used by Rossman (2008). A potential assessment task, where students set up an investigation to consider difference in reaction time for two groups of students was also included. Limited descriptions of the lessons are provided in the Appendix. Except for the first, detailed lesson plans were provided, as well as worksheets with instructions and questions for students. An example of a worksheet originally supplied to teachers, with the space for recording responses removed, for the Swimming with Dolphins investigation, is shown in Fig. 30.1.

Swimming with Dolphins

Name:

This lesson uses data from an experiment presented in a two-way table to investigate whether for people with mild to moderate depression swimming 4 hours per day with dolphins produces more improvement in depression than swimming 4 hours per day without dolphins.

MAKE A HYPOTHESIS TO TEST

1. The data from the experiment, where 30 patients were randomly allocated to the two treatments, Control (swimming 4 hours per day in the Caribbean for four weeks) and Dolphins (swimming 4 hours per day in the Caribbean for four weeks in the presence of dolphins), are shown in the TinkerPlots plot below, where the Results of the experiment were judged as Improvement or No Improvement in depression.

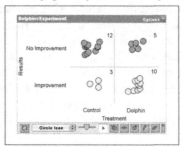

Looking at the data make a hypothesis about the effect of the treatment on depression. Why do you say this?

DATA ANALYSIS

2. There are two TinkerPlots files to provide instructions, if you need them, to carry out the resampling to make a decision on your hypothesis. **Dolphins1.tp** helps set up the Sampler to randomly reallocate of the Results to the Treatments for the 30 patients. Collect 3 resamples and record the number of patients whose depression improved after swimming with dolphins.

    i _____

    ii _____

    iii _____

    How do these values compare with the 10 observed in the experiment?

3. The second file **Dolphins2.tp** continues and sets up the History tool to collect the number in the Dolphins-Improvement cell. This can be done many times and the outcomes plotted to compare with the 10 from the experiment.

    How many resamples did you collect? _____

    How many outcomes were 10 or larger for Dolphin-Improvement? _____

    What percentage is this of the total number of random resamples your collected? _____

    Are you convinced that swimming with dolphins is more successful than the control in reducing depression? Explain why or why not.

4. Compare your results with others in your class. Do you all agree on the conclusion in your analyses? Write a summary of what the class found.

5. In carrying out your investigation you have made an informal inference.

    What was your *evidence*?

    What was your *generalisation*?

    How *uncertain/certain* were you of your inference?

Fig. 30.1 Student worksheet for one lesson (space for writing removed)

Data were collected in the form of saved *TinkerPlots* files or filled-in worksheets for each lesson for each student whose parent had signed an ethics consent form. Due to the teachers' preferences, Class A answered questions in text boxes in *TinkerPlots* and Class B answered questions in space provided on the worksheets. All students answered the same questions, regardless of format. As well two researchers attended all class periods making notes and interacting with students, one in each class. There was no protocol for the interactions and they consisted of answering questions about using the software or asking questions about what was seen on the screen as the researchers walked past. For this report a descriptive account is presented of the aspects of the project that evolved over the period of implementation in the light of the nature of teaching in real time in an Australian high school. The factors that played a role in influencing the results included the time restrictions on the teachers for preparation, the choices made by teachers from the lessons provided and amendments to them, the availability of computer labs, technical difficulties with files on the school's computer system, the relationship between teachers and students, and student absences due to other school activities. Although these factors can be considered limitations from the ideal design of the study, they encompass actual issues in every Australian high school. Later in the sequence of lessons the feedback of teachers to the researchers resulted in changes to the worksheets that were provided to the students.

3 Outcomes

Although the expectation of the researcher was that the lessons would take the allotted time within the teachers' unit of work on Statistics and Probability, in fact the content was too much to allow for reinforcement to be completely effective. The teachers did not feel it necessary to have an extended time with the introductory data sets in Lesson 1, and introduced *TinkerPlots* with the lesson from *Digging into Australian Data with TinkerPlots* (Watson et al. 2011) on two football codes. The lesson and worksheet were from the sixth section of the book and made assumptions about the students' skills that they did not have, due to lack of exposure to the software. Although students did achieve the outcome of distinguishing the two football codes based on characteristics of the players, they required considerable help from teachers and researchers in working with *TinkerPlots*. For the later lessons, instructions for implementing procedures in *TinkerPlots* were included within the software files but there were times when the students waited to be helped rather than working through the instructions.

The next three lessons progressed roughly according to plan. It was important for students to appreciate that in the context of probability, the Sampler operated randomly "with replacement," allowing each trial outcome selected to be replaced and have the same chance of being chosen in the next round of trials. Lesson 2 considered the number of random trials with the Sampler before simulated tosses of a single die approached the expected percentage. One pair of students reported 2000 trials and an outcome of 18%, 18%, 16%, 16%, 15%, 17%. Generally the students were surprised at how many tosses were required to approach the expected distributions.

Lesson 3 was more challenging in testing students' appreciation of sample size. When initially asked whether tossing 10 coins or 30 would be more likely to result in more than 60% heads, most students suggested the two scenarios would have the same chance "because there is a 50/50 chance to be heads," with a few suggesting 30 "because there are more tosses" and a few suggesting 10 because "the lower number of samples will make it less accurate." Although their simulations generally showed twice as many times there were more than 60% heads with 10 coins, some students were reluctant to give up their belief in equality of the two contexts, claiming it was "random." As well, some students needed to be reminded several times of the two sample size issues: the number of times the coins were tossed (10 or 30) and the number of simulations that were carried out (decided by the students). This was an issue also considered by Maxara and Biehler (2010) but with older students in a more sophisticated setting.

Overall students found Lesson 4 more motivating because they were aware of the issue of China wanting a one-child policy but families wanting to ensure they have a boy (Konold 1994). The lesson helped students set up the Sampler to randomly select a boy or girl until a boy was born and then keep track of the number of children in the family. A History collection was set up to collect family size for many families. When asked to predict, most students suggested families would have more girls and there would be more girls overall. The explanation shown in Fig. 30.2 was typical of those who completed the lesson. As students did more trials, the average family size became closer to 2 and the students realized that this implied a boy and a girl.

4. The value that it was approaching when we tested it was around 2.0597. So that means that for most families you would have at least 2 children and that would come down to having a boy and a girl. (Because you have to stop when you have a boy)

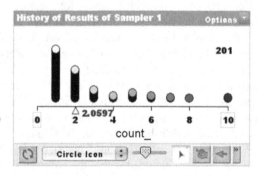

Fig. 30.2 Plot and explanation for the One-Son lesson simulation

Lesson 5 based on random sampling without replacement from the First Fleet data set was the first encounter students had with random sampling from a large finite population that could be placed in the Sampler. The lesson was of particular interest to one of the teachers. Some students collected many random samples to show that one particular ship was typical of the entire fleet for one attribute, the "value of crime" for which the convicts were transported. This sampling procedure was accepted relatively easily by students.

Lesson 6 incorporated all of the *TinkerPlots* skills introduced so far, and one new tool, the Ruler, which was used to measure the difference in medians for two different treatments. Other skills included using the Sampler, sampling without replacement, plotting random samples, measuring the difference of medians with the Ruler, keeping a History collection of the differences, and plotting them to compare with the difference in the original experiment. This was the lesson based on memorizing meaningful and nonsense words and the teachers decided that there would be ownership of the data and more interest in the analysis if students carried out the experiment for themselves. The researcher hence created two lists of 3-letter meaningful words such as AGO, HIM, WON, and nonsense words such as MZA, EOC, XCS. The design of randomly assigning half of the class to each treatment was modified by the teachers who thought it would be better for the students to memorize each set of words. Hence the randomization was amended to randomly assign the order in which the students memorized the words, half starting with meaningful and half with nonsense. Students recorded their data on four sticky notes, two for each treatment. They then placed one for each treatment on a stacked plot on the whiteboard and the other two were placed in a box to be randomly selected later. After a discussion of the shape of the data and the difference in medians for the two types of words, the second data values from the set of data were drawn out (randomly) and assigned to the two treatments. In Class A the random allocation was to the meaningful treatment first and in Class B the allocation alternated between the two treatments. Discussion was held about the expected difference in medians for the randomly allocated data and comparison with the experiment carried out by the students. In Class A the difference in the students' experimental data was 4 words remembered. In Class B the difference in the original data was 3. (There was some confusion about correct values but this was checked by the researcher and amended before data were entered into *TinkerPlots*.) The random allocation for Class A had a difference of 3, whereas for Class B it was 2. The display of the original data and the random allocation from Class B is shown in Fig. 30.3. Discussion took place about carrying out many random resamplings and what would be required for the students to be convinced that it was easier to memorize meaningful than nonsense words.

The following class period the students were provided with *TinkerPlots* files with their class's data and the instructions for setting up the Sampler to carry out the resampling procedure. These instructions were included within *TinkerPlots* files. Although the instructions built upon procedures used in previous lessons, except for the introduction of the Ruler, some students found the process tedious and were distracted by other students or links available on the computers. The few students who completed the worksheet activities in Class B recorded the results for four runs of the Sampler, obtaining differences in medians of, e.g., ¯2, 1, 0, and ¯1. In response to the question of whether the results were more or less than her class had obtained, one student wrote, "The differences in these medians are much less than our original median, because it is all random data." One group of students completed 50 random resamples with 4% equal to or greater than the class result. In response to the question, "Do you believe the data you collected in your class would have happened by chance?" the group said, "No, it wasn't by chance but we needed more people to make the results more accurate."

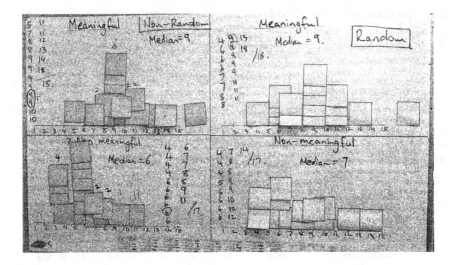

Fig. 30.3 Display of class data (left) and random data (right) from memorizing meaningful and nonsense words

In Class A, students had the opportunity to analyse their class results or a comparison of the two classes, an additional lesson prepared by the researcher because of the different sample sizes represented in the two classes. Some groups struggled to complete all of the instructions to keep track of the history of the resamples. A group of three boys whose results are shown in Fig. 30.4, measured "nonsense words remembered – meaningful words remembered" and hence referred to the result for ¯4 in their explanation. Although their explanation could have been more explicit, it conveyed the message of the appropriate outcome. Another group of three provided the outcomes and explanation in Fig. 30.5 with one boy writing the text in the first person for the group.

9. The difference in the original class one was 4 but in the sampler 4 was an extreme. This would of been unlikely to have occurred by chance and shows that it was caused rather than just having randomly occurred. This show that meaningful words can be remembered more easily then nonsense words.

11. We collected 402 random samples less than 0% were the same as the original samples. This is unlikely that it occured via chance.

12. We are 99% sure that this didn't occur by chance. This very much confirms our suspicion that meaningful words are more easier to remember than nonsense words.

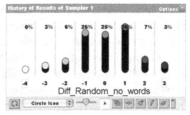

Fig. 30.4 Analysis of meaningful-nonsense word experiment

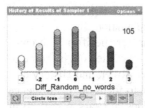

After 50 random samples, the average difference between meaningful and non-meaningful words is 0.

I believe the original data was not entirely random for it is easier to remember words with a meaning. The most recent, entirely random, data did not contain any median differences of 4, as the class memory test did. This provides solid ground for an inference to be made that meaningful words are easier to remember than non-meaningful words.

due to the major difference in random and the actual results, it is fairly safe to assume that the rest of Australia would see similar averages as this test did.

A random outlier or multiple outliers in a random data set may be significant enough to influence the entire average/median and therefore prove the inference incorrect however after a large number of collections, the average would prove the inference a likely case.

Fig. 30.5 Analysis of the meaningful-nonsense word experiment

The final lesson using *TinkerPlots* for Class A was the lesson on Swimming with Dolphins. Because they had carried out the instructions for the earlier lessons more successfully than Class B, they were given the option of having a worksheet with or without *TinkerPlots* instructions. About half chose each option. In this class the students filled in text boxes in *TinkerPlots* rather than completing worksheets. One pair of students collected 411 resamples of the data and reported their results as shown in Fig. 30.6, with the first point being their hypothesis (see Fig. 30.1) and the second, their conclusion after resampling.

1. The people swimming with dolphins show more improvement than the ones that swam on their own. The results tell us that the more people who swam with the dolphins improved. We are 95% certain with this outcome.

2. To test whether this was due to the dolphins or was just chance we could create a plot with 30 different people. We found that getting 10 people cured by dolphins (if it was just by chance) was extremely unlikely and happened only 2% of the time, this indicates the original data shows that it was the dolphins and not just chance that cured 10 out of 15 people.

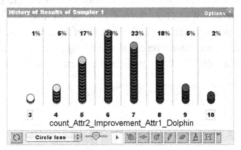

Fig. 30.6 Analysis of the Swimming with Dolphins experiment

In Class B, the filled-in worksheets for the Swimming with Dolphins lesson demonstrated that the students could create the Sampler and collect and keep track of the data but some had difficulty in moving from verbal discussion of the unlikely outcome from the experiment to the random allocations and seeing this as a favourable outcome supporting their hypothesis about the beneficial outcome of swimming with dolphins. The outcome being tested was 10 of 30 people improving when swimming with dolphins compared to random allocations. Students were asked to collect and report 3 random resamples, make a comment, and then collect and keep track of "many" more (a number of their choice). One student reported values of 5, 7, and 7 to compare with 10 and commented, "They are all lower making the experiment significant." The student then did 100 resamples with one result greater than or equal to 10 and when asked if convinced that swimming with dolphins was more successful than the control in reducing depression wrote, "Yes because the data we are given confirms this." Another student who found 1% of 303 resamples equal to or more extreme than the experiment, reasoned differently,

"I think so because swimming with dolphins would add to the experience and feel relaxing," moving back from the data for a contextual explanation.

Class B finished with the lesson written by the researcher specifically to compare the resampling process for the two classes, an activity carried out by some of the students in Class A earlier. In this lesson the Sampler, Ruler, and History collection had been set up for the students, in order for them to focus on the influence of the different sample sizes in the two classes, 38 values in Class A and 30 in Class B. Fig. 30.7 shows the two class sets of data that the students were investigating. Each class had a median difference of 4 for the difference in meaningful and nonsense words remembered. The difference in sample size of 8 meant that it was less likely for there to be a random difference greater than 4 based on the data from the Class A data. This was in fact the result that all students obtained, rarely finding a random difference greater than 4 from Class A data but often finding between 5% and 10% of random differences greater than 4 for Class B. When asked which class's data gave them more confidence that meaningful words were easier to remember, many students were confused and influenced by the larger percentages (e.g., 7% for Class B rather than 0% for Class A) and said that Class B provided more confidence. It was agreed by the teachers and researcher that this extension was too much for most of the students to absorb alongside the other concepts that were the focus of resampling, for example being able to produce more and more resamples. Doing a larger number of trials to approximate probability distributions had been a focus of Lessons 2, 3, and 4.

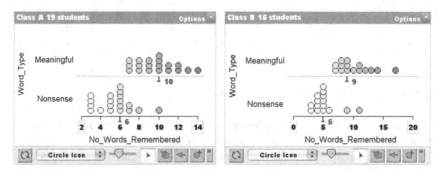

Fig. 30.7 Data from the two classes when memorizing meaningful and nonsense data

One student, who worked in a pair producing the outcomes in Fig. 30.8 for 500 resamples for each class shown in Fig. 30.7, answered a question about "which data set gives you more confidence about whether the difference in medians observed in class was unusual?" with "[Class A] data is slightly more reliable because it had more results than [Class B]."

At the request of the teachers, a written assessment was devised for their use. As part of the assessment students were asked to explain some of the outcomes obtained in the earlier lessons. They were not specifically asked to set up or explain *TinkerPlots* procedures but were asked why the process was used to draw inferences from the two experiments (memorizing words or swimming with dolphins). Some students had difficulty stating the reasons clearly but others appeared to have assimilated the process.

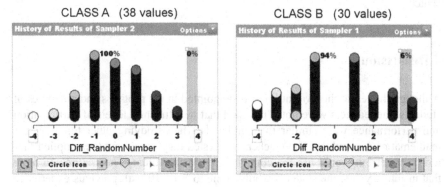

Fig. 30.8 Plots of 500 resamples for the data sets from the two classes

> We did this to test if it [the experiment] was simply due to chance or if the results were caused by something in the experience e.g., it is unlikely that 10 people would be cured by chance so this shows it was probably as a result of being with dolphins.

Finally students were given two data sets where resampling could be applied and asked to suggest a procedure based on the data presented for deciding in the first case if there were a difference between at rest and active-heart rates (12 of each) and in the second case to decide if there were an association between time spent on a mobile phone per week (<10 hours or $\geq$ 10 hours) and difficulty sleeping (Yes, No). The first setting mirrored the memorizing words experiment and the second, swimming with dolphins. A number of students made appropriate suggestions based on their experiences throughout the lessons, for example, suggesting how to set up a Sampler. Others, however, instead focussed on the design of the experiments making suggestions such as the following.

- [heart rates] First of all you would have to get them to set at their normal heart rate before ascending the stairs. Collect their RHR (resting heart rate), get them to ascend the stairs and rest, again take their RHR. You would want to repeat this at least 10 times.
- [mobile phones] I would collect more data so that the results would be more accurate and so that there is something to compare them against.
- [mobile phones] I would ask the students which use their mobile phone for less than 10 hours to use it for 10 and see if their sleeping is affected by this change.

Others made suggestions for dealing with the existing data.

- [heart rates] I would write a hypothesis, make a graph showing the results, and then set up a table – like the first table for the dolphins effect on patients – so that the before and after results could be easily displayed and compared.

These responses indicate the influence of previous learning and at times a mis-interpretation of the questions, which asked for a consideration of the data as presented.

4 Discussion

Although the researcher was at times disappointed in the progress and outcomes of students, the teachers were not. They said that the students' levels of participation and performance were similar to their levels when studying other topics in the mathematics curriculum. The teachers suggested they would cut the supplied material further but would use the content again in the future. They expressed hope that in future years, the students would come to Year 10 with previous experience with *TinkerPlots* from the earlier years in school.

The teachers expressed confidence in their own understanding of inference as an important idea underlying the school statistics curriculum but regretted the fact that the word did not explicitly appear in the curriculum. They felt that their students' increased appreciation of the concept of random and of the importance of choosing larger sample sizes when possible, were noticeable and sustainable, given the emphases during the activities.

The students expressed mixed views on the use of *TinkerPlots*. Some thought it was too juvenile and too easy, whereas others felt it was too complex to set up the necessary procedures. Several students said "the unit didn't feel like maths." The teachers felt this view was likely due to less emphasis on statistics in earlier years and the lack of fixed "correct" answers throughout the unit. As well there was mixed opinion of the students on their interest in the contexts chosen. Almost all contexts were the "most engaging" to some students and "least engaging" for others, although no one picked Swimming with Dolphins as the least engaging. The meaningful words experiment was, however, claimed to be boring by some. It would appear important for teachers to choose contexts that interest them and that they can use to motivate their students.

As noted by one of the teachers the word "inference" is not used in the Foundation to Year 10 *Australian Curriculum: Mathematics* (ACARA 2012b). Students are asked to compare box plots, histograms, and dot plots and "draw conclusions" in grade 10 (p. 63) but no criteria are given for making judgments. Random samples are introduced in grade 8 but not mentioned specifically again. In this study there were several classroom issues, as noted in the Methodology, that kept the outcomes from reflecting what the researcher considered an ideal learning context. Conducting research in the perfect classroom with students on task 100% of the time and teachers following plans exactly, is every researcher's dream; but if outcomes are to be translated to the wider context of typical schools and classrooms, perhaps it is not responsible to claim that the ideal outcomes should be expected there. Whether resampling is a realistic topic to introduce in grade 10 awaits further research in other school contexts.

The model suggested by Biehler and Prömmel (2010) would be ideal for the opportunity to work intensely initially with students to build up the skills of simulation and an appreciation of the importance of sample size but at this time the curriculum in Australia does not free enough time for such an endeavor. The issue of whether instructions are best given on paper handouts, within *TinkerPlots* files, or orally by the teachers, to increase time-on-task and efficiency is one topic to be considered in future research. Situations where students have been using *TinkerPlots* for several years previously in the high school with more straightforward investigations both in data and chance would save much effort at the beginning of a unit such as this. As recognition increases of the General Capability for Information and Communication Technology (ICT) in the *Australian Curriculum*, where students "employ their ICT capability to perform calculations, draw graphs, collect, manage and interpret data" (p. 11), perhaps more time will become available.

Acknowledgments This project was funded in association with the University of Tasmania Excellence in Research Medal for 2011. The author gratefully acknowledges the assistance of Research Fellow, Dr Sue Stack.

References

Arnholt, A. T. (2007). Resampling with R. *Teaching Statistics, 29*(1), 21-26.

Australian Curriculum, Assessment and Reporting Authority (ACARA) (2012a). *Draft senior secondary curriculum – Specialist mathematics*. Sydney: Author. Retrieved from www.acara.edu.au/verve/_resources/specialist+mathematics.pdf#xml.

Australian Curriculum, Assessment and Reporting Authority (ACARA) (2012b). *The Australian Curriculum: Mathematics, Version 3.0, 23 January 2012*. Sydney, NSW: ACARA.

Bakker, A. (2002). Route-type and landscape-type software for learning statistical data analysis. In B. Phillips (Ed.), *Proceedings of the Sixth International Conference of Teaching Statistics: Developing a statistically literate society, Cape Town, South Africa* [CD-ROM]. Voorburg, The Netherlands: International Statistical Institute.

Biehler, R. (1997). Software for learning and for doing statistics. *International Statistical Review, 65*(2), 167-189.

Biehler, R. (2006). Working styles and obstacles: Computer-supported collaborative learning in statistics. In A. Rossman, & B. Chance (Eds.), *Proceedings of the Seventh International Conference on Teaching Statistics: Working cooperatively in statistics education, Salvador, Brazil*. [CD-ROM]. Voorburg, The Netherlands: International Association for Statistical Education and the International Statistical Institute.

Biehler, R., Ben-Zvi, D., Bakker, A., & Makar, K. (2013). Technology for enhancing statistical reasoning at the school level. In M. A. Clement, A. J. Bishop, C. Keitel, J. Kilpatrick, & A. Y. L. Leung (Eds.), *Third international handbook on mathematics education* (pp. 643-689). New York: Springer.

Biehler, R., & Prömmel, A. (2010). Developing students' computer-supported simulation and modelling competencies by means of carefully designed working environments. In C. Reading (Ed.), *Data and context in statistics education: Towards an evidence-based society. Proceedings of the 8th International Conference on the Teaching of Statistics, Ljubljana, Slovenia*. [CD-ROM] Voorburg, The Netherlands: International Statistical Institute.

Christie, D. (2004). Resampling with Excel. *Teaching Statistics, 26*(1), 9-14.

Clements, C., Erickson, T., & Finzer, B. (2007). *Exploring statistics with Fathom Dynamic Data Software Version 2.* Emeryville, CA: Key Curriculum Press and Key College Press.

Cobb, G. (2007). The introductory statistics course: A Ptolemaic curriculum? *Technology Innovations in Statistics Education, 1*(1), Article 1. Retrieved from http://escholarship.org/uc/item/6hb3k0nz.

Common Core State Standards Initiative (2010). *Common Core State Standards for Mathematics.* Retrieved from http://www.corestandards.org/the-standards.

Kahneman, D., & Tversky, A. (1972). Subjective Probability: A judgement of representativeness. *Cognitive Psychology, 3*, 430-454.

Key Curriculum Press. (2005). *Fathom. Dynamic Data Software, Version 2.01.* [Computer software]. Emeryville, CA: KCP Technologies.

Konold, C. (1994). Teaching probability through modeling real problems. *Mathematics Teacher, 87*(4), 232-235.

Konold, C., & Miller, C. D. (2011). *TinkerPlots: Dynamic data exploration* [computer software, Version 2.0]. Emeryville, CA: Key Curriculum Press.

Maxara, C., & Biehler, R. (2006). Students' probabilistic simulation and modeling competence after a compter-intensive elementary course in statistics and probability. In A. Rossman, & B. Chance (Eds.), *Proceedings of the Seventh International Conference on Teaching Statistics: Working cooperatively in statistics education, Salvador, Brazil.* [CD-ROM]. Voorburg, The Netherlands: International Association for Statistical Education and the International Statistical Institute.

Maxara, C., & Biehler, R. (2010). Students' understanding and reasoning about sample size and the law of large numbers after a computer-intensive introductory course on stochastics. In C. Reading (Ed.), *Data and context in statistics education: Towards an evidence-based society. Proceedings of the 8th International Conference on the Teaching of Statistics, Ljubljana, Slovenia.* [CD-ROM] Voorburg, The Netherlands: International Statistical Institute.

National Council of Teachers of Mathematics (2009). *Focus in high school mathematics: Reasoning and sense making.* Reston, VA: Author.

Rossman, A. J. (2008). Reasoning about informal statistical inference: One statistician's view. *Statistics Education Research Journal, 7*(2), 5-19. Retrieved from http://www.stat.auckland.ac.nz/serj.

Rossman, A., & Chance, B. (2008). *Concepts of statistical inference: A randomization-based curriculum.* San Luis Obispo, CA; California Polytecnic University. Retrieved from http://statweb.calpoly.edu/csi.

Scheaffer, R., & Tabor, J. (2008). Statistics in the high school mathematics curriculum: Building sound reasoning under uncertain conditions. *Mathematics Teacher, 102*(1), 56-61.

Shaughnessy, J. M., Chance, B., & Kranendonk, H. (2009). *Focus in high school mathematics reasoning and sense making: Statistics and probability.* Reston, VA: National Council of Teachers of Mathematics.

Taffe, J., & Garnham, N. (1996). Resampling, the bookstrap and Minitab. *Teaching Statistics, 18*(1), 24-5.

Tintle, N., VanderStoep, J., Holmes, V-L., Quisenberry, B., & Swanson, T. (2011). Development and assessment of a preliminary randomization-based introductory statistics curriculum. *Journal of Statistics Education, 19*(1). Retrieved from http://www.amstat.org/publications/jse/v19n1/tintle.pdf.

Watson, J. M. (2013). Resampling with TinkerPlots. *Teaching Statistics, 35*(1), 32-36.

Watson, J., Beswick, K., Brown, N., Callingham, R., Muir, T., & Wright, S. (2011). *Digging into Australian data with TinkerPlots.* Melbourne: Objective Learning Materials.

Watson, J., & Chance, B. (2012). Building intuitions about statistical inference based on resampling. *Australian Senior Mathematics Journal, 26*(1), 6-18.

Appendix

Summary of Lessons Developed for Study

Lesson 1: The goal was a rapid introduction to *TinkerPlots* using extracts from lessons and data sets in *Digging into Australian data with TinkerPlots* (Watson et al., 2011). Data on Australian venomous snakes or prime ministers would introduce data cards, attributes, and plots; body measurements could reinforce box plots and scatterplots, both in the Australian curriculum; and data on two different football codes would provide mystery data and use other *TinkerPlots* skills and introduce creating a formula.

Lesson 2: This lesson introduced the Sampler in *TinkerPlots*, the object upon which resampling is based. Here it was used to simulate the rolling of a single die or of two dice together with the sum of their outcomes, with more trials achieving results closer to the theoretical expectation.

Lesson 3: Again the Sampler was used in the probability context, reinforcing the importance of sample size in simulating the tossing of 10 coins or 30 coins many times to decide which setting was likely to have more than 60% heads. This was a problem developed from Kahneman and Tversky's (1972) Hospital problem.

Lesson 4: The modelling process was extended to introduce the History button to keep track of repeated samples collected by the Sampler, a process required for resampling. The context for this lesson was determining family size with a population policy allowing families to have children until the first son was born.

Lesson 5: This lesson demonstrated the use of the Sampler for random sampling from a finite population, the task required for resampling from a randomized experiment. The context was the population of 780 convicts arriving on the First Fleet to Australia in 1788. Random sampling took place to determine if a particular ship was typical of the entire fleet.

Lesson 6: This lesson, the first to introduce resampling, had two versions: one if the teacher wanted the class to collect its own data on memorizing meaningful or nonsense words and another if the data from Shaughnessy et al. (2009) were used instead. The Sampler was used to randomly reallocate the number of words remembered to each treatment, the Ruler measured the difference in medians for the two conditions, and the History collection kept track of the difference in medians for each resampling. These could be compared to the actual difference observed in the experiment to see if the actual difference appeared to be random (like others) or unusual.

Lesson 7: The second setting for resampling was the Swimming with Dolphins study (Rossman 2008) based on data in a two-way table for 30 participants suffering mild to moderate depression. Again the Sampler and History collection were the basis of decision-making.

Lesson 8: This lesson was meant to provide the opportunity to assess student skills with *TinkerPlots* and understanding of resampling by asking for an analysis of two pairs of data sets on student reaction times. In one case there was expected to be a likely difference, whereas in the other, there was no difference. [In the end, this lesson was not used.]

Chapter 31
Math-Bridge: Closing Gaps in European Remedial Mathematics with Technology-Enhanced Learning

Sergey Sosnovsky, Michael Dietrich, Eric Andrès

CeLTech, German Research Center for Artificial Intelligence (DFKI)

George Goguadze

Leuphana-Universität Lüneburg

Stefan Winterstein

CeLTech, German Research Center for Artificial Intelligence (DFKI)

Paul Libbrecht

CERMAT, Martin-Luther University of Halle Wittenberg

Jörg Siekmann, Erica Melis[†]

CeLTech, German Research Center for Artificial Intelligence (DFKI)

Abstract Math-Bridge is an e-Learning platform for online courses in mathematics. It has been developed as a technology-based educational solution to the problems of bridging courses taught in European universities. Math-Bridge has a number of unique features. It provides access to the largest in the World collection of multilingual, semantically annotated learning objects (LOs) for remedial mathematics. It models students' knowledge and applies several adaptation techniques to support more effective learning, including personalized course generation, intelligent problem solving support and adaptive link annotation. It facilities a direct access to LOs by means of semantic search. It provides rich functionality for teachers allowing them to manage students, groups and courses, trace students'

[†] Dr. Erica Melis was the initiator and the coordinator of the Math-Bridge project. She passed away during the implementation of Math-Bridge.

progress with the reporting tool, create new LOs and assemble new curricula. Math-Bridge offers a complete solution for organizing technology-enhanced learning (TEL) of mathematics on individual-, course- and/or university level.

1 Introduction

Many students enrolling into European colleges and universities lack mathematical competencies necessary for their studies, especially, in math-intensive engineering and science disciplines (ACME 2011). This often leads to serious learning problems, and even causes students to drop out of their learning programs. For instance, drop-out rates for most engineering disciplines in Germany grew about 10% over the last 15 years and now stay at 25-35%. Drop-out rates for math-intensive science study programs in German universities and colleges have also grown across the board to 15-40% (Heublein et al. 2008). Similar figures apply for several other countries of EU, such as the Netherlands, Germany, Spain, UK, etc.

One source of the problem is that many students simply underestimate the requirements to their mathematical background when selecting a study program. This often results in confusion, frustration and lack of motivation to continue the study. On the other hand, the schools themselves cannot always provide skilful teachers and enough content to prepare their pupils for the university-level mathematics courses. Although high drop-out rates cannot only be solved by focusing on mathematics education, an early opportunity to close the competency gap and increase students' awareness about the of university programs requirements can help them to make an informed choice of future studies, prepare properly and avoid potential detrimental effects on motivation.

In order to facilitate pragmatic efforts in this direction, and take a significant step towards improving European educational practices in the field of remedial mathematics, in 2009, a consortium of educators, mathematicians and computer scientists from nine universities and seven countries initiated the project Math-Bridge[1]. This paper presents the technical side of this project. Section 2 underlines the problems of existing remedial courses offered by individual European universities. Section 3 briefly summarizes the most important aspects of the approach implemented by Math-Bridge. Sections 3 to 6 provide the details of the Math-Bridge technology including the design and implementation of the developed e-learning platform, characteristics of the accumulated collection of digital mathematical content and the main functionality available to the user of Math-Bridge, both students and teachers. Section 7 concludes the paper.

[1] This work has been supported by the EU eContentplus project "Math-Bridge: European Remedial Content for Mathematics" (ECP-2008-EDU-428046).

2 Problems of existing remedial courses

Educators across EU have long realized this set of challenges. The common solution is offering to students a dedicated bridging (or remedial) course in the beginning of their study program. Although these courses might help to remedy the problem in the universities administering them, there are several drawbacks that prevent implementation of a broader, cheaper and more effective solution.

Accessibility
To improve the situation, access to bridging mathematical content has to be provided not only on paper during remedial sessions at the university, but constantly on the Web. When (or even before) choosing a university/college, students should be able to prepare for the requirements of their study subjects.

Support for cross-cultural and multi-lingual access
The target competencies that students need to master often vary in description and names across countries, study programs and universities. It is not easy for prospective students, to get access to such requirements and learn them in advance. A solution is necessary that would provide a mapping between different institutions' curricula/requirements to a more generic set of competencies.

At the same time, mathematical notations themselves often vary across countries, which makes understanding of math content much more difficult for international students. Naturally, this problem also occurs when the content is available only in one language. With student mobility increasing and Bologna process gaining momentum, the number of international students in European universities grows. For international students, transition to the university-level mathematics is aggravated by the necessity to learn it in a foreign language that they did not use when taking the mathematics courses in a school (Setati et al. 2011).

A related issue is the lack of appropriate metadata that would otherwise facilitate the discovery of learning content in a desired language.

Content reuse and interoperability
Even when the remedial content for mathematics is digitized, it lacks the most basic support for content discoverability and interoperability. This means, the content cannot be easily found, altered and reused by anybody but its authors. Content authoring for educational systems is a tedious and expensive procedure. Therefore, it is crucial to ensure that the high-quality remedial content becomes easily available for the third-party users.

The problems often start with the formats in which this content exists. Many teachers simply distribute their course materials as Word or TeX-documents, thus supporting no basic means for content reuse. Moreover, a large part of the content is compartmentalized by publishers or universities. It is not annotated with proper metadata, i.e. does not carry the semantic (mathematical) information that would be necessary to discover and retrieve the appropriate LOs easily.

In order to realize the full potential of WWW as the infrastructure for content delivery, the remedial mathematical content should be:

1. Implemented using standard-based semantic representation (XML and RDF-based formats provide the basic means for content reuse across systems and contexts) (Buswell et al. 2004; Carlisle et al. 2010; Kohlhase 2006);
2. Dissected into individual LOs (this way, meaningful pieces of learning material can be discovered, and reused, the content can be re-assembled according to the curricular of different courses, intelligent e-learning applications can present to the students only the most appropriate learning resources);
3. Provided with open-standard metadata, e.g. (IEEE 2002; IMS 2003; ADL 2009) (which will facilitate content discovery, and enable adaptive access to the content based on its characteristics and the learning needs of the student);
4. Authored in such a way that the meaning of mathematical symbols is decoupled from their rendering in the browser (this is essential for supporting multi-lingual and multi-cultural access to the content).

Interactivity
To support meaningful learning experiences and enhance students' engagement, remedial content collections must have interactive LOs that can react to students' actions by providing proper feedback. Unfortunately, the dominant type of learning content available online is instructional texts and lecture slides.

Infrastructures for students, teachers, and authors
Students' motivation, self-assessment, and performance can be improved by implementing a range of TEL approaches supporting adaptive access to LOs, timely assessment of student knowledge, self-organized and exploratory learning. This functionality is not available for them within the current bridging courses.

An adequate infrastructure for effective authoring of learning resources is also missing. Developing an individual LO or assembling an entire course remains a very complex task poorly supported by the majority of existing authoring tools.

A teacher of a bridging course must be able to easily manage all key aspects of the course, including student, course material, assessment tests, etc. A teacher should be also provided with rich facilities for monitoring students' progress, detecting potential learning problems and intervening if necessary.

3 Math-Bridge: the Approach

The Math-Bridge project has addressed the outlined set of problems by applying a range of techniques from the fields of Intelligent Tutoring Systems, Adaptive Hypermedia, Semantic Web and TEL. The result of these efforts is the e-Learning platform for bridging courses with a number of unique features. It provides multilingual and multi-cultural semantic access to remedial mathematics content, which adapts to the requirements of a learner and his/her subject of study. It brings together content from different European sources and offers it in a unified way. This access is provided through an online Pan-European learning service for remedial mathematics, which is built by collecting appropriate learning resources, extending them in terms of structure and multi-linguality and making them useful and easy-to-find. The extended formats of the content makes a wider use of standards and, hence, enables content reusability and transferability between different learning environments. In order to achieve this, an analysis of the target bridging competencies required for target subjects of study has been performed, the semantic and multi-lingual search software has been developed, the assessment tools have been implemented, and a range of remedy-scenario has been authored, which includes specific diagnostic means and decisions for the transition from school to higher education. The solution implemented by Math-Bridge has been based on achieving several operational objectives:

1. To collect and harmonize high-quality remedial content developed by experts in bridging-level mathematics and make this content Web-accessible;
2. To enable cross-cultural and multi-lingual presentation of this content, thus, promoting its reuse across the borders;
3. To motivate the technological reuse of the content by implementing it in a shareable format and enriching it with metadata based on open standards;
4. To offer different types of personalized access to the content, thus supporting multiple usage scenarios of the platform: from individual exploratory e-Learning to classroom-based knowledge training and testing;
5. To foster the platform adoption by increasing its usability not only for students, but also for other stakeholders, including teachers and university officials.

4 Math-Bridge: Content and Knowledge Base

Mathematical Content Collections
The Math-Bridge content base consists of several collections of learning material covering the topics of secondary and high school mathematics. They were originally developed for teaching bridging courses by mathematics educators from several European universities participating in the project.

The content collections have been transformed through a sequence of operations in order to enable discoverability, interoperability and adaptability of LOs constituting them. Fig. 31.1 presents the complete procedure of content transformation, step-by-step. The resulting database of remedial content is available as a collection of individual LOs, transformed into the OMDoc format for mathematical documents (Kohlhase 2006) and provided with metadata.

Compared to the majority of adaptive e-Learning applications, Math-Bridge supports a multitude of LO types. The OMDoc language used for representing content in Math-Bridge defines a hierarchy of LOs to describe the variety of mathematical knowledge. On the top level, LOs are divided into concept objects and satellite object. Satellite objects are the main learning activities, they structure the learning content, which students practice with: exercises, examples, and instructional texts. Concept objects have a dualistic nature: they can be physically presented to a student, and s/he can browse them and read them; at the same time they are used as elements of domain semantics, and, as such, employed for representing knowledge behind satellite objects and modelling students' progress.

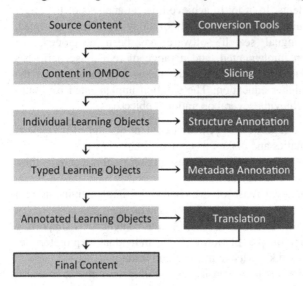

Fig. 31.1 Content transformation procedure (left: evolution of the content; right: stages of the procedure)

There are five main types of concept object available in Math-Bridge (see Fig. 31.2). Symbol is a special kind of concept objects. They represent the most abstract entities in Math-Bridge, atomic mathematical concepts, which do not have content of their own. A symbol has a representation – a physical manifestation, which can be shown to a student and used in equations, but there is no actual learning content behind a symbol. Math-Bridge symbols are combined into an ontology, which is described on page 444.

The rest of the concept objects model the most typical mathematical notions:

- Definition is a statement, indicating meaning of one or several symbols;
- Axiom is a postulate about one or several symbols;
- Assertion is a statement about symbols; there can be several types of assertions, such as theorem, lemma, and corollary;
- Proof represents a formal inference of an assertion and is always connected to the assertion it proves.

The total number of LOs in the Math-Bridge content base is almost 11,000. Table 31.1 presents the details of this content base broken down by LO type.

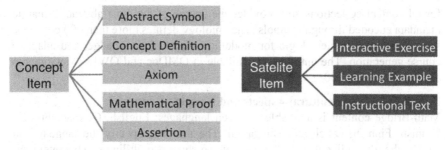

Fig. 31.2 Hierarchy of LO types in Math-Bridge

Table 31.1 Different types of LOs in the Math-Bridge content collection

Type	Count
Interactive Exercise	5060
Instructional Text	2142
Learning Example	1563
Concept Definition	950
Assertion	634
Mathematical Proof	334
Axiom	36

Metadata Schema

The content collections of Math-Bridge consist of multiple LOs of different types. Both, the collections and the individual LOs are described in terms of their properties and attributes. The LOs link to each other by multiple relations. Both, the attributes describing LOs and the relations linking them together are specified via various metadata. The metadata elements can be divided into the following three categories:

- descriptive metadata used for administrative, cataloguing and licensing purposes; represented mainly using the Dublin Core standard (DCMI 1999).
- pedagogical metadata helping authors to specify multiple educational properties of LOs; adopted and extended from IEEE LOM and IMS LD standards (IEEE 2002; IMS 2003).
- semantic metadata connecting different LOs to one another; partially relaying on OMDoc and SKOS standards (Kohlhase 2006; Miles and Bechhofer 2009).

Math-Bridge metadata plays the core role in the overall architecture of the platform. It enables LOs discovery, course composition, students' knowledge tracing and subsequent adaptation of the learning content.

Math-Bridge Ontology

In order to model the domain of bridging mathematics, an ontology for the target subset of mathematical knowledge has been created. It serves as a reference point for all content collections and provides the source of the most abstract semantic metadata encoded through symbols. The ontology defines more than 600 concepts. It is used by the system logic for modelling students' knowledge, and adaptive course generation. The ontology is available in OMDoc and OWL[2] (McGuinness and van Harmelen 2004).

Multilingual/Cross-cultural Aspects and Notation Census
Math-Bridge content is available in seven languages: English, German, French, Spanish, Finnish, Dutch and Hungarian. The user can specify the language, in which s/he would like to read the content. To support multilingual students, individual LOs can be translated on the fly. It is important to mention, that Math-Bridge translates not only the text but also the presentation of formulæ. Although mathematics is often called a "universal language", this is not fully true. In many countries, the same mathematical concepts use very different symbols (Libbrecht 2010). In order to address this challenge, Math-Bridge separates the semantic and the presentation layer of math symbols. Inside the content, symbols are encoded using unambiguous entities, and when presented to the user, a correct notation is chosen based on the current language (Melis et al. 2009). A public "notation census" has been conducted to document different notations of all symbols in all languages[3]. Table 31.2 presents the quantities of LO translations available for each language.

Table 31.2 Different types of LOs in the Math-Bridge content collection

Language	Count
English	9792
German	9792
Spanish	5099
Dutch	5484
French	4391
Finnish	5149
Hungarian	5905

5 Math-Bridge: Technology-Enhanced Learning of Mathematics

The Math-Bridge platform provides students with multilingual, semantic and adaptive access to mathematical content. It has been developed based on the ActiveMath technology, and can be considered the next phase in the evolution of the ActiveMath intelligent tutoring system (Melis et al. 2001; Melis et al. 2006; Melis

---

[2] http://www.math-bridge.org/content/mathbridge.owl

[3] http://wiki.math-bridge.org/display/ntns/Home

et al. 2009). Fig. 31.3 presents the dashboard of Math-Bridge. This is the entry point to the platform that every user sees after the login. Its interface consists of several widgets that provide access to different facilities available within Math-Bridge: regular and personalized courses, questionnaires, tests, and bookmarks.

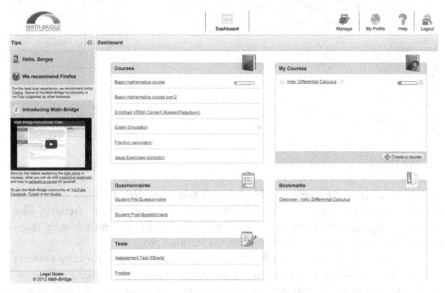

Fig. 31.3 Math-Bridge dashboard

If a user clicks on any of the courses, the student interface of Math-Bridge will launch (Fig. 31.4). It consists of three panels. The left panel is used for navigation through learning material using the topic-based structure of the course. The topics can have subtopics and can be folded and unfolded.

When one of the bottom-level topics is chosen, the content page associated with this topic is presented in the central panel. Each page can consist of multiple LOs that a student can read. Exercise LOs have a button "Start Exercise", which launches this particular interactive exercise in a separate tab. Students can also open new tabs when opening the results of the search and browsing through the LO metadata.

Whenever a LO is clicked on the content page, it gains focus. The right panel provides access to the details of the LO currently in focus, as well as some additional features, such as semantic search and social feedback toolbox. It can also be used for on-demand individual translation of the LO in focus to any language available for it, while the system interface and the rest of the content remain in the original language. This is one of the ways Math-Bridge supports international students. Alternatively, the entire system interface and course material can be accessed in any of the supported languages; the language can be specified at login.

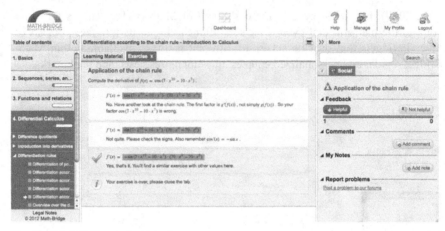

Fig. 31.4 Math-Bridge student interface

Tracing Students' Progress and Modeling Their Knowledge

Math-Bridge logs every student interaction with learning content. Actions, like loading a page or answering an exercise are stored in the student's history database and help tracing and interpreting her/his learning progress.

The results of interactions with exercises (correct/ incorrect/ partially correct) are used by the student-modelling component of Math-Bridge to produce a meaningful estimation of the student's progress. The student model of Math-Bridge supports multi-layered representation of student's masteries. For every concept in the domain the model maintains a set of values estimating the probabilities that the student has mastered individual mathematical competencies associated with this concept. Every exercise in Math-Bridge is linked with one or several concepts (symbols, theorems, definitions etc.) and the competencies that the exercise is training for these concepts. A correct answer to the exercise is interpreted by the system as evidence that the student advances towards mastery and will result in the increase of probabilities for the corresponding concepts and competencies in this student's model. (Faulhaber and Melis 2008) provides the architectural and implementation details of this student modelling mechanism.

Personalized Courses

The course generator component of Math-Bridge allows students to automatically assemble a course optimized for their needs and adapted to their knowledge and competencies based on the current states of their student models. To generate a course, students need to select the target topics and a learning scenario. Several scenarios are available within Math-Bridge: a student can choose to explore a new topic, train a particular competency, prepare for an exam, master a previous topic or a assemble a course that will focus on the current gaps in student's knowledge (Ullrich 2008; Biehler et al. 2012). Each course type is generated based on a set of pedagogical rules defining the top-level structure of the course and the learning goals. The generation tool queries the student model and the metadata storage in order to assemble a didactically valid sequence of LOs. Pedagogical metadata (such as exercise difficulty) and semantic metadata (such as prerequisite-outcome relations) play the central role in this process. On the interface level, generation of a single course is a simple 4-step process (see Fig. 31.5). After that, the generated course appears in the "My Courses" widget, and the student can access it the same way s/he accesses standard pre-defined courses.

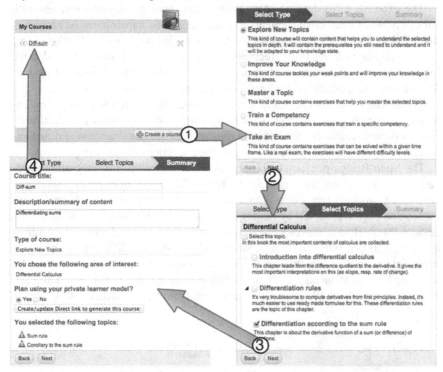

Fig. 31.5 Course generation in Math-Bridge: 1) Click the button "Create a Course" in the "My Courses" widget of the Dashboard; 2) Choose one of the six course generation scenarios; 3) Select one or more target topics; 4) Name the course.

Adaptive Navigation Support

The amount of content available within Math-Bridge is massive. Some of the pre-defined bridging courses consist of thousands of LOs. Math-Bridge helps students finding the right page to read and/or the right exercise to attempt by implementing a popular adaptive navigation technique – adaptive annotation (Brusilovsky et al. 2009). The annotation icons show the student how much progress s/he has achieved for the corresponding part of learning material. Math-Bridge computes annotations on several levels: each course in the Math-Bridge dashboard (Fig. 31.3, widget "Courses"), each topic within a course table of contents (Fig. 31.4, left panel) and each content page under a topic are provided with individual progress indicators aggregating student's learning activity on the corresponding level.

Interactive Exercises and Problem Solving Support

Interactive exercises play two important roles in Math-Bridge. First of all, they maintain constant assessment of students' knowledge thus providing the input for the student-modelling component. Second, they give students the opportunity to train mathematical competencies and apply in practice theoretical knowledge acquired by reading the rest of the content.

The exercise subsystem of Math-Bridge can serve multi-step exercises with various types of interactive elements and rich diagnostic capabilities. At each step, Math-Bridge exercises can provide students with instructional feedback ranging from mere flagging the (in)correctness of given answers to presenting adaptive hints and explanations (the central panel of the student interface shown by Fig. 31.4 presents an example of feedback produced by a Math-Bridge exercise).

Math-Bridge can automatically generate interactive exercises powered by external domain reasoner services. Currently, Math-Bridge employs a collection of IDEAS domain reasoners for stepwise diagnosis of students' actions and generation of advanced feedback on every step of their solutions (Heeren et al. 2010).

The Math-Bridge platform also implements functionality for integrating third-party exercise services that maintain the full cycle of student-exercise interaction. As a result, students can access within Math-Bridge both, native Math-Bridge exercises and exercises served by remote systems. The integration is seamless for the student (Math-Bridge makes no difference in how native and external exercises are launched) and fully functional (Math-Bridge makes no difference in how students' interactions with native and external exercises are logged and interpreted by its modelling components). Currently Math-Bridge integrates two external exercise systems: STACK (Sangwin 2008) and mathe-online (Embacher 2006).

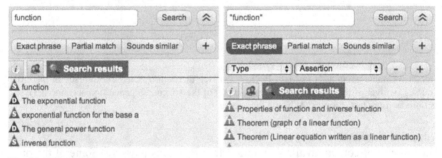

Fig. 31.6 Math-Bridge search tool

Semantic Search of LOs

In addition to navigating through the course topics, students have a more direct way to find LOs of their interest – by using the Math-Bridge search tool. They can use default search based on simple string matching, advanced search that allows more precise specification of general search parameters (exact or practical matching, lexical or phonetic matching) and semantic search. The semantic search mode fully utilizes the advanced metadata schema of Math-Bridge. Students can specify the type of the desired LO (e.g. only exercises), its difficulty (e.g. only easy exercises), its target field of study (e.g. only easy exercises designed for physics students), etc. Fig. 31.6 presents the interface of the Math-Bridge search tool. The left part shows the results of simple search with the word "function"; the right part shows the results of semantic search with the same word and the target type of LOs restricted to assertions.

6 Technology-Enhanced Teaching of Mathematics

Math-Bridge offers teachers and university IT specialists a complete arsenal of tools necessary to setup, administer and teach online courses.

Content and Course management
Teachers can create their courses from scratch or reuse one of the existing tables of contents. They can design assessment tests, exams and questionnaires, and author individual LOs and collections of new material.

User and Group Management
There are three categories of users in the system. Students can access learning content individually or as a part of a course. Teachers can manage their courses, including content visibility and student roster. Administrators have access to all aspects of Math-Bridge user management. They can change user parameters and rights, modify group membership, and assign a teacher to a course. Naturally, administrators can also do everything that other users can.

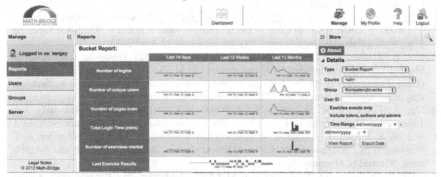

Fig. 31.7 Math-Bridge teacher interface: reporting tool

Course Monitoring with the Reporting Tool
It is easy for teachers to monitor students' progress within Math-Bridge: a dedicated reporting tool allows them to trace individual student's performance or results of the entire class. The reporting tool can also help in discovering potentially problematic LOs (e.g. an exercise that nobody has solved correctly). Fig. 31.7 presents an outcome example of an aggregated report on the system usage. Overall, Math-Bridge provides teachers with about 10 different reports.

7 Conclusion

Math-Bridge is a full-fledged e-Learning platform developed to help individual learners, classes of students, as well as entire schools and universities to achieve their real-life educational goals. Math-Bridge implements a number of advanced technologies to support adaptive and semantic access to learning content. Fostering the adoption of these technologies by the general public is the primary goal of Math-Bridge. The platform has been evaluated in several experiments with more than 3000 students from nine universities and seven European countries. Detailed results of two of these experiments can be found in (Biehler et al. 2013) and (Kangas et al. 2012). Math-bridge has been used under different scenarios: as the main learning platform in a distant course, as an online component in a blended course, and as a supplementary tool in a traditional course. Under such diversity of usage scenarios and educational settings, it has been confirmed that students learn with Math-Bridge and that they in general feel very positive about using the tool. Further experiments are required to detect other effects of using Math-Bridge for learning and teaching remedial mathematics.

References

ACME. (2011). *Mathematical needs: the mathematical needs of learners. Report. Advisory Committee on Mathematics Education.*
http://www.acme-uk.org/media/7627/acme_theme_b_final.pdf. Accessed 12 April 2013.
ADL. (2009). SCORM 2004 4th Edition. Documentation suite. Advanced Distributed Learning. http://www.adlnet.gov/scorm/scorm-2004-4th. Accessed 12 April 2013.
Biehler, R., Fischer, P. R., Hochmuth, R., & Wassong, T. (2012). Designing and evaluating blended learning bridging courses in mathematics. In M. Pytlak, T. Rowland, & E. Swodoba (Eds.), *Proceedings of the 7th congress of the European society for research in mathematics education, Rzeszow, Poland* (pp. 1971-1980). University of Rzeszów.
Biehler, R., Fischer, P. R., Hochmuth, R., & Wassong, T. (2013). Eine Vergleichsstudie zum Einsatz von Math-Bridge und VEMA an den Universitäten Kassel und Paderborn. In I. Bausch, R. Biehler, R. Bruder, P. R. Fischer, R. Hochmuth, W. Koepf, S. Schreiber, & T. Wassong (Eds.), *Mathematische Vor- und Brückenkurse: Konzepte, Probleme und Perspektiven*. Wiesbaden, Germany: Springer Spektrum.
Brusilovsky, P., Sosnovsky, S., & Yudelson, M. (2009). Addictive links: the motivational value of adaptive link annotation. *New Review of Hypermedia and Multimedia, 15*(1), 97-118.
Buswell, S., Caprotti, O., Carlisle, D. P., Dewar, M. C., Gaëtano, M., & Kohlhase, M. (Eds.). (2004). *The OpenMath standard. Standard specification*. The OpenMath society. http://www.openmath.org/standard/om20-2004-06-30/omstd20.pdf. Accessed 12 April 2013.
Carlisle, D., Ion, P., & Miner, R. (Eds.). (2010). *Mathematical markup language (MathML) version 3.0: W3C recommendation*. http://www.w3.org/TR/MathML3/. Accessed 12 April 2013.
DCMI. (1999). *Dublic core metadata initiative. Standard specifications*. http://dublincore.org/. Accessed 12 April 2013.
Embacher, F. (2006). The didactical significance of interactive animations. In *Proceedings of Dresden international symposium on technology and its integration into mathematics education. Dresden, Germany.*

Faulhaber, A., & Melis, E. (2008). An efficient student model based on student performance and metadata. In M. Ghallab, C. Spyropoulos, N. Fakotakis, & N. Avouris (Eds.), *Proceedings of 18th European conference on artificial intelligence*, (pp. 276-280). IOS Press.

Heeren, B., Jeuring, J., van Leeuwen, A., & Gerdes, A. (2010). Specifying Rewrite Strategies for Interactive Exercises. *Mathematics in Computer Science, 3*(3), 349–370.

Heublein, U., Schmelzer, R., & Sommer, D. (2008). *Die Entwicklung der Studienabbruchquote an den deutschen Hochschulen: Ergebnisse einer Berechnung des Studienabbruchs auf der Basis des Absolventenjahrgangs 2006. Projektbericht.* HIS Hochschul-Informations-System GmbH. http://www.his.de/pdf/21/20080505_his-projektbericht-studienabbruch.pdf. Accessed 12 April 2013.

IEEE. (2002). *IEEE standard for learning object metadata. Approved publication of IEEE (Designation: 1484.12.1-2002).* IEEE Computer Society/Learning Technology Standards Committee. http://ltsc.ieee.org/wg12/. Accessed 12 April 2013.

IMS. (2003). *IMS learning design specification. Version 1.0 final specification The IMS Global Learning Consortium.* http://www.imsglobal.org/learningdesign/. Accessed 12 April 2013.

Kangas, J., Miilumäki, T, & Pohjolainen, S. (2012). Mathematics remedial instruction with Math-Bridge elearning system. In *Proceedings of 16th SEFI MWG Seminar on mathemerics education of engineers. June 28-30, 2012, Slamanca, Spain.*

Kohlhase, M. (2006). *OMDoc – An open markup format for mathematical documents [version 1.2].* Berlin/Heidelberg, Germany: Springer Verlag.

Libbrecht, P. (2010). Notations around the world: census and exploitation In S. Autexier, J. Calmet, D. Delahaye, P. Ion, L. Rideau, R. Rioboo, & A. Sexton (Eds.), *Intelligent computer mathematics* (pp. 398-410). Berlin/Heidelberg, Germany: Springer Verlag.

McGuinness, D., & van Harmelen, F. (Eds.). (2004). *OWL Web ontology language overview. W3C recommendation.* http://www.w3.org/TR/owl-features/. Accessed 12 April 2013.

Melis, E., Andres, E., Büdenbender, J., Frischauf, A., Goguadze, G., Libbrecht, P., & Ullrich, C. (2001). ActiveMath: a generic and adaptive Web-based learning environment. *International Journal of Artificial Intelligence in Education, 12*(4), 385-407.

Melis, E., Goguadze, G., Homik, M., Libbrecht, P., Ullrich, C., & Winterstein, S. (2006). Semantic-aware components and services of ActiveMath. *British Journal of Educational Technology, 37*(3), 405-423.

Melis, E., Goguadze, G., Libbrecht, P., & Ullrich, C. (2009). ActiveMath - a learning platform with Semantic Web features. In D. Dicheva, R. Mizoguchi, & J. Greer (Eds.), *Semantic Web technologies for e-Learning* (pp. 159-177). IOS Press.

Miles, A., & Bechhofer, S. (Eds.). (2009). *SKOS: simple knowledge organization system reference. W3C recommendation.* http://www.w3.org/TR/skos-reference/. Accessed 12 April 2013.

Sangwin, C. J. (2008). Assessing Elementary Algebra with STACK. *International Journal of Mathematical Education in Science and Technology, 38*(8), 987-1002.

Setati, M., Nkambule, T., & Goosen, L. (Eds.). (2011). *Proceedings of the ICMI Study 21 Conference: Mathematics and language diversity.* São Paulo, Brazil: ICMI.

Ullrich, C. (2008). *Pedagogically founded courseware generation for Web-based learning - an HTN-planning-based approach implemented in PAIGOS.* Berlin/Heidelberg, Germany: Springer Verlag.

Kapitel 32
Mathematik als Werkzeug: Sicht- und Arbeitsweisen von Studierenden am Anfang ihres Mathematikstudiums

Michael Liebendörfer

Leuphana Universität Lüneburg

Laura Ostsieker

Universität Paderborn

Abstract Die Rolle von mathematischen Weltbildern für das Lernen von Mathematik an der Universität wurde bisher nirgendwo eingehend untersucht. Gleichwohl deutet einiges darauf hin, dass bei der Wahrnehmung von Hochschulmathematik und der Interpretation von Aufgaben falsche Orientierungen eine große Hürde darstellen können. Ausgehend von Beobachtungen an der Universität beschreiben wir Fehler und problematische Ansichten von Studierenden, die mutmaßlich auf schematisch orientierten Beliefs beruhen. Wir skizzieren mögliche Zusammenhänge aus der Literatur und illustrieren das Phänomen dann anhand von Interviewdaten und Aufgabenbearbeitungen.

Einleitung

Im Gegensatz zur Schuldidaktik ist die Hochschuldidaktik in der Mathematik ein breites, wenig beforschtes Gebiet. Speziell bezogen auf Bachelor Mathematik und gymnasiales Lehramt versucht die „Arbeitsgemeinschaft Mathematik im Lehramt Gymnasium und Bachelor Mathematik" (BaGym) im khdm das Dickicht auf diesem Feld zu lichten. Zwangsweise tauchen dabei immer wieder Beobachtungen in Untersuchungen auf, die mit der eigentlichen Forschungsabsicht nichts zu haben. Obwohl die allermeisten solcher Phänomene einem erfahrenen Hochschullehrer bekannt sein dürften, sind nicht viele beforscht oder wissenschaftlich diskutiert. Der vorliegende Artikel will so ein Phänomen umreißen: Bei verschiedenen Projekten der AG BaGym konnten wir beobachten, wie sich im Sprechen und Handeln der Studierenden epistemologische Überzeugungen zeigten, die Mathematik als Anwenden von Werkzeugen sehen und erfolgreichem Lernen z.T. entgegenstehen. Wir schauen dabei auf Äußerungen, die sich nicht bewusst auf das eigene Weltbild beziehen. Unsere Absicht ist dabei weniger, Antworten zu geben, als vielmehr das Problemfeld zu illustrieren. Da diese Beobachtungen nach unserem Kenntnisstand noch nicht eingehend wissenschaftlich untersucht wurden, skizzieren wir die Literatur dort, wo wir mögliche Zusammenhänge sehen. Wir untermauern dieses Bild dann mit Daten aus einigen Interviews, und beschreiben dann die Auswirkungen dieser Haltung anhand von einigen Bearbeitungen konkreter Aufgaben.

Forschungsstand und Hintergrund

Mathematische Weltbilder wurden unter dem Begriff Beliefs ausführlich untersucht, siehe etwa (Grigutsch et al. 1998). Wir greifen dabei eine Vorstellung heraus, die in der Regel „schematisch" genannt wird. Mathematiktreiben ist demnach vorrangig das Anwenden von bekannten Regeln und Formeln.

Aus der Schule ist u.a. durch die Ergebnisse der TIMS-Studie bekannt, dass am Ende der gymnasialen Oberstufe den meisten Schülern vorrangig diese Bedeutung von Mathematik bekannt und wichtig ist: „Zentrales Moment des mathematischen Weltbildes von Oberstufenschülern ist die schematisch-algorithmische Ausrichtung von Mathematik" (Baumert et al., 2000). Diese Erkenntnis stützt sich auf quantitative Erhebungen. Konkret gaben z.B. 5 von 6 Schülern an, dass die Aussage „Mathematik betreiben heißt: allgemeine Gesetze und Verfahren auf spezielle Aufgaben anwenden." stimmt, oder eher stimmt. Dieser Befund korrespondiert z.T. mit den Zielen der Schule, die als einen Schwerpunkt den Aufbau von Prozessfähigkeiten umfassen (vgl. Fischer et al. 2009).

Die Hochschule setzt aber andere Schwerpunkte. In Deutschland wird vor dem Hintergrund hoher Abbruchquoten im Studiengang Mathematik (Heublein et al. 2010) als eine mögliche Ursache in verschiedenen Arbeiten eine Veränderung des Charakters der Mathematik beim Übergang von der Schule zur Hochschule genannt. So schreiben Fischer et al. über den schulischen Mathematikunterricht in Abgrenzung zur Hochschulmathematik: „Ziel des Mathematikunterrichts ist dabei nicht allein, dass Schülerinnen und Schüler analog den Mathematikern eine schulmathematische ‚Theorie' entwickeln und somit für sie neues Wissen konstruieren und akzeptieren, sondern es geht vor allem auch um den Erwerb einer mathematischen Grundbildung, die insbesondere mathematische Prozessfähigkeiten umfasst" (Fischer et al. 2009). Auch Rach und Heinze (2013) beschreiben Unterschiede zwischen der universitären und der schulischen Mathematik, sie sprechen von dem „Werkzeug Mathematik in realen Kontexten im Sinne von Modellierungsaktivitäten", das im Gegensatz zur Schule an der Hochschule wesentlich weniger bedeutsam sei. Im Gegenzug ist an der Hochschule der Beweis von zentraler Bedeutung. Dies wird aber nicht von allen Studierenden erkannt. Ein möglicher Zusammenhang wird ebenfalls bei Rach und Heinze (2013) gegeben: „Wenn man dagegen Mathematik in der Schule eher als instrumentelles Anwendungsfach betreibt, haben mathematische Beweise eine geringe Bedeutung für die Evidenzgenerierung, da hier auch andere Autoritäten (z.B. Schulbuch, Lehrkraft) ausreichen, um die Gültigkeit von Aussagen sicherzustellen".

Die Rolle von Beliefs im Fachstudium wurde in Deutschland trotz dieser vermuteten Bedeutung bisher nicht eingehender untersucht. Aus England sind z.B. in (Daskalogianni und Simpson 2001) Probleme durch unpassende Beliefs und deren problematische Anpassungsprozesse berichtet. Beispielhaft wurde dort der Fall einer Studentin Katherine skizziert, die ihre Beliefs nicht anpassen konnte, obwohl sie den Widerspruch zur Hochschulmathematik aktiv erlebte. Sie konnte in Folge nicht richtig mit neuen Konzepten arbeiten, insbesondere nicht mit Beweisen.

Methode

Interviews

Im Rahmen einer Vorstudie zum Mathematikinteresse wurde am Ende des Wintersemesters 2011/12 an der Universität Kassel eine Gruppendiskussion zum Studienerleben im ersten Semester veranstaltet. Die Teilnehmer waren fünf Freiwillige aus der Vorlesung zur Linearen Algebra 1, denen wir hier neue Namen geben wollen: Anna studierte Physik Bachelor im 3. Semester, Ben gymnasiales Lehramt im 1. Semester, Cora Mathematik Bachelor im 1. Semester, Dirk Mathematik Bachelor im 3. Semester und Erik Mathematik Bachelor im 1. Semester, der ein verwandtes Studium bereits erfolglos abgebrochen hatte.

Zusammen mit zwei Forschern wurde ca. 120 Minuten diskutiert, wobei das Thema nur sehr moderat gelenkt wurde. So haben die Studierenden von sich aus die Frage aufgegriffen, worum es in der Mathematik und auch im Studium nach ihrer Meinung geht. Die Diskussion wurde aufgenommen, transkribiert und unter anderem Fokus (Studienerleben, Interesse) induktiv und deduktiv codiert. Mit der Teilnehmerin Anna folgte ein Einzelinterview.

Aufgabenbearbeitungen

Ein Teil der vorliegenden Aufgabenbearbeitung stammt aus Analysis-1-Klausuren am Ende des Wintersemesters 2010/11 an der Universität Kassel. Die Teilnehmer waren Bachelorstudierende der Fächer Mathematik und Physik und Lehramtsstudierende mit Lehrberechtigung für die Sekundarstufe II. Die erste Klausur wurde von 68 Studierenden geschrieben, an der Nachklausur nahmen 47 Studierende teil.

Die übrigen Aufgabenbearbeitungen stammen aus der Analysis-1-Klausur am Ende des Wintersemesters 2011/12 an der Universität Paderborn. An dieser Klausur haben 96 Studierende der Studiengänge Bachelor Mathematik, Bachelor Technomathematik, Gymnasiales Lehramt, Lehramt Berufskolleg und Informatik mit Nebenfach Mathematik teilgenommen. Die Studierenden waren entweder in ihrem ersten Semester oder sie hatten die Veranstaltung Analysis 1 bereits in einem vorherigen Semester besucht.

Zusätzlich wurden in einer qualitativen Untersuchung ebenfalls in Paderborn Studienanfänger in der fünften Semesterwoche des Wintersemesters 2011/12 beobachtet, wie sie in fünf Kleingruppen von jeweils zwei bis drei Studierenden Aufgaben zum Thema Konvergenz bearbeiten. Sie wurden dazu aufgefordert, die Aufgaben gemeinsam zu lösen und über die Aufgabenbearbeitung zu diskutieren. Dadurch erhielten wir nicht nur das Resultat, die Aufgabenbearbeitung, sondern gewannen zusätzlich Einblick in den Lösungsprozess.

Ergebnisse

Eindrücke aus Interviewdaten

Während des Interviews skizzierten vier der fünf Teilnehmer ein Bild, das die Anwendungs- und Prozessseite der Mathematik sehr betont. Ben schildert z.B.:

> Also ich bin eher so der Praktische, der eher so Zahlen, sehr gut mit Zahlen umgehen kann [...] Das Problem ist nur hier an der Uni war es erst mal so dieser Unterschied zwischen dem selber Rechnen und selber Ergebnisse Erzielen und nur Zuhören, erst mal NICHTS verstehen, das Ganze dann Zuhause nacharbeiten zu müssen, und dann diese Übungsblätter, also es war mir viel zu theoretisch.

Er führt später aus, er habe schon in der Schule „stochastische Probleme über einfache Konsolenanwendungen gelöst" und vermisst an der Uni solche praxisnahen Angebote: „Meinetwegen irgendeine Verschlüsselungssache, die sie dann auch privat irgendwie nutzen könnten". Der Praxisbegriff umfasst dabei nicht nur Tätigkeiten mit sofortigem außermathematischem Nutzen, sondern z.B. auch Rechnen nach vorgegebenen Regeln. Wenn das (wie bei den Übungsaufgaben oft) nicht der Fall ist, sind die Studierenden verunsichert und abgeschreckt. Gerade Beweisaufgaben sind oft nicht mit einem klaren Verfahren zu bearbeiten. Emil beschreibt das so: „die Aufgabe, was da steht ist klar. Aber nicht was jetzt zu tun ist." und erklärt, dass Abschreiben in solchen Situationen wesentlich näher liegt, als sich mit der Aufgabe auseinanderzusetzen. Als erfolgreich wurden Momente beschrieben, in denen klar wird, was zu tun ist, etwa bei Anna: „Da dachten wir, was wollen die von uns, dass man den Körper nachweist, bis man dann raus hatte, man muss die und die Eigenschaften nachweisen."

Die Gegenposition nimmt Cora ein. An der Schule kritisiert sie, dass man „[...] viel so rechnet, auch immer die gleichen Aufgabentypen. Und das fand ich immer langweilig, ich wollte eher immer die Strukturen dahinter verstehen und das äh kommt ja in der Schule ziemlich kurz oft". Glücklich beschreibt sie die Erfahrung an der Universität: „Man hat erst mal eine allgemeine Definition gehabt, das heißt man wusste überhaupt, wovon man eigentlich redet." Es ist überraschend, dass sie, obwohl sie auch Mühen mit dem Stoff hatte, im Gegensatz zu den anderen den Stoff letztlich beherrschte, und z.B. als einzige die Übungsaufgaben selbstständig lösen konnte.

Im nachfolgenden Interview wurde die Position von Anna noch weiter ausgeführt. Es bestätigte sich ihre starke Kalkülorientierung. Erfreut schilderte sie „Du musst nur das wissen: Integral u'v = uv- Integral uv'. Und sofort konnte ich alle partiellen Integrationen rechnen!", oder auch „Diese Kurvenintegrale, von vorne bis hinten, alles schön ausrechnen, das finde ich gut. Und immer noch, ich rechne immer noch gerne. Aber Beweise; man sieht noch nicht wirklich, wie man diese Beweise macht". Das Bild, das sie von Beweisen zeichnet, ist dann auch nicht das von Evidenzgenerierung, sondern eher das von einem neuen Aufgabentyp. Das hatte sich auch bei anderen Studierenden gezeigt (etwa bei Bens „Lösung" stochastischer Probleme durch Simulation, s.o.).

Insgesamt zeigt sich, dass die Mehrheit der Studierenden dieser (sehr kleinen) Studie Mathematik vor allem als Werkzeug verstanden hat, aber nicht als Methode zu (relativer) Wahrheitsfindung. Die einzige Studentin, die davon abweicht, hat die Rolle von Definitionen und Sätzen reflektiert und betont das Ziel des Aufbaus von Wissen, nicht von Rechen- oder Programmierfähigkeiten. Sie ist gleichzeitig die einzige, die in der Lage war, die Übungsaufgaben selbstständig zu lösen und zuversichtlich auf die Klausur schaute.

Im nächsten Teil werfen wir nun einen Blick auf die Bearbeitung von Aufgaben mit typischen Fehlern – auch hier wird sich die Sichtweise auf Mathematiktreiben als Werkzeuganwendung problematisch erweisen.

Bearbeitung von Aufgaben

Die Theorieentwicklung ist ein wesentlicher Bestandteil der Hochschulmathematik. Ein großer Teil dieser Theorie wird in Form von Sätzen formuliert. Beim Lösen von Aufgaben können häufig Aussagen von Sätzen benutzt werden, die zuvor bewiesen wurden. Die Betrachtung verschiedener Bearbeitungen zeigt, dass viele Studierende dabei regelmäßig die Voraussetzungen der Sätze vernachlässigen. So wenden sie Sätze an, ohne zu begründen, warum die Voraussetzungen erfüllt sind, oder wenden sie sogar dann an, wenn die Voraussetzungen nicht erfüllt sind. Dies soll an einigen Beispielen demonstriert werden.

In der qualitativen Untersuchung sollten die Studierenden unter anderem die folgende Aufgabe bearbeiten:

Untersuche die Folge $(a_n)_{n \in \mathbb{N}}$ gegeben durch $a_n = \frac{1-n}{4n-1}$ auf Konvergenz. Gib gegebenenfalls den Grenzwert begründet an und bestimme für ein beliebiges vorgegebenes $\varepsilon > 0$ ein mögliches von ε abhängiges $N \in \mathbb{N}$ im Sinne der Definition der Konvergenz.

Für die Untersuchung des Konvergenzverhaltens betrachtet eine Kleingruppe zunächst die Folge, die durch den Term im Zähler definiert ist, und die Folge, die durch den Kehrwert des Terms im Nenner definiert ist, und möchte den Grenzwertsatz für Produkte darauf anwenden. Das ist hier jedoch nicht zulässig, da die Zählerfolge nicht konvergiert und somit die Voraussetzungen des Grenzwertsatzes nicht erfüllt sind. Die Studierenden hingegen wählen letztendlich lediglich deshalb einen anderen Lösungsweg, weil sie nicht wissen, ob die Folge, die durch den Term im Zähler definiert ist, stärker gegen minus Unendlich strebt als die Folge, die durch den Kehrwert des Terms im Nenner definiert ist, gegen Null, oder umgekehrt.

Gerade in Bezug auf die Grenzwertsätze lässt sich eine solche Herangehensweise bei zahlreichen Studierenden beobachten. Die Grenzwertsätze werden oft lediglich als Rechenregeln verstanden, die Existenzaussage wird vernachlässigt. Dies ist nicht nur zu Beginn des Semesters der Fall, sondern zeigt sich auch noch in der Klausur am Ende des ersten Semesters. Die folgende Aufgabe war dort gestellt:

Sei $(a_n)_{n \in \mathbb{N}}$ eine konvergente Folge reeller Zahlen mit Grenzwert $a \in \mathbb{R}$, sei $(b_n)_{n \in \mathbb{N}}$ eine weitere Folge in $\mathbb{R}$, so dass $b_n \neq 0$ für alle $n \in \mathbb{N}$ und $\lim_{n \to \infty} \frac{a_n}{b_n} = 1$. Zeige, dass $(b_n)_{n \in \mathbb{N}}$ gegen a konvergiert.

Die Bearbeitung in Abbildung 32.1 ist beispielhaft für zahlreiche weitere.

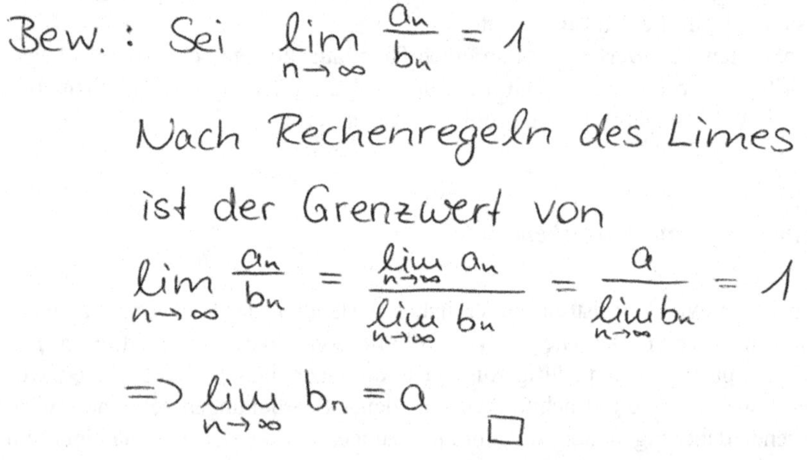

Bew.: Sei $\lim_{n \to \infty} \frac{a_n}{b_n} = 1$

Nach Rechenregeln des Limes ist der Grenzwert von

$$\lim_{n \to \infty} \frac{a_n}{b_n} = \frac{\lim_{n \to \infty} a_n}{\lim_{n \to \infty} b_n} = \frac{a}{\lim_{n \to \infty} b_n} = 1$$

$$\Rightarrow \lim_{n \to \infty} b_n = a \qquad \square$$

Abb. 32.1 Falsche Bearbeitung: die zu zeigende Konvergenz wird vorausgesetzt.

58 von 96 Studierenden, die an der Klausur teilgenommen haben, haben den Grenzwertsatz angewendet, obwohl die Folge $(b_n)_{n \in \mathbb{N}}$ nicht als konvergent vorausgesetzt war und somit nicht alle Voraussetzungen erfüllt waren.

Auch in weiteren Themengebieten lassen sich ähnliche Verhaltensmuster finden. In einer anderen Analysis-1-Klausur war die folgende Aufgabe gestellt:

Zeigen Sie: Ist (E, d) ein metrischer Raum, $K \subset E$ kompakt, und $F \subset E$ abgeschlossen, dann ist $F \cap K$ kompakt.

Auch hierbei haben einige Studierende Sätze benutzt, deren Voraussetzungen nicht erfüllt waren. So hat in der Bearbeitung in Abbildung 32.2 ein Studierender den Satz von Bolzano-Weierstraß benutzt, der besagt, dass jede beschränkte Folge von komplexen Zahlen eine konvergente Teilfolge enthält. Die Voraussetzung, dass es sich um eine Folge komplexer Zahlen handeln muss, wurde missachtet.

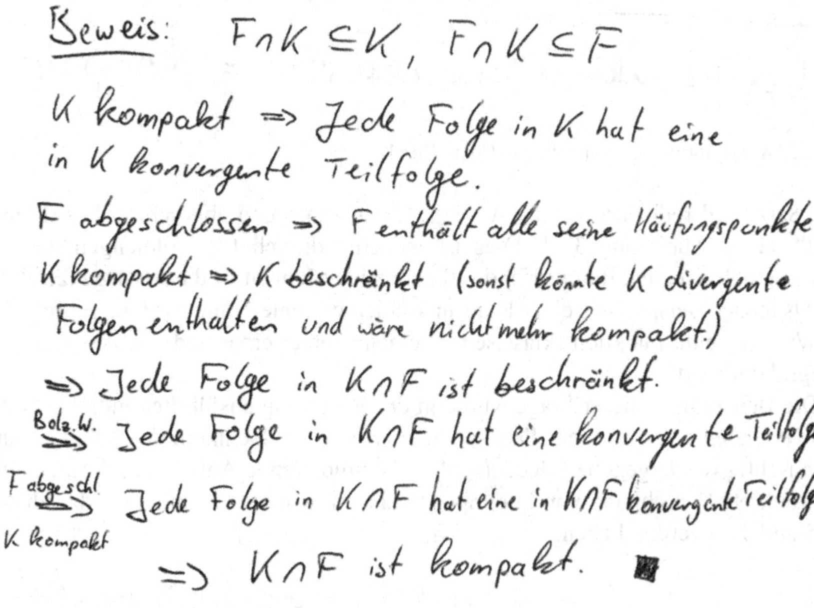

Beweis: $F \cap K \subseteq K, \quad F \cap K \subseteq F$

K kompakt $\Rightarrow$ Jede Folge in K hat eine in K konvergente Teilfolge.

F abgeschlossen $\Rightarrow$ F enthält alle seine Häufungspunkte.

K kompakt $\Rightarrow$ K beschränkt (sonst könnte K divergente Folgen enthalten und wäre nicht mehr kompakt.)

$\Rightarrow$ Jede Folge in $K \cap F$ ist beschränkt.

Bolz.W. $\Rightarrow$ Jede Folge in $K \cap F$ hat eine konvergente Teilfolge

F abgeschl $\Rightarrow$ Jede Folge in $K \cap F$ hat eine in $K \cap F$ konvergente Teilfolge

K kompakt $\Rightarrow$ $K \cap F$ ist kompakt. ∎

Abb. 32.2 Anwendung eines Satzes trotz fehlender Voraussetzungen.

Andere Studierende haben wie in Abbildung 32.3 mit dem Satz von Heine-Borel argumentiert. Dieser sagt aus, dass Teilmengen des n-dimensionalen reellen Raums genau dann kompakt sind, wenn sie abgeschlossen und beschränkt sind. In dieser Aufgabenstellung ist jedoch die Voraussetzung, dass es sich um eine Teilmenge des n-dimensionalen reellen Raums handeln muss, nicht erfüllt.

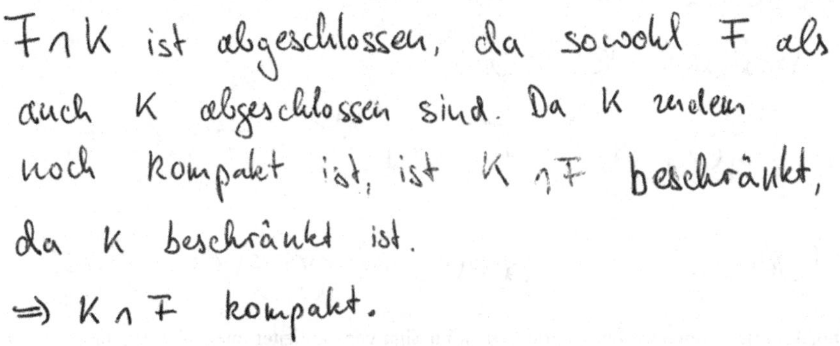

$F \cap K$ ist abgeschlossen, da sowohl F als auch K abgeschlossen sind. Da K zudem noch kompakt ist, ist $K \cap F$ beschränkt, da K beschränkt ist.

$\Rightarrow$ $K \cap F$ kompakt.

Abb. 32.3 Weitere Anwendung eines Satzes trotz fehlender Voraussetzungen.

Ganz deutlich zeigt sich dieses Verhaltensmuster auch noch bei einem weiteren Studenten, der zunächst aufschreibt, was der Satz von Heine-Borel seiner Meinung nach aussagt.

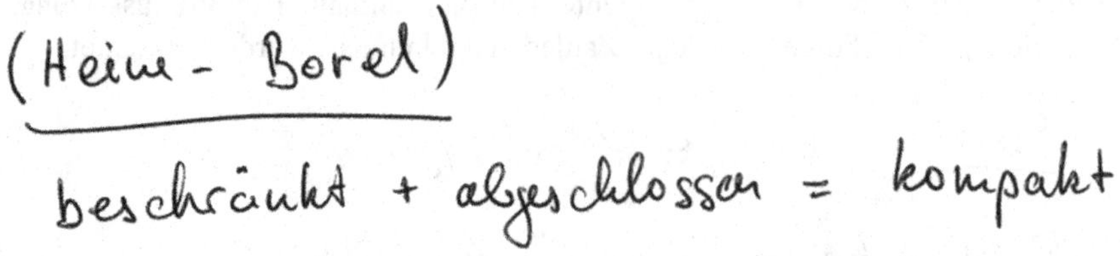

$$(\text{Heine - Borel})$$

$$\underline{\hspace{4cm}}$$

$$\text{beschränkt} + \text{abgeschlossen} = \text{kompakt}$$

Abb. 32.4 Kurzformel für den Satz von Heine-Borel.

Der Satz wird reduziert auf die Aussage „beschränkt und abgeschlossen = kompakt", siehe Abbildung 32.4. Dies ist jedoch lediglich für Teilmengen des n-dimensionalen reellen Raums richtig, für sonstige Mengen ist die Aussage falsch.

Als letztes Beispiel aus einer Klausur wählen wir eine Frage zur Integration:

Wann ist eine Funktion (klassisch) Riemann-Integrierbar und wie ist dann das Integral definiert?

Die Behandlung dieser Frage wurde in der Vorlesung ausführlich mittels Ober- und Untersummen diskutiert. Das hatten 16 von 47 Teilnehmern der Nachklausur auch richtig wiedergegeben. Jedoch gaben 12 Studierende Antworten, die sich auf die konkrete Berechnung von Integralen beziehen, wie die Beispiele in Abbildung 32.5 und 32.6 verdeutlichen.

Das Integral ist als Umkehrung des Differentials definiert.

Abb. 32.5 Falscher Merksatz zum Integral.

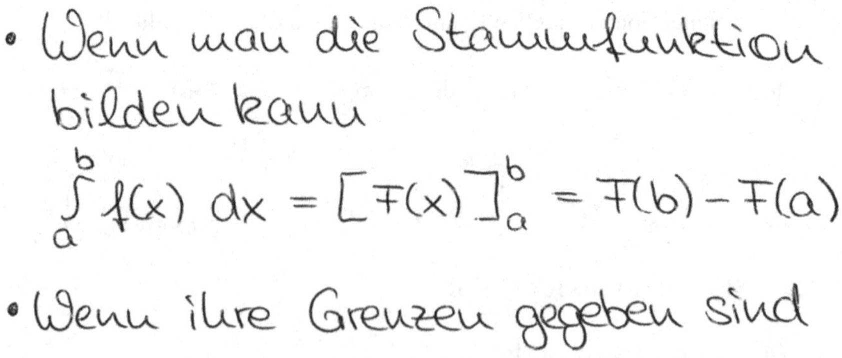

• Wenn man die Stammfunktion bilden kann

$$\int_a^b f(x)\, dx = \left[F(x) \right]_a^b = F(b) - F(a)$$

• Wenn ihre Grenzen gegeben sind

Abb. 32.6 Beschreibung von Integrierbarkeit im Sinn von expliziter Integralberechnung.

Drei weitere Studierende bezogen sich auf das Kalkül und die Definition und weitere 12 Studierende gaben gar keine Antwort.

Die verschiedenen Aufgabenbearbeitungen zeigen, dass viele Studierende Sätze anwenden, ohne zu begründen, warum deren Voraussetzungen erfüllt sind, oder Sätze auch dann anwenden, wenn deren Voraussetzungen nicht erfüllt sind. Möglicherweise sind die Studierenden noch zu sehr durch schulische Herangehensweisen geprägt. Die Studienanfänger sind mit der Theorieentwicklung, die in der Hochschulmathematik eine zentrale Rolle spielt, nicht vertraut, daher greifen sie auf die ihnen bekannten Vorgehensweisen zurück. Einschränkend muss man für die Klausurbearbeitungen festhalten, dass die Studierenden von ihren Antworten nicht unbedingt überzeugt sein müssen. Möglicherweise wird lieber eine mutmaßlich falsche Antwort gegeben, als gar keine. Man müsste sich dann trotzdem fragen, warum sie ihre Antwort nicht offensichtlich falsch fanden und sich an die erforderten Inhalte nicht erinnern konnten.

Diskussion

Die dargelegten Beispiele illustrieren die Auswirkungen eines schematisch orientierten mathematischen Weltbildes. Auch wenn die Interviewdaten und die Aufgabenbearbeitungen von unterschiedlichen Personen stammen, ergeben sie ein kohärentes Bild von einem Studierenden-Typus, der mathematische Aussagen vor allem als Handlungsoptionen interpretiert. Mathematik treiben heißt für ihn, Kalküle zu erweitern und auszuführen. Der Stellenwert von Beweisen und Voraussetzungen von Sätzen wird stark unterschätzt. Das steht im Widerspruch zu der Mathematik, die an der Hochschule präsentiert wird. Ausgehend von einer falschen Erwartungshaltung könnte es sein, dass Studierende nur Aussagen als relevant bewerten, die für Rechnungen (als Haupttätigkeiten ihres mathematischen Weltbildes) im weitesten Sinne subjektiv nutzbar sind. Es stellt sich also die Frage, warum der Anpassungsprozess der Beliefs nicht gelingt. Dafür sind viele Ursachen denkbar. Ein Ansatz könnte sein, dass Wenn-Dann-Aussagen mit einer Vielzahl an Prämissen und mathematischen Objekten das Arbeitsgedächtnis Vieler überfordern. Die Reduktion auf kurze Regeln (beschränkt + abgeschlossen = kompakt) wäre hier als Ausweg aus der Überforderung zu interpretieren. Man kann sich auch vorstellen, dass viele Studierende kein Problembewusstsein entwickeln, was die Anwendbarkeit von Sätzen und die Existenz von Objekten (z.B. Grenzwerten, insbesondere Integralen) betrifft. Hier könnte die ausführliche Betrachtung von Gegenbeispielen hilfreich sein.

Wie eingangs bereits bemerkt, hat sich die AG BaGym (noch) nicht auf diese Schwierigkeiten konzentriert, obwohl wir erkannt haben, dass eine Auseinandersetzung damit dringend erforderlich wäre.

462   Michael Liebendörfer, Laura Ostsieker

Literaturverzeichnis

Baumert, J., Bos, W., & Lehmann, R. (Eds.). (2000). *TIMSS/III: Dritte Internationale Mathematik- und Naturwissenschaftsstudie- Mathematische und Naturwissenschaftliche Bildung am Ende der Schullaufbahn.* Opladen: Leske + Budrich.
Daskalogianni, K., & Simpson, A. (2001). Beliefs overhang: the transition from school to university. In J. Winter (Ed.), Proceedings of the British Congress of Mathematics Education. In collaboration with the British Society for Research into Learning Mathematics (Vol. 21, Number 2, S. 97–108). http://www.bsrlm.org.uk/IPs/ip21-2/BSRLM-IP-21-2-Full.pdf.
Fischer, A., Heinze, A., & Wagner, D. (2009). Mathematiklernen in der Schule – Mathematiklernen an der Hochschule: die Schwierigkeiten von Lernenden beim Übergang ins Studium. In A. Heinze & M. Grüßing (Eds.), *Mathematiklernen vom Kindergarten bis zum Studium. Kontinuität und Kohärenz als Herausforderung beim Mathematiklernen* (S. 245–264). Münster: Waxmann.
Grigutsch, S., Raatz, U., & Törner, G. (1998). Attitudes of mathematics teachers towards mathematics. [Einstellungen gegenüber Mathematik bei Mathematiklehrern.]. *Journal für Mathematik-Didaktik, 19*(1), 3–45.
Heublein, U., Hutzsch, C., Schreiber, J., Sommer, D., & Besuch, G. (2010). *Ursachen des Studienabbruchs in Bachelor-und in herkömmlichen Studiengängen. HIS Forum Hochschule, Band 2/2010.* http://www.his-hf.de/pdf/pub_fh/fh-201002.pdf.
Rach, S., & Heinze, A. (2013). Welche Studierenden sind im ersten Semester erfolgreich? *Journal für Mathematik-Didaktik, 34*(1), 121–147. doi:10.1007/s13138-012-0049-3.

Kapitel 33
Der operative Beweis als didaktisches Instrument in der Hochschullehre Mathematik

Leander Kempen

Universität Paderborn

Abstract Dieser Beitrag berichtet von einer Studie, in der 64 Hausaufgabenbearbeitungen von Erstsemesterstudierenden zum operativen Beweis analysiert und kategorisiert wurden. Im Kontext einer neu entwickelten Brückenkursvorlesung sollten Betrachtungen von beispielgebundenen Beweisen den Übergang zum formalen Beweis bereiten. Die Ergebnisse zeigten jedoch, dass für die Studierenden die Unterscheidung zwischen einem beispielgebundenen Beweis und einer Verifikation anhand einiger Beispiele problematisch ist. Am Ende des Beitrags werden Folgerungen für den Einsatz operativer Beweise in der Lehrveranstaltung dargestellt.

Einleitung

Im Rahmen der Arbeitsgruppe 5 (eLearning und Mathematik im Übergang Schule/ Hochschule) des Kompetenzzentrums Hochschuldidaktik Mathematik (khdm) wurde eine neue obligatorische Erstsemesterveranstaltung für Lehramtsstudierende des Bachelor-Studiengangs für Haupt- und Realschulen entwickelt, die den Übergang zur universitären Mathematik erleichtern soll. In dieser Veranstaltung wird die Mathematik nicht als ein axiomatisches, deduktives System dargestellt, vielmehr soll die universitäre Mathematik direkt mit der zuvor vermittelten Schulmathematik verbunden werden. Im Wintersemester 2011/12 wurde diese Veranstaltung „Einführung in die Kultur der Mathematik" von Rolf Biehler zum ersten Mal durchgeführt. Um die Studierenden an die formale Sprache der Mathematik und das formale Beweisen heranzuführen, wurden im Gebiet der elementaren Zahlentheorie sog. operative Beweise (etwa Wittmann 1985) als eine valide und alternative Argumentationsform eingeführt. Sie sollten den Studierenden dabei helfen, eine Beweisidee zu finden, einen Beweisanfang zu schaffen und damit den Übergang zum formalen Beweis bereiten, indem die Beweisidee beibehalten und u.a. mit Variablen formalisiert wird. Allerdings ließen sich bei der Korrektur der ersten abgegeben Hausaufgabe erhebliche Probleme der Studierenden mit dem Prinzip des operativen Beweises erkennen.

In diesem Artikel sollen die Ergebnisse einer Untersuchung von studentischen Lösungen zu einem in den Hausaufgaben geforderten operativen Beweis beschrieben und mögliche Folgerungen dargestellt werden.

Theoretischer Rahmen

In Anlehnung an die *intuitive Beweisstufe* von Branford (1913) und das Prinzip der Invarianz von Piaget (1967) wurde das *operative Programm der Erkenntnisgewinnung* seit den 1970er Jahren insbesondere von E. C. Wittmann für die deutschsprachige sowie die internationale Mathematikdidaktik fruchtbar gemacht. Das vor diesem Hintergrund ausgearbeitete Prinzip des *operativen Beweises* charakterisiert Wittmann wie folgt: „Inhaltlich-anschauliche, operative Beweise stützen sich [...] auf Konstruktionen und Operationen, von denen intuitiv erkennbar ist, daß sie sich auf eine ganze Klasse von Beispielen anwenden lassen und bestimmte Folgerungen nach sich ziehen." (1988, S. 249). Des Weiteren schreibt er hierzu: „Dadurch, dass aber nicht auf einzelne Beispiele, sondern *auf allgemein ausführbare Operationen* und deren „*Wirkungen*" zurückgegriffen wird, ist die Allgemeingültigkeit gesichert. Man nennt Beweise dieser Art deshalb *operative Beweise*" (2007, S. 38; Hervorhebungen im Original).

Padberg (1997) greift in seinen verschiedenen Begründungsniveaus zum Beweisen das *beispielgebundene Beweisen* auf, um zu einer Beweisidee zu gelangen, die die Studierenden im Übergang zum formalen Beweis beibehalten sollen. Auch Sandefur et al. (2012), Mason et al. (1982/2010) und Schoenfeld (1985) weisen ausdrücklich auf die Bedeutung von Beispielkonstruktionen und Beispielbetrachtungen für das formale Beweisen hin. In der Hochschuldidaktik der Mathematik wurden bisher verschiedene Vorgehensweisen bei der Erarbeitung formaler Beweise erprobt (vgl. hierzu Selden, 2012). In der englischsprachigen Forschung hat sich vor allem der Begriff des *generic proof* etabliert, mit dem das generische Moment der Betrachtungen an konkreten (Zahlen-)Beispielen betont wird. Dazu Leron und Zaslavsky: „They [the generic proofs] enable students to engage with the main idea of the complete proof in an intuitive and familiar context, temporarily suspending the formidable issues of full generality, formalism and symbolism." (2009, S. 56). Auch Rowland plädiert aufgrund seiner bisherigen Lehrerfahrungen: „In effect, I am saying that the potential of the generic example as a didactic tool is virtually unrecognized and unexploited in the teaching of number theory, and I am urging a change in this state of affairs." (2002, S. 157).

Ein *formaler Beweis* wird in diesem Kontext, in Anlehnung an Stylianides und Ball (2008), als eine mathematische Argumentation zur Verifikation einer Behauptung verstanden, die die folgenden drei Kriterien erfüllt: (i) es werden Argumente benutzt, die von der Community (hier: den Veranstaltungsteilnehmern) als wahr akzeptiert werden und welche ohne weitere Verifikationen zugänglich sind, (ii) es werden Formen des Begründens und logischen Schließens angewendet, die gültig und bekannt sind und (iii) die Argumentation wird in einer angemessenen

Sprache kommuniziert, die auch formale mathematische Symbole (Variablen u.a.) beinhaltet.

Forschungsfragen

Die Forschungsfragen dieser Untersuchung beziehen sich auf das didaktische Potenzial von operativen Beweisen in der Hochschullehre:

1. Wie argumentieren Erstsemesterstudierende, wenn sie aufgefordert werden, einen operativen Beweis zu führen?
2. Benutzen Sie das generische Argument aus dem operativen Beweis auch im darauf aufbauenden formalen Beweis?

Die Lernsequenz

Als die Studierenden ihre erste Hausaufgabe abgaben haben, hatten bereits zwei Vorlesungen und ein Tutorium stattgefunden.

Zu Beginn wurden die Ziele des ersten Vorlesungsabschnitts („Beweisen und Entdecken in der Arithmetik") benannt: An einfachen Beispielen den Prozess des Beweisens und Entdeckens kennen lernen und zwischen Überprüfungen einer allgemeinen Behauptung (i) an einigen Zahlenbeispielen, (ii) mit operativen Beweisen und (iii) mit allgemeinen formalen Beweisen mit Variablen unterscheiden. Durch die Aussage: „Jemand behauptet: Die Summe von drei aufeinanderfolgenden natürlichen Zahlen ist immer ungerade", wurde ein „Forschungsprozess" initiiert, in dem die Behauptung zunächst an einigen Beispielen getestet wurde und man zu der Vermutung gelangte, dass die Summe immer gleich dem Dreifachen der mittleren Zahl sei. Als vorläufige Begründung für diese Vermutung wurde ein operativer Beweis in Form einer (fiktiven) Schülerlösung präsentiert (s. Abb. 33.1) und von den Studierenden diskutiert. Es wurde festgehalten, dass die vorgenommenen algebraischen Operationen und die damit verbundenen Argumentationen in allen zu betrachtenden Fällen möglich sind, diese folglich einen allgemeingültigen beispielgebundenen Beweis darstellen. Folgende Zusammenfassung wurde formuliert:

In diesem Beispiel vollführen wir Operationen mit konkreten Zahlen, die aber auch mit allen anderen natürlichen Zahlen so möglich sind. Damit weicht diese Argumentation von den bisherigen Überlegungen zu unserem Problem ab. Es ist ein „operativer" Beweis und allgemeingültig zu bewerten. Weiter wird neben der Verifikation der Behauptung auch eine Erklärung geliefert, warum die Summe immer das Dreifache der mittleren Zahl sein muss. Hiernach wurde das allgemeingültige Argument aus dem operativen Beweis auch im formalen Beweis verwendet (s. Abb. 33.1).

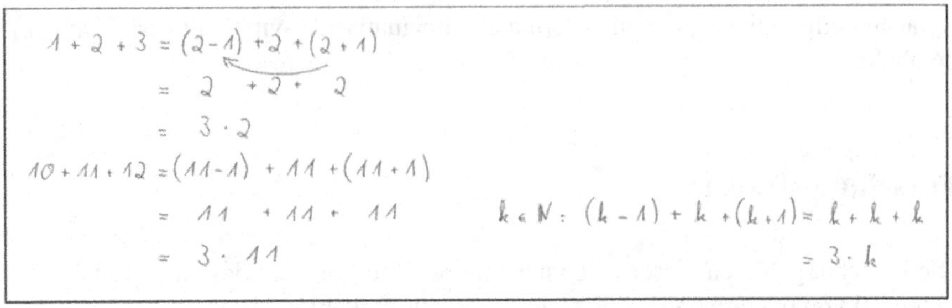

$$1 + 2 + 3 = (2-1) + 2 + (2+1)$$
$$= 2 + 2 + 2$$
$$= 3 \cdot 2$$
$$10 + 11 + 12 = (11-1) + 11 + (11+1)$$
$$= 11 + 11 + 11 \qquad k \in \mathbb{N}: \; (k-1) + k + (k+1) = k + k + k$$
$$= 3 \cdot 11 \qquad\qquad\qquad\qquad = 3 \cdot k$$

Abb. 33.1 Der operative Beweis (links) und die formalisierte Beweisidee (rechts)

Nach diesem „Forschungsprozess" gelangte man schließlich zu der Behauptung, dass die Summe von k aufeinanderfolgenden natürlichen Zahlen genau dann durch k teilbar ist, wenn k eine ungerade Zahl ist. Die Summe von k aufeinanderfolgenden natürlichen Zahlen mit Startzahl n wurde weiter als $S_{n,k}$ definiert, und zum Abschluss die Behauptung formal bewiesen.

In den Tutorien wurde vor allem die formale Repräsentation von geraden und ungeraden Zahlen besprochen, da diese in der ersten Hausaufgabe zu benutzen war. In der folgenden Übungsaufgabe sollten die Studierenden die Repräsentation einer ungeraden Zahl durch *2n-1* innerhalb eines Beweises per Widerspruch anwenden.

Aufgabe

Die Teilnehmer der Veranstaltung mussten wöchentlich Hausaufgaben in Einzelarbeit lösen und abgeben, um an der abschließenden Klausur teilnehmen zu dürfen. Für die Analyse der folgenden Aufgabe wurde ein Kategoriensystem bzgl. der Argumentationsqualität der Bearbeitungen entwickelt (s. Abschnitt „Ergebnisse") und die Lösungen von 64 Studierenden aus vier Übungsgruppen eingescannt und den Kategorien zugeordnet. Damit sollte die Akzeptanz des operativen Beweises bei den Studierenden und sein Potenzial als didaktisches Instrument erforscht werden.

> Beweisen Sie die nachfolgende Behauptung mit einem operativen Beweis und einem formalen Beweis. Formulieren Sie vor dem formalen Beweis zunächst die Behauptung mit Variablen:
> „Die Summe aus einer ungeraden natürlichen Zahl und ihrem Doppelten ist immer ungerade."

Aufgabenanalyse und erwartete Ergebnisse

Zunächst besteht ein operativer Beweis aus verallgemeinerungsfähigen (algebraischen) Operationen an konkreten (Zahlen-)Beispielen. Hier muss eine Argumentation entwickelt werden, warum die Behauptung in den konkreten Beispielen und auch in allen anderen möglichen Fällen wahr ist.

- Argumentationsmethode (1):

$$1 + 2 \times 1 = 3 \times 1 = 3 \qquad 3 + 2 \times 3 = 3 \times 3 = 9 \qquad 5 + 2 \times 5 = 3 \times 5 = 15$$

Man erkennt, dass das Ergebnis immer das Dreifache der Ausgangszahl ist. Da das Produkt zweier ungerader natürlicher Zahlen ungerade ist, muss das Ergebnis immer ungerade sein.

- Argumentationsmethode (2):

$$1 + 2 \times 1 = 1 + 2 = 3 \qquad 3 + 2 \times 3 = 3 + 6 = 9 \qquad 5 + 2 \times 5 = 5 + 10 = 15$$

Man erkennt, dass die Summe immer aus einem ungeraden und einem geraden Summanden besteht, da das Doppelte einer ungeraden Zahl gerade ist. Daher muss die Summe ungerade sein.

Ergebnisse

Die betrachtete Aufgabe enthält mehrere Aufgabenteile. Im Folgenden soll allerdings nur auf die Bearbeitungen des operativen Beweises eingegangen werden.
Um die Argumentationsqualität der Bearbeitungen untersuchen zu können wurde das folgende Kategoriensystem entwickelt:

- E0: Der „operative Beweis" beinhaltet Beispiele, die nicht zu der Behauptung passen (s. Abb. 33.2 links oben).
- E1: Der „operative Beweis" besteht nur aus einer Verifikation durch verschiedene Beispiele, ohne dass allgemeingültige Prinzipien benannt werden (s. Abb. 33.2 rechts oben).
- P1: In den Beispielen innerhalb des „operative Beweises" werden allgemeingültige Operationen und Umformungen deutlich, welche allerdings nicht expliziert werden (s. Abb. 33.2 links unten).
- P2: In den operativen Beweisen werden allgemeingültige Prinzipien deutlich, die benannt und in der folgenden Argumentation zum Beweisen der Behauptung genutzt werden (s. Abb. 33.2 rechts unten).

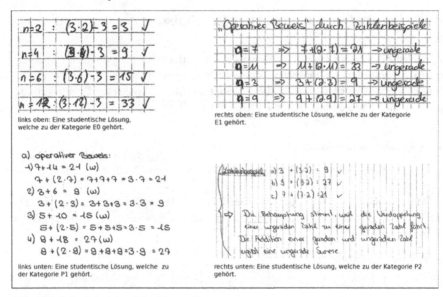

links oben: Eine studentische Lösung, welche zu der Kategorie E0 gehört.

rechts oben: Eine studentische Lösung, welche zu der Kategorie E1 gehört.

links unten: Eine studentische Lösung, welche zu der Kategorie P1 gehört.

rechts unten: Eine studentische Lösung, welche zu der Kategorie P2 gehört.

Abb. 33.2 Ankerbeispiele zu den Kategorien

Die Häufigkeiten sind dabei wie folgt:
E0: 3 (5,6%), E1: 36 (67,9%), P1: 8 (15,1%) und P2: 6 (11,3%). Somit präsentierten 39 Studierende (73,5%) in ihren „operativen Beweisen" ausschließlich Beispiele ohne Nennung von Beobachtungen oder weiterführenden Argumentationen (E0 + E1). Augenscheinlich ist ihnen der Unterschied zwischen dem Konzept des operativen Beweises und der bloßen Verifikation einer Behauptung durch Beispiele nicht bewusst. Von den 14 Studierenden, deren Bearbeitungen in den Beispielen verallgemeinerbare Zusammenhänge erkennen ließen (P2 + P1), nutzten nur sechs Studierende diese Erkenntnisse, um in einer folgenden Argumentation die Behauptung zu „beweisen" (P2). In diesen sechs Lösungen mit einem vollständigen operativen Beweis wurde die Argumentationsmethode (1) zweimal und die Argumentationsmethode (2) viermal genannt (s.o.). Von den 14 Studierenden, deren Bearbeitungen in die Kategorie P1 oder P2 fallen, versuchten elf die Konstruktion des formalen Beweises. Acht dieser Studierenden nutzten im formalen Beweis dieselbe Argumentation, wie in ihrem vorherigen operativen Beweis.

Ausblick

Die hier genannten Ergebnisse, auch wenn sie nicht als repräsentativ gelten können, geben Anlass zu einer kritischen Beleuchtung des didaktischen Instruments „operativer Beweis". Die Studierenden hatten überraschende Schwierigkeiten mit diesem Beweisprinzip, obwohl sich eine gesamte Vorlesungs- und Übungssitzung mit dieser Thematik beschäftigt hatten. Die dargestellte Problematik deckt sich mit der bisherigen Forschung: So ist bekannt, dass Erstsemesterstudierende für das Grundschullehramt Probleme mit der Unterscheidung zwischen einem Beweis und der Verifikation durch Beispiele haben (siehe z.B. Martin und Harel 1989; Recio und Godino 2001). Auch Rowland (2002) bezeichnet das Spannungsfeld zwischen dem erklärenden Moment und der allgemeingültigen Argumentation in einem generic proof, besonders für Studienanfänger der Mathematik, als problematisch. Es bleibt hier zu betonen, dass diejenigen Studierenden, welche einen generischen Aspekt in ihren Beispielbetrachtungen identifiziert haben, diesen auch auf ihren formalen Beweis übertragen konnten.

Erste Konsequenzen wurden aus diesen Ergebnissen bereits gezogen. Die beschriebene Lehrveranstaltung wurde vor ihrer wiederholten Durchführung inhaltlich überarbeitet: Innerhalb der Vorlesung wurde auf typische Fehlvorstellungen und Probleme hinsichtlich des operativen Beweises eingegangen und eine explizite Abgrenzung der Argumentationen anhand der vier beschriebenen Kategorien gemacht. Des Weiteren wurden die Tutoren der Veranstaltung speziell im Hinblick auf das beispielgebundene Beweisen fachlich geschult und neue Übungsaufgaben eingesetzt. Weitere quantitative und qualitative Studien zum operativen Beweis und dessen Anwendung in der Hochschuldidaktik sollen folgen.

Literatur

Branford, B. (1913). *Betrachtungen über mathematische Erziehung vom Kindergarten bis zur Universität*. Leipzig: Teubner.

Leron, U., & Zaslavsky, O. (2009). Generic proving: Reflections on scope and method. In F.-L. Lin, F.-J. Hsieh, G. Hanna, & M. de Villiers (Hrsg.), *Proceedings of the ICMI Study 19 Conference: Proof and Proving in Mathematics Education*, Vol. 2 (S. 53-58). Taipei, Taiwan: The Department of Mathematics, National Taiwan Normal University.

Martin, W. G., & Harel, G. (1989). Proof frames of preservice elementary teachers. *Journal for Research in Mathematics Education, 20*(1), 41-51.

Mason, J., Burton, L., & Stacey, K. (1982). *Thinking mathematically* (2nd extended edition). Harlow: Pearson Prentice Hall.

Padberg, F. (1997). *Einführung in die Mathematik 1 - Arithmetik*. Heidelberg/ Berlin: Spektrum Akademischer Verlag.

Piaget, J. (1967). *Psychologie der Intelligenz* (3rd ed.). Zürich: Rascher.

Recio, A. M., & Godino, J. D. (2001). Institutional and personal meanings of mathematical proof. *Educational Studies in Mathematics, 48*(1), 83-99.

Rowland, T. (2002). Generic proofs in number theory. In S. R. Campbell, & R. Zazkis (Hrsg.), *Learning and Teaching Number Theory* (S. 157-183). Westport, Connecticut: Ablex Publishing.

Sandefur, J., Mason, J., Stylianides, G. J., & Watson, A. (2012). Generating and using examples in the proving process. *Educational Studies in Mathematics, 83*(3), 323-340. doi: 10.1007/s10649-012-9495-x.

Schoenfeld, A. (1985). *Mathematical problem solving*. New York: Academic.

Selden, A. (2012). Transitions and proof and proving at tertiary level. In G. Hanna, & M. de Villiers (Hrsg.), *Proof and Proving in Mathematics Education: The 19th ICMI Study* (S. 391-422). Heidelberg: Springer Science + Business Media.

Stylianides, A. J., & Ball, D. L. (2008). Understanding and describing mathematical knowledge for teaching: Knowledge about proof for engaging students in the activity of proving. *Journal of Mathematics Teacher Education, 11*(4), 307-332. doi: 10.1007/s10857-9077-9

Wittmann, E. C. (1981). Beziehungen zwischen operativen Programmen in Mathematik, Psychologie und Mathematikdidaktik. *Journal für Mathematik-Didaktik, 2,* 83-95.

Wittmann, E. C. (1985). Objekte–Operationen–Wirkungen: Das operative Prinzip in der Mathematikdidaktik. *Mathematik lehren, 11,* 7-11.

Wittmann, E. C., & Müller, G. (1988). Wann ist ein Beweis ein Beweis? In P. Bender (Hrsg.), *Mathematikdidaktik: Theorie und Praxis* (S. 237-257). Berlin: Cornelsen.

Wittmann, E. C., & Ziegenbalg, J. (2007). Sich Zahl um Zahl hochangeln. In G. Müller, H. Steinbring, & E. C. Wittmann (Hrsg.), *Arithmetik als Prozess* (S. 35-53). Seelze: Kallmeyer.

Kapitel 34
Werkstattbericht der Arbeitsgruppe "Mathematik in den Ingenieurwissenschaften"

Markus Hennig, Axel Hoppenbrock, Jörg Kortemeyer, Bärbel Mertsching, Gudrun Oevel

Universität Paderborn

Abstract In diesem Beitrag werden drei Teilprojekte aus der Arbeitsgruppe „Mathematik in den Ingenieurwissenschaften" des Kompetenzzentrums Hochschuldidaktik Mathematik vorgestellt. Das Teilprojekt „Situierter Erwerb von Mathematikkenntnissen in den Ingenieurwissenschaften" fokussiert auf den Erwerb mathematischer Kompetenzen im Rahmen von ingenieurwissenschaftlichen Einführungsveranstaltungen. Exemplarisch wird hier die Lehrveranstaltung *Grundlagen der Elektrotechnik A* an der Universität Paderborn untersucht, in der ein abgestimmtes Blended Learning Szenario unter Verwendung eines Wikis umgesetzt wurde. In dem Beitrag wird insbesondere die Konzeption des Szenarios unter didaktischen Gesichtspunkten beschrieben und aufgezeigt, wie der Ansatz der hohen Heterogenität der Studierenden Rechnung trägt.

Das Teilprojekt „Mathematik für Maschinenbauer: Integration des Modellierens in ingenieurwissenschaftlichen Zusammenhängen" setzte als Ausgangspunkt seiner Aktivitäten die Analyse von Problemsituationen bei der Vermittlung und dem Erwerb von Mathematik in „klassischen" Mathematik für Maschinenbauer–Veranstaltungen. Schwerpunkt ist derzeit die Entwicklung von kontextgebundenen und anwendungsorientierten Aufgaben zur Unterstützung des Lehr-/Lernprozesses sowie deren empirische Evaluation.

Das Paderborner Teilprojekt des BMBF-geförderten Projekts KoM@ING hat das Ziel, Beiträge zur Kompetenzmodellierung zu liefern. Es werden Studien zur Kompetenzentwicklung bezogen auf die Mathematikveranstaltungen für Ingenieure und die Verwendung der Mathematik in Einführungsveranstaltungen zur Elektrotechnik erstellt. Dabei stellt die an Hochschulen übliche curriculare Trennung zwischen mathematischen und ingenieurwissenschaftlichen Veranstaltungen eine besondere Herausforderung dar. Zur Beschreibung der Kompetenzen soll durch Analyse von Aufgabenbearbeitungen ein Modell entstehen, welches typische Lösungsstrategien Studierender bei der Bearbeitung mathematikhaltiger Elektrotechnik-Aufgaben beschreibt.

Einleitung[1]

Nach einer Studie der HIS brachen im Jahr 2010 insgesamt 48 % der Ingenieurstudenten an Universitäten ihr Studium ab. In den Studiengängen Maschinenbau
und Elektrotechnik waren es sogar über 50 % (s. Abb. 34.1). Dieser Wert liegt
weit über der durchschnittlichen universitären Abbrecherquote von 35 % (Heublein et al. 2010). Die Ursachen für diese hohen Quoten sind vielfältig und bisher
nur wenig untersucht.

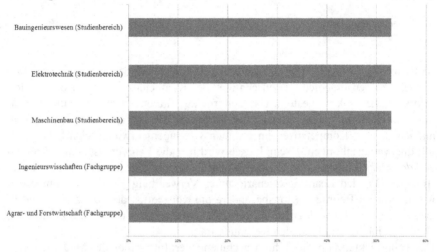

Abb. 34.1 Studienabbruchquoten in Bachelorstudiengängen an Universitäten. Bezugsgruppe Ab-
solventen 2010 (Daten aus: Heublein et al. 2010, S.17)

Eine der wenigen Studien in Deutschland, die sich mit diesem Thema beschäftigt haben, stammt von Derboven und Winkler. In dieser Online-Studie wurden
680 Studienabbrecher zu ihren Gründen befragt. 56,4 % der Befragten – Mehr-
fachnennungen waren möglich – gaben als einen Abbruchgrund an, dass es im
Studium überwiegend darum ging, Formeln anzuwenden, ohne sie zu verstehen
(Derboven und Winkler 2009).

Dieses Ergebnis deckt sich mit den Erfahrungen vieler Mathematikdozenten.
„Mathe ist das Hauptproblem, der überwiegende Teil der Studienabbrecher gibt
deswegen auf" klagt Aloys Krieg, Prorektor der RWTH Aachen in der Süddeut-
schen Zeitung vom 24.03.2012 (vgl. Demmer 2012).

Neben der Frustration und der verlorenen Zeit für die Studenten ist diese hohe
Abbrecherzahl vor dem Hintergrund des großen Mangels an Ingenieuren (s.
Tab. 34.1) von besonderer Bedeutung.

Die Universität Paderborn stellt sich diesem Problem. Rolf Biehler ist es gelungen, vorhandene Initiativen zu bündeln und drei Projekte zu initiieren, die sich
zum Ziel gesetzt haben, die Schwierigkeiten von Ingenieurstudenten mit der Ma-

---

[1] Autor: Axel Hoppenbrock

thematik genauer zu analysieren sowie Lehrinnovationen zur Reduktion dieser Schwierigkeiten zu erproben und zu evaluieren.

Tab. 34.1 Ausgewählte Fachkräftelücken in den MINT-Berufen (2008) (Daten aus Hetze 2011, S. 17)

	Fachkräftelücke	In % der entsprechenden Gruppe an sozialver- sicherungspflichtig Beschäftigten 2007
Maschinen- und Fahrzeugbau	36 556	25,6
Maschienenbautechniker	25 242	25,2
Datenverarbeitunsfachleute	22 426	4,8
Elektroingenieure	17 310	10,7
Architekten, Bauingenieure	5 631	4,8
Physiker, Mathematiker,	333	1,4
sonst. Naturwissenschaftler	0	0
Insgesamt	143 741	

Zwei Teilprojekte des Kompetenzzentrums Hochschuldidaktik Mathematik (khdm) beschäftigen sich mit der Ingenieurmathematik; sie sind im Bereich des Maschinenbaus (Projekt: *Mathematik für Maschinenbauer unter Leitung von Prof. Dr. Gudrun Oevel*) und der Elektrotechnik (Projekt: *Situierter Erwerb von Mathematikkenntnissen in den Ingenieurwissenschaften unter Leitung von Prof. Dr. Bärbel Mertsching*) angesiedelt. Beide Teilprojekte gehen davon aus, dass in der Ausbildung der Ingenieure neben dem Verständnis der Mathematik „an sich" zusätzlich die Notwendigkeit und Schwierigkeit besteht, nicht-mathematisch formulierte technische Sachverhalte in eine mathematische Form zu übersetzen, diese dann mit den gebotenen mathematischen Methoden zu bearbeiten und anschließend die „Rückübersetzung" vorzunehmen, also die technische Relevanz der mathematischen Resultate zuverlässig zu beurteilen. Die beiden Teilprojekte sind in ihren Ansätzen komplementär angelegt: Während in der „Mathematik für Maschinenbauer" exemplarisch untersucht wird, wie die Mathematikvorlesung durch Elemente des Modellierens aus den Ingenieurwissenschaften angereichert werden kann, wird im zweiten Teilprojekt der alternative Weg erprobt, nämlich mathematische Kompetenzen im Kontext einer ingenieurwissenschaftlichen Vorlesung zu vermitteln.

Das neueste Projekt heißt „Kompetenzmodellierung und Kompetenzentwicklung, integrierte IRT[2]-basierte und qualitative Studien bezogen auf Mathematik und ihre Verwendung im ingenieurwissenschaftlichen Studium" (*KoM@ING*). Neben der Kompetenzmodellierung und Kompetenzentwicklung geht es in dem Projekt auch um integrierte IRT-basierte und qualitative Studien bezüglich Mathematik und ihren Anwendungen in ingenieurwissenschaftlichen Studiengängen.

Zusammenfassend illustrieren die genannten Beispiele die Breite der Forschungsthemen im Bereich der Mathematikausbildung für Ingenieure sowie ihre

---

[2] IRT = Item Response Theory

Relevanz über die Universität Paderborn hinaus. Rolf Biehler kommt das Verdienst zu, die Aktivitäten zusammengeführt und in den Fokus der Mathematikdidaktik nicht nur an der Universität Paderborn gerückt zu haben. Die Beteiligten in den Teilprojekten haben in vielfältiger Weise von seinen Erfahrungen und seinem didaktischen Wissen profitiert.

Situierter Erwerb von Mathematikkenntnissen in den Ingenieurwissenschaften am Beispiel der Grundlagen der Elektrotechnik[3]

Die Mathematik-Kompetenzen von Studienanfängerinnen und -anfängern in den Ingenieurwissenschaften reichen häufig zur Bewältigung der Fachthemen nicht aus, was bei ihnen Frust erzeugt, der bis zum Studienabbruch führen kann, siehe z.B. Mustoe (2006) oder Bamforth et al. (2010). Darüber hinaus werden schnell weitere mathematische Verfahren benötigt, die in den begleitenden Veranstaltungen der höheren Mathematik zum Teil erst in den Folgesemestern behandelt werden. Diese Situation stellt die Lehrenden der ingenieurwissenschaftlichen Eingangsveranstaltungen vor große Herausforderungen. Exemplarisch wird in diesem Projekt untersucht, wie fehlende mathematische Fähigkeiten in der Veranstaltung *Grundlagen der Elektrotechnik A (GET A)*, die seit inzwischen 10 Jahren vom Lehrstuhl von Bärbel Mertsching im Bachelorstudiengang Elektrotechnik an der Universität Paderborn im ersten Semester durchgeführt wird, ausgeglichen werden können. Die Ergebnisse des Projektes sollen auf andere ingenieurwissenschaftliche Eingangsveranstaltungen übertragbar sein.

Aktuell sind über 200 Studierende für die Veranstaltung GET A registriert. Aus einer Vorlesung und einer Übung (4+2 SWS bzw. 8 ECTS-Punkte) bestehend vermittelt die Veranstaltung ein fundamentales Verständnis der vielfältigen Erscheinungen des Elektromagnetismus. Im Gegensatz zu den - in der Regel anschaulichen - Gesetzen der Mechanik, die starke Bezüge zu alltäglichen Vorgängen haben, sind die elektrischen und magnetischen Vorgänge bedeutend abstrakter. Für die Beschreibung dieser physikalischen Phänomene werden beispielsweise die zentralen Begriffe des *elektromagnetischen Feldes* sowie des *elektrischen Dipols* eingeführt, die keine Entsprechung im Alltag besitzen. Für ihre mathematische Formulierung werden mathematische Fertigkeiten und Kenntnisse benötigt, die teilweise deutlich über die Schulmathematik hinausgehen, aber erst später in der universitären Mathematiklehre behandelt werden.[4]

---

[3] Autoren: Bärbel Mertsching und Markus Hennig

[4] Speziell in Paderborn besteht eine weitere Herausforderung darin, dass die Veranstaltung auch von Studierenden weiterer Studiengänge (z.B. Informatikstudierende mit dem Nebenfach Elektrotechnik) in zum Teil höheren Fachsemestern besucht wird. Weiterhin gliedert sich die Hörerschaft in Studierende mit Fachhochschulreife und allgemeiner Hochschulreife.

Die Lehrveranstaltung beginnt mit der Behandlung der grundlegenden Phänomene des statischen elektrischen Feldes. Hierzu werden die Grundzüge der Differential- und Integralrechnung als bekannt vorausgesetzt. Doch gleich am Anfang wird die Beherrschung mathematischer Konzepte verlangt, die deutlich über das hinaus gehen, was in den Mathematik-Kursen der gymnasialen Oberstufe vermittelt wird: Zur Beschreibung der Coulomb-Kräfte zwischen Punktladungen wird ein sicherer Umgang mit der Vektorrechnung sowie den grundlegenden orthogonalen Koordinatensystemen benötigt. Die Berechnung der Feldgrößen *elektrische Feldstärke* und *elektrische Flussdichte* erfordert die Verwendung von Linien- und Hüllflächenintegralen, siehe z.B. Ida (2004).

Diesen Herausforderungen wurde zunächst mit einfachen Ansätzen zum situierten Erwerb von Mathematikkenntnissen begegnet: Das Vorlesungsskript erhielt einen Anhang mit Kurzeinführungen zu den relevanten mathematischen Themen, auf die in der Veranstaltung - zur besseren Veranschaulichung auch unter Zuhilfenahme multimedialer Anwendungen - zurückgegriffen werden konnte. Darüber hinaus wurden Sonderübungen zu den Themen *mehrdimensionale Analysis, Koordinatensysteme* sowie zu *linearen Gleichungssystemen* angeboten. Die Tatsache, dass die Durchfallquote in den letzten Jahren teilweise trotzdem über 50 % lag, motivierte die Erprobung weiterer Maßnahmen.

Zur Evaluierung und Systematisierung dieser Maßnahmen wurden zunächst die in der Veranstaltung vorkommenden mathematischen Inhalte erfasst, analysiert und ihren jeweiligen mathematischen Themengebieten zugeordnet. Darauf basierend wurde dann ein Fragebogen entwickelt, der von den Studierenden am Ende des Wintersemesters 2010/2011 ausgefüllt und zur Identifizierung derjenigen mathematischen Inhalte herangezogen wurde, die besondere Probleme bereiteten. Einen Überblick über die Ergebnisse gibt Abb. 34.2. Daraus resultierten folgende Schwerpunkte in großer Übereinstimmung mit den Themen, die bereits zuvor als problematisch eingestuft und deswegen im Anhang des Skripts und in den Sonderübungen behandelt wurden:

- Verwendung von Polar-, Zylinder- und Kugelkoordinaten, z. B. zur Beschreibung elektromagnetischer Felder;
- Formulierung infinitesimaler Linien-, Flächen- und Volumenelemente in verschiedenen Koordinatensystemen;
- Umgang mit mehrdimensionalen Linien- und Ringintegralen in verschiedenen Koordinatensystemen;
- Umgang mit mehrdimensionalen Flächen-, Hüllflächen- und Volumenintegralen in verschiedenen Koordinatensystemen, etwa im Zusammenhang mit dem Satz von Gauß;
- Formulierung von Größenabhängigkeiten mit Hilfe des Differentialquotienten; sowie
- Umgang mit Matrizen und den zugehörigen Begriffen wie Determinante oder Matrizenmultiplikation.

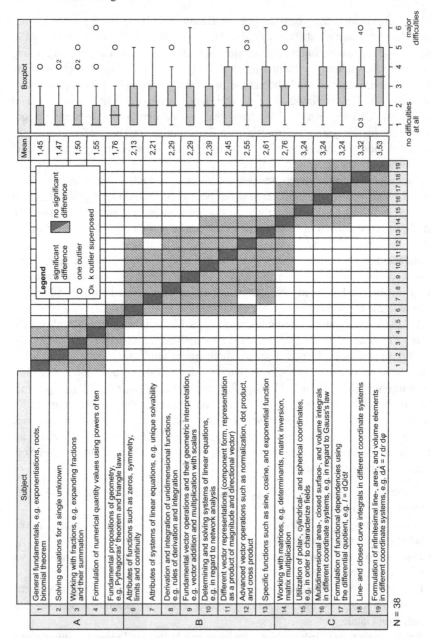

Abb. 34.2 Ergebnisse der Umfrage zu den mathematischen Schwierigkeiten der Studierenden der Veranstaltung GET A aus (Hennig und Mertsching 2012). In den Boxplots zeigen die breiten Balken die Medianwerte, die Rechtecke werden durch die 25 % und 75 % Quantile begrenzt, während die „Antennen" die größten Werte beschreiben, die nicht als Ausreißer behandelt wurden. Zur Bestimmung der signifikanten Unterschiede (p<0.05) wurden „Post-hoc Pairwise Comparisons" (unkorrigiert) durchgeführt.

In einem nächsten Schritt wurde auf Basis wissenschaftlicher Veröffentlichungen zur Didaktik der Ingenieurwissenschaften nach Maßnahmen im Hinblick auf die eingangs beschriebenen Probleme recherchiert. Mögliche Lösungsvorschläge umfassen Änderungen der Lehrpläne (so dass die Mathematik- und Fachlehre besser aufeinander abgestimmt sind), die Durchführung zusätzlicher Mathematik- und Brückenkurse zu Anfang oder parallel zur Lehrveranstaltung sowie die Einrichtung von Unterstützungszentren und Lerntreffs ("Mathematics Support Centers"), siehe z.B. Lawson et al. (2008) oder Korenko et al. (2010). In dem oben genannten Fragebogen wurden weiterhin demografische Daten, eine Bewertung der Veranstaltungselemente, das Arbeitsverhalten der Studierenden sowie Aspekte bezüglich der Studienmotivation erfragt. Die Untersuchungsergebnisse bestätigen die heterogene Zusammensetzung der Hörerschaft und lassen in Zusammenhang mit den engen Lehrplänen im ersten Studienabschnitt (Tetour et al. 2010) den Schluss zu, dass *zusätzliche losgelöste Mathematikangebote nicht wahrgenommen* werden. Eine Umstrukturierung der Lehrpläne kommt aufgrund der fortgeschrittenen Themen ebenfalls nicht in Frage. Weiterhin wurde durch Recherchen deutlich, dass sowohl national als auch international keine wissenschaftlichen Projekte zur Vermittlung mathematischer Kompetenzen im Kontext der technischen und physikalischen Fachlehre existieren. Grundsätzlich lässt sich aber auf Theorien aus dem Bereich der situierten Kognition (Brown et al. 1989) aufbauen, die sich beispielsweise mit dem problemorientierten Lernen ("Problem Based Learning") oder sogenannten Ankern ("Anchored Instruction") zur Interessens- und Motivationssteigerung befassen. Daraus lassen sich einige grundlegende Forderungen ableiten, die im Rahmen der Konzeption des Teilprojekts berücksichtigt wurden:

- Verwendung komplexer Ausgangsprobleme: Als Ausgangspunkt des Lernprozesses sollte ein interessantes und intrinsisch motivierendes Problem dienen ("Das-Problem-lösen-wollen").
- Authentizität und Situiertheit: Die Lernumgebung sollte den Studierenden den Umgang mit realistischen und authentischen Problem ermöglichen.
- Multiple Perspektiven: Die Lernumgebung sollte verschiedene Kontexte anbieten, damit das Wissen auf andere Problemstellungen übertragen werden kann.
- Artikulation und Reflexion: Zur Förderung der Abstrahierung des Wissens sollten Problemlöseprozesse artikuliert und reflektiert werden.
- Lernen im sozialen Austausch: Lernumgebungen sollten dem sozialen Kontext einen wichtigen Stellenwert zuweisen (zum Beispiel in Form von Diskussionen).

Auf Basis dieser Forderungen und der oben angeführten Erkenntnisse und Untersuchungsergebnisse wird die Lehrveranstaltung wie folgt angepasst: Statt wie bisher auf den Anhang zurückzugreifen, werden knapp gehaltene mathematische Exkurse in eine webbasierte Plattform ausgelagert, welche aus den Vorlesungs- und Übungsunterlagen per Hyperlink direkt aufgerufen werden können. Die Inhalte stehen damit in unmittelbarer Relation zu einem komplexen und authentischen Ausgangsproblem (z. B. dem Konzept des Einheitsvektors bei der Einführung der Coulomb-Kraft). Darüber hinaus können weiterführende mathematische Themen,

Bezüge zwischen verschiedenen Gegenständen, Übungsaufgaben zum Selbstlernen, etc. integriert werden.

Zur Entwicklung des Online-Angebots wurden verschiedene Lernplattformen („Virtual Learning Environments") geprüft, resultierend wurde nun ein Wiki (vgl. http://www.mediawiki.org) aufgebaut. Die Zweckmäßigkeit dieser Plattform ergibt sich insbesondere aus der hohen Verbreitung und der dementsprechend niedrigen Hemmschwelle zur Nutzung des Angebots durch die Studierenden. Hinzu kommt, dass die Plattform unter anderem durch die Bereitstellung eines Vorlagensystems eine schnelle und komfortable Erstellung von Inhalten ermöglicht. Das System dient in erster Linie zur Bereitstellung von und zum Umgang mit Informationen, das heißt die Inhalte sollen zunächst nicht von Studierenden bearbeitet oder erstellt werden können. Das Wiki behandelt die oben genannten mathematischen Schwerpunktthemen und bietet aufgrund der Wichtigkeit des Themas zusätzlich eine Einführung in die Vektorrechnung.

Das Angebot wird durch zusammengetragene Referenzen auf hilfreiche Zusatzinformationen (z. B. andere Vorlesungsskripte) sowie auf multimediale Lehrmaterialen ergänzt. Die Förderung des mathematischen Verständnisses durch solche Hilfsmittel lässt sich aus wissenschaftlichen Studien ableiten. Ein wichtiger Mehrwert des Wikis besteht in der Begrenzung des Umfangs (der Tiefe), in der mathematisches Fachwissen behandelt wird. Dank des Einsatzes des Wikis müssen sich die Studierenden nicht durch ein mathematische Lehrbuch arbeiten: Durch die Strukturierung der Inhalte können sie schnell die gerade benötigten Informationen abrufen. Da nämlich mathematische Begriffe direkt auf erklärende Artikel verweisen, können die Studierenden mathematische Themen gezielt und in Bezug auf die individuellen Vorkenntnisse aufarbeiten. Weiterhin werden auf der Plattform Aufgaben eingebunden, die eine Reflexion der Inhalte ermöglichen und außerdem zur selbständigen Diagnose von Schwierigkeiten mit speziellen Themen geeignet sind. Durch diese Eigenschaften des Online-Angebots wird zudem der Heterogenität der Studierenden Rechnung getragen: das Wiki erlaubt den Fortgeschrittenen ein rasches Nachschlagen, während sich die Anfänger gezielt in eine Thematik einarbeiten können. Da alle Studierenden hierbei selbstständig mathematische Methoden im Zusammenhang mit einer technischen Problemstellung anwenden, erhalten sie laufend individuelle Rückmeldung zu ihrem eigenen Leistungsstand. Das Wiki ist weiterhin mit einem Forum ausgestattet, so dass sich die Studierenden austauschen und Fragen stellen können.

Das GET A Wiki, dessen Hauptseite Abb. 34.3 zeigt, wurde im Wintersemester 2012/13 erstmalig in der Vorlesung und in den Übungen zu den Grundlagen der Elektrotechnik A eingesetzt. Die Zugriffe wurden mit Hilfe von Google Analytics aufgezeichnet, wobei eine rege Benutzung des Angebots nachgewiesen werden konnte. Die Akzeptanz des Online-Angebots wurde durch drei Umfragen während des Semesters überprüft. Erste Ergebnisse des Projekts wurden in (Hennig und Mertsching 2012) veröffentlicht.

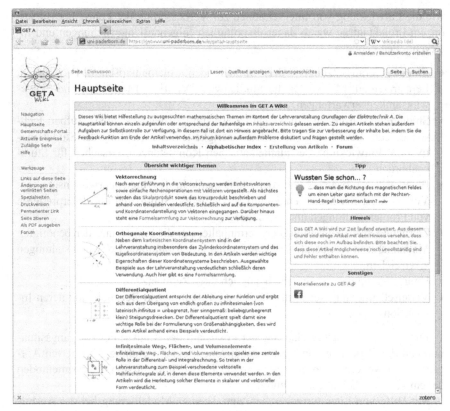

Abb. 34.3 Ausschnitt der Startseite des GET A Wikis.

Insgesamt wird mit dem GET A Wiki ein generalisierbares und auf didaktische Erkenntnisse gestütztes Konzept zur Vermittlung mathematischer Kompetenzen im situativen Kontext der Fachlehre entwickelt. Die Nutzung des Zusatzangebots wird dabei einerseits durch die enge Verzahnung der Veranstaltungselemente sowie andererseits durch wöchentliche Bonusübungen und drei Zwischentests, in denen die Studierenden zusätzliche Punkte für die Abschlussklausur sammeln können, gefördert.

Mathematik für Maschinenbauer: Integration des Modellierens in ingenieurwissenschaftlichen Zusammenhängen[5]

Im Fokus des Forschungsinteresses dieses Projekts stehen die Untersuchung von Problemen bei der Mathematikausbildung für Ingenieure des Maschinenbaus so-

---

[5] Autorin: Gudrun Oevel

wie die Entwicklung von Interventionen zur Verbesserung der Lehr-/ Lernsituation. Im Projektantrag waren daher ursprünglich folgende Untersuchungen und Interventionen geplant:

- Eingangsvoraussetzungen der Studierenden: mathematisches Fachwissen, Lernstrategien, Motive für die Studienwahl, Volition, Motivation und Selbstwirksamkeitserwartungen,
- Studierende im ersten Studienjahr: Lern- und Arbeitsverhalten, Aufbau fachspezifischer Lernstrategien, Lernschwierigkeiten, Entwicklung der Studienmotivation, Entwicklung der mathematischen Kompetenzen, Ursachen für Studienabbruch,
- Betonen des Einsatzgebietes der Mathematik in den Ingenieurwissenschaften: Vorbereitung der Studierenden auf Modellieren, Simulieren und Interpretieren von Problemstellungen und Lösungen,
- Zeitliche Umstrukturierung der Lerninhalte, um die Anpassung der benötigten Mathematik in den parallel stattfindenden grundständigen Lehrveranstaltungen zu erreichen,
- Umgestaltung der Lerninhalte bezüglich ihrer Relevanz,
- Veranschaulichung der Mathematik durch Real-World-Beispiele und deren Integration in den Stoff.

Im bisherigen Projektverlauf lag der Fokus auf der Analyse der Ausgangssituation und dem Erstellen von kontextgebundenen und anwendungsorientierten Aufgaben. Ein weiterer Schwerpunkt war die Einarbeitung in Evaluationsmethoden der empirischen Bildungsforschung.

Mit Hilfe von Literaturrecherchen zum Thema „Modellierung bei der Ausbildung von Ingenieuren" und der Evaluation des Core Curriculum der SEFI Mathematical Working Group (Barry und Steele 1993) wurde analysiert, welche Ergebnisse im Bereich der Hochschulausbildung für Ingenieure vorliegen. Aus den dort geschilderten Problemen bei der Arbeit mit Modellierungs-Aufgaben hat sich für das Teilprojekt ein pragmatischerer Ansatz ergeben, der sich nun in Richtung der folgenden Forschungshypothesen konkretisiert hat:

1. „sinnvolle" Anwendungsaufgaben erhöhen die Kompetenzen

  - Mathematisierung eines Anwendungsproblems (Konstruktion eines passenden mathematischen Modells);
  - Probleme mathematisch lösen;
  - Mathematisch kommunizieren.

Diese Kompetenzen sind notwendig, um die Kompetenz „mathematisch modellieren" zu erwerben, die wiederum für Ingenieure für wichtig gehalten werden (u.a. Biembengut und Hein 2007 und exemplarische Befragungen im Bereich Maschinenbau an der Universität Paderborn).
2. Es lassen sich Kriterien für „sinnvolle" Anwendungsaufgaben entwickeln.
3. Studierende im Maschinenbau finden Anwendungsaufgaben sinnvoll und motivierend. Sie nutzen sie weiterhin als „Merkhilfe" für mathematische Verfahren

und können so leichter den Transfer von mathematischen Methoden auf unterschiedliche Anwendungsszenarien herstellen.

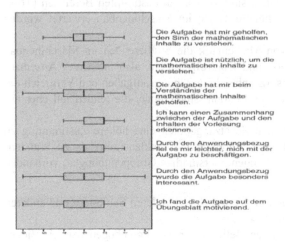

Abb. 34.4 Ergebnisse der Befragung

Bei der qualitativen Untersuchung mit Hilfe eines Fragebogens im WS 2010/11 ergaben sich folgende Befunde:

1. Im Gegensatz zur ersten Forschungshypothese sind die Studierenden im Fach Maschinenbau bezüglich der Mathematik stark motiviert und sehen die Notwendigkeit von Mathematik in ihrem Fach ein.
2. Die Studierenden halten die Beispiele und Simulationen für sehr sinnvoll. Eine Ergebnisanalyse ist beispielhaft in Abbildung 34.4 dargestellt.
3. Die Studierenden kritisieren sehr stark die Korrektur von Übungsaufgaben sowie die aus ihrer Sicht schlechte Vorbereitung der studentischen Tutoren.

Die Ergebnisse konnten mit Hilfe der studentischen Veranstaltungskritik validiert werden. In Sachen Motivation decken sie sich mit den Befunden aus der BMBF-Langzeitstudie zur Studienqualität und Attraktivität der Ingenieurwissenschaften (BMBF 2007).

Im Vergleich mit anderen qualitativen Untersuchungen im Rahmen des khdm mit teilweise identischen Frageblöcken konnte darüber hinaus festgestellt werden, dass

- die Motivation in der Veranstaltung konstant hoch blieb;
- die Motivation zur Bearbeitung der Übungsblätter deutlich stärker extrinsischen als intrinsischen Charakter hatte;
- die Relevanz der zu erwerbenden Mathematik-Kompetenzen deutlich höher für spätere Veranstaltungen als für die parallel stattfindende Technische Mechanik und Technische Darstellung eingeschätzt wird;

- bei der Einstellung zur Mathematik im Vergleich mit Lehramtsstudierenden (für Gymnasium, Haupt–, und Real- und Gesamtschule; HRGGy) keine deutlichen Unterschiede zu erkennen sind, obwohl in bestimmten Bereichen (Toolbox, praktische Relevanz) höhere Werte im Maschinenbau erwartet worden sind;
- beim Üben und Lernen von Mathematik die Studierenden im Maschinenbau tendenziell eher Elaboration und Kontrollstrategie, also intensives Auseinandersetzen mit dem Stoff bspw. anhand von Übungsaufgaben und Vergleich der eigenen Ergebnisse, nutzen als Lehramtsstudierende, die tendenziell eher auf Memorisation bauen;
- bei der Frage nach sinnvollen Zusatzangeboten eine höhere Zustimmung zu freiwilligen Ergänzungen (Sonderübungen, Probeklausuren) als zu Stundenplanerweiterungen (Zentralübungen, vierstündige Gruppenübungen) vorhanden war.

Diese Ergebnisse sollen nun im Vergleich mit anderen Semestern und anderen Kontrollgruppen weiter überprüft werden.

Zusätzlich zu den quantitativen Befragungen per Fragebögen wurden Videoaufzeichnungen erstellt, in denen Gruppen von Studierenden zum einen Aufgaben lösen und zum anderen über ihre Meinungen zu und den Erfahrungen mit den Anwendungsaufgaben befragt wurden. Eine erste Analyse dieser Aufzeichnungen zeigte, dass die Studierenden den Anwendungsbezug der Aufgaben als Hilfsmittel zur Erinnerung an die dahinterliegende Mathematik verwenden. Der Transfer dieser Mathematik auf einen anderen Anwendungskontext konnte geleistet werden. Zusätzlich betonten die Studierenden die Wichtigkeit einer Visualisierung (Skizze des Sachverhalts, siehe Abb. 34.5).

Beide Ergebnisse werden derzeit im Zusammenhang mit der Fragestellung nach „sinnvollen" Anwendungsaufgaben genauer untersucht.

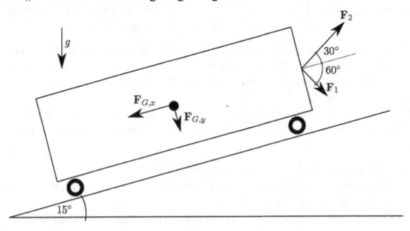

Abb. 34.5 Skizze zu einer Übungsaufgabe

KoM@ING[6]

Die Abkürzung KoM@ING steht für „Kompetenzmodellierung und Kompetenz-entwicklung, integrierte IRT-basierte und qualitative Studien bezogen auf Mathematik und ihre Verwendung im ingenieurwissenschaftlichen Studium". An dem Teilprojekt von KoM@ING an der Universität Paderborn sind neben Rolf Biehler auch Niclas Schaper und als wissenschaftlicher Mitarbeiter Jörg Kortemeyer beteiligt. Es gibt eine enge Zusammenarbeit mit Prof. Dr. Reinhard Hochmuth und Dr. Stephan Schreiber an der Leuphana Universität Lüneburg. Weitere Teilprojekte befinden sich an der Technischen Universität Dortmund, an der Ruhr-Universität Bochum, am IPN in Kiel und an der Universität Stuttgart.

Das KoM@ING-Projekt ist Teil des BMBF-Forschungsprogramms KoKoHs, wobei diese Abkürzung für Kompetenzmodellierung und Kompetenzerfassung im Hochschulsektor steht. Das Ziel des Paderborner Teilprojekts ist Kompetenzmodellierung, Kompetenzerfassung und Kompetenzentwicklung bezogen auf Studierende von elektrotechnikhaltigen Studiengängen in den ersten drei Semestern.

Hierbei spielen bei ingenieurwissenschaftlichen Studiengängen einerseits hochschulbezogen metakognitive, volitionale und motivationale Kompetenzherausforderungen, aber andererseits auch Spannungsverhältnisse zwischen den „Grundlagen der Elektrotechnik" und der „Mathematik für Ingenieure" eine große Rolle. Hierbei gibt es im Wesentlichen zwei Kritikpunkte:

1. Die Asynchronität zwischen mathematischer und ingenieurwissenschaftlicher Ausbildung: So ist bei einem Zugang über Felder in der Grundlagenvorlesung zur Elektrotechnik bereits in den ersten Wochen ein Verständnis von Integralsätzen nötig, die in der Mathematik für Ingenieure üblicherweise erst innerhalb des zweiten Semesters behandelt werden.
2. Die Mathematik kann in der ingenieurwissenschaftlichen Ausbildung nicht ausschließlich situiert oder begleitend erlernt werden. Die Mathematik hat einen eigenen logischen Aufbau, der auch bei der Konzeption einer ingenieur-wissenschaftlichen Mathematikveranstaltung eine zentrale Rolle spielt, um Verständnis bei den Studierenden zu sichern.

Daraus folgt, dass bei der Kompetenzmodellierung immer beide Seiten berücksichtigt sein müssen. Hierzu werden die Teilkompetenzen ermittelt, die einerseits in der „Mathematik für Ingenieure" und andererseits in den „Grundlagen der Elektrotechnik" curricular erwartet. Hieraus wird ein Gesamtmodell entwickelt, das die Schnittstelle zwischen „Mathematik für Ingenieure" und „Mathematik in den Ingenieurwissenschaften" besser gestalten helfen soll.

In dem Bereich „Mathematik in den Ingenieurwissenschaften" gibt es bislang keine Kompetenzmodellierungen bzw. Kompetenzmessinstrumente. Das Projekt schließt außerdem an die Kompetenzauffassungen der empirischen Bildungsforschung und die Methoden der Kompetenzmodellierung und –messung in tertiären

---

[6] Autor: Jörg Kortemeyer

484   Markus Hennig et al.

Bildungsbereichen an, zu denen der Leser Näheres in den Literaturangaben findet (vgl. Klieme und Hartig 2007).

In dem Projekt KoM@ING sind die drei großen ingenieurwissenschaftlichen Bereiche Maschinenbau, Elektrotechnik und Bauingenieurwesen abgedeckt. Hierdurch kann dieser Bereich in der nötigen Breite erfasst werden.

Für das Paderborner Teilprojekt wurden bereits viele Materialien unterschiedlicher Universitäten, einschließlich der TU9-Universitäten Stuttgart und Hannover, gesichtet. So konnten die unterschiedlichen inhaltlichen Dimensionen der Kompetenzmodelle entwickelt werden:

- „Mathematik für Ingenieure": schulmathematische Grundlagen, Analytische Geometrie, Lineare Algebra, Analysis in einer Veränderlichen, Analysis in mehreren Veränderlichen, Differentialgleichungen
- „Grundlagen der Elektrotechnik": Gleichstrom, Wechselstrom, Felder

Ein weiterer zentraler Punkt sind die kognitiven Prozesse beim Lösen von mathematikhaltigen Elektrotechnik-Aufgaben. Hier ergeben sich – angelehnt an den Lösungsplan von Blum und Leiß (vgl. Blum 2006) - drei Schritte:

1. Mathematisieren der elektrotechnischen Aufgabenstellung
2. „Mathematik machen", also das innermathematische Lösen des Problems
3. Rückübersetzung und Validierung des Ergebnisses

Mathematikhaltige Aufgaben in der Elektrotechnik stellen eine besondere Herausforderung dar, da sie neben den in der Grundlagenvorlesung zur Elektrotechnik vermittelten Inhalten häufig auch intuitive Vorstellungen zu physikalischen Sachverhalten voraussetzen. Um dieses zu berücksichtigen, haben wir ein Modell aus der amerikanischen Physikdidaktik (siehe Redish und Tuminaro 2007), welches sich mit dem Lösen mathematikhaltiger Physikaufgaben auseinandersetzt und das obige Phasenmodell ergänzt.

In den nächsten Projektphasen werden diese Ergebnisse mit Hilfe von Experten validiert, um so das Gesamtmodell zu erhalten. Hierzu werden aktuell Klausurbearbeitungen aus Elektrotechnik-Klausuren zu den Grundlagen der Elektrotechnik B („GET B") aus Paderborn hinzugezogen.

Literaturverzeichnis

Bamforth, S. E., Robinson, C. L., Croft, T., & Crawford, A. (2007). Retention and progression of engineering students with diverse mathematical backgrounds. *Teaching Mathematics and its Applications*, 26(4), 156-166.

Barry, M.D.J., & Steele, N.C. (1993). A core curriculum in mathematics for the European engineer: an overview. *International Journal of Mathematical Education in Science and Technology*, 24(2), 223-229.

Biembengut, M.S., & Hein, N. (2007). Modeling in engineering: Advantages and difficulties. In C. Haines, P. Galbraith, W. Blum, and S. Khan (Eds.), *Mathematical Modeling ICTMA 12: Education, Engineering and Economics* (S. 415-423). Chichester: Horwood Publishing.

Blum, W. (2006). Modellierungsaufgaben im Mathematikunterricht–Herausforderung für Schüler und Lehrer. In A. Büchter, H. Humenberger, S. Hußmann, & S. Prediger (Hrsg.), *Realitätsnaher Mathematikunterricht–vom Fach aus und für die Praxis* (S. 8-23). Hildesheim: Franzbecker.

BMBF (2007). *Studienqualität und Attraktivität der Ingenieurwissenschaften.* http://www.bmbf.de/pub/qualitaet_attraktivitaet_ingenieurswissenschaften.pdf. Abgerufen am 26.01.2012.

Brown, J. S., Collins, A., & Duguid, P. (1989). Situated Cognition and the Culture of Learning. *Educational Researcher*, 18(1), 32-42.

Demmer, C. (2012, 24./25. März). Das Kreuz mit der Mathematik: Die hohe Abbrecherquote im Maschinenbaustudium geht leicht zurück. Die Hochschulen nehmen das Problem inzwischen ernst. Süddeutsche Zeitung. S. V2/12. http://pix.sueddeutsche.de/app/szbeilagen/nas/orm_sonderthemen/pdf/karriere-im-maschinenbau--2012-03-24.pdf. Abgerufen am 17.10.2012.

Derboven, W., & Winkler, G. (2009). *Ingenieurswissenschaftliche Studiengänge attraktiver gestalten.* New York: Springer Verlag.

Hennig, M., & Mertsching, B. (2012). Situated Acquisition of Mathematical Knowledge - Teaching Mathematics within Electrical Engineering Courses. In A. Avdelas (Ed.), *Proceedings of the 40th SEFI Annual Conference 2012: Engineering Education 2020: Meet the Future.* (S. 264-265). Brüssel, Belgien: SEFI.

Hetze, P. (2011). *Nachhaltige Hochschulstrategien für mehr MINT-Absolventen.* http://www.stifterverband.info/publikationen_und_podcasts/positionen_dokumentationen/mint_hochschulstrategien_2011/mint_hochschulstrategien_2011.pdf. Abgerufen am 23.04.2013.

Heublein, U., Richter, J., Schmelzer, & R., Sommer, D. (2012). *Die Entwicklung der Schwund- und Studienabbruchquoten an den deutschen Hochschulen – Statistische Berechnungen auf der Basis des Absolventenjahrgangs 2010. HIS: Forum Hochschule, Bd. 3.* http://www.his.de/pdf/pub_fh/fh-201203.pdf. Abgerufen am 8.10.2012.

Ida, N. (2004). *Engineering Electromagnetics* (2nd Edition). New York: Springer Verlag.

Klieme, E., & Hartig, J. (2008). Kompetenzkonzepte in den Sozialwissenschaften und im erziehungswissenschaftlichen Diskurs. In M. Prenzel, I. Gogolin, & H. Krüger (Hrsg.), *Kompetenzdiagnostik* (S. 11-29). Wiesbaden: VS Verlag für Sozialwissenschaften.

Korenko, B., Cervenova, J., & Janicek, F. (2010). Teaching of Electromagnetism using E-Courses. In *Proceedings of the Joint International IGIP-SEFI Annual Conference*. Trnava, Slovakia. http://www.sefi.be/wp-content/papers2010/papers/1274.pdf. Abgerufen am 16.05.2013.

Lawson, D., Croft, A. C., & Carpenter, S. L. (2008). Mathematics Support Real, Virtual and Mobile. *The International Journal for Technology in Mathematics Education*, 15(2), 73-78.

Mustoe, L. (2006). Coming to terms with change: mathematics for engineering undergraduates. In M. Demlová, & D. Larson (Eds.), *Proceedings of the 13th European Seminar on Mathematics in Engineering Education* (S. 29-37). Kongsberg, Norway: Buskerud University College.

Tetour, Y., Richter, T., & Boehringer, D. (2010). Integration of Virtual and Remote Experiments into Undergraduate Engineering Courses. *Proceedings of the Joint International IGIP-SEFI Annual Conference*, Trnava, Slovakia. http://www.sefi.be/wp-content/papers2010/papers/1295.pdf. Abgerufen am 16.05.2013.

Kapitel 35
Das Deutsche Zentrum für Lehrerbildung Mathematik (DZLM)

Ziele und Fortbildungsprogramme

Jürg Kramer, Thomas Lange

Deutsches Zentrum für Lehrerbildung Mathematik (DZLM), Institut für Mathematik, Humboldt-Universität zu Berlin

Abstract   Ein essentieller Baustein zu einer besseren mathematischen Bildung ist die dritte Phase der Lehrerbildung, das Lernen im Beruf. Hier soll sich das Wissen und Können der Lehrkräfte kontinuierlich fort- und weiterentwickeln. Die nationalen Bildungsstandards, der Umgang mit Heterogenität innerhalb der Schülerschaft, neue Informations- und Kommunikationsmedien sind nur einige aktuelle Anforderungen, die an Lehrkräfte gestellt werden. Diese Herausforderungen erfordern ein verstärktes Bemühen um eine kontinuierliche Fort- und Weiterbildung. Im Oktober 2011 wurde deshalb auf Initiative der Deutsche Telekom Stiftung das Deutsche Zentrum für Lehrerbildung Mathematik (DZLM) gegründet, dessen Ziele und Fortbildungsprogramme in diesem Beitrag vorgestellt werden sollen. Dabei soll insbesondere auf die wissenschaftlichen Grundlagen, die Qualitätsstandards und die Evaluation der Fortbildungsaktivitäten eingegangen werden. Fort- und Weiterbildungen für Lehrkräfte sind im Sinne des Titels dieses Festbandes „Hilfsmittel", um das Lehren und Lernen von Mathematik, und hier insbesondere von Stochastik, zu unterstützen und zu verbessern.

Einleitung

Die professionelle Kompetenz von Lehrkräften ist ein bedeutsamer Einflussfaktor für guten Unterricht und erfolgreiche Lehr-Lern-Prozesse, die wiederum eine zentrale Voraussetzung für eine erfolgreiche Kompetenzentwicklung bei den Schülerinnen und Schülern sind. Wie Kompetenzen von Lehrkräften modelliert und erfasst werden können, war Gegenstand mehrerer Studien wie TEDS-M (Blömeke et al. 2010) und COACTIV (Kunter et al. 2011). In der COACTIV-Studie konnte gezeigt werden, dass insbesondere das fachdidaktische Wissen von Lehrkräften mit der kognitiven Aktivierung und der individuellen Unterstützung im Unterricht zusammenhängt (Baumert et al. 2010). Diese beiden Merkmale sind wesentliche Prädiktoren für den Lernzuwachs bei den Schülerinnen und Schülern.

Die Förderung von Lehrerkompetenzen ist also ein wichtiger Ansatzpunkt zur Verbesserung des Lernerfolgs von Schülerinnen und Schülern. Ein zentrales Anliegen des Deutschen Zentrums für Lehrerbildung Mathematik (DZLM) ist die Förderung von Lehrerkompetenzen durch Fortbildungen und Unterstützung von professionellen Lerngemeinschaften.

Die Deutsche Telekom Stiftung hat Anfang 2009 die Expertengruppe „Mathematik entlang der Bildungskette" berufen. Eine Empfehlung der Expertengruppe war die Einrichtung eines nationalen Zentrums für Lehrerbildung im Fach Mathematik (Tenorth et al. 2010). Die Deutsche Telekom Stiftung folgte dieser Empfehlung mit einer Ausschreibung für Einrichtung und Betrieb eines solchen Zentrums an einer deutschen Hochschule. Das Konzept eines Konsortiums unter Leitung der Humboldt-Universität zu Berlin erhielt dann den Zuschlag und gründete im Oktober 2011 das Deutsche Zentrum für Lehrerbildung Mathematik (DZLM). Dem Konsortium gehören neben der Humboldt-Universität zu Berlin, die Freie Universität Berlin, die Deutsche Universität für Weiterbildung, die Pädagogische Hochschule Freiburg und die Universitäten in Bochum, Dortmund, Duisburg-Essen und Paderborn an.

Rolf Biehler ist Vorstandsmitglied des DZLM und leitet die Abteilung „Sekundarstufe I" des DZLM. Seine Abteilung konzipiert und entwickelt Fort- und Weiterbildungen für Mathematiklehrkräfte der Sekundarstufe I, insbesondere im Bereich des kompetenzorientierten Mathematikunterrichts, der Didaktik der Stochastik und methodischer Aspekte beim Lehren und Lernen mit digitalen/neuen Medien. Die Kursentwicklungen basieren auf aktueller Forschung zum professionellen Lehrerwissen und die Fort- und Weiterbildungen sind selbst wieder Gegenstand der Forschungsarbeit seiner Arbeitsgruppe an der Universität Paderborn.

Ausgangslage im Bereich der Fortbildung von Mathematiklehrkräften

Bisherige Angebote zur Fortbildung von Mathematiklehrkräften sind kaum zwischen staatlichen Instanzen und anderen Trägern aufeinander abgestimmt, wenig nachhaltig und zumeist befristet. Zudem werden Lehrerinnen und Lehrer für Fortbildungen nicht regelmäßig freigestellt und nicht überall haben Schulen ausreichend Mittel für Fortbildungen zur Verfügung (Richter et al. 2012). Viele Fortbildungsprogramme sind nicht wissenschaftlich fundiert, werden nicht ausreichend öffentlich dokumentiert und geeignet evaluiert. Es gibt wenig Kooperationen zwischen den Bundesländern und zwischen Fortbildungsprogrammen.

Die Fortbildung von Lehrkräften im Fach Mathematik ist in den Bundesländern unterschiedlich organisiert. Verschiedene Multiplikatorinnen und Multiplikatoren sind verantwortlich für die Fortbildung von Lehrerinnen und Lehrern: (Fach)berater, Fach(bereichs)leiter, Fortbildner, Mentoren, Moderatoren, Referenten usw. Multiplikatorinnen und Multiplikatoren „rutschen" oft wenig vorbereitet in ihre Tätigkeit hinein und müssen diese neben ihren anderen Aufgaben ausführen. Es existieren für sie kaum Qualifizierungsprogramme, die eine fachspezifische Qualifizierung adressieren.

Der Mathematikunterricht in Deutschland leidet außerdem unter einem erheblichen Fachlehrermangel, insbesondere an den Grund- und Hauptschulen, zunehmend mehr Lehrerinnen und Lehrer unterrichten daher fachfremd. Hinzu kommt eine steigende Zahl an Quer- bzw. Seiteneinsteigern.

Klassische Fortbildungen für alle Lehrerinnen und Lehrer im Fach Mathematik sind meist einmalige halb- oder ganztägige Veranstaltungen, die schulübergreifend angeboten werden. Mehrteilige Fortbildungen mit der Gelegenheit neue Impulse in der Praxis zu erproben oder eine Begleitung und Unterstützung von Lehrkräften einer Schule bei der Unterrichtsentwicklung sind selten.

Neben Fortbildungen stehen Lehrerinnen und Lehrern eine große Anzahl von Foren und Informationsnetzen zur Verfügung; dennoch ist es für sie schwierig gesuchte Informationen zu finden, geeignete Materialien für den Unterricht zu erhalten und seriöse von weniger hilfreichen Informationen zu trennen.

In Deutschland fehlte bislang eine zentrale Anlaufstelle für die Lehrerbildung in Mathematik – obwohl gerade in diesem Fach hoher Bedarf besteht, wie Studien immer wieder belegen.

Ziele und Programmatik des DZLM

Das übergeordnete Ziel des DZLM ist es, eine umfassende Organisation von kontinuierlicher, professioneller Fortbildung für das Lehren von Mathematik zu etablieren. Das Zentrum soll ein Ort der Information, Dokumentation, Qualitätssicherung, Programmentwicklung und Fortbildung sein. Dabei arbeitet das Zentrum zusammen mit allen Akteuren im Feld der mathematischen Bildung: den Hochschulen in der Fachwissenschaft Mathematik, der Mathematikdidaktik und der Bildungsforschung, den Lehrkräften, Schulen und Netzwerken, den Bildungsadministrationen und Fortbildungseinrichtungen.

Insbesondere Qualifizierungsprogramme für Multiplikatorinnen und Multiplikatoren sowie fachfremd unterrichtende Lehrerinnen und Lehrer stehen im Fokus. Die Fortbildungskurse werden bundesweit organisiert und durchgeführt. Die DZLM-Aktivitäten ergänzen dabei bestehende Fortbildungsangebote in den Bundesländern. Das mittelfristige Ziel ist die Entwicklung des DZLM zu einem Beispiel für ein erfolgreiches deutschlandweit wirkendes Lehrerbildungszentrum.

Um diese Ziele zu erreichen sieht das DZLM-Konzept drei wesentliche Programmlinien vor.

Programme für Multiplikatorinnen und Multiplikatoren

Ein Grundgedanke des DZLM-Konzepts ist die Implementierung einer Kaskade von Professionalisierungsmaßnahmen. Dafür sollen Multiplikatorinnen und Multiplikatoren qualifiziert werden, damit sie ihr vertieftes Wissen und ihre erweiterten Kompetenzen an Lehrerinnen und Lehrer weitergeben können.

Es hat sich herausgestellt, dass die Bedarfsformulierungen und die Voraussetzungen für Multiplikatorenfortbildungen in den verschiedenen Bundesländern sehr unterschiedlich sind. Die beiden ersten und am weitesten entwickelten Multiplikatorenkurse in Nordrhein-Westfalen für die Primarstufe und die Sekundarstufe I beinhalten die Themen: Kompetenzorientierung und Heterogenität (Grundschule) sowie kompetenzorientierter Mathematikunterricht aus inhaltsbezogener (Stochastik) und prozessbezogener Perspektive (Sekundarstufe). Sie umfassen 240 bis 250 Stunden über ein Schuljahr verteilt. Der Stundenumfang teilt sich auf 13 bzw. 15 Präsenztermine (ganztägige Seminare) und dazwischen auf praxis-basiertes Arbeiten, kollaboratives Lernen (online) und Selbststudium (online) auf. Die Multiplikatorinnen und Multiplikatoren in Nordrhein-Westfalen sind Mathematikmoderatoren der regionalen Kompetenzteams. Die Teilnehmenden werden am Ende des Kurses ein DZLM-Zertifikat für aktive Mitarbeit erhalten. Die aktive Mitarbeit umfasst die Erstellung eines Portfolios von Fortbildungsmaterialien und die Durchführung einer eigenen Fortbildung. Der Kurs für die Sekundarstufe wird von Rolf Biehler und den Mitarbeitern seiner DZLM-Abteilung entwickelt und durchgeführt.

Ausgebildete Multiplikatorinnen und Multiplikatoren sollen bei ihrer weiteren Fortbildungsarbeit in den Regionen durch eine spezielle Programmlinie unterstützt werden, bspw. in Form von professionellen Lerngemeinschaften. Auf Grundlage der Erfahrungen mit den Multiplikatorenkursen soll ein bundesweit angebotener Masterstudiengang für Lehrkräfte konzipiert werden.

Programme für fachfremd Unterrichtende und den Vorschulbereich

Lehrkräfte werden teilweise auch in Fächern eingesetzt, die sie nicht studiert haben. Dies liegt zum einen daran, dass praktische Begebenheiten an den Schulen dies erfordern (bspw. Lehrermangel in dem Fach), zum anderen vor allem in der Grundschule, aber teilweise auch in der Sekundarstufe, am Klassenlehrerprinzip, bei dem die Klassenlehrkraft möglichst viele Fächer in ihrer Klasse unterrichtet. Über die Anzahl von Lehrkräften, die in Deutschland im Fach Mathematik fachfremd unterrichten, ist bisher wenig bekannt. Ebenso ist bislang nur wenig erforscht, wie sich fachfremder Unterricht auf die Kompetenzen von Schülerinnen und Schülern auswirkt.

In der kürzlich veröffentlichten Bundesländer-Vergleichsstudie zu Kompetenzen von Schülerinnen und Schülern am Ende der vierten Jahrgangsstufe in den Fächern Deutsch und Mathematik wurde auch diesen beiden Fragen nachgegangen (Richter et al. 2012). 27 Prozent der 1200 befragten Mathematiklehrkräfte aus allen Bundesländern gaben an, Mathematik zu unterrichten, ohne Mathematik studiert zu haben. Dabei schwankt der Anteil zwischen 1 Prozent in Thüringen und 48 Prozent in Hamburg. Die durchschnittlich erreichten Kompetenzen von Schülerinnen und Schülern, die durch fachfremd unterrichtende Lehrkräfte unterrichtet werden, ist dabei geringer als bei solchen, die nicht fachfremd unterrichtet werden. Besonders deutliche Unterschiede gibt es bei den leistungsschwächsten Schülerinnen und Schüler. Diese profitieren offenbar sehr deutlich von fachlich qualifizierten Lehrerinnen und Lehrern.

Erste Qualifizierungskurse für fachfremd unterrichtende Lehrerinnen und Lehrer sind im Rahmen des DZLM angelaufen bzw. für das Schuljahr 2012/13 geplant. In Schleswig-Holstein wird ein Primarstufenkurs des IQSH (Institut für Qualitätsentwicklung an Schulen Schleswig-Holstein) mit dem Thema „Mathematik mit Kompetenz" wissenschaftlich begleitet. In Bayern wird ein Kurs für fachfremd unterrichtende Haupt- und Mittelschullehrer durchgeführt sowie ein Multiplikatorenkurs für Fachkoordinatorinnen und -koordinatoren, die Ansprechpartner für fachfremd unterrichtende Lehrerinnen und Lehrer sind. In Baden-Württemberg werden Multiplikatorinnen und Multiplikatoren im Grundschulbereich qualifiziert, um fachfremd Unterrichtende fortzubilden.

Eine Empfehlung der Expertengruppe „Mathematik entlang der Bildungskette" ist bereits im Kindergartenbereich mit Qualifizierungsmaßnahmen anzusetzen. In den Bundesländern wurden Bildungsprogramme für den Kindergartenbereich eingeführt, und es gibt viele Initiativen, die im MINT-Bereich aktiv sind, wie bspw. die Stiftung „Haus der kleinen Forscher". Im nächsten Jahr sollen Qualifizierungsprogramme für diesen Bereich im DZLM konzipiert werden.

Programme für alle Lehrerinnen und Lehrer

Für alle Lehrerinnen und Lehrer werden schulübergreifende und schulinterne Fortbildungskurse angeboten (einschließlich des Aufbaus und der Unterstützung professioneller Lerngemeinschaften), um dezentralisierte regionale Aktivitäten zu unterstützen, die sich unmittelbar nach den Bedürfnissen von Schulen richten. Dabei wurde zunächst auf das Kursangebot des Vorgängerprojekts „Mathematik Anders Machen" zurückgegriffen und dieses sukzessive ergänzt und weiterentwickelt. Zunächst wurde das Format der Kurse ausgeweitet, wie im nächsten Kapitel vorgestellt wird.

Qualitätsstandards

Eine zentrale Aufgabe des DZLM ist die Entwicklung bundesweiter Standards für Lehrerfortbildungen im Fach Mathematik. Dies ist ein wichtiges Alleinstellungsmerkmal des Zentrums. Die Qualität von Lehrerfortbildungen wird anhand von Indikatoren bestimmt. Dies sind: die Relevanz der Themen, die geförderten Kompetenzen bei den Teilnehmern, die Gestaltung und die Formate der Kurse. Je höher die Qualität der Fortbildungskurse ist, desto wirkungsvoller sollten diese die Unterrichtspraxis der Lehrerinnen und Lehrer und letztlich die Leistungen der Schülerinnen und Schüler verbessern.

Thematisch konzentrieren sich die Angebote des DZLM auf folgende Kategorien:

- Themenkategorie 1: Mathematik mit Blick auf Fachwissenschaft und -didaktik
- Themenkategorie 2: Kompetenzorientierung im Mathematikunterricht
- Themenkategorie 3: Lehr- und Lernprozesse in der Mathematik
- Themenkategorie 4: Fortbildungsmanagement und -didaktik

Alle Maßnahmen des DZLM lassen sich mindestens einer der vier DZLM-Themenkategorien zuordnen. Für Fortbildungen für Multiplikatorinnen und Multiplikatoren gilt darüber hinaus: Es wird immer Themenkategorie 4 sowie mindestens eine der Themenkategorien 1 bis 3 angeboten.

Gestaltungsprinzipien und Kompetenzrahmen

Die Gestaltungsprinzipien sind verbindliche Kriterien, nach denen alle DZLM-Fortbildungsangebote (für Multiplikatoren, fachfremd unterrichtende Lehrer, alle Lehrer etc.) gestaltet werden. Sie sind sowohl Selbstverpflichtung bei der Durchführung eigener Kurse, als auch Verpflichtung für externe Anbieter, die im Auftrag oder in Kooperation mit dem DZLM Kurse anbieten. Gleichzeitig sind die Gestaltungsprinzipien Indikatoren für die Evaluation von DZLM-Fortbildungsangeboten.

Die folgenden sechs Prinzipien wurden aus der Literatur identifiziert, um möglichst wirksame Fortbildungsangebote für Multiplikatoren und Lehrkräfte zu gestalten. Die Fortbildungen sind *teilnehmerorientiert* (Franke et al. 2001), d.h., sie beziehen die individuellen Voraussetzung, aber auch die individuellen Bedürfnisse der Teilnehmenden ein. Sie fördern und fordern die aktive und eigenverantwortliche Teilhabe bei der Gestaltung und Durchführung, um eine Veränderung des eigenen Handelns zu fördern. Die Fortbildungen sind *fallbezogen* (Timperley et al. 2007; Lipowsky und Rzejak 2012), indem sie Bezug zu typischen Situationen aus Unterrichts- und Fortbildungspraxis nehmen, um den Teilnehmenden Folgen veränderten Handelns in Unterricht bzw. Fortbildung deutlich werden zu lassen. Die Fortbildungen sind *kompetenzorientiert* (Garet et al. 2001; Timperley et al. 2007;

Lipowsky und Rzejak 2012), d.h. sie sind ausgerichtet an inhaltlichen und methodischen Kompetenzen, die die Teilnehmenden erwerben sollen. Diese Kompetenzen sind die Zielkategorien der Fortbildungen, die evaluiert werden und die in einem Kompetenzrahmen eingebettet sind, der unten ausführlicher vorgestellt wird. Die Fortbildungen sind *vielfältig* durch unterschiedliche Zugangs- und Arbeitsweisen (Deci und Ryan 1993; Lipowsky und Rzejak 2012). Diese unterschiedlichen Vermittlungsformate tragen zur Vertiefung, Verstetigung und Vernetzung bei und werden im Unterabschnitt „Kursformate" noch ausführlicher dargestellt. Die Fortbildungen sind *kooperationsfördernd* und *(selbst)reflexionsanregend* (Bonsen und Hübner 2012; Lipowsky und Rzejak 2012). Durch gemeinsames Arbeiten an Problemlösungen und deren Umsetzung wird die Kooperation der Teilnehmenden gefördert und langfristige Zusammenarbeit angeregt. Die gemeinsame Reflexion wird als wichtige Grundlage der Erkenntnisgewinnung und Kompetenzentwicklung angesehen.

Als Zielkategorien für DZLM-Fortbildungsangebote hat das DZLM Kompetenzmodelle und Kompetenzbeschreibungen entwickelt, die sich an die in der Lehrerbildungsforschung verwendeten Modelle anlehnen (Shulman 1986; Ball et al. 2008; Blömeke et al. 2010) und diese um weitere Aspekte ergänzen, die sich aus dem spezifischen Auftrag des DZLM ergeben. Professionswissen, mathematik-bezogene Vorstellungen (Beliefs) und – angesichts des zu erheblichen Anteilen online-gestützten DZLM-Fortbildungsangebots – technische Fertigkeiten sind die drei zentralen Zielkategorien der Kompetenz für alle Teilnehmenden an einer DZLM-Fortbildung. Für Multiplikatorinnen und Multiplikatoren sind zudem Fertigkeiten in der Fortbildungsdidaktik sowie im Fortbildungsmanagement von großer Bedeutung. Zusammen stellen diese Konstrukte den Kompetenzrahmen der Fortbildungsmaßnahmen des DZLM dar.

Der Aspekt Professionswissen beinhaltet die Kompetenzdimensionen: fachbezogenes Wissen, fachdidaktisches Wissen und pädagogisches Wissen. Für die Ausdifferenzierung dieser Kompetenzdimensionen wird auf die Empfehlungen der gemeinsamen Kommission der Deutschen Mathematiker-Vereinigung (DMV), der Gesellschaft für Didaktik der Mathematik (GDM) sowie des Lehrerverbandes Mathematisch Naturwissenschaftlicher Unterricht (MNU) „Standards für die Lehrerbildung im Fach Mathematik" aus dem Jahr 2008 zurückgegriffen. Das fachbezogene Wissen ist insbesondere für die Fortbildung von fachfremd unterrichtenden Lehrerinnen und Lehrern relevant, aber immer eingebettet in fachdidaktische Fragestellungen.

Kursformate

Die Kursformate des DZLM sind eng verknüpft mit den Gestaltungsprinzipien. Insbesondere sollen die Formate das „Gestaltungsprinzip der Vielfältigkeit" ermöglichen. Das bedeutet, dass unterschiedliche Zugangs- und Arbeitsweisen (Vermittlungsformate) zum Einsatz kommen, die aufeinander bezogen und ver-

netzt sind. Die verschiedenen Vermittlungsformate ermöglichen eine aktive Mit-
gestaltung und begünstigen damit eine höhere Nachhaltigkeit. Neben klassischen
Präsenzseminaren zur intensiven Zusammenarbeit und Reflexion ist praxis-
basiertes Arbeiten zur Erprobung von Konzepten in der Unterrichtspraxis ein
wichtiger Bestandteil der Formate, unterstützt durch kollaboratives Lernen online
und einem Selbststudium online. Die Kursformate werden durch das Profil, also
den Wechsel der Vermittlungsformate, und den Gesamtumfang bestimmt.

DZLM-Fortbildungskurse sollen aus mindestens einer Impuls-, Erprobungs-
und Reflexionsphase bestehen. In der Erprobungsphase werden die Teilnehmenden durch E-Learning-Angebote unterstützt. Die abschließende gemeinsame Reflexion und Evaluation der im Unterricht gewonnen Erkenntnisse und Erfahrungen
soll die Nachhaltigkeit der Fortbildung fördern. Halbtägige oder eintägige Impuls-
kurse sind nur als niederschwellige Einstiegskurse gedacht. Intensivkurse stellen
eine Ausweitung der Standardkurse dar, bei denen die Unterrichtsentwicklung einer Schule unterstützt werden soll. Bei umfangreicheren Kursformaten für Multiplikatorinnen und Multiplikatoren sowie fachfremd unterrichtende Lehrerinnen
und Lehrer wechseln sich Präsenzphasen und andere Vermittlungsformate mehr-
mals ab.

Qualitätssicherung und Evaluation

Der erste Schritt der Qualitätssicherung ist ein Antragsverfahren für die Entwicklung und/oder Durchführung von Fortbildungskursen. Jede Fortbildungsmaßnah-
me, die von DZLM-Mitarbeitern oder externen Anbietern/Entwicklern initiiert
wird, durchläuft diesen Antrags- und Begutachtungsprozess. Der Antrag umfasst
insbesondere Angaben, inwieweit das Angebot den DZLM-Qualitätsstandards gerecht wird, also zu den Zielen (angestrebte Kompetenzentwicklung), den gewählten Themen, der Umsetzung der Gestaltungsprinzipien und dem Format.

Die Erreichung der Qualitätsstandards bei der Durchführung der Fortbildungen
und ob damit die Ziele des DZLM erreicht werden, wird evaluiert. Wie in der Einleitung bereits beschrieben, ist letztlich für die Evaluation der Wirksamkeit von
Fortbildungsmaßnahmen die Erhöhung von Schülerkompetenzen entscheidend.
Da die Wirkungskette lang und komplex ist, insbesondere bei Multiplikatorenfortbildungen, sollte nicht nur das Endergebnis, die Schülerleistungen evaluiert werden, sondern auch die verschiedenen Ebenen der Wirkungskette. Lipowsky (2010)
unterscheidet vier Ebenen:

1. Unmittelbare Reaktionen der teilnehmenden Lehrkräfte: Zufriedenheit, Veran-
   staltungsqualität, Kompetenz der Fortbildner, Relevanz der Inhalte;
2. Lernzuwachs der teilnehmenden Lehrkräfte: Erweiterung der Lehrerkognition
   (Wissen, Überzeugungen) und affektiv-motivationale Dimensionen;
3. Veränderungen im unterrichtlichen Handeln der teilnehmenden Lehrkräfte:
   bspw. Intensivierung kognitiv herausfordernden und aktivierenden Unterrichts;

4. Entwicklung der Schülerinnen und Schüler: Lernleistungen und affektiv-motivationale Entwicklung.

Der Fokus in der ersten Phase des DZLM liegt auf nutzertypischen Befragungen der Teilnehmenden. Diese sollen die Umsetzung und Zielerreichung der Fortbildung bewerten. Für die Umsetzung wurde ein Instrument entwickelt, das – neben Fragen zum demographischen Hintergrund, schulischen Rahmenbedingungen und verschiedenen Zufriedenheitsmerkmalen – die sechs Gestaltungsprinzipien des DZLM erfasst (siehe oben). Für die Zielerreichung wird der DZLM-Kompetenzrahmen herangezogen (siehe ebenfalls vorherigen Abschnitt). Dabei wird der eigene Lernzuwachs von den Teilnehmenden bewertet (Ebene 2 nach Lipowsky).

Verwendet wird ein Fragenkatalog, der übergreifende Fragen enthält, die jeder Teilnehmerin und jedem Teilnehmer gestellt werden, und weitere Fragen, die zuvor im Hinblick auf die spezifischen Ziel- und Umsetzungsmerkmale für jede Fortbildung individuell angepasst werden. Die Teilnehmenden werden in jedem Fall im Anschluss an die Veranstaltungen befragt. Diese Befragung beinhaltet stets eine Bewertung der Teilnehmenden, wie sie selbst ihre Kompetenzen vor und nach der Fortbildung einschätzen (retroperspektivischer Pretest).

Das allgemeine Design wird individuell auf jede Fortbildung angepasst. Bspw. beschränkt sich die wissenschaftliche Begleitung bei Ein-Tages-Fortbildungen auf die Abstimmungen der Zielindikatoren mit den Referentinnen und Referenten sowie eine einmalige Teilnehmerbefragung. Bei längeren Maßnahmen, wie bspw. den derzeitigen Multiplikatorenfortbildungen in Nordrhein-Westfalen, gibt es neben einer Vorabbefragung, die den Referentinnen und Referenten zur Vorbereitung dient, auch Zwischenbefragungen für jedes Modul.

Ausblick

Das DZLM hat eine Reihe von ersten Aktivitäten gestartet, um schnell Wirkung entfalten zu können, Erfahrungen zu sammeln und das DZLM bekannt zu machen. Insbesondere die ersten Fortbildungsmaßnahmen für Multiplikatorinnen und Multiplikatoren in Nordrhein-Westfalen sind als Pilotprojekte bereits an einem möglichen Regelbetrieb orientiert. Sie dienen auch der Konzipierung des geplanten Masterstudiengangs, bzw. könnten als Bausteine eines solchen Studiengangs dienen. Weitere Multiplikatorenkurse in Baden-Württemberg, Bayern, Berlin und Schleswig-Holstein werden noch in diesem Schuljahr 2012/13 folgen. In Brandenburg, Bremen, Hessen, Mecklenburg-Vorpommern, Niedersachen, Rheinland-Pfalz, Sachsen und Thüringen laufen Verhandlungen mit den Kultusministerien.

Für fachfremd unterrichtende Lehrerinnen und Lehrer und alle anderen Lehrerinnen und Lehrer wurden Kurse nach dem oben beschriebenen Muster entwickelt und in die Implementationsphase gebracht. Auch die ersten Pilot-Maßnahmen zur Unterstützung von Netzwerken und Lehrergruppen wurden beschlossen.

Der Ausbau der Kooperationen mit weiteren Hochschulen und Fortbildungsanbietern wird die effektive und effiziente Verbreitung von DZLM-Fortbildungskursen unterstützen. Die Erfahrungen und Evaluationen aus den ersten Fortbildungsmaßnahmen werden dem DZLM helfen, sein Angebot zu reflektieren, es weiter auszubauen, zu verbessern und zu verfeinern, um dem Anspruch eines bundesweiten Zentrums für Lehrerbildung Mathematik gerecht zu werden. Das aktuelle Fortbildungsprogramm kann auf http://www.dzlm.de eingesehen werden.

Literaturverzeichnis

Ball, D. L., Thames, M. H., & Phelps, G. (2008). Content Knowledge for Teaching: What Makes It Special? *Journal of Teacher Education, 59*(5), 389–407.

Baumert, J., Kunter, M., Blum, W., Brunner, M., Voss, T., Jordan, A., Klusmann, U., Krauss, S., Neubrand, M., & Tsai, Y. M. (2010). Teachers' mathematical knowledge, cognitive activation in the classroom, and student progress. *American Educational Research Journal, 47*(1), 133–180.

Blömeke, S., Kaiser, G., & Lehmann, R. (2010). *TEDS-M2008: Professionelle Kompetenz und Lerngelegenheiten angehender Primarstufenlehrkräfte im internationalen Vergleich.* Münster: Waxmann.

Bonsen, M., & Hübner, C. (2012). Unterrichtsentwicklung in Professionellen Lerngemeinschaften. In J. Bauer, & J. Logemann (Hrsg.), *Effektive Bildung* (S. 55–76). Münster: Waxmann.

Deci, E. L., & Ryan, R. M. (1993). Die Selbstbestimmungstheorie der Motivation und ihre Bedeutung für die Pädagogik. *Zeitschrift für Pädagogik, 39,* 223–238.

Franke, M. F., Carpenter, T. P., Levi, L., & Fennema, E. (2001). Capturing teachers' generative change: A follow-up study of professional development in mathematics. *American Educational Research Journal, 38*(3), 653–689.

Garet, M., Porter, A., Desimone, L., Birman, B.F., & Suk Yoon, K. (2001). What makes professional development effective? Results from a national sample of teachers. *American Educational Research Journal, 38*(4), 915–945.

Kunter, M., Baumert, J., Blum, W., Klusmann, U., Krauss, S., & Neubrand, M. (Eds.). (2011). *Professionelle Kompetenz von Lehrkräften: Ergebnisse des Forschungsprogramms COACTIV.* Waxmann Verlag GmbH.

Lipowsky, F. (2010). Lernen im Beruf. Empirische Befunde zur Wirksamkeit von Lehrerfortbildung. In F. H. Müller, A. Eichenberger, M. Lüders, & J. Mayr (Hrsg.), *Lehrerinnen und Lehrer lernen – Konzepte und Befunde zur Lehrerfortbildung* (S. 51–77). Münster: Waxmann.

Lipowsky, F., & Rzejak, D. (2012). Lehrerinnen und Lehrer als Lerner – Wann gelingt der Rollentausch? Merkmale und Wirkungen effektiver Lehrerfortbildungen. *Schulpädagogik heute, 5*(3), 1–17.

Richter, D., Kuhl, P., Reimers, H., & Pant, H. A. (2012). Aspekte der Aus- und Fortbildung von Lehrkräften in der Primarstufe. In P. Stanat, H. A. Pant, K. Böhme, & D. Richter (Hrsg.), *Kompetenzen von Schülerinnen und Schülern am Ende der vierten Jahrgangsstufe in den Fächern Deutsch und Mathematik* (S. 237–250). Münster: Waxmann.

Shulman, L. S. (1986). Those who understand: knowledge growth in teaching. *Educational Researcher, 15*(2), 4–31.

Tenorth, H. E., Blum, W., Heinze, A., Peter-Koop, A., Post, M., Selter, C., Tippelt, R., & Törner, G. (2010). *Mathematik entlang der Bildungskette: Empfehlungen einer Expertengruppe zur Kompetenzentwicklung und zum Förderbedarf im Lebenslauf.* Bonn: Deutsche Telekom Stiftung.

Timperley, H., Wilson, A., Barrar, H., & Fung, I. (2007). *Teacher professional learning and development. Best Evidence Synthesis Iteration.* Wellington, New Zealand: Ministry of Education.